Life of Mine | Mine Waste and Tailings Conference 2025

LIFE OF MINE VOLUME

29–30 July 2025
Brisbane, Australia

The Australasian Institute of Mining and Metallurgy
Publication Series No 3/2025

AusIMM

Published by:
The Australasian Institute of Mining and Metallurgy
Ground Floor, 204 Lygon Street, Carlton Victoria 3053, Australia

ISBN 978-1-922395-53-5

Advisory Committee

Life of Mine Committee

Claire Cote
Conference Advisory Committee Chair

Alex Thin
FAusIMM(CP)

Brad Radloff
MAusIMM

Ingrid Meek
MAusIMM(CP)

Jason Dunlop

Michael McLeary
MAusIMM

Sarah McConnell
MAusIMM

Sibasis Acharya
FAusIMM

Todd Bell
MAusIMM

Vanessa MacDonald

Maggie Ng

Mine Waste and Tailings Committee

David Williams
FAusIMM
Conference Advisory Committee Chair

Angélica Amanda Andrade

Lis Boczek

Edgar Salas

Hernan Cifuentes

Theo Gerritsen
MAusIMM(CP)

Symon Jackson
FAusIMM(CP)

Fernanda Maluly Kemeid
MAusIMM

Marcelo Llano

Allan McConnell
MAusIMM

Kathy Tehrani
MAusIMM

Chenming Zhang

Andy Fourie

Arun Muhunthan

Kristy Ellis

Kevin Spencer
MAusIMM

AusIMM

Julie Allen
Head of Events

Fiona Geoghegan
Senior Manager, Events

Raha Karimi
Program Coordinator, Events

Reviewers

We would like to thank the following people for their contribution towards enhancing the quality of the papers included in this volume:

Sibasis Acharya

Angelica Amanda Andrade

Todd Bell

Lis Boczek

Pascal Bolz

Kate Brand

Mark Chapman

Robynne Chrystal

Hernan Cifuentes

Amelia Corzo

Claire Cote

Manuel de Membrillera Ortuño

Jason Dunlop

Mansour Edraki

Andy Fourie

Philippe Garneau

Theo Gerritsen

Morteza Ghamgosar

Babak Hedayatifar

Stuart Henderson

Joe Hinton

Symon Jackson

Sina Kazemian

Fernanda Maluly Kemeid

Anjan Kundu

Scott Lines

Marcelo Llano

Vanessa MacDonald

Allan McConnell

Sarah McConnell

Karen McKenzie

Michael McLeary

Mauricio Medina Florez

Ingrid Meek

Liane Millington

Jack Moorhead

Camilo Moreno

Arun Muhunthan

Maggie Ng

Louisa Nicolson

Ina Prinsloo

James Purtill

Brad Radloff

Louisa Rochford

Edgar Salas

Mandana Shaygan

Kevin Spencer

Gideon Style

Kathy Tehrani

Alexander Thin

Ed Tuplin

Alex Walker

David Williams

Joseph Wu

Lauren Zappala

Chenming Zhang

Foreword

On behalf of AusIMM and the Sustainable Minerals Institute at The University of Queensland, welcome to the seventh Life of Mine (LOM) Conference.

Since its inception, LOM has grown to become a cornerstone of thought leadership in mine rehabilitation and closure, championing the importance of holistic thinking across the mine life cycle.

For the first time ever, LOM will run concurrently with the Mine Waste and Tailings Conference. This provides delegates a unique opportunity for cross-disciplinary dialogue, and a chance to explore strategic decisions shaping the future of mine waste management.

We are proud to present LOM's most extensive technical program to date, featuring 110 presentations across two streams. Attendees will benefit from a rich array of interactive panels and case studies, designed to encourage collaborative learning. A defining strength of this conference is the participation of AusIMM's traditional core professional disciplines – geology, mining engineering, and mineral processing – alongside those working in environmental, social, and sustainability-focused roles. The program has been carefully curated to foster cross-disciplinary engagement and learning.

A highlight of this year's conference is our opening keynote panel exploring the critical role of regional context in shaping effective closure strategies. Our keynote panellists will discuss how circular economy principles can turn environmental challenges into business opportunities, and the importance of engaging with supply chain actors to meet growing demands for transparency in environmental and social performance.

There is also a dedicated session on innovative rehabilitation practices, alongside a panel discussion and two must-attend technical sessions on building resilient landforms. Notably the LOM program includes a session dedicated to the Global Industry Standard for Tailings Management, reflecting the growing relevance of this topic for all professionals.

Importantly, the conference recognises that mine planners play a vital role in closure planning, and the perspectives of communities and Traditional Owners are also woven throughout the program, with a dedicated panel on strategies for building thriving, sustainable communities in mining regions.

I extend my heartfelt thanks to our presenting authors for their keen insights and our abstract reviewers for their time and diligence. We are also deeply grateful to our sponsors and exhibitors for their continued support.

Finally, I wish to acknowledge the Advisory Committee and AusIMM Management Team, whose dedication and expertise have been instrumental in bringing this event to life.

Welcome to the conference – we hope you find the program enriching and enjoy the opportunity to connect with your peers.

Yours faithfully,

Claire Côte

Life of Mine Conference Chair

The University of Queensland

Foreword

On behalf of the Advisory Committee, welcome to the 2025 Mine Waste and Tailings Conference, the fifth in this highly successful series first launched in 2015.

Co-hosted by AusIMM and The University of Queensland, this conference aims to continue as an industry-orientated benchmark for sharing knowledge and experience on all aspects of mine waste and tailings management, sustainable practice and closure. The co-location with Life-of-Mine will enrich the delegate experience and provide new learning opportunities.

As the demand for minerals and mine waste volumes continues to increase exponentially in the face of diminishing ore grades, the mining industry faces sustained threats to its financial and social licences to operate. Over the last decade, tailings dam failures have come to dominate discussion, with the longer term threat of acid and metalliferous drainage and erosion ever-present.

The mining industry accepts that the majority of the world's future minerals will come from ever more low-grade, high-tonnage, ultra-mechanised operations.

The industry is recognising that we are in the mine waste business, and this conference will present the latest ideas and tools to ensure we can better manage the responsibilities we have as professionals. With four pre-eminent keynote speakers, five industry expert panel discussions and numerous papers, I am confident that this conference will build on the noted success of its predecessors, as well as promoting networking to foster new connections.

I would like to thank the Advisory Committee, the AusIMM Management Team, authors, paper reviewers, attending delegates and all our sponsors and exhibitors, as well as our supporting partner, QTG.

We are delighted you are joining us at the conference, and we hope that you will find it an enjoyable and rewarding event.

Yours faithfully,

Emeritus Professor David John Williams FAusIMM

Mine Waste and Tailings Conference Chair

The University of Queensland

Sponsors

Technical Session Sponsors

Ausenco

 **ERM**

xylem

Name Badge and Lanyard Sponsor

WEIR
Mining technology for a sustainable future

Coffee Cart Sponsor

McCOSKER
SAFETY FIRST. PRODUCTION ALWAYS.

Conference Proceedings Sponsor

MINE EARTH

Conference App Sponsor

 **kurloo**

Supporting Partner

QTG

Contents

Life of Mine — volume 1

Innovations in rehabilitation and closure

Integrated strategic mine planning

Life of mine waste and tailings management

Post-mining land use

Standards and reporting

Mine Waste and Tailings — volume 2

Case studies on operational aspects and closure

Design loadings and parameter selection, including BAT and BAP

Managing mine-affected water

MCA, risk assessment, governance and compliance

Mine closure and rehabilitation to accommodate site settings

Minimising and managing mine wastes, including dewatering tailings and comingling

Miscellaneous

Monitoring waste storages during operation and post-closure

Tailings and foundation characterisation

Tailings dam breach and runout analysis

Innovations in rehabilitation and closure

Soil amelioration with lignite as a means to improve rehabilitation outcomes

T Baumgartl[1], N Baghbani[2] and V Filipovic[3]

1. Director Future Regions Research Centre, Federation University, Churchill Vic 3842.
   Email: t.baumgartl@federation.edu.au
2. Associate Lecturer, La Trobe University, Bundoora Vic 3083.
   Email: a.baghbani@latrobe.edu.au
3. Senior Lecturer, University of Queensland, St Lucia Qld 4072. Email: v.filipovic@uq.edu.au

ABSTRACT

This study provides new insights into optimising lignite application for large-scale mine rehabilitation, offering a cost-effective and sustainable approach for improving post-mining landscapes.

Successful rehabilitation and revegetation often lack suitable topsoil substrate capable of sustaining vegetation. Challenges arise from detrimental chemical properties (eg salinity, dispersive tendencies) or physical characteristics (eg high clay content, high bulk density, and root penetration resistance) of available materials.

This study investigates the effects of lignite as a soil amendment on the mechanical and hydrological properties of a clay-rich subsoil from a coalmine in Victoria. Lignite was added to soil in concentrations of 2 per cent, 5 per cent, 9 per cent and 17 per cent. Multiple wetting and drying cycles were applied to consolidate the samples before measuring soil properties. An additional set of samples was prepared to evaluate the response of vegetation (grass cover) to coal-amended soil under water stress conditions. The following soil properties were assessed: pore size distribution, bulk density, hydraulic conductivity, plant available water, grass cover under drought conditions. Results showed, that adding up to 17 wt per cent of lignite increased the total pore volume by more than 10 per cent and plant available water by up to 8 per cent, along with an increase in saturated hydraulic conductivity. While high lignite concentrations improved soil properties, low concentrations resulted in either no or rather negative outcomes compared to the unamended control soil, such as higher bulk density, reduced total pore volume and no change in hydraulic conductivity. Vegetated samples subjected to drought stress exhibited prolonged water availability in soils with higher lignite concentrations. The improvement of physical and mechanical soil properties may also allow soil chemical improvement due to increased salt leaching capacity.

These findings support the use of lignite as a viable amendment in mine rehabilitation efforts, potentially reducing reliance on imported topsoil and improving long-term soil sustainability.

BACKGROUND

Rehabilitation of mining environments is often challenged by a lack of sufficient amounts of topsoil or the need to place topsoil on landforms unsuitable for stable landscape development. Failure risks may stem from physical and chemical limitations such as high bulk density, low permeability, and poor water retention, sodicity, poor soil structure – factors that significantly hinder successful revegetation. Common practice is the chemical and/or physical amelioration using gypsum, fertiliser application or deep ripping among others. While these measures improve conditions in the short-term, the benefits often diminish over time due to the consumption and leaching of chemical amendments or the re-compaction of mechanically loosened soil. This also applies to the use of fresh organic matter like compost. Ideally, soil amelioration should provide benefits lasting years or decades. Amongst a range of various amendments tested as soil improver, lignite has been trialled frequently primarily as an amendment influencing the chemical conditions or for enhancing fertility (eg Abraha, Tesfamariam and Truter, 2019; Paterson, Mushia and Mkula, 2019), as a component to create artificial soils (eg Walmsley *et al*, 2020) or tested as option for biological improvement (Rumpel, 2004). The authors are not aware of studies specifically focusing on the effect of soil amendment with lignite on soil physical properties. Background of this approach is the knowledge from soil mechanics literature, that mixing of two substrates of different particle size distributions can lead to a change of the pore system and to a decrease or increase of bulk densities and hence affect soil physical and soil hydrological characteristics.

OBJECTIVE

The objective of this study was to assess the effect of lignite coal addition on physical soil property improvement. A clayey-silty soil with a strong tendency to compact and limit water flow was amended with lignite to improve soil hydrological properties. Lignite was considered beneficial due to its long-term stability and ability to sustain soil functions over time and its neutral effect on plants. The added amounts of lignite span over a wide range and included concentrations well above what lignite applications reported in literature for eg chemical amendments commonly report.

METHODS

The material used in this study was a clayey-silty subsoil sourced from a coalmine in Victoria, Australia, selected for its high bulk density and low permeability. The subsoil was airdried and crushed. Crushed lignite was incorporated into the soil at varying concentrations of 2 per cent, 5 per cent, 9 per cent, and 17 per cent (by weight). Lignite was thoroughly mixed with the subsoil to ensure uniform distribution before conducting any tests. To evaluate the effect of the particle size, two lignite fractions were used: one with 60 per cent of coal particles >2 mm and another one with only <2 mm particles.

After mixing, the prepared soil-lignite samples and a control sample (soil only) underwent multiple wetting and drying cycles in a glasshouse to simulate field-like conditions and ensure consolidation. In total four sets of samples were prepared in trays (600 mm × 400 mm × 200 mm): two with the coarse lignite fraction and two with the fine fraction. One set of each of the replicates was seeded with grass and, after establishment, subjected to drought to assess the grass response.

The following tests were conducted to quantify soil functional and capacity properties.

Soil hydraulic properties were measured using intact soil cores (250 cm^3, 80 mm diameter and 50 mm height), collected from the soil consolidated trays. The HYPROP device (Meter Group Inc, USA/Germany) was used to determine accurately the soil water retention characteristics (SWRC) and derive soil hydraulic conductivity values based on the principle of evaporation, change of water potential and weight loss. The SWRC was fitted using the Fredlund/Xing approach to derive total pore volume, plant available water and saturated hydraulic conductivity were derived.

Grass response to drought conditions was assessed through image analysis and the use of AI tools quantifying the pixel greenness and yellowness of the grass and its distribution across the tray.

RESULTS AND DISCUSSION

HYPROP test data and SWRC fitting yielded values for total pore volume, saturated hydraulic conductivity and the plant available water capacity (PAWC). Table 1 presents the mean values from four test replications. Results show a slight decrease in total pore volume with the addition of 2 per cent lignite, but an increase to 50.3 per cent at 17 per cent lignite, 10 per cent higher than the control. The PAWC increases slightly for the 2 per cent and 5 per cent lignite mixing ratio, but is up to 6.3 per cent higher in the 17 per cent lignite mixing ratio compared to the control. Likewise the saturated hydraulic conductivity ksat increases from very low values of <3 cm/d for the control and the two low lignite concentrations to higher levels in higher concentrations (Table 1).

The findings illustrate that lignite addition modifies the pore system by increasing the pore volume – and lowering bulk density. Soils not only gain porosity overall, but also altered the pore size fraction, which can hold water. The addition of 17 per cent lignite raises the PAWC by >6 per cent, which for a 0.5 m deep soil equates to an increase of the available water amount by more than 30 mm. While it can be argued, the increase of the total pore volume and PAWC is attributed to the properties of lignite, only, as lignite has a very high pore volume and very low bulk density, the HYPROP simulated saturated hydraulic conductivity strongly indicates that the increase of ksat with increasing lignite content is indeed due to changing the pore system to larger pore volume and lower bulk density and in consequence therefore higher flow capacity.

TABLE 1

Hydraulic parameters of tested soil-lignite mixtures.

	Lignite mixing ratio	0%	stdev	2%	stdev	5%	stdev	9%	stdev	17%	stdev
Total pore volume	[%]	40.3	3.6	39.3	1.4	44.7	2.1	48.7	2.1	50.3	1.2
Difference of PAWC to control	[%]			2.1	4.4	2.6	6.1	5.1	3.8	6.3	3.2
Saturated hydraulic conductivity	[cm/d]	2.7	-	0.3	-	3.0	-	325.0	-	161.3	-

Results of soil and soil-lignite mixtures are based on four replicates.

At a point in time following multiple wetting and drying cycles and sufficient grass establishment, all trays were subjected to identical drought conditions. Over three weeks, the grasses increasingly turned yellow. Photos were taken, and the greenness of the pixels analysed and classified using image analysis. The results showed that by week three 80 per cent of grass in the 17 per cent lignite treatment remained green, compared to less than 20 per cent in the control and 2 per cent treatments (Table 2).

As nutrient conditions were identical across treatments, grass survival appeared primarily dependent on water availability. As shown from the hydrological tests, the soils with the higher lignite concentrations created pore distributions with increased water holding capacity. This seems to directly translate into prolonged plant survival under drought for high lignite concentrations.

TABLE 2

Grass vitality under drought conditions over time (classified as % green versus yellow).

Mixture	Percentage of green grass (%)			Percentage of yellow grass (%)		
%	week 1	week 2	week 3	week 1	week 2	week 3
0	29	18	12	62	76	81
2	31	29	18	63	67	74
5	52	47	44	46	51	54
9	68	67	64	26	27	30
17	89	82	80	3	12	12

SIGNIFICANCE

The objective of the study was to test whether the addition of lignite as a largely neutral substrate could positively influence the properties of a clayey-silty subsoil substrate, which otherwise by its properties would not be considered as a suitable substrate for revegetation. Unlike to many tests reported in literature, the tests presented here also included high lignite concentrations. Results showed that mixing of the soil substrate with coal at low concentrations did not have any beneficial effect, even reducing the total pore volume compared to the control. In contrast, high concentrations of lignite created substrates with higher porosity and plant available water capacity. Glasshouse trials confirmed that grasses in higher lignite treatments survived longer during drought conditions and confirmed the finding of higher plant water availability at high concentrations of lignite (9 per cent, 17 per cent) from soil hydrological tests.

While the vegetation trials did not show obvious signs of nutrient deficiency or other adverse impacts, further studies will be required to verify the potential of leaching of dissolved organic carbon, relative nutrient dilution by the addition of large quantities of lignite, the positive role of humate for nutrient storage or the impact on microbial activity to list a few.

Overall, the addition of lignite (or potentially black coal) offers the opportunity to improve hydrological properties, but also increase soil volume. This may be beneficial where shortage of topsoil availability in mine rehabilitation is an issue. In case of the addition of 17 per cent lignite the total soil volume would increase by about 30 per cent. Increased porosity and hydraulic conductivity may also be beneficial for the amelioration of saline substrates by enhancing the leaching capacity and, hence, leaching of salts.

For future application, a verification of the long-term viability and sustainability of achieved soil benefits will be required from large scale field trials.

ACKNOWLEDGEMENTS

The authors like to acknowledge AGL Loy Yang and Australian Coal Innovation for the financial support for this study and Wayne Powrie, Federation University, for the diligent execution of laboratory tests.

REFERENCES

Abraha, A B, Tesfamariam, E H and Truter, W F, 2019. Can a Blend of Amendments Be an Important Component of a Rehabilitation Strategy for Surface Coal Mined Soils?, *Sustainability,* 11:4297.

Paterson, D G, Mushia, M N and Mkula, S D, 2019. Effects of stockpiling on selected properties of opencast coal mine soils, *S Afr J Plant Soil,* 36:101–106.

Rumpel, C, 2004. Microbial use of lignite compared to recent plant litter as substrates in reclaimed coal mine soils, *Soil Biology and Biochemistry,* 36:67–75. https://doi.org/10.1016/j.soilbio.2003.08.020

Walmsley, A, Mundodi, L, Sederkenny, A, Anderson, N, Missen, J and Yellishetty, M, 2022. From spoil to soil: utilising waste materials to create soils for mine rehabilitation, in *Mine Closure 2022: Proceedings of the 15th International Conference on Mine Closure* (eds: A B Fourie, M Tibbett and G Boggs), pp 1237–1248 (Australian Centre for Geomechanics: Perth). https://doi.org/10.36487/ACG_repo/2215_92

Driving innovation in mine closure monitoring through automation and an integrated digital knowledge base

K Burkell[1], M Adams[2], A Zahradka[3], A-M Dagenais[4] and M Lato[5]

1. Senior Product Manager, Civil Engineer, Cambio Earth, Calgary, AB T2P 3L8, Canada. Email: katie.burkell@cambioearth.com
2. Principal Geoenvironmental Engineer, BGC Engineering, Halifax NS B3J 3N6, Canada. Email: madams@bgcengineering.ca
3. Data Scientist, Geological Engineer, Cambio Earth, Victoria BC V8W 3G9, Canada. Email: aron.zahradka@cambioearth.com
4. Principal Tailings and Mine Waste Engineer, BGC Engineering, Montreal QC H3B 5L1, Canada. Email: amdagenais@bgcengineering.ca
5. Chief Innovation Officer, Principal Geotechnical Engineer, BGC Engineering, Ottawa ON K1Z 6X6, Canada. Email: mlato@bgcengineering.ca

ABSTRACT

Effective mine closure monitoring is crucial for environmental protection, regulatory compliance, and long-term stability of post-mining landscapes. Current techniques and approaches, while robust, often lack integration and automation, leading to inefficiencies and missed opportunities for proactive management. This paper explores how a geospatial digital knowledge base can transform mine closure monitoring by integrating diverse data sets, automating analysis, and providing actionable insights.

The proposed approach includes the automatic ingestion, processing, and visualisation of airborne LiDAR or drone photogrammetry-derived change detection data to identify surface deformation and erosion patterns. Coupling this with satellite-based InSAR data, mine design information, risk assessments, and geological models enhances understanding of site conditions and adherence to closure criteria. Additionally, satellite imagery for vegetation monitoring offers a cost-effective method to assess ecological health and recovery.

Integrating real-time instrumentation data, such as piezometers, shape acceleration arrays, soil moisture sensors, tensiometers, and flow metres alongside climate data, provides a dynamic view of site conditions, enabling early detection of potential risks. Summary dashboards consolidate these data sets and compare them against closure criteria, streamlining reporting processes and facilitating regulatory compliance. Finally, access to historical site information through the knowledge base accelerates root cause analysis in the event of anomalies or failures, saving time and resources while improving decision-making.

This paper highlights how a geospatial digital knowledge base could be used to enhance monitoring at closed mine sites by showcasing how various operating mine sites and other sectors, such as linear infrastructure, utilise these data sets and the Cambio platform. By leveraging the Cambio platform, complex earth science data can be transformed into actionable insights, enabling better decision-making and risk management. For mine closure, this approach provides a holistic, proactive, and cost-effective solution for long-term site management.

INTRODUCTION

Overview of mine closure monitoring

Mine closure objectives vary for each project; but are considered to commonly include achieving physical, chemical and ecological stability, effectively managing risks, and attaining long-term care of the facility. The International Council on Mining and Metals (ICMM) has provided a Good Practice Guide for Integrated Mine Closure (ICMM, 2025) that describes common objectives and how monitoring fits into the closure project. A mine closure plan requires a comprehensive understanding of site conditions, as well as an understanding of how the established high-level objectives translate to site-specific objectives. Site specific objectives include a defined post-closure land use, meeting criteria for success, achieving social transition, and defining what relinquishment of the site looks like – all of this aligned with the expectations of engaged stakeholders. Closure monitoring provides information on the condition of the site and the performance toward meeting the criteria for success.

A robust knowledge base is needed to inform the design of closure; but also, to support monitoring, maintenance and management during execution. As described in ICMM (2025) the knowledge base supports the domain model as well as monitoring, measurement and inspections. The domain model includes not only technical information, but also historical information and local/traditional knowledge (when available). Such a broad knowledge base is necessary in the management of a closed mine site; however, that broad nature of the knowledge base creates challenges in the management of, and accessibility of, relevant information to inform decisions. Tones, Howe and du Plooy (2021) provide a useful summary of the challenges of building and maintaining a knowledge base that provides coherent and timely access to data that informs closure planning but is also directly applicable to closure execution. Incorporating closure monitoring data from various sources and instrument types into the knowledge base adds to the challenge, particularly when the post-closure reduction of site resources is considered.

Current data management and software strategies

Historically, data management and project information have been maintained through reports, drawings, and fragmented geographic information systems (GIS). This traditional approach, while necessary for documentation and regulatory reporting, often results in static interpretation of data which may be difficult to interpret and understand and would benefit from being viewed and analysed dynamically as new information becomes available. Relying on consultants or computer-aided design (CAD) teams to produce drawings or models can lead to significant delays in decision-making, with turnaround times extending to weeks or months. By then, the data may be outdated, obsolete, and unable to impact real-time decisions effectively. By digitising the data, it becomes possible to manipulate and interrogate it in real-time, facilitating informed decisions.

Engineers and owners often still require multiple applications to achieve a comprehensive understanding of on-site activities and main software applications are limited to office use due to network connectivity requirements. A comprehensive digital knowledge bases should continuously integrate near-real-time data with historical data, providing a dynamic and evolving understanding of the mine site. This continuous data integration allows operators to have a holistic understanding of current conditions at every stage of the mine life cycle. As the project progresses and more data becomes available, managing and analysing this growing data set becomes increasingly crucial. Automated data integration empowers teams to assess hazards, changing conditions, or potential non-compliance issues in near-real-time. This capability enhances the understanding of site conditions by coupling data sets for comprehensive analysis. This includes adherence to closure criteria, integration of historical site information, and providing a dynamic view of site conditions for early detection of potential risks. A software platform that can provide comprehensive data management capability, enhances learning and adaptation, ultimately leading to more informed and effective decision-making throughout the mine's life cycle is a huge asset for operators during mine closure and aligns with the recommendations from ICMM. This paper describes such an approach, utilising a geospatial digital knowledge base, Cambio™, which integrates diverse data sets, provides automation in support of analysis, and results in actionable insights.

The Cambio platform, integrates data sources, including remote sensing, instrumentation, field observations, historical project information (designs, reports, site characterisation data, baseline and operational monitoring data), and makes them accessible and viewable together in a single platform which is available on web and mobile applications (Figure 1). Having the data all viewed together in context with one another makes it that much easier to gain actionable insights. The cloud-based platform leverages over 20 years of earth science engineering expertise and is API (application program interface) driven so it can be integrated with a variety of existing platforms. This type of software architecture allows for other systems to remain the source of truth for certain data types while still allowing for a comprehensive geospatial consolidation of data, enhancing the visualisation and understanding of site conditions.

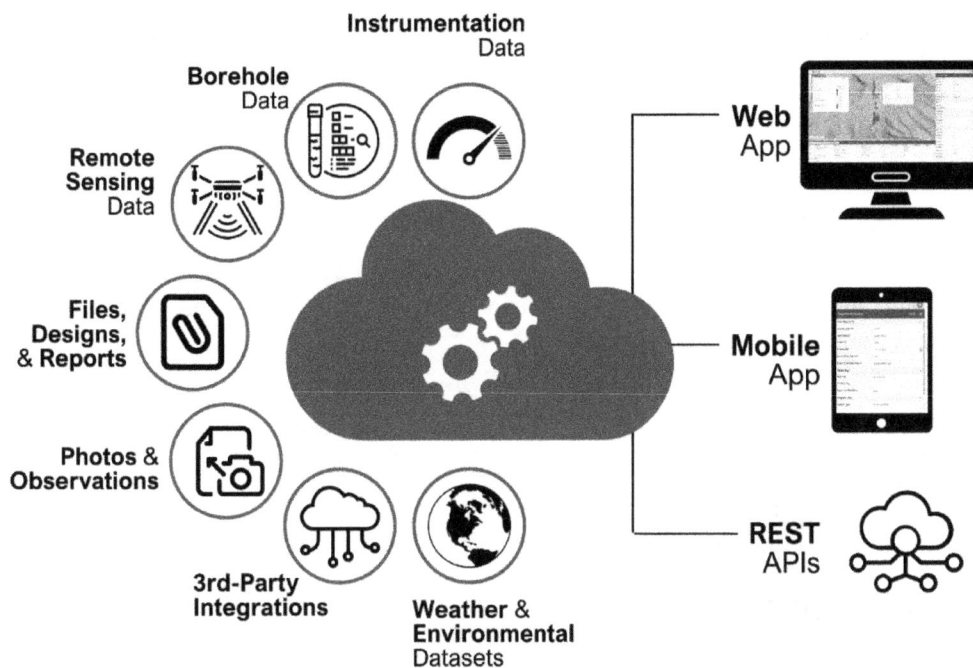

FIG 1 – Schematic of an integrated platform model with automated ingestion pathways and various data accessibility options.

The following sections discuss closure landforms and the associated monitoring requirements in closure, then highlight key aspects of the Cambio platform that has been used by various projects and industries, and how it could benefit and change the monitoring of closed mining projects.

CLOSURE LANDFORMS

Closure landforms resulting from mining, whether they are waste rock piles, mine pits, various stockpiles or tailings storage facilities, will require some form of monitoring to inform closure performance and for regulatory compliance. The main components of a closure landform include vegetation, cover systems, geometry, and surface water management infrastructure.

Vegetation can be a control of both physical stability (controlling erosion and pore water pressure) and chemical stability (controlling water ingress and the creation of contact water). Landform covers themselves (upon which the vegetation may be established) also contribute to control of these same stabilising processes. Covers can isolate and divert clean water from disturbed (source) materials and keep it clean. They can limit infiltration and mitigate the creation of contact water, as well as reduce the driving forces for migration of that contact water. Limiting infiltration can also lower pore pressures to enhance physical stability. Oxygen ingress might also be controlled to limit conditions that degrade the contact water.

The geometry of the landforms themselves can also limit the creation of contact water and limit the extent of impact of that contact water. They can contribute to water management such that recharge is limited, and clean water is directed around areas with risk of chemical impacts.

The physical stability of landforms must be monitored to confirm those landforms are performing as the design intended. The health of the vegetative component (if present) of cover systems can also be monitored to confirm those covers are performing as the design intended. Discharging water (seepage and run-off) from the site can be monitored to confirm it is meeting the performance objective of promoting ecological stability of the receiving environment. Quantification of water within the water management system can confirm the water balance is meeting the design intent.

Following from the above description of closure landscape components and their function are monitoring parameters that arise to assess the performance of the closure landscapes:

- Vegetation health and coverage.

- Landform geometry and drainage pattern disruption (ie settlement, slope movement, gully formation).

- Cover system integrity (eg erosion, gully formation).

- Water management system imbalance (ie changes in water balance components).

- Infiltration rates (eg lysimeters) and pore water pressures (eg vibrating wire piezometers and monitoring wells).

- Water quality at internal receptor points (eg ponds), within the subsurface (eg at monitoring wells) and at discharge points.

- Monitoring the ecological health of the site and the receiving environment, based on the land use of those areas (eg flora and fauna health, contaminant loading, water quality at consumption points).

Some of the above parameters may be monitored for compliance to regulatory requirements. Others might be monitored to identify maintenance needs prior to regulatory compliance issues.

CLOSURE MONITORING

Current monitoring techniques and challenges

A typical closure program includes an active closure phase, followed by a passive closure phase. Each of these phases has its own monitoring requirements:

- Active closure phase – typically requires more frequent and denser measurements against performance criteria during the evolution of conditions from operations, through closure construction, and into closure. An updated conceptual model of the site based on the conditions changed during closure is needed.

- Passive closure phase – to be informed by critical indicators of the updated conceptual site model. Monitoring may be limited to confirming the physical stability of landforms and health of receiving environments.

Using the same monitoring program components previously listed, following is a description of how they might be addressed in closure:

- Vegetation establishment – approaches to monitoring this range from on-site inspection with ground level surveys and in some cases sample analysis, to light detection and ranging (LiDAR) surveys and interferometric synthetic aperture radar (InSAR) remote sensing techniques. Sage et al (2022) present an example of the latter application.

- Landform geometry monitoring – typically involves topographic surveys, sometimes with comparison to (and calibration of) settlement models as well as inspections. Characteristics of tailings, for example, are assessed throughout operation to inform tailings management planning. These characteristics can then be applied to closure planning and design and further inform the performance monitoring program. Settlement or other geometry changes can be identified through surveys and can trigger maintenance to preserve the function of the landform. Brink and Heymann (2024) discuss post-closure fill settlement and a risk management framework to address it. Monitoring observations (as part of a broad knowledge base) are an integral part of that framework.

- Surface water (SW) flows/drainage pattern disruption – typically monitored through inspection and surveys. Observations of stagnation points (ie ponding that can enhance infiltration) or drainage patterns that depart from designed drainage network. Departures from the design drainage pattern can result in unanticipated concentration of flows, causing erosion that can further disrupt drainage patterns and degraded SW quality, and in some cases can degrade the cover system. Observations and surveys can be supported by flow quantity instrumentation that can trigger site inspections as SW flow patterns vary from historic or modelled values. Crisp et al (2024) present example of the application of LiDAR to monitor stability and erosion of rehabilitated landforms.

- Water balance updates can be applied to assess changes in water movement at a closed site. The flow pattern changes discussed above can be precursors to larger shifts in the water balance that can have a significant impact on the overall performance of the closed landform's physical and chemical stability. For example, this shift can give rise to changes in the volume of a pond that forms part of the closure landscape, which can lead to shifts in infiltration quantities, changes in discharging water quality, etc.

- Cover performance can be inferred by analysis of surface water flows, water balance updates and pore pressure measurements, or it can be directly assessed by monitoring the infiltration rates or the soil moisture profile of the cover system. Ultimately, it might also be assessed by measurement of groundwater levels, or by the quality of the underlying or downgradient groundwater (see below).

- Water quality monitoring is typically done at a minimum at the receiving environment and other compliance points (potentially upgradient points that may allow remedial responses prior to receiving environment impact). A rising trend of contaminants in the site discharge point without any apparent integrity deficiencies in the landform may indicate that the design is not achieving chemical stability of the waste rock, tailings or other waste within the landform. This is a direct and comprehensive measure of performance of the closure controls. However, it has limitations in that it does not immediately identify the location of the performance gap and is reactive to performance shortcomings that may have been years in the making.

- Monitoring the ecological health of the receiving environment is a step that is further along the migration pathway than water quality monitoring, as it identifies the impacts of the quality exceedances that have already occurred; therefore, incorporating toxicological effects and any natural mitigating factors that may arise in the pathway-receptor-exposure chain. Responses to these monitoring triggers are reactive and require an understanding of the root cause of the observed trigger.

The monitoring components described above are at different placements along the pathway from source to receptor. Early identification of performance gaps provides more time to address issues while maintaining compliance, often more efficiently and easily remediated than if addressed later and is likely nearer to the root cause of the performance gap; for example, deficiency in the integrity of the cover or in the diversion of clean water can be more efficiently addressed than remediation of a fully developed contact water plume.

Effective mine closure monitoring including all the monitoring aspects described above is crucial to achieving the objectives of closure. There are many approaches to mine closure monitoring that strike individual balances between effectiveness and efficiency. An opportunity exists to apply an approach that integrates the disparate aspects of mine closure monitoring, allows automation of data gathering or management, and supports proactive management of the site.

Geospatial digital knowledge base for closure monitoring

Safe closure and monitoring of a mine facility requires the combined efforts and attention of multidisciplinary groups and effective communication of complex subjects to a wide range of stakeholders with varying levels of technical understanding. Large quantities of data are produced throughout a mine's life cycle, and as data collection becomes easier, the effective management of the data becomes more challenging. With an exponential growth in available data both in terms of different sources and quantity of information that can now be used to monitor sites, more advanced data management, processing, and analysis has become increasingly more important. Otherwise, the value of the data is underutilised (Johnson *et al*, 2022).

Additionally, the regulatory landscape is evolving, and conformance to standards such as the Global Industry Standard on Tailings Management (GISTM; Global Tailings Review, 2020) requires operators to develop and maintain an interdisciplinary knowledge base of social, environmental, economic, and technical data to inform decisions throughout the tailings facility life cycle, including closure and post-closure phases. The development of an integrated knowledge base allows project teams to access key data via a central location which can be used to inform decision-making during mine closure and long-term monitoring.

Recently there has been an emerging trend and ongoing advancements in software platforms that are being used for monitoring earth science data. One of which is the concept of a geospatial knowledge base which is typically designed for the comprehensive management of earth science and project specific data. The platforms function as virtual representations of project sites, whether it be a network of pipeline data, other linear infrastructure data (such as a railways or highways), communities or mine sites. These 'digital twins' integrate a multitude of data sets such as geological, environmental, operational, remote sensing, instrumentation, and weather data into a unified system. The integration of these diverse data sets facilitates a holistic understanding of the project area, enabling efficient and informed decision-making.

The preceding discussion has outlined a framework of closure monitoring, the drivers for each component, and some examples of the approaches used to monitor those components. Research is abundant and the body of experience in application of general monitoring approaches is extensive. However, many challenges can be encountered in applying monitoring to closed sites.

- As the body of data to inform closure management become larger, accessibility and the ability to navigate through the monitoring history to develop a comprehensive assessment of site performance becomes more challenging.

- Managing the closure monitoring data sets, there is a very large body of data from the operation phase of the mine life. These data are crucial in assessing the closure monitoring data to develop repair and maintenance actions in response to monitoring triggers.

- Remote sensing approaches might be used as an alternative or enhancement to site inspections. There has been much recent research in the application of these technologies to mine sites (eg Braimbridge *et al*, 2019; Kelcey *et al*, 2019; Jones and Franklin, 2019; Morel *et al*, 2021; Sage *et al*, 2022; Crisp *et al*, 2024). However, these approaches are also data intensive. Photogrammetry, LiDAR surveys and other remote sensing approaches involve large data sets.

PROPOSED SOLUTION – DATA INTEGRATION

Data accessibility

Configurable workspaces in Cambio offer significant benefits for project team collaboration, review meeting preparation, and discussions. A workspace is a customisable environment that anyone can configure and save their preferred project views, allowing quick access to relevant data and tools. Workspaces can be personal or shared amongst project teams, allowing for streamlined collaboration. A key aspect of mine closure includes stakeholder engagement and workspaces could be a way to help convey complex concepts or visual changes to the site over time. With the ability to set specific layers, filtering options, and map locations, these workspaces ensure that all team members are viewing the same data, which is necessary for effective communication and decision-making.

During review meetings, having a standardised workspace helps streamline discussions, as everyone can easily reference the same information. This consistency also enhances monitoring and reporting efficiency, as predefined workspaces reduce the time spent on setting up views so that reports are based on uniform data sets. For example, an Engineer of Record (EOR) and team used a Cambio workspace to organise and present design layers that illustrated complex sequencing and geometry in response to a review board action item. The workspace allowed everyone to toggle layers on and off, view surfaces in 2D or 3D, and apply colour coding for clarity. It was specifically configured to provide external review board members with controlled access to relevant data and contextual information for independent technical review board meetings.

Additionally, configurable workspaces serve as a comprehensive tool for viewing changing site conditions. By regularly updating the data layers and filters, teams can monitor real-time changes in the environment, infrastructure, or other relevant factors. This dynamic capability allows for timely responses to new developments and allows all stakeholders to be informed about the latest site conditions. Overall, configurable workspaces foster a more cohesive and productive team environment by providing a reliable and shared platform for geospatial analysis.

The Cambio platform is set-up with dashboards that provide a rolled-up summary of critical data, so that important information is brought to the team's attention promptly. These dashboards offer a heads-up view of site conditions, allowing teams to quickly identify areas that require prioritised resources or attention. One example of standard dashboards would be in the space of instrumentation monitoring – including current reading levels versus site thresholds, number of instruments above thresholds, instruments that are no longer receiving data and prioritised list of instruments needing to be manually read or require maintenance (Figure 2). Another example would be site performance monitoring – given certain compliance criteria, is the site or site personnel meeting standards including number of outstanding actions, dam assessment recommendations, number of compliance specifications passing or failing. By including compliance criteria within the system, dashboards can also show relative compliance, aiding in meeting reporting requirements efficiently.

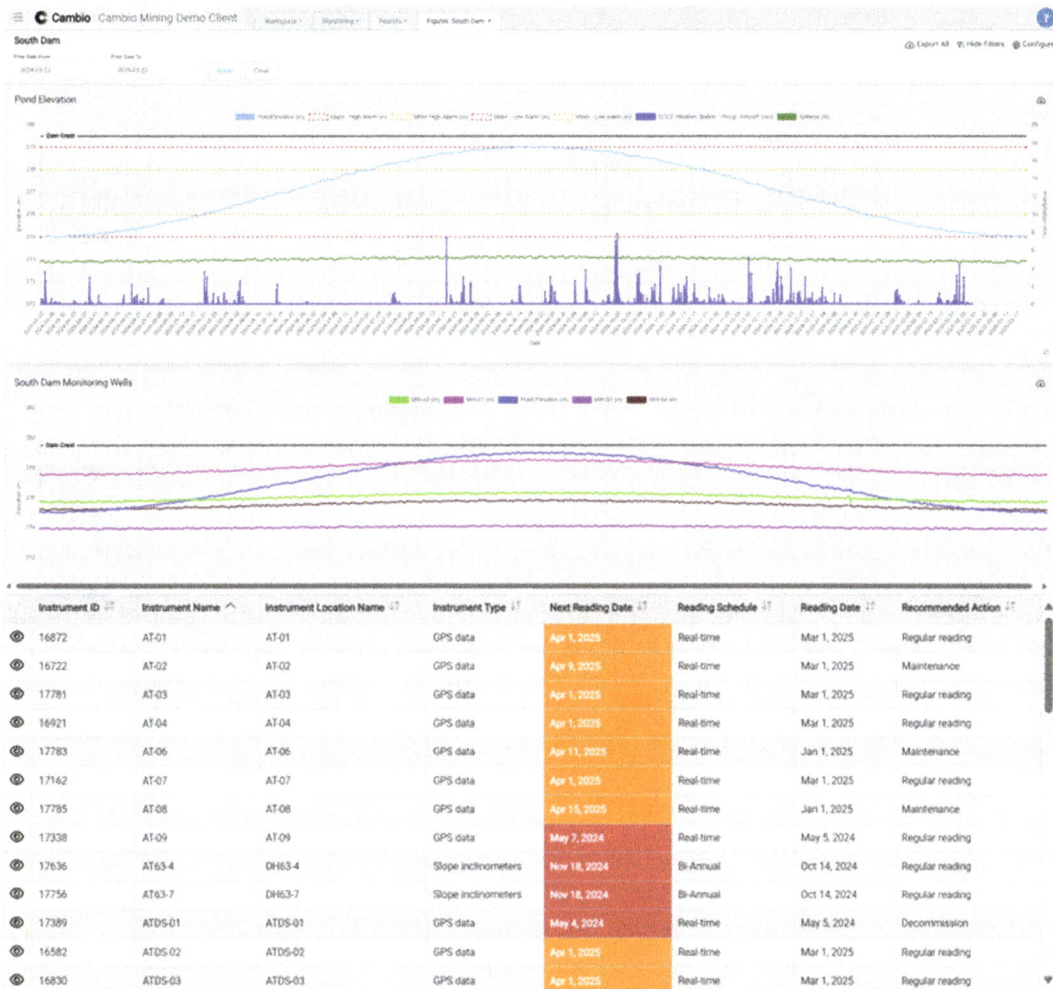

FIG 2 – Example of a monitoring dashboard in Cambio used to show instruments performance and summarise maintenance and monitoring requirements with colours to highlight time frames and recommendations outlined in the table. Each row can be expanded to show more detail.

As described in the opening sections of the paper, an important aspect of mine closure is the mine closure plan which requires an understanding of site conditions, as well as status of site-specific objectives. Dashboards as described above could play an important role in the digital transformation of mine closure plans. The connective and geospatial benefits of these dashboards are particularly valuable. Users can identify spatial trends and because the dashboards are linked to full data sets users can drill down into detailed information as needed. This summary view helps teams monitor performance parameters in relation to closure criteria and respond to changing conditions effectively.

Monitoring

Mine closure monitoring is used to help decommissioned sites achieve long-term stability and meet environmental and societal expectations. The process encompasses two primary aspects: performance monitoring, which assesses site conditions against predefined success criteria, and information gathering, which involves documenting and maintaining a database of site conditions throughout the life cycle of the mine, spanning from baseline data collection through operations and into closure. A digital geospatial knowledge base serves as a tool to streamline closure monitoring and the assessment of data arising from those monitoring efforts.

Selecting success criteria for closure using the SMART approach (specific, measurable, achievable, relevant and timely) is recommended (ICMM, 2025). The application of the SMART framework to a geospatial digital knowledgebase can help define the configuration to align with closure objectives and site-specific requirements. It provides a structured and actionable methodology to select, prioritise, and implement the data and functionalities that the digital twin should encompass. The following sections will highlight technical advances in the monitoring space – remote sensing and instrumentation, that can be applied to mine closure monitoring and how these data sources are only enhanced by the use and integration with a digital knowledgebase.

Cambio could improve mine closure monitoring projects by providing real-time data integration and dynamic analysis capabilities. By pairing historical data, closure criteria, and automatically integrating real-time data, owners have a comprehensive and up-to-date understanding of site conditions, facilitating timely and informed decision-making. By enhancing data management and analysis, project efficiency and compliance can be significantly improved, ultimately leading to more effective mine closure outcomes.

Remote sensing for settlement, erosion, and vegetation monitoring

One of the most significant advances in the earth sciences profession has been the evolution and availability of remote earth sensing technologies, including photogrammetry, LiDAR and satellite-based InSAR. These technologies have allowed for rapid and accurate assessments of ground movement. Remote sensing data can be used to monitor erosion, settlement, or changes to mining facilities that could be hard (or impossible) to see with the human eye. With fewer people on-site the use of drone-based photogrammetry or LiDAR, or satellite based InSAR, paired with automated processing techniques can identify changes and potential issues sooner, at scales that were previously impossible to achieve. Tailing facilities can be assessed to determine where millimetres of movement may be occurring, and this can be repeated to understand not just the locations of change, but also whether it is accelerating or slowing down.

LiDAR

The use of LiDAR or photogrammetry data has become increasingly popular at mine facilities (Schmidt *et al*, 2015). By comparing multiple LiDAR data sets, the positional change in the data sets can be assessed over time in a process known as LiDAR change detection (LCD). Changes analysed between LiDAR data sets are reported as positive or negative relative to the baseline or initial LiDAR survey. Positive model differences can be interpreted as material accumulation or bulging, and negative model differences can be interpreted as a loss of material, erosion or settlement. Availability of on-site data collection and LCD processing have increased both efficiency and quality of results. These advancements have enabled high-accuracy LCD to be performed in automated ways. One of the primary advantages of employing a digital knowledge base is the automation of data analysis. Utilising sophisticated algorithms and machine learning techniques, the platform continuously processes incoming data, identifying patterns and anomalies that may not be immediately discernible to the human eye. This automated analysis generates actionable insights, such as issuing triggers based on changing site conditions or deviation from set closure criteria.

The integration of LiDAR data also provides significant advantages for vegetation management in mine closure monitoring. Unlike traditional imagery, LiDAR captures detailed three-dimensional structural information about vegetation, including canopy height, density, and overall biomass. By performing repeated LiDAR surveys over rehabilitated areas, mine operators can accurately track vegetation growth and health over time, confirming compliance with closure criteria related to ecological recovery and vegetation density or height and confirming the function of the cover system or erosion control requiring the vegetative component. Additionally, automated LiDAR analysis within the digital knowledge base streamlines the monitoring process, rapidly identifying areas that are lagging in recovery, and providing precise data that informs remediation activities.

InSAR

In addition to LiDAR, InSAR provides a valuable complementary method for monitoring settlement and surface deformation during mine closure, particularly for tailings dams and impoundments. InSAR works by transmitting radar signals from satellites towards the Earth's surface and measuring the phase differences between signals collected at different times. These phase differences indicate slight shifts or movements on the ground surface, enabling the detection of subtle ground movements, often with detection limits as precise as a few millimetres per annum over extensive areas. Regularly acquired satellite radar data, such as from the freely available Sentinel-1 constellation or the upcoming NISAR mission, enable continuous (sub-weekly) tracking of slow-moving processes like subsidence, creep, or settlement across tailings facilities, waste rock dumps, and reclaimed mine surfaces. Additionally, commercial InSAR data providers offer higher-resolution imagery and more frequent revisit times, making them preferable for detailed, rapid-response monitoring. A major benefit of spaceborne radar imagery is that since it is continuously collected globally, historical data extending back to at least 2014 is often available for analysis at a mine site.

The potential effectiveness of InSAR is demonstrated by the Brumadinho Tailings Dam collapse in Brazil. Advanced analysis of satellite data following the failure event demonstrated that ground deformation precursors were detectable months prior to the disaster (Grebby et al, 2021). This highlights how routine InSAR monitoring could have provided critical early warning, potentially averting the failure or earlier engagement of the emergency response plan. While aerial LCD provides precise, high-resolution 3D measurements ideal for identifying localised changes and quantifying erosion or structural deformation typically greater than a few centimetres, InSAR excels in detecting broader-scale, subtle ground movements over time. Automated InSAR processing within digital platforms facilitates early detection of unexpected deformation patterns, allowing site operators to proactively respond to potential stability concerns. By combining the strengths of both LCD and InSAR, owners gain a comprehensive understanding of site stability. Thus, integrating InSAR into the digital knowledge base significantly enhances the ability to monitor post-mining landforms and infrastructure, promoting long-term safety and compliance with closure objectives.

In addition to monitoring ground deformation, Synthetic Aperture Radar (SAR) can also be used to estimate soil moisture conditions. Radar backscatter is sensitive to changes in surface moisture, particularly in the top few centimetres of soil, and variations in moisture content can be observed through time-series analysis. Monitoring soil moisture is especially important in the context of mine closure, as it influences vegetation recovery, erosion potential, and slope stability (through the identification of seepage zones) and is an indicator of infiltration conditions and other aspects of cover performance. SAR-based soil moisture estimates can support early identification of overly dry or saturated conditions, helping inform revegetation efforts, deviations from drainage design, cover performance and risk assessments related to run-off or slope failure.

A recent addition to the space-based radar toolkit is the Surface Water and Ocean Topography (SWOT) satellite, launched as a joint mission by NASA and CNES. SWOT provides a new capability for monitoring water covers on tailings facilities and pit lakes in mine closure settings. Using advanced radar interferometry, SWOT measures surface water extent, elevation, and volume changes with unprecedented precision. Its global coverage and ability to measure water bodies as small as 250 m wide make it a powerful tool for large-scale mine closure monitoring. By integrating SWOT data with other remote sensing and on-site hydrological data, mine operators can gain deeper insights into long-term water balance, weather- and climate-driven changes, potential risks associated with water-covered mine closure strategies, and other aspect of water management system performance.

Multispectral and hyperspectral imagery

In addition to LiDAR and InSAR, multispectral and hyperspectral satellite imagery provides important complementary capabilities for remote monitoring of environmental conditions at closed mine sites.

These remote sensing technologies capture reflected solar energy across multiple spectral bands, allowing detailed assessments of vegetation health, moisture conditions, seepage, erosion, and water quality issues around tailings facilities. Multispectral sensors typically measure reflectance in a limited number of discrete bands (eg visible, near-infrared, short wave infrared), enabling effective monitoring of vegetation health, soil moisture content, and erosion patterns. In contrast, hyperspectral sensors measure reflectance across dozens or hundreds of narrow spectral bands, providing detailed spectral signatures that can identify specific minerals, contaminants, or subtle variations in vegetation stress, soil conditions, and water quality.

Multispectral imagery is widely available from public satellites such as Sentinel-2 (European Space Agency) and Landsat (NASA/USGS), both of which offer free access to regular sub-weekly imagery at resolutions of 10–30 m. These platforms are highly effective for routine monitoring of vegetation health through indices like Normalized Difference Vegetation Index (NDVI), detecting moisture anomalies indicative of seepage or ponding, and tracking erosion through changes in exposed soil and sediment transport patterns. Commercial providers offer higher-resolution multispectral imagery, often at daily revisit intervals and sub-meter spatial resolution, enabling more precise monitoring of localised issues such as erosion gullies and seepage zones.

Hyperspectral imagery, while less widely available, provides additional capabilities for detailed environmental monitoring. This type of imagery can detect subtle variations in water quality downstream of tailings sites, including elevated turbidity, sediment plumes, acid mine drainage, and other contaminants based on their unique spectral signatures. Additionally, hyperspectral data can precisely differentiate erosion-affected areas from stable soil by detecting specific mineral signatures associated with disturbed or freshly exposed substrates. Public access to hyperspectral data is currently limited, but missions such as NASA's forthcoming Surface Biology and Geology (SBG) mission aim to improve availability. Meanwhile, commercial providers offer satellite and airborne hyperspectral collection tailored to specific monitoring needs.

Remote sensing data integration

Integrating remote sensing data sets – such as LiDAR, InSAR, and multispectral/hyperspectral imagery – with other mine closure design and monitoring data significantly enhances the quality and speed of decision-making processes. Each remote sensing method provides unique and complementary insights: LiDAR delivers high resolution deformation, erosion and vegetation measurements, InSAR continuously tracks subtle movements, and multispectral or hyperspectral imagery monitors vegetation, moisture conditions, erosion, and contaminants. When these diverse data streams are combined and analysed alongside traditional mine closure data sets (eg closure design, geological models, hydrogeological data, geotechnical instrumentation readings, climate data, and historical site records including operations and construction records), owners gain a holistic understanding of evolving site conditions.

For example, integrating InSAR-derived ground deformation data with piezometer readings and weather data enables early detection and analysis of potential instability in tailings dams (Figure 3), such as the correlations identified by Grebby *et al* (2021) following the Brumadinho tailings dam failure. Similarly, combining LiDAR-derived vegetation structure measurements with multispectral or hyperspectral vegetation health indices (eg NDVI) and weather data provides a comprehensive view of vegetation growth, health, and ecological recovery, quickly identifying areas requiring additional rehabilitation or maintenance efforts and potentially identifying distress indicators in the receiving environment. Combining high-resolution digital elevation models derived from LiDAR surveys with soil moisture and ponding assessments from freely available multispectral imagery enables up-to-date mapping of current drainage paths and ponding areas.

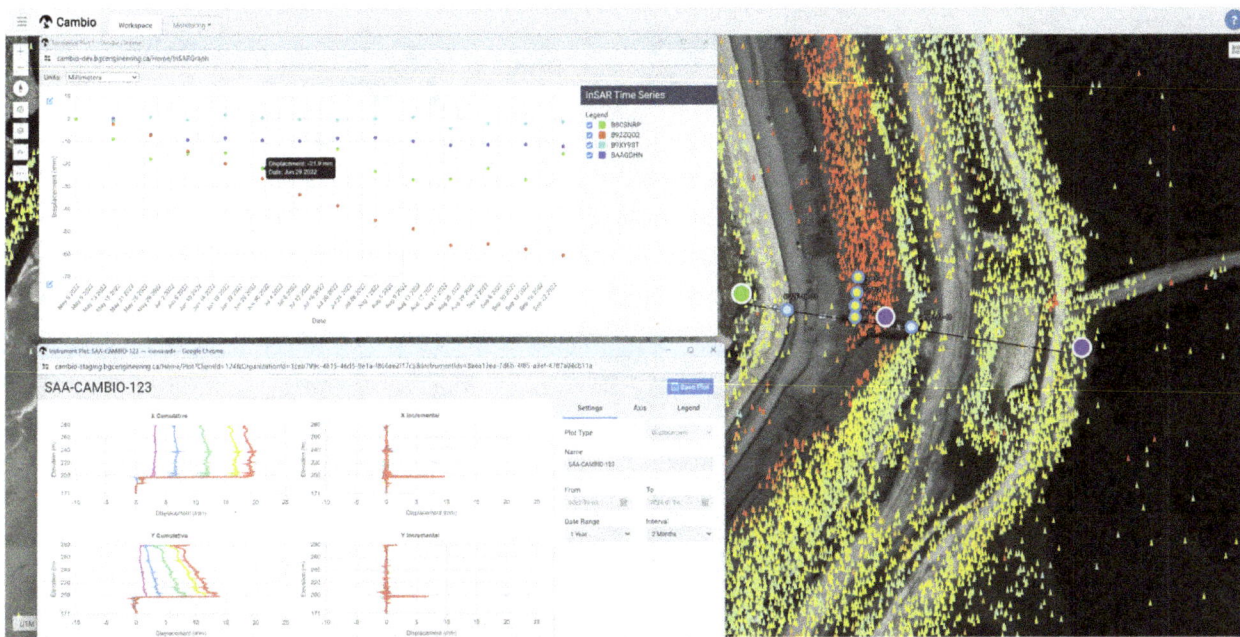

FIG 3 – Example of InSAR, instrumentation (SAA) data, standard cross-sections and project imagery all viewed together in one platform.

Finally, the combination of remote sensing data collection with digital knowledge bases allows for fully remote collection and analysis of data, allowing for more efficient ways of allocating resources. This capability enables experts to monitor and analyse mine site conditions from anywhere, so that issues are identified and addressed promptly. When fieldwork is necessary, it can be more targeted and efficient, as personnel can access the project data while in the field, even without the need for internet connectivity. This approach not only saves time but also enhances the effectiveness of field operations.

Instrumentation

An integrated monitoring platform that supports a broad range of instrument types could address the challenges associated to disparate monitoring data, and enable centralised management of installation records, spatial context, historical and real-time readings, and associated documentation such as photos or maintenance logs. A system with this level of data integration can improve traceability, support more rigorous quality control, and make it easier to interpret and communicate monitoring results.

One example of such a system is Cambio, which is designed to manage detailed metadata associated with each instrument, including configuration parameters, calibration factors, thresholds, and maintenance history, alongside geospatial and temporal data. Both manual and automated data acquisition methods are supported, allowing integration of field-collected measurements with real-time sensor feeds.

Visualisation tools within the platform facilitate the interpretation of instrument data by enabling the configuration of plots and dashboards tailored to specific project needs (Figure 4). These can be used to track readings against defined thresholds, supporting early detection of conditions that may warrant further investigation. Visual outputs can be exported in a variety of formats to aid communication with project teams, regulators, and other stakeholders.

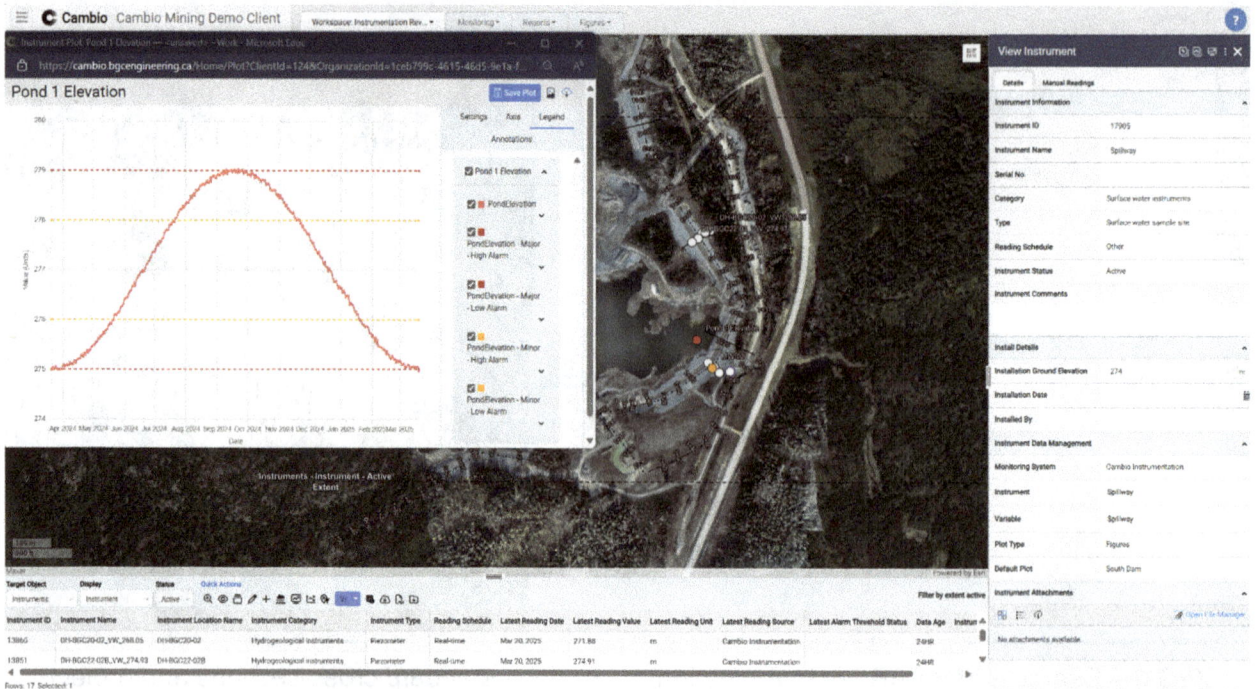

FIG 4 – Image of instruments in alarm status can be viewed in summary tables and map views, providing a comprehensive overview of site conditions.

The ability to generate cross-sections dynamically, using up-to-date instrument readings, site designs, historical topography, borehole logs, and other spatial data sets (Figure 5) reduces the time typically required to produce such visualisations and allows for more iterative, on-demand analysis. A proactive monitoring approach like this helps identify potential issues and enables them to be addressed promptly, reducing the risk of non-compliance and environmental hazards.

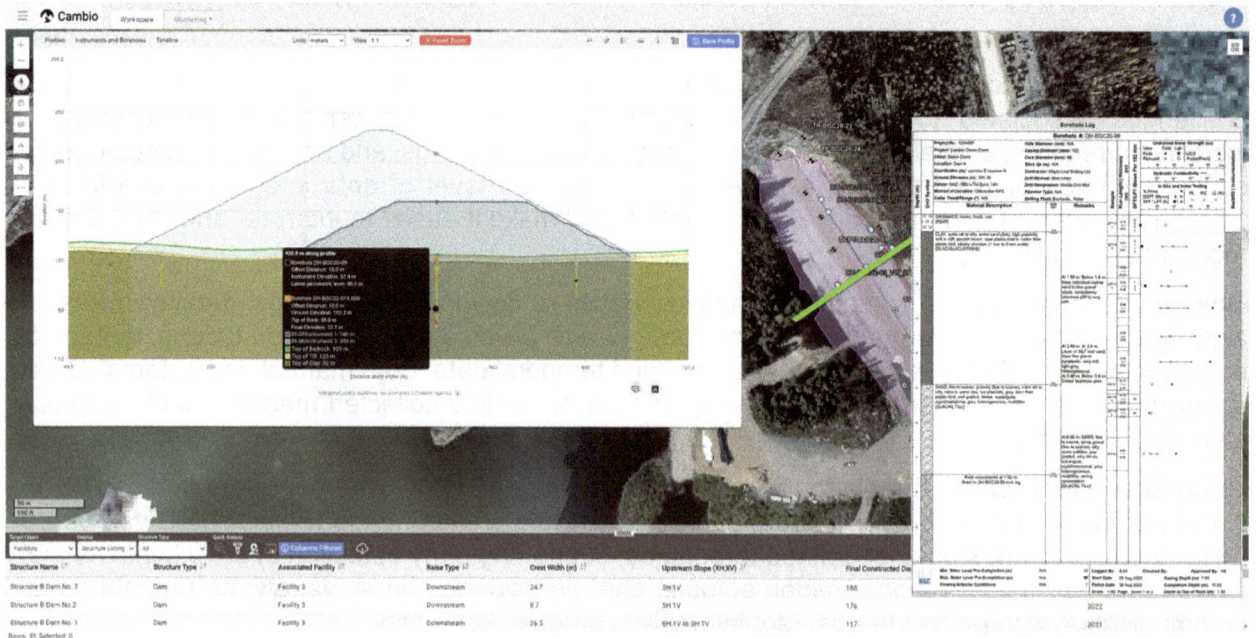

FIG 5 – Image of the Cambio platform highlighting the ability to view multiple data sources (borehole location and logs, instruments, cross-sections) at once to better understand site conditions.

The integration of diverse data sources, combined with configurable visualisation and alerting capabilities, supports more efficient and responsive monitoring workflows. For mine closure projects, where maintaining long-term site stability and compliance is critical, this type of system has the potential to improve closure monitoring.

CASE STUDIES AND APPLICATIONS

Innovative monitoring approaches at tailing construction projects

Case study #1

The safe management of tailings facilities necessitates the combined efforts of multidisciplinary teams and the clear communication of complex subjects to a diverse range of stakeholders. A mine site in British Columbia, Canada, has adopted Cambio to support tailings dam construction and performance monitoring.

Effective construction and operations management at the mine site required a comprehensive understanding of the project site and historical conditions. Cambio supports the construction teams by integrating design lines, borrow area polygons, and construction imagery, enhancing communication of construction progress. It also aids in visualisation and communication between on-site and off-site staff by combining field photographs with recent imagery and topography data.

As noted previously, Cambio can ingest and display a variety of data sets and interact with other software platforms. Project teams analyse and visualise data in plan or three-dimensional views or in profile by creating sections through topography, design surfaces, drill hole stratigraphy, and instrumentation. The platform is used by the owner and EOR teams for day-to-day communication, such as reviewing historical data including as-built information and tailings deposition locations, projecting future tailings facility configurations at various design stages, and planning site investigation and instrumentation installation programs. This integration facilitates communication among multidisciplinary teams regarding design considerations and the incorporation of new data into existing design bases.

Examples of daily tasks completed in Cambio during annual dam construction include, tracking weekly fill placement volumes with the automated change detection tool to monitor construction progress. In the field, the Cambio mobile application is used by construction quality assurance and quality control teams to log testing and sample data, which is then summarised in a dashboard to track compliance with design specifications and further utilises the fill placement volumes to calculate required test frequencies. The owner and EOR teams use Cambio to monitor and review dam performance using the instrumentation module. Instrumentation is essential for performance-based design and risk management, with readings compared to predefined thresholds to guide dam performance assessments.

Case study #2

The implementation of standard dam inspection forms in Cambio, equipped with georeferenced information, has significantly enhanced reporting efficiency and quality for a mine in Manitoba, Canada (Figure 6). The forms provide a structured and repeatable checklist for field personnel, so that all necessary observations are consistently captured during inspections. The georeferenced data allows inspectors to precisely locate and document their findings, which is crucial for tracking changes over time and maintaining accurate records. This systematic approach not only streamlines the reporting process but also improves the reliability and comprehensiveness of the data collected, facilitating better decision-making and risk management.

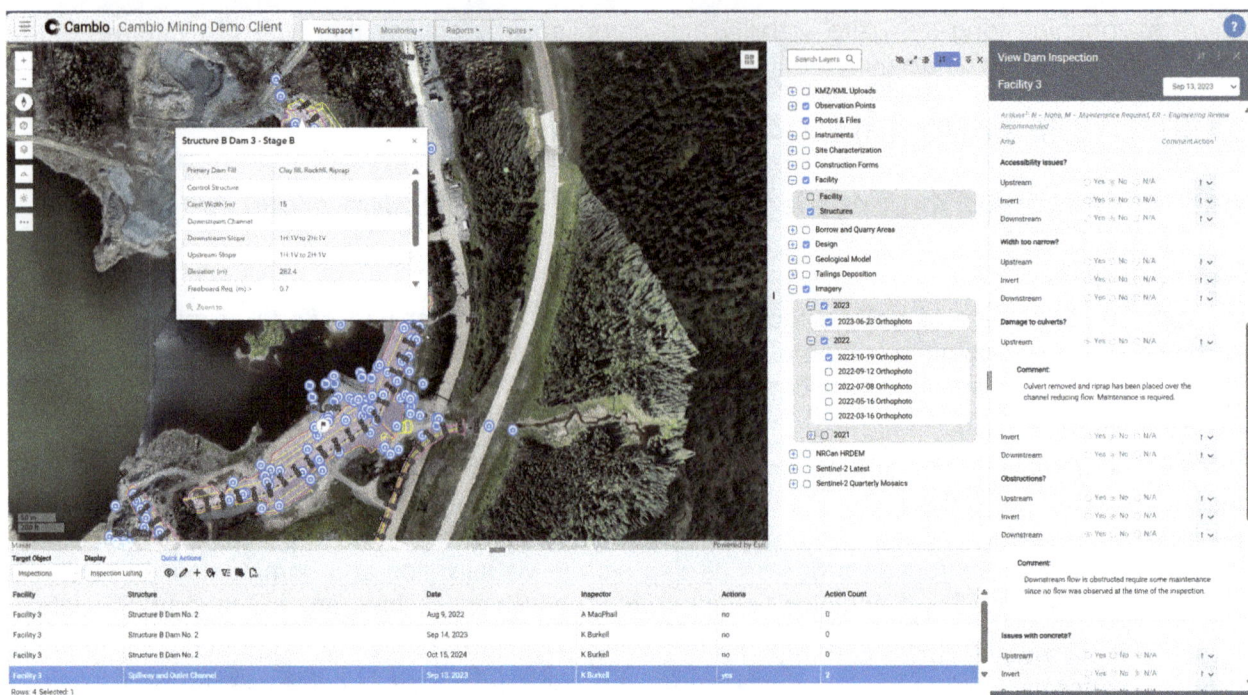

FIG 6 – screenshots of a digital knowledge base containing digital dam inspection forms, key structure information and georeferenced photo location and images.

Moreover, the use of these standardised forms has greatly aided in onboarding new engineers onto the project and site. New team members can quickly familiarise themselves with the inspection protocols and site conditions by referring to the detailed notes, recommendations, photos, and observation locations recorded by previous engineers. This promotes continuity between project team members since they have access to all historical data and insights necessary to perform thorough inspections. The ability to review past observations and recommendations helps maintain consistency in monitoring and addressing dam performance issues, thereby upholding the integrity of the inspection process.

On both of the above mining project examples the owners have benefited from having easy access to borrow locations and drill hole logs found in Cambio. This information is invaluable for planning ahead of the field season, as it allows for efficient resource allocation and preparation. By having detailed records of borrow areas and drill hole data, operators can make informed decisions about material sourcing and site logistics, ultimately optimising project timelines and reducing costs. The integration of these data sets into the Cambio platform allows all stakeholders to have the necessary tools and information to support effective project management and execution. This could be analogous to having access to closure criteria information or closure tasks that need to be executed on a routine basis. With having less personnel available to do the work, having access to the historical information as well as current and upcoming actions leads to more effective use of resources

Asset monitoring and geohazard management

Case study #3 – linear infrastructure

Cambio is used by over 150 organisations to digitally transform the collection, analysis, visualisation of earth science data and help clients address the increased regulatory requirements of the pipeline and transportation industries, similar to those currently seen in the mining industry. Historically, pipeline geohazard management systems were unique to each operator, leading to inconsistent and subjective assessments. These programs relied on reports summarising individual geohazards or provided a static overview of pipeline systems, lacking critical geospatial information and dynamic data integration due to changing environmental conditions. In response, in the early 2000s, a platform was developed to provide a centralised repository for data, enabling repeatability, scalability, and consistency through the use of factual data and algorithms. These industry-leading algorithms are used to assess the probability of failure (PoF) at locations with known geohazards. These algorithms consider various asset specific priorities, site conditions, and hazard/ environmental characteristics. The system includes varying levels of assessments or inspections, from basic screenings or photo inspections, field inspections on the mobile app to detailed evaluations. The standardisation of forms makes reporting more repeatable and efficient, while a central database allows for summary dashboards and the application of machine learning algorithms to identify trends and gain valuable insights. These insights aid in proactive rather than reactive monitoring. This inspection system is only one portion of the overall management program for each operator, with outcomes and timings influenced by the operator's risk tolerances. The outcomes of the inspections are paired with real-time weather monitoring, instrumentation readings, and remote sensing data to identify and changes to site conditions to aid in proactive monitoring of the overall system. Engineering recommendations are embedded, and sites are assigned priorities, actions and due dates to meet regulatory requirements. Cambio for Critical Infrastructure is used by operators to characterise, assess, and monitor over 300 000 geohazards across more than 500 000 km of linear infrastructure. Over the last ten years, the software has enabled a ten times reduction in risk through proactive hazard management (Newton *et al*, 2022).

There are parallels that can be drawn between this inspection, reporting, recommendation and required action workflow to ones we see in both operating and mine closure projects and fits with how the ICMM described the role of a knowledge base – 'supports the domain model as well as monitoring, measurement and inspections' (ICMM, 2025). A platform solution, like Cambio, could be used to enhance the execution plan for mine closure. Through the digitation of action plans and closure criteria the steps toward closure can be clearly tracked, monitored, and communicated amongst all project personnel, including stakeholders. Various communication tools within the platform could be used to leverage the information in the digital knowledge base and make it accessible and actionable.

Case study #4 – reservoir

A reservoir construction project has used Cambio as an innovative tool to monitor reservoir conditions, specifically focusing on landslides and shoreline erosion before, during, and after dam construction and reservoir filling. By developing a comprehensive framework for managing hydro reservoir projects through Cambio, the project has significantly improved asset management. This project involved assessing potential hazards resulting from the reservoir's creation, including landslides, landslide-generated waves, and wind and wave-generated shoreline erosion.

A reservoir slope and shoreline monitoring program were established that aimed to gather information on slope stability and shoreline erosion processes to manage public safety and infrastructure risks. Initiated in 2015, the monitoring program has implemented remote sensing tools such as terrestrial, airborne, and UAV-based LiDAR, photogrammetry, and three types of satellite InSAR. Additionally, geotechnical drilling and the installation of new instrumentation, along with the retrofit of historical instruments, has been carried out. Cambio has been central to this effort, serving as the primary data management tool for all monitoring activities. It enables the visual monitoring of conditions against specific thresholds, analogous to mine closure monitoring. By processing, analysing, and visualising remote sensing data, Cambio helps closely monitor shoreline erosion and landslide activity. This, combined with a robust instrumentation program featuring real-time data, enhances fieldwork efficiency by targeting specific areas for inspection and reducing unnecessary fieldwork. This approach is highly applicable to the mine closure space, demonstrating the versatility and effectiveness of Cambio in managing geohazard risks.

CONCLUSIONS

The process of mine closure is inherently complex, requiring long-term monitoring, adaptive management, and the integration of diverse data sets to attain compliance with environmental and societal expectations. Maintaining a comprehensive knowledge base and monitoring are crucial components of integrated and effective mine closure (ICMM, 2025). The knowledge base contains a broad range of information in many forms, ranging from the domain model to monitoring data, as well as historical information and local/traditional knowledge. Specifically, regarding mine closure monitoring, it can be comprised of traditional methods (surveys, inspections, instrumentation readings etc) with ever improving automation capabilities as well as new remote sensing technologies such as photogrammetry, LiDAR and InSAR. An integrated knowledge base that supports holistic assessment, and includes automated ingestion, processing and visualisation of the many complex data streams is becoming a necessity given this growth in the number of data sources and the size of the data sets.

Cambio is a proposed solution that addresses many of the challenges traditionally associated with mine closure monitoring. By providing centralised access to real-time data, configurable workspaces, interactive dashboards, and advanced geospatial tools, Cambio transforms the way mine closure data is collected, analysed, shared, and acted upon. Future advancements could include the implementation of automated alerting for remote sensing sources like InSAR, similar to what has been implemented on instrumentation readings. These alerts associated TARPS can help identify subtle trends before they escalate into major issues, enabling proactive management.

The primary focus of this paper has been monitoring and maintaining the physical stability of mine closure landforms. But it should be noted that the capabilities of the Cambio platform can be extended to the monitoring of chemical and ecological stability that are also alluded to in the paper. This addition would allow for a fully comprehensive management of risks and long-term care of the facility.

REFERENCES

Braimbridge, M, Mackenzie, S, Lyons, M, Clarke, T and Bow, B, 2019. Whole-of-landform erosion assessment using unmanned aerial vehicle data, in *Mine Closure 2019: Proceedings of the 13th International Conference on Mine Closure* (eds: A B Fourie and M Tibbett), pp 397–406 (Australian Centre for Geomechanics: Perth). https://doi.org/10.36487/ACG_rep/1915_32_Braimbridge

Brink, G and Heymann, E, 2024. A conceptual risk management framework for post-closure settlement of fill, in *Mine Closure 2024: Proceedings of the 17th International Conference on Mine Closure* (eds: A B Fourie, M Tibbett and G Boggs), pp 487–502 (Australian Centre for Geomechanics: Perth). https://doi.org/10.36487/ACG_repo/2415_35

Crisp, H, Mackenzie, S, Gregory, S, Sprenkels, T and Slabber, A, 2024. Application of remote sensing data to measure erosion on rehabilitated landforms at the Abydos mine, in *Mine Closure 2024: Proceedings of the 17th International Conference on Mine Closure* (eds: A B Fourie, M Tibbett and G Boggs), pp 1005–1018 (Australian Centre for Geomechanics: Perth). https://doi.org/10.36487/ACG_repo/2415_72

Global Tailings Review, 2020. Global Industry Standard on Tailings Management (GISTM), co-convened by International Council on Mining and Metals, United Nations Environment and Principles for Responsible Investment.

Grebby, S, Sowter, A, Gluyas, J, Toll, D, Gee, D, Athab, A and Girindran, R, 2021. Advanced analysis of satellite data reveals ground deformation precursors to the Brumadinho Tailings Dam collapse, *Commun Earth Environ*, 2:2. https://doi.org/10.1038/s43247-020-00079-2

International Council on Mining and Metals (ICMM), 2025. *Integrated Mine Closure – Good Practice Guide*, third edn.

Johnson, C, Schmidt, S, Taylor, J and de La Chapelle, J, 2022. Geospatial Database Development: Supporting Geohazard Risk Assessments Through Real-Time Data and Geospatial Analytics, Proceedings of the ASME 2022, 13th International Pipeline Conference, Canada: ASME.

Jones, P L and Franklin, C, 2019. Relinquishment criteria verification: quality assurance/quality control using unmanned aerial vehicles, in *Mine Closure 2019: Proceedings of the 13th International Conference on Mine Closure* (eds: A B Fourie and M Tibbett), pp 1461–1476 (Australian Centre for Geomechanics: Perth). https://doi.org/10.36487/ACG_rep/1915_114_Jones

Kelcey, J, Blaxland, D, Smith, B and Gove, A, 2019. The analysis and validation of landform stability using unmanned aerial vehicles, in *Mine Closure 2019: Proceedings of the 13th International Conference on Mine Closure* (eds: A B Fourie and M Tibbett), pp 1127–1138 (Australian Centre for Geomechanics: Perth). https://doi.org/10.36487/ACG_rep/1915_90_Kelcey

Morel, J, Foumelis, M, Raucoules, D and Lemal, S, 2021. InSAR assets in ground movements survey on abandoned coalfields, in *Mine Closure 2021: Proceedings of the 14th International Conference on Mine Closure* (eds: A B Fourie, M Tibbett and A Sharkuu), QMC Group, Ulaanbaatar. https://doi.org/10.36487/ACG_repo/2152_47

Newton, S, Van Hove, J, Porter, M and Ferris, G, 2022. Assessing geohazard probability of pipeline failure: Lessons and improvements from the last 10 years, Proceedings of the ASME 2022, 14th International Pipeline Conference (IPC2022), Canada.

Sage, E, Holley, R, Carvalho, L, Miller, M, Magnall, N and Thomas, A, 2022. InSAR monitoring of a challenging closed mine site with corner reflectors, in *Mine Closure 2022: Proceedings of the 15th International Conference on Mine Closure* (eds: A B Fourie, M Tibbett and G Boggs), pp 779–788 (Australian Centre for Geomechanics: Perth). https://doi.org/10.36487/ACG_repo/2215_56

Schmidt, B, Malgesini, M, Turner, J and Reinson, J, 2015. Satellite monitoring of a large tailings storage facility, in *Proceedings of the Tailings and Mine Waste 2015 Conference*, 10 p. Available from: <https://www.photosat.ca/wp-content/uploads/2020/06/Golder-paper-satellite-monitoring-large-tailings-storage-facility.pdf>

Tones, A, Howe, L and du Plooy, J, 2021. Knowledge makes the world go around: Knowledge management in mine closure planning, in *Mine Closure 2021: Proceedings of the 14th International Conference on Mine Closure* (eds: A B Fourie, M Tibbett and A Sharkuu), QMC Group, Ulaanbaatar. https://doi.org/10.36487/ACG_repo/2152_06

Case study – using drone-based NDVI imagery to monitor groundcover performance of a creek diversion at a mine

A Costin[1], B Silverwood[2], B Wehr[3] and G Dale[4,5,6]

1. Environmental Engineer, Verterra Ecological Engineering, Brisbane Qld 4001.
   Email: adam.costin@verterra.com.au
2. Senior GIS and Systems Engineer, Verterra Ecological Engineering, Brisbane Qld 4001.
   Email: ben.silverwood@verterra.com.au
3. Principal Scientist, Verterra Ecological Engineering, Brisbane Qld 4001.
   Email: bernhard.wehr@verterra.com.au
4. Chief Technical Officer, Verterra Ecological Engineering, Brisbane Qld 4001.
   Email: glenn.dale@verterra.com.au
5. Adjunct Associate Professor, Faculty of Engineering, School of Civil and Environmental Engineering, Queensland University of Technology.
6. Adjunct Associate Professor, Centre for Sustainable Agricultural Systems, University of Southern Queensland.

EXTENDED ABSTRACT

Comprehensive yet efficient quantitative monitoring of revegetation performance in post-mining landscapes presents significant challenges. Traditional monitoring approaches are usually time-consuming, struggle to identify underperforming areas, and only provide extrapolated estimates of groundcover. This paper presents a methodology developed for monitoring revegetation performance of a coalmine creek diversion through drone-based multispectral imagery, that successfully provided evidence for compliance with completion criteria, while also providing opportunities for early intervention.

METHODOLOGY

Our approach uses Normalised Difference Vegetation Index (NDVI), a well-established vegetation health indicator, calculated from multispectral imagery captured via unmanned aerial vehicles (UAVs), to assess groundcover percentage (Hamylton et al, 2020; de la Iglesia Martinez and Labib, 2023; Prasetya Nugraha and Hariyanto, 2024). While NDVI itself is widely used in vegetation monitoring, our novel approach lies in calibrating site-specific NDVI thresholds at the pixel level to estimate groundcover percentage across an entire rehabilitation area. This method was specifically designed to evaluate compliance with a threshold of ≥75 per cent groundcover required for diversion opening approval, while also identifying underperforming areas requiring targeted maintenance.

The methodology incorporates several key features that distinguish it from traditional monitoring approaches:

- Frequent data collection: Site personnel were trained to collect multispectral imagery using UAVs, creating data sets that allowed for groundcover estimates and growth rate assessments to identify areas responding differently to the average. For this project, the mine needed to purchase a multispectral drone specifically for this purpose, as standard RGB drones commonly available at mine sites are insufficient for the required NDVI calculations.

- Automated processing: Processing scripts were developed to standardise imagery analysis, reducing the time between data collection and actionable insights while maintaining consistency across monitoring periods.

- Zone-specific calibration: Rather than applying uniform NDVI thresholds across the entire site, high-resolution images were used to 'ground-truth' NDVI values and calibrate them to specific areas, accounting for variations in soil type, aspect and vegetation communities. This calibration process involved collecting high-resolution images at representative locations throughout each revegetation zone, correlating NDVI values with actual groundcover percentages (by manual visual assessment) and developing zone-specific threshold values that could be applied across similar areas (Hamylton et al, 2020).

- Complete spatial coverage: Unlike traditional transect-based monitoring methods that assess only a small percentage of the rehabilitation area, this approach enabled complete assessment of the entire diversion footprint, eliminating sampling bias and providing a comprehensive spatial understanding of revegetation success.

- Field validation: Traditional field-based groundcover measurements (transects and quadrat plots) were used to verify the accuracy of NDVI-derived groundcover percentages.

The team encountered several practical challenges during implementation, including issues with image quality due to cloud cover and time of day affecting shadows. These factors sometimes necessitated repeat flights or careful scheduling to ensure optimal conditions for image capture.

RESULTS

The method successfully quantified groundcover across the entire diversion footprint with good confidence (varying with image quality), providing the required evidence for diversion opening criteria sign-off, while also enabling identification of underperforming areas requiring more intensive monitoring or intervention (Figure 1a).

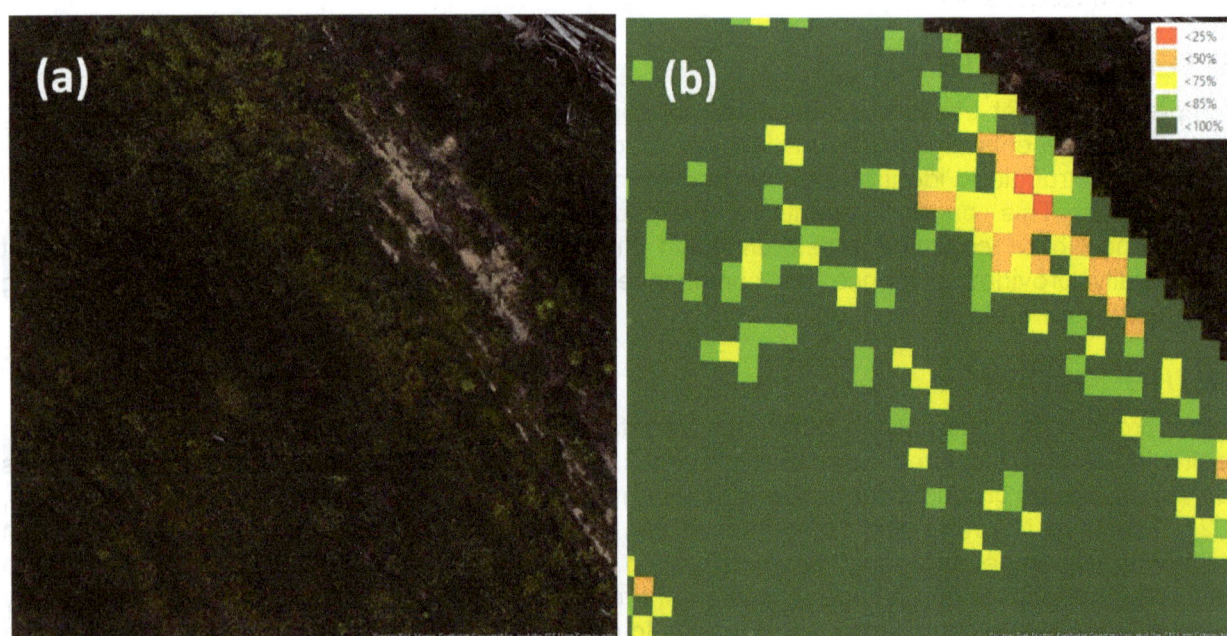

FIG 1 – (a) Drone image of a creek diversion and (b) corresponding processed NDVI image showing estimated groundcover percentage per m^2 pixel.

The regular frequency of monitoring enabled quantification of groundcover growth rates for varying areas (Figure 1b). This data was primarily used for assessing area performance, helping in early identification of lagging or leading sections, and was used to roughly estimate when specific areas were expected to achieve compliance thresholds. This information supported proactive planning of maintenance activities and regulatory submissions.

A key strength of the approach was its ability to rapidly identify emerging issues before they became significant problems. Areas showing early signs of erosion or poor groundcover compared with surrounding areas were flagged for immediate inspection. This early intervention capability substantially reduced overall maintenance costs and improved rehabilitation outcomes by addressing problems before they required more extensive remediation. While visual on-site identification was still needed to identify specific issues like weed infestation, the multispectral imagery could be used to monitor treatment effectiveness and weed growth patterns if required.

Compared to traditional monitoring methods (Almalki *et al*, 2022; Juanda, Martono and Saria, 2021), our approach demonstrates significant advantages in terms of spatial coverage, ease of implementation and cost-effectiveness. While conventional transect-based monitoring typically samples less than 1 per cent of a rehabilitation area, a drone-based approach can achieve

100 per cent coverage at a comparable or lower cost while reducing the time required for data collection and analysis.

APPLICATIONS

The approach developed for this specific creek diversion has potential for broader application across various revegetation scenarios, including other types of mine rehabilitation, linear infrastructure corridors, gully and streambank rehabilitation and natural ecosystem restoration.

The increasing availability and decreasing cost of drone technology creates an opportunity to establish a new approach to rehabilitation monitoring. While many mining operations don't yet have multispectral drone capabilities, the effectiveness may justify the investment.

CONCLUSION

The drone-based NDVI monitoring methodology developed for this coalmine creek diversion represents a practical application of existing technology in a novel way to address rehabilitation monitoring challenges. By providing comprehensive spatial and temporal coverage, enabling early intervention, and generating complete data sets for compliance reporting, the approach addresses key limitations of traditional monitoring methods.

This methodology offers a cost-effective framework for supporting timely management decisions, ensuring compliance with rehabilitation targets, and contributing to successful establishment of post-mining landscapes. As drone technology becomes more accessible, this approach has potential to become more widely adopted for revegetation monitoring.

REFERENCES

Almalki, R, Khaki, M, Saco, P M and Rodriguez, J F, 2022. Monitoring and Mapping Vegetation Cover Changes in Arid and Semi-Arid Areas Using Remote Sensing Technology: A Review, *Remote Sens*, 14. https://doi.org/10.3390/rs14205143

de la Iglesia Martinez, A and Labib, S M, 2023. Demystifying normalized difference vegetation index (NDVI) for greenness exposure assessments and policy interventions in urban greening, *Environ Res*, 220:115155. https://doi.org/10.1016/j.envres.2022.115155

Hamylton, S M, Morris, R H, Carvalho, R C, Roder, N, Barlow, P, Mills, K and Wang, L, 2020. Evaluating techniques for mapping island vegetation from unmanned aerial vehicle (UAV) images: Pixel classification, visual interpretation and machine learning approaches, *Int J Appl Earth Obs Geoinformation*, 89:102085. https://doi.org/10.1016/j.jag.2020.102085

Juanda, E T, Martono, D N and Saria, L, 2021. Analysis of vegetation change on coal mine reclamation using Normalized Difference Vegetation Index (NDVI), *IOP Conf Ser Earth Environ Sci*, 716:012035.

Prasetya Nugraha, A and Hariyanto, T, 2024. Evaluation of Vegetation Health in the PT X Reclamation Area Using the NDVI Method Based on Unmanned Aerial Vehicle (UAV) Multispectral Orthophoto Data, *IOP Conf Ser Earth Environ Sci*, 1418:012010. https://doi.org/10.1088/1755-1315/1418/1/012010

The Life of Mine | Mine Waste and Tailings Conference 2025 | Brisbane, Australia | 29–30 July 2025

GIS-based optimisation of fill material transport and allocation for progressive mine closure

J Diana[1] and S Dressler[2]

1. Closure Planner, WSP, Brisbane Qld 4006. Email: julia.diana@wsp.com
2. Senior Landform Design Engineer, WSP, Newcastle NSW 2300. Email: sven.dressler@wsp.com

ABSTRACT

Material transport accounts for a major share of operating expenditure (OPEX) in open pit mining, often representing over half of total operational costs. As a result, it is a strategic area for optimisation, particularly when integrated with mine closure planning. Recent studies highlight the benefits of mathematical models such as fuzzy and linear programming in optimising haulage operations, boosting fleet efficiency and reducing costs. In parallel, mine closure practices are evolving to emphasise early integration with operational planning, as recommended by the International Council on Mining and Metals (ICMM), 2019. Progressive closure—conducting rehabilitation during active mining—has emerged as a best practice, offering operational, financial, and reputational advantages by aligning material movement with long-term closure goals. This study investigates the intersection of transport optimisation and closure planning, using a bauxite mine in Northern Australia as a case study. The site is undergoing a transition to closure and faces a shortage of growth media required for rehabilitation. Available material is scattered across the site, presenting a logistical challenge for efficient allocation. A linear programming (LP) transportation model was developed using Python's PuLP library to minimise haulage distances while meeting demand under three topsoil placement scenarios (0.4 m, 0.3 m and 0.2 m). A dummy source was added to account for potential deficits. The model successfully identified optimal material allocations, prioritised nearby sources, and flagged high-risk shortage areas. It also provided operational insights, such as haul distance distribution and equipment selection. The approach supports strategic stockpile planning, reinforces the value of progressive closure, and illustrates how optimisation tools can improve closure outcomes. The results have practical implications for mine planning, risk management, and long-term cost control. This case study contributes to the growing body of research on applying operations research techniques to mine rehabilitation, highlighting the benefits of integrated, data-driven approaches for sustainable closure planning.

INTRODUCTION

Material transport represents a significant portion of operating expenditure (OPEX) in open pit mining, making it a strategic target for optimisation. According to Losaladjome Mboyo et al (2025), transport can account for 43–70 per cent of total operational costs in mining. Similarly, Moradi-Afrapoli, Osanloo and Rahmannejad (2021) found that material handling—including transport—can represent up to 60 per cent of OPEX in truck-and-shovel operations. Their study introduced a fuzzy linear programming model for truck dispatching, resulting in notable gains in fleet efficiency and overall productivity.

These findings underscore the importance of transport efficiency—through optimised logistics or advanced mathematical modelling—as a component of mine economic planning. In the context of closure, the ICMM's Integrated Mine Closure Guide (2019) emphasises that closure should be embedded from the earliest stages of mine planning, spanning short-, medium- and long-term horizons. Integrating closure within the mine life cycle (LoM) enables more effective risk, cost, and stakeholder management.

The concept of progressive closure—carrying out closure activities during operations—is particularly strategic. Material movement, such as waste rock placement, capping, and revegetation, can be coordinated with mining operations to deliver technical, social, and financial benefits. These include gradually reducing disturbed areas, lowering long-term liabilities, improving stakeholder relationships, cost predictability, and even tax advantages in some jurisdictions (ICMM, 2019).

Implementing progressive closure during the high cash flow phase can also ease the financial burden typically concentrated in the post-operational period. Although Net Present Value (NPV) analysis may suggest higher upfront costs, the tangible and intangible benefits—such as risk reduction, improved reputation, and potential adjustments to financial assurance—reinforce the strategic value of this approach (ICMM, 2019).

Combining transport optimisation with progressive closure planning creates a clear opportunity to enhance value throughout the mine's life and mitigate challenges at decommissioning. Linear Programming (LP) methods—particularly the classical transportation problem—are powerful tools for modelling the optimal allocation of materials from sources to destinations while respecting supply and demand constraints. In closure planning, these models can also help identify areas at risk of material shortages, supporting proactive stockpile planning during operations.

This study applied this approach to a bauxite mine in Australia, aiming to allocate growth media stockpiles across multiple rehabilitation areas efficiently. The methodology involved: (a) a material balance across different topsoil thickness scenarios; and (b) a transportation model designed to minimise haulage distances. The model generated valuable insights into optimal allocation strategies, directly supporting decision-making and the logistical optimisation of mine closure.

CONTEXT AND LITERATURE REVIEW

Site context

The target site for this project is a bauxite mine located in northern Australia, currently undergoing transition to closure. While progressive rehabilitation has been implemented across much of the site, several areas—particularly internal roads—will require rehabilitation at final closure. The mine currently holds approximately 1.5 million cubic metres (Mm^3) of growth media suitable for these remaining areas. However, these stockpiles are widely scattered across the site, and the combination of fixed structures and extensive haul road rehabilitation creates a highly complex source-to-destination allocation matrix. The currently stockpiled growth media is insufficient to rehabilitate all the current and future disturbed areas. As the mining operation progresses more growth media will be stripped and stockpiled, which will close this deficit. In this context, it is essential not only to plan transport routes that minimise total haulage costs but also to accurately identify areas most vulnerable to material shortages. Doing so will support more strategic stockpiling during the operational phase, helping to avoid logistical bottlenecks and reduce costs at closure. This spatial and logistical complexity calls for a highly efficient allocation strategy—one that ensures optimal resource use and successful rehabilitation outcomes at the lowest operational cost

Algorithms and possible solutions

Transportation problems can be addressed using various approaches depending on the primary goal—whether to quickly obtain a feasible solution or to find the global optimum directly. Classical heuristics such as the Northwest Corner Method, Least Cost Method, and Vogel's Approximation Method (VAM) are commonly used for generating initial solutions with low computational cost. These are particularly useful in investigative contexts or when computational resources are limited. However, heuristics do not guarantee optimality and can lead to significantly suboptimal allocations.

To ensure optimal solutions, refinement techniques like the Modified Distribution Method (MODI) or the Potential Method (U-V Method) are often applied (Srinivasan, 2010). In contrast, tools like the PuLP library take a complete linear programming (LP) approach, solving the problem using proven optimisation algorithms such as the Simplex method or interior-point techniques.

In the LP formulation, continuous variables represent the quantity transported between each source-destination pair. The objective function minimises total cost, subject to constraints that meet supply limits and demand requirements. Unlike heuristics, which follow localised, rule-based logic, LP enables a global optimisation search performed automatically by specialised solvers. This makes LP far more robust and precise, especially in large-scale problems or scenarios with multiple constraints.

PuLP, a Python library, provides an intuitive interface for modelling and solving linear and mixed-integer problems and integrates with solvers. It is a powerful tool for automated decision-making in logistics, industrial operations, and planning contexts.

The Simplex method, introduced by Dantzig (1951), remains widely used for solving LP problems. It relies on the fact that the feasible region—known as the simplex—is convex and that the optimal solution, when it exists, lies at one of the region's vertices (for problems where costs and constraints are linear). The algorithm iteratively evaluates these vertices to identify the minimum-cost solution. In computational environments, Simplex is implemented across various solvers. It is accessible through libraries like PuLP, enabling the resolution of complex models with multiple constraints, penalties, integer variables, or multi-objective structures (Winston, 2004).

A simplified graphical example of Simplex is illustrated by Carvalho Jnr, Koppe and Costa (2012) in Figures 1 and 2, where problems with two decision variables form a polygonal feasible region. The solution process involves evaluating the corner points (vertices) of this region iteratively until the optimal value of the objective function is found.

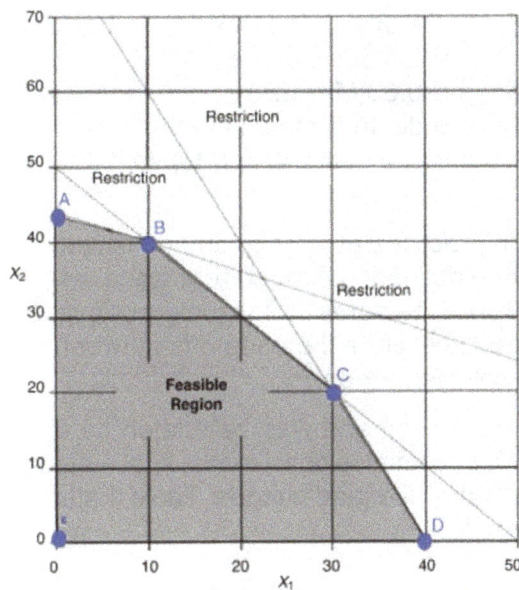

FIG 1 – Graphical representation of the algorithm restriction setting and feasible region within restrictions boundaries (adapted from Carvalho Jnr, Koppe and Costa, 2012).

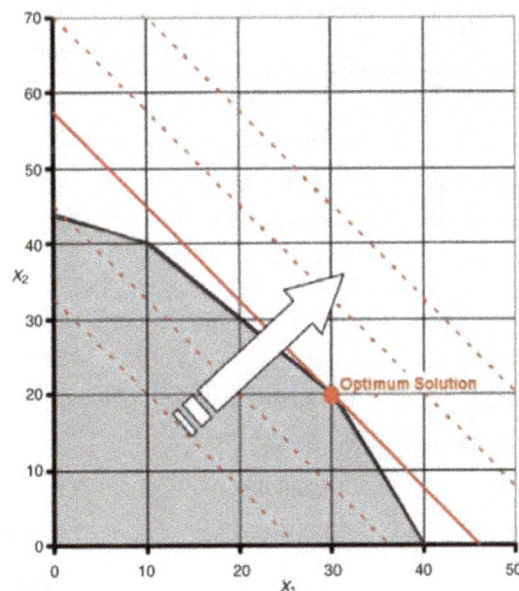

FIG 2 – Graphical representation of the point of optimum solution (adapted from Carvalho Jnr, Koppe and Costa, 2012).

Traditional transportation models are based on the balanced condition, where total supply equals total demand, ensuring feasibility without adjustments. However, in real-world applications, this balance is rarely achieved. As in the case study presented in this article, one common workaround is to convert the model into a balanced version by introducing a dummy row with additional artificial supply (Srinivasan, 2010).

This dummy source can be assigned either zero cost to remain neutral or prohibitively high transport cost, as done in this model. By doing so, the solver is discouraged from allocating to the dummy unless strictly necessary. This approach ensures mathematical feasibility and highlights which demand points are most sensitive to material shortages, supporting more informed and strategic stockpiling decisions during operational planning.

METHODOLOGY

Material balance

The first step involved quantifying the growth media required to rehabilitate all disturbed areas at mine closure and comparing this to the available stock. Using LiDAR surveys and stockpile data, the team identified approximately 1.3 million cubic metres (Mm^3) of growth media stored across various dispersed locations on-site. Around 0.63 Mm^3 of red clay soil was also identified, designated for structural backfilling of a specific industrial area ('chemical pad'). Due to previous revegetation failures, this red soil was deemed unsuitable for topsoil use and excluded from ecological rehabilitation calculations.

Based on existing rehabilitation practices at the site and consultations with the operating company, three cover-depth scenarios were defined: 0.4 m (current standard), 0.3 m and 0.2 m. Thinner layers would allow broader coverage with the same soil volume but might affect ecological performance. These scenarios were developed to explore the trade-offs between maintaining the current standard and reducing soil thickness in selected areas.

For each scenario, the total topsoil demand was calculated by summing the requirements for all relevant closure areas, including decommissioned roads, final mined-out floors left, decommissioned infrastructure zones, stockpile areas, and land bridges. Table 1 shows the material balance amounts per relevant closure area assessed.

TABLE 1

Material balance amounts.

Category	Area (ha)	Scenario with 0.4 m placement depth	Scenario with 0.3 m placement depth	Scenario with 0.2 m placement depth
		GM volume needed for rehab (m^3)		
Roads to be rehabilitated	540.2	2 160 620	1 620 465	1 080 310
Retained roads	10.1	40 400	30 300	20 200
Growth media stockpiles footprint	102.2	408 615	306 461	204 307
Final mined floor remaining	327.8	1 311 269	983 452	655 635
Chemical floor	79.7	318 739	239 054	159 369
Infrastructure areas	27.4	109 433	82 074	54 716
Redsoil stockpiles footprint	18.3	73 122	54 842	36 561
Land bridges	7.7	30 989	23 242	15 494
Conveyor area	44.1	35 242	26 431	17 621
Total growth media requirement (m^3)		**4 488 428**	**3 366 321**	**2 244 214**
Current stockpiled material available (m^3)		1 301 501	1 301 501	1 301 501
Material balance considering current stockpiled material (m^3)		-3 186 927	-2 064 820	-942 713

In the 0.4 m scenario (the thickest cover layer), the total soil volume required for rehabilitation was estimated at approximately 4 488 000 m³. With only 1 301 500 m³ currently stockpiled, this would result in a deficit of around 3.19 Mm³ if no additional sources are considered. In the 0.3 m scenario, the demand decreases to approximately 3 366 000 m³, and then we see in the 0.2 m scenario it decreases to about 2 244 000 m³. These reductions indicate that if thinner layers are accepted in some areas, the available material becomes more than sufficient to meet demand.

Transport mathematical modelling

With source and destination volumes and locations defined, the next step was to optimise the transport routes. This was formulated as a classic linear programming transportation problem, where a set of sources i (stockpiles of growth media) each have a supply S_i (in m³), and a set of destinations j (rehabilitation areas, grouped by proximity or type) each have a demand D_j (in m³).

For each source-destination pair, the shortest transport distance d_{ij} was calculated using GIS tools (eg Dijkstra's algorithm or ArcGIS Network Analyst) based on the internal road network. This resulted in a complete origin-destination distance matrix, which d_{ij} can be converted to monetary cost using a unit transport rate. The objective function minimises the total transport cost, defined as:

$$C = \sum_i \sum_j d_{ij} \cdot x_{ij}$$

where x_{ij} represents the amount of soil transported from source i to destination j. Constraints ensure that: (a) supply is not exceeded; and (b) demand is met:

$$\sum_i x_{ij} \geq D_j$$

While standard transportation problems assume full demand satisfaction, a dummy source with high cost was added here to allow partial fulfillment and flag potential shortfalls.

The model was implemented using the open-source PuLP library in Python. The workflow included: loading input data (supply S_i, demand D_j and distance matrix from CSV files), building the model in PuLP, solving it, and exporting the results. A preliminary data validation step ensured dimensional consistency to avoid indexing errors.

Model balancing (dummy-source addition)

Given that the material balance indicated a potential shortfall of growth media in some scenarios (particularly with a 0.4 m application depth), an artificial 'dummy' source was added to the model. This source was assigned virtually unlimited supply but an extremely high transport cost. This is a standard technique in transportation problems to ensure feasibility when not artificially generating material unless necessary.

In this case, the dummy source was configured to cover an estimated deficit of approximately 1 458 000 m³ — the expected shortfall if only current stockpiles are considered (excluding future stripping operations). A prohibitively high distance was assigned (several orders of magnitude greater than real mine distances) to discourage its use, ensuring the solver resorts to it only when no real source can fulfill a given demand.

Any allocation from the dummy source in the results indicates an unmet demand and helps pinpoint areas with critical shortages. This also highlights regions that could benefit from future *stockpiling* or reallocation of excess material from surplus scenarios. Thus, the dummy source maintains mathematical feasibility (ensuring 100 per cent of demand is addressed in the model) while revealing material deficits' spatial extent and location.

The model was then solved for the three cover depth scenarios (0.4 m, 0.3 m, 0.2 m), adjusting demands D_j accordingly. Initially, only current stockpile volumes S_i were considered. Alternatively, future supply (eg material expected to be generated through stripping by 2029) can be incorporated as additional sources. Both approaches offer valuable insights: one highlights immediate gaps, while the other simulates optimal conditions with full resource availability.

In this study, emphasis was placed on current stock availability to identify shortfalls and optimise spatial allocation. The model produced optimal solutions in seconds for all scenarios, minimising total haulage distance.

RESULTS AND DISCUSSION

The transport model provided the optimal allocation of growth media from each source stockpile to each demand location across all scenarios. Overall, the minimised transport cost significantly reduces travel distances compared to non-optimised distributions. In all simulations, the majority of soil was allocated from sources located close to the target areas — with substantial volumes moved along short routes, particularly under 1 km.

A clear inverse relationship was observed between transport distance and volume: the greater the distance, the smaller the volume transported, as the model prioritises exhausting nearby sources before resorting to more distant ones. Notable peaks in transported volume were found at very short distances, especially around 0.2 km, 0.8 km and 1.2 km, corresponding to stockpiles situated near key rehabilitation areas. In contrast, only modest volumes were assigned to longer hauls (over 2–3 km) and only when local sources could not meet the demand. Figures 3, 4 and 5 show the model results for the 0.2 m, 0.3 m and 0.4 m scenarios on a map.

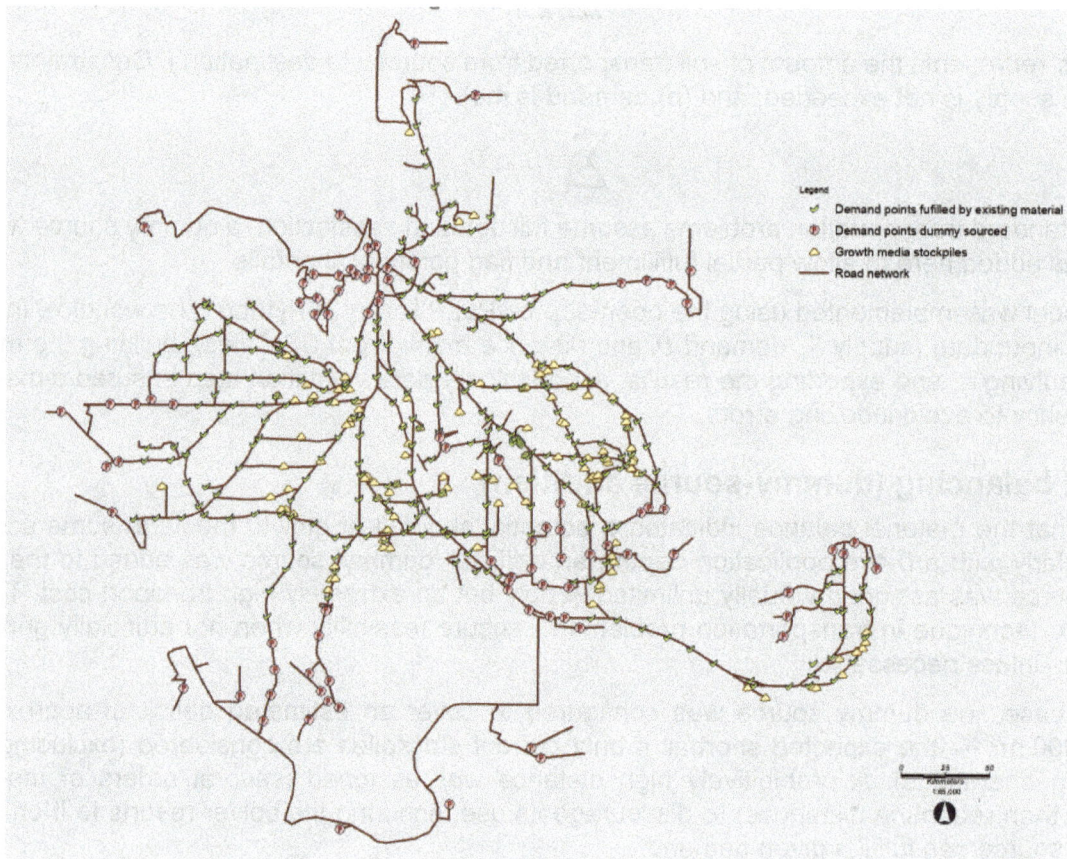

FIG 3 – A 0.2 m depth placement scenario, showing points of demand (fulfilled and dummy sourced) and the growth media stockpiles locations (source: J Diana).

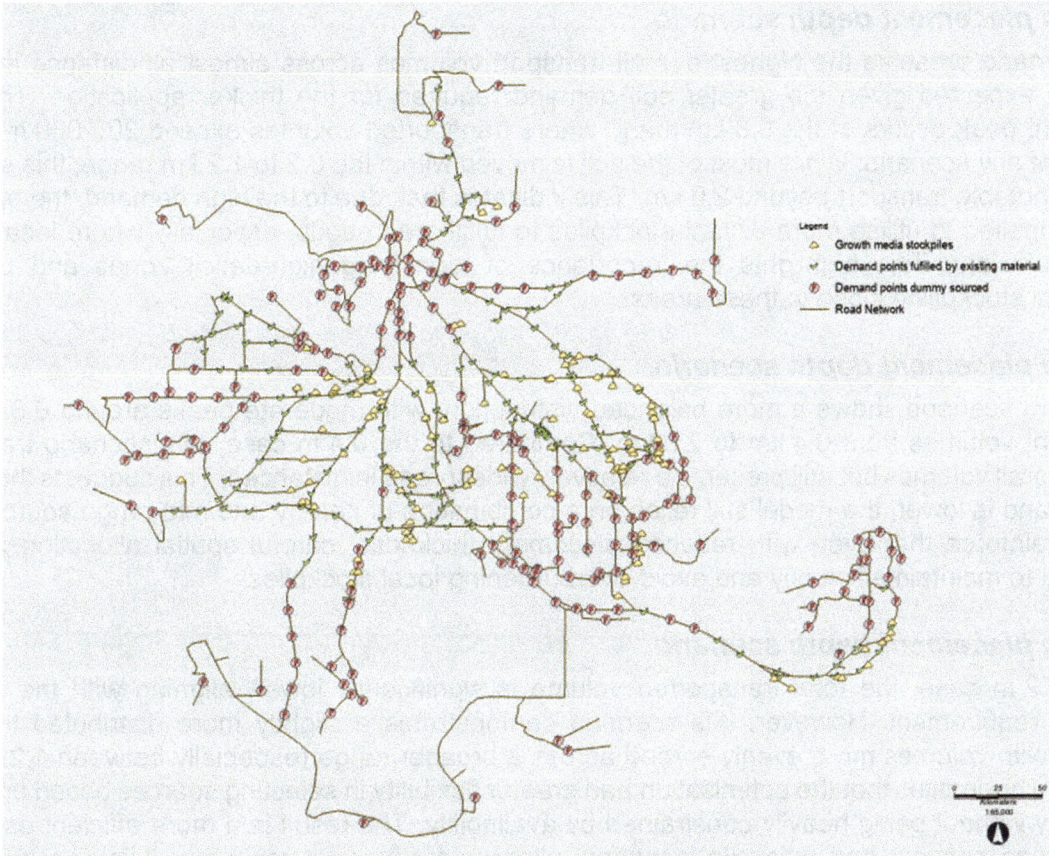

FIG 4 – A 0.3 m depth placement scenario, showing points of demand (fulfilled and dummy sourced) and the growth media stockpiles locations (source: J Diana).

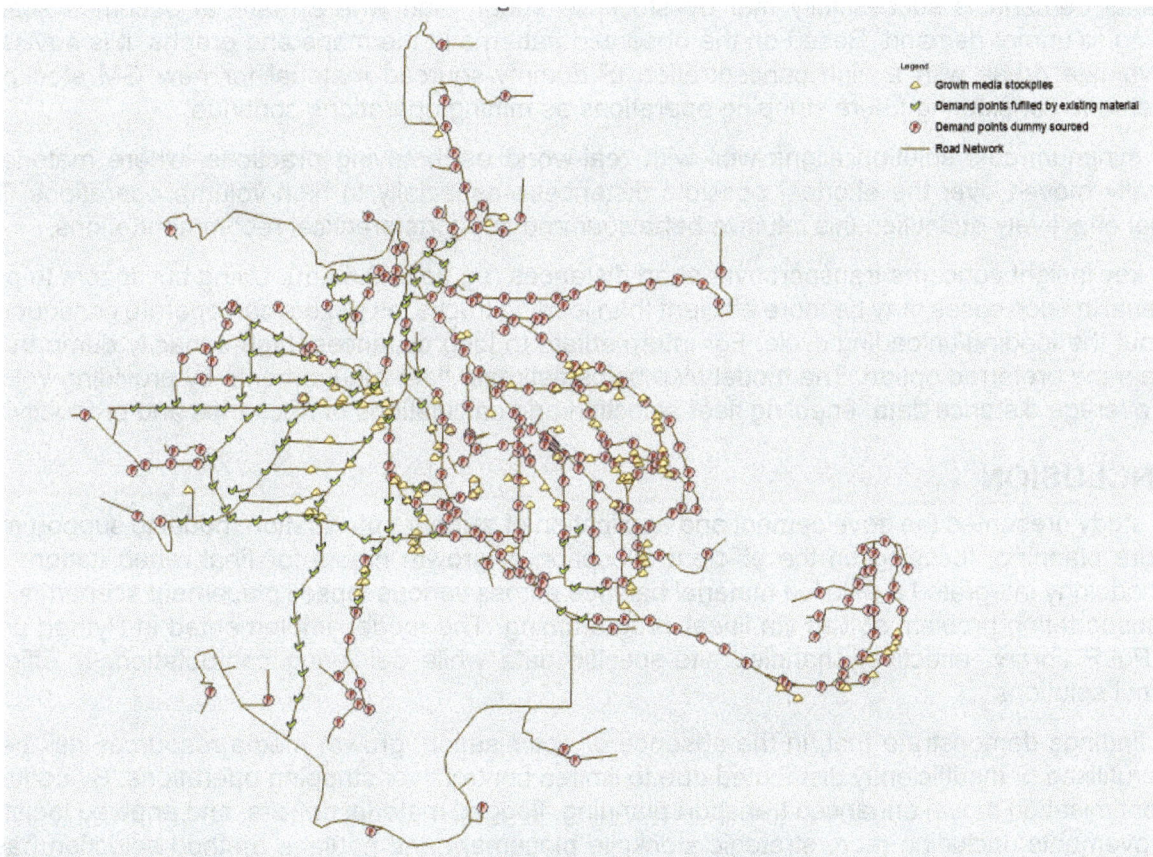

FIG 5 – A 0.3 m depth placement scenario, showing points of demand (fulfilled and dummy sourced) and the growth media stockpiles locations (source: J Diana).

A 0.4 m placement depth scenario

This scenario presents the highest overall transport volumes across almost all distance intervals, which is expected given the greater soil demand required for the thicker application. The most prominent peak occurs at the 0.8 km mark, where transported volumes exceed 200 000 m³ — the highest of any scenario. While most of the soil is moved within the 0.2 to 1.2 km range, this scenario exhibits notable transport beyond 2.0 km. This indicates that, due to the high demand, the optimiser was compelled to utilise more distant stockpiles to fulfill area needs, especially where local supply was insufficient. This highlights the importance of identifying high-deficit zones and planning additional stockpiling close to these areas.

A 0.3 m placement depth scenario

The 0.3 m scenario shows a more balanced distribution, with moderate peaks around 0.8 km and consistent volumes from 0.4 km to 2.0 km. Compared to the 0.4 m case, this scenario transports lower overall volumes but still presents a relatively widespread in distances. This suggests that, while the demand is lower, the model still relies on a combination of nearby and mid-range sources. The results reinforce that even with reduced placement thickness, careful spatial allocation remains essential to maintain efficiency and avoid overburdening local stockpiles.

A 0.2 m placement depth scenario

In the 0.2 m case, the total transported volume is significantly lower, aligning with the minimal material requirement. However, this scenario demonstrates a slightly more distributed transport pattern, with volumes more evenly spread across a broader range (especially between 1.2 km and 2.2 km). This implies that the optimisation had greater flexibility in selecting sources based on spatial efficiency without being heavily constrained by availability. The result is a more efficient use of the overall road network and stockpile locations, allowing for broader resource utilisation and lower overall pressure on individual sources.

By cross-referencing this data with Figures 3, 4 and 5, it is possible to identify the regions where material demand is successfully met by stockpile supply and where material becomes scarce, leading to unmet demand. Based on the observed patterns in the maps and graphs, it is advisable to evaluate areas with a high concentration of dummy-sourced material for new GM stockpiles placement, considering future stripping operations as mining operations continue.

The minimum-cost solution aligns well with real-world earthmoving practices, where material is typically moved over the shortest possible distances—especially in high-volume operations. The model effectively quantifies this intuitive behaviour and supports practical recommendations.

One key insight concerns transport over short distances (eg up to 0.5 km). Using bulldozers to push material in such cases may be more efficient than loading trucks, as dozers can operate continuously without the loading/unloading cycle. For intermediate to long distances, high-capacity dump trucks remain the preferred option. The model also helps estimate fleet requirements by providing volume and average distance data, enabling fleet selection and calculations of truck trips and productivity.

CONCLUSION

This study presented the development and application of a linear optimisation model to support mine closure planning, focusing on the efficient allocation of growth media for final rehabilitation. The methodology integrated a detailed material balance across various topsoil placement scenarios with a transportation problem solved via linear programming. The model, implemented in Python using the PuLP library, effectively handled site-specific data while delivering computationally efficient optimal solutions.

The findings demonstrate that, in the absence of optimisation, growth media resources risk being underutilised or insufficiently distributed due to limited control over stripping operations. By contrast, the optimisation model enhanced transport planning, flagged material deficits, and enabled logistical improvements, including more strategic stockpile placement and haulage method selection based on distance and quantity.

From both cost and operational perspectives, adopting an optimised plan reduces expenses and enhances confidence in meeting closure targets without material shortfalls. Importantly, the approach reinforces the need to integrate mine planning and closure from early stages, as recommended by ICMM's Integrated Mine Closure guidance (2019), by ensuring stripped materials are stored with final land use in mind.

Academically, this case study underscores the practical value of operations research tools—particularly linear programming—in addressing complex logistical challenges in the mining sector. The modelling framework developed here is adaptable beyond this site-specific application and holds promise for broader use in other mine closure contexts, particularly those facing similar constraints in material availability, spatial variability, or haulage logistics.

Moreover, the model's structure is flexible enough to accommodate operational differences among mines extracting various commodities. For example:

- *Bauxite and coalmines*, which typically involve shallow, strip mining with high volumes of overburden, can directly benefit from optimising short-distance haulage and progressive rehabilitation strategies.

- *Iron ore and base metal operations*, often characterised by deeper pits and more complex material segregation (eg NAF versus PAF), can adapt the model by integrating geochemical constraints and selective placement zones for capping and encapsulation.

- *Uranium or rare earth sites*, which require careful management of radiologically impacted or chemically sensitive materials, could extend the optimisation framework by including material classification rules and exclusion zones in the transport matrix.

While this study focused on current growth media stockpiles, incorporating projected stripping volumes through 2029 could significantly shift material allocation dynamics. Future modelling efforts may investigate this aspect to assess whether anticipated material supply can offset current shortfalls, reducing reliance on dummy-source assumptions. Further research may also explore integrating monetary costs (eg haulage method cost differentials), testing model sensitivity to key parameters (stockpile volume, distance, stripping rates), and introducing multi-objective optimisation to balance cost, emissions, and ecological priorities.

Ultimately, this modelling approach offers a versatile and data-driven framework that can help a wide range of mine sites—regardless of commodity type—enhance closure outcomes, reduce long-term liabilities, and promote more sustainable land use transitions.

REFERENCES

Dantzig, G B, 1951. The simplex method for linear programming, *Linear Programming and Extensions*, pp 3–15 (Princeton University Press: Princeton).

De Carvalho Júnior, J A, Koppe, J C and Costa, J F C L, 2012. A case study application of linear programming and simulation to mine planning, *Journal of the Southern African Institute of Mining and Metallurgy*, 112(6):477–484.

International Council on Mining and Metals (ICMM), 2019. Strategic planning for mine closure, in *Integrated mine closure: Good practice guide*, 2nd edn, pp 1–25 (International Council on Mining and Metals: London).

Losaladjome Mboyo, A, Smith, R, Chen, X and Wang, Y, 2025. Optimising haulage operations in open pit mining using integrated simulation approaches, *International Journal of Mining Science and Technology*, 35(2):123–134.

Moradi-Afrapoli, M, Osanloo, M and Rahmannejad, H, 2021. Truck dispatching using fuzzy linear programming in open pit mines, *Resources Policy*, 70:101917.

Srinivasan, G, 2010. Transportation problem and solution methods, *Operations Research: Principles and Applications*, 2nd edn, pp 120–145 (PHI Learning Pvt. Ltd.: New Delhi).

Winston, W L, 2004. The simplex algorithm, in *Operations Research: Applications and Algorithms*, 4th edn, pp 85–114 (Thomson Brooks/Cole: Belmont).

Looking backward to look forward – palaeoclimate recreations and potential applications for mine closure

A Goto[1] and A Karrasch[2]

1. Senior Water Engineer, Red Earth Engineering – A Geosyntec Company, Brisbane Queensland 4000. Email: anna.goto@redearthengineering.com.au
2. Principal Water Engineer, Red Earth Engineering – A Geosyntec Company, Brisbane Qld 4000. Email: alex.karrasch@redearthengineering.com.au

ABSTRACT

Effective life-of-mine planning is increasingly facing the challenge of designing for perpetuity. To allow effective mine closure, landforms must accommodate a closure state that is safe, stable, non-polluting, self-sustaining and supports a post-mining land use. This challenge is magnified by the requirement to consider various future conditions, such as uncertain climates in the medium term, and the possibility of substantially modified regional climatic conditions.

For hydrologic and hydraulic landform design in the medium term to perpetuity, the challenge has resulted in substantial challenges in establishing the design climatic conditions that should be considered across a mine's life cycle, such as the Probable Maximum Precipitation and Flood (PMP and PMF) or continuous climatic conditions. These assumptions are key in establishing hydrologic, hydraulic, landform evolution, cap and surface design aspects.

In Australia, recent national guidance has stressed the need to consider adaptive management in landform design and management. However, methods to estimate potential future conditions primarily involve the introduction of additional uncertainty within estimates with ensuing outcomes difficult to communicate to key stakeholders.

Palaeoclimates are geologically studied conditions, which represent a potential alternative approach for design in perpetuity by considering the potential envelope of conditions that could be experienced in post-closure, in addition to traditional approaches. Palaeoclimates are informed by proxies, such as tree growth rings, corals, to Antarctic ice cores, and can allow us to reconstruct historical climate conditions.

This paper proposes a simple framework for the application of palaeoclimate and palaeohydrology to mine closure designs. Published palaeohydrology case studies have been reviewed and contextualised for potential impacts to mine closure applications.

Palaeoclimate recreations will provide mine owners and designers with an appreciation for natural climate variability – if it has happened before, can it happen again? This provides an opportunity for palaeoclimate recreations to provide confidence in closure designs, as palaeoclimate recreations can establish a more robust benchmark for natural climate variability, and envelope the quagmire of uncertainty in future climate condition estimates.

INTRODUCTION

The design and implementation of effective mine closure strategies is critical to ensuring that post-mining landscapes are non-polluting, stable, and capable of supporting post closure land uses. Design and analysis, including the hydrological, hydraulic, geomorphological, landform evolution performance of the post-closure design in a long-term, or in-perpetuity design horizon, is a critical consideration to achieving the closure objectives.

Current hydrotechnical closure design practices rely on estimates derived from historical climate data which is then adjusted to account for non-stationarity. The adjustment typically assumes non-stationarity is a function of global surface temperatures, often as studied by the International Panel on Climate Change (IPCC) (2021a) and subsequent literature. The reliabilities of adjustment procedures varies, and may not adequately capture the full range of climate variability and extreme weather events that could impact mine closure outcomes.

Palaeoclimatology, the study of past climates using geological and biological proxies, offers an independent approach for reconstructing historical climate patterns. These palaeo-climate recreations provide insight into the frequency and intensity of extreme events (such as floods) and potential climatic conditions. Coupled with recorded historical data, it can reduce uncertainties associated with rainfall and flood estimations for mine closure design.

This paper presents a literature review into palaeoclimate and palaeohydrology reconstructions and explores potential applications to inform mine closure design. Current design methodologies are examined and the incorporation of palaeoclimate data in the context of rare to extreme design floods. Integration of palaeoclimate insights into continuous climate data is explored. Based on the review, a draft framework is proposed.

PALAEOCLIMATOLOGY AND PALAEOHYDROLOGY

Palaeoclimatology is the study of past climates. It involves reconstructing and developing an understanding of Earth's climate history over geological timescales, ranging from decades to millennia ago. Palaeohydrology, specifically focuses on the study of past hydrological systems – the movement, distribution, and quality of water during periods predating modern hydrologic records. These two fields are closely related and relies on geological and biological records as indirect evidence, or proxies, to reconstruct historical climate and hydrological conditions.

Modern climate records, based on instrumental observation, reliably began in the 1880s (NASA, nd). In Australia, direct rainfall measurements began in the 1870s (National Climate Centre, 2000). River gauge records, if they exist at all, rarely extend more than a century into the past. Palaeoclimatology can extend climate records far beyond the reach of instrumental data, offering insights into natural climate variability, enhancing our understanding of the frequency and intensity of extreme weather events (like floods and droughts), and providing greater confidence in potential climate futures.

However, it is also crucial to interpret notable palaeohydrologic periods or events in the broader palaeoclimate context. Atmospheric greenhouse gas concentrations, global temperatures, and continental configurations have also varied throughout history and have influenced hydrologic processes.

PROXIES USED FOR PALAEOHYDROLOGIC RECONSTRUCTIONS

Reconstructing palaeofloods and palaeorainfall relies on a combination of geological and biological proxies, dating techniques, and climate modelling. The following sections present different types of palaeo proxies and discusses their uses for reconstruction of palaeohydrology.

A summary of palaeoclimate proxies is presented in Table 1.

TABLE 1

Summary of palaeoclimate proxies.

Proxy data	Measurement	Climate parameter	Span limits (years)	Resolution	Limitations
Geomorphic features (Baker, 2008)	Physical properties	Flood extent	>1 000 000	Individual events, or clusters of individual events.	Limited to individual flood events, evidence typically represents minimum peak flood stage.
Lake/ terrestrial sediments (Queensland Government, 2018)	Physical and chemical properties, isotopes (diatoms, ostracods, leaf material), sedimentation rate, ecological assemblage	Lake levels, chemistry, temperature, drought, flooding	50 – >1 000 000	Decadal – centennial	Diverse limitations depending on depositional environment
Marine sediments (Queensland Government, 2018)	Physical and chemical properties, isotopes (foraminifera, shells, leaf material), sedimentation rate, ecological assemblage	Rainfall, salinity, temperature, greenhouse gas levels	5000 – >1 000 000	Centennial to millennial	Reduced potential to capture high-frequency climate signals
Ice Cores (Queensland Government, 2018; Vance et al, 2013)	Stable isotopes, accumulation of snow/dust, greenhouse gases	Air temperature, precipitation rate, greenhouse gas levels, global ice volume, volcanic eruption, zonal wind strength, atmospheric and/or ocean system changes	500 – >1 000 000	Sub-annual to annual	Limited to polar areas, long-term planetary scale factors, potential underrepresentation of extreme events due to smoothing, regional specificity
Tree rings (Queensland Government, 2018)	Tree ring width	Rainfall, snowfall, temperature,	200–2000	Annual	Reduced potential to capture long-term climate signals, biased to temperate areas.
Coral (Queensland Government, 2018)	Isotopes, chemical properties, growth rate	Temperature, salinity, river inputs, sea surface temperatures	100–500	Sub-annual to annual	Spatially limited to the tropics
Speleothem (Queensland Government, 2018)	Isotopes, chemical properties, growth rate	Rainfall, temperature, cyclones	100–10 000	Annual to centennial	Site-specific calibration required to understand complex soil-groundwater interactions

Geomorphic features and sediment deposits

Extreme floods leave behind distinctive geomorphic and sedimentary signatures that serve as proxies for reconstructing past hydrological events (US Army Corps of Engineers, nd). Geologic records show evidence of catastrophic floods and intense rainfall episodes have occurred throughout Earth's history. Sometimes far exceeding anything observed in modern times. Deep scabland channels, giant current ripples with wavelengths of tens of metres, plunge pools, and streamlined erosional islands are evidence of historically large floods.

Geomorphic features such as slackwater deposits, high-level gravel bars and boulder deposits, erosional scarps and terraces are also indicators of the occurrence of palaeofloods. Slackwater deposits, typically found in bedrock canyons and tributary mouths, can form layered sequences that record multiple flood events (Baker, 2008; Harden *et al*, 2021). Geomorphic and sedimentary features can indicate the maximum extent of past floods, which can form the basis of estimating envelope maximum flood magnitudes (Baker, 2008; Mike *et al*, 2020). The presence of no evidence of flood at higher elevations, such as a stable terrace with developed soils and established vegetation, have also been used to define non-exceedance bounds (Harden *et al*, 2021). The spatial scale and morphology of these landforms, when analysed with hydraulic modelling, allow scientists to estimate palaeoflood velocities and discharge volumes. A visual representation of the above geomorphic features are presented in Figure 1.

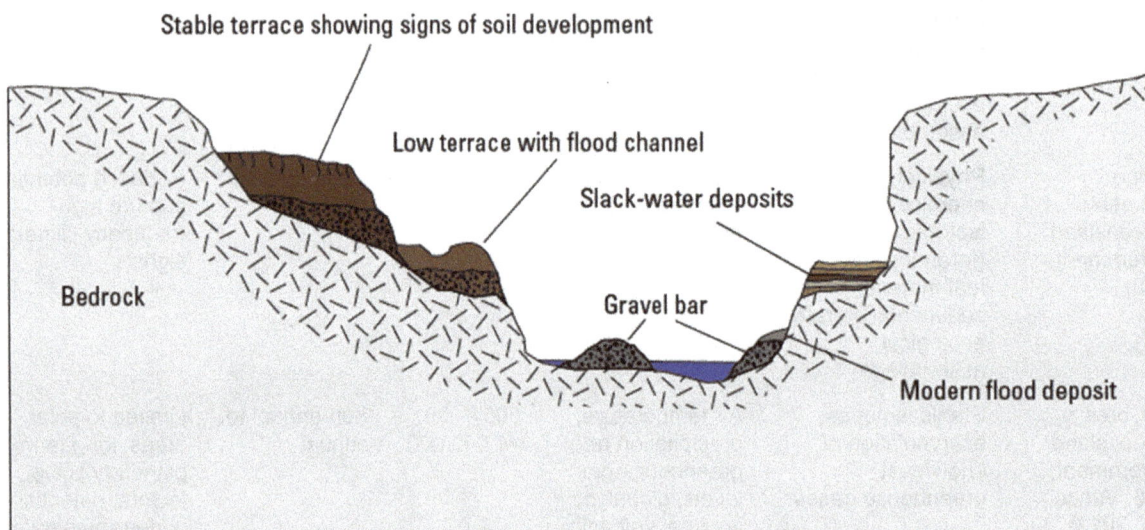

FIG 1 – Schematic diagram showing older stable terrace with developed soils (evidence of no recent floods), a younger terrace with well-developed overflow channel, the active channel with gravel bars, and a series of slack-water deposits along the channel margin on the bank opposite the terraces (Harden *et al*, 2021).

A snapshot of previous studies which have identified palaeoflood indicative sites globally is presented below in Figure 2.

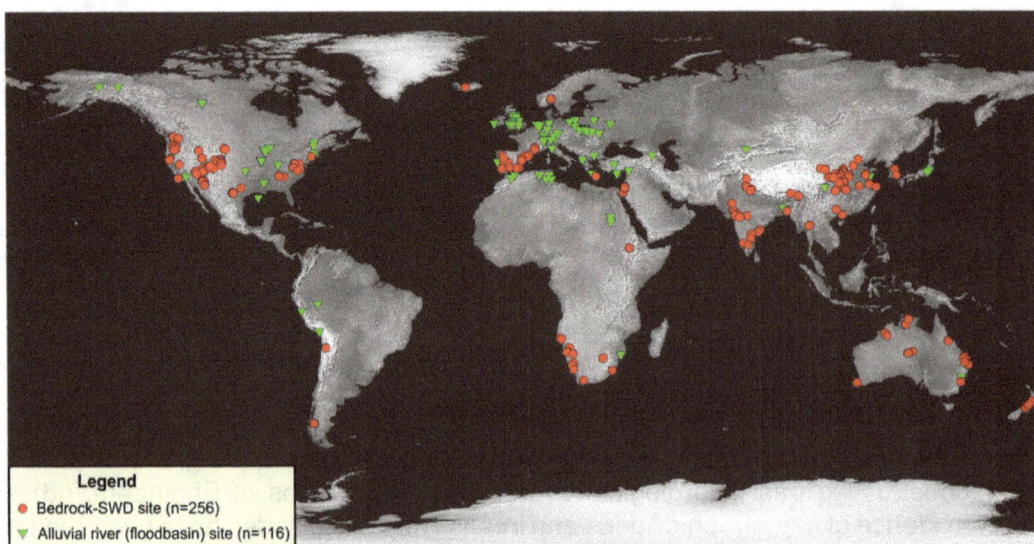

FIG 2 – Global distribution of selected sites with palaeoflood stratigraphy. Database modified from PAGES (Past Global Changes) working group (Baker *et al*, 2022).

Complementing these geomorphic indicators are stratigraphic sedimentary records preserved in terrestrial, freshwater, and marine environments. Large floods often deposit thick layers of coarse sediment in valleys, lakes, and sheltered slackwater zones – areas where floodwaters slow and drop their suspended load. In lake sediment cores, flood events are marked by distinct coarse-grained layers interbedded within finer background sediments. Grain-size analysis and geochemical scanning help distinguish flood layers from other depositional events like landslides. Sediments and their time of deposition can be dated using methods discussed in sections below.

Together, geomorphic and sedimentary evidence, coupled with channel geometry and hydraulic modelling can provide a robust framework for understanding the frequency, magnitude, and climatic drivers of extreme floods across geologic timescales.

Ice cores

Ice cores can serve as proxies for past atmospheric conditions. They consist of layers of ice that can contain trapped air bubbles, chemical compounds, dust and sea salt particles, which have been studied to yield correlations to past atmospheric conditions (Vance et al, 2013).

For example, summer sea salt concentrations transported by winds contained in Antarctic ice cores were used to infer large-scale atmospheric circulation patterns (Tozer et al, 2018). Coupled with a statistically significant correlation between the presence of salt in the ice cores and annual rainfall for eastern Australia, >1000 year rainfall records were reconstructed. Further, other climate data can be inferred such as temperature and greenhouse gas concentrations (Tozer et al, 2018; Vance et al, 2013).

Biologic proxies

For more recent prehistoric events (late Holocene), tree-ring records and palaeoecology can capture flood signals. Extremely wide or narrow rings, or isotopic anomalies in wood, may reflect flood-induced water availability or stress (Harden et al, 2021).

Pollen and plant remains in sediment cores can show shifts to more moisture-loving vegetation during pluvial periods. However, these proxies usually indicate prolonged climate conditions rather than single flood events.

Dating of proxies

All the proxies above are underpinned by robust dating to place the palaeo-events on a timeline. Dating techniques include:

- Radiometric dating – determines age based on the decay of naturally occurring radioactive isotopes.
- Optically stimulated luminescence (OSL) – determines when the last time mineral grains were exposed to sunlight before burial.
- Dendrochronology – tree ring dating.
- Tephrochronology – layers of volcanic ash are use as time markers to date and correlate geological, archaeological, and palaeo-environmental records.

Where possible multiple, independent dating techniques are employed to improve chronological accuracy of palaeoevents (Baker, 2008).

Databases of palaeoclimate proxy and reconstruction data

Palaeoclimate proxy and reconstruction data around the world has been made available through a number of online databases such as, National Oceanic and Atmospheric Administration (NOAA) Palaeo Data Search, hosted by the National Centers for Environmental Information (NCEI, nd) and PAGES (Past Global Changes), an international project provides a library of data and research related to palaeoscience (PAGES, nd). A representation of palaeoflood study data available on PAGES is shown in Figure 2.

For palaeoclimate proxy records relevant to Australia's climate, the PalaeoWISE Database is available (Croke et al, 2021a). See Figure 3 for a visual representation of the types of proxies, spatial and temporal spread of data available.

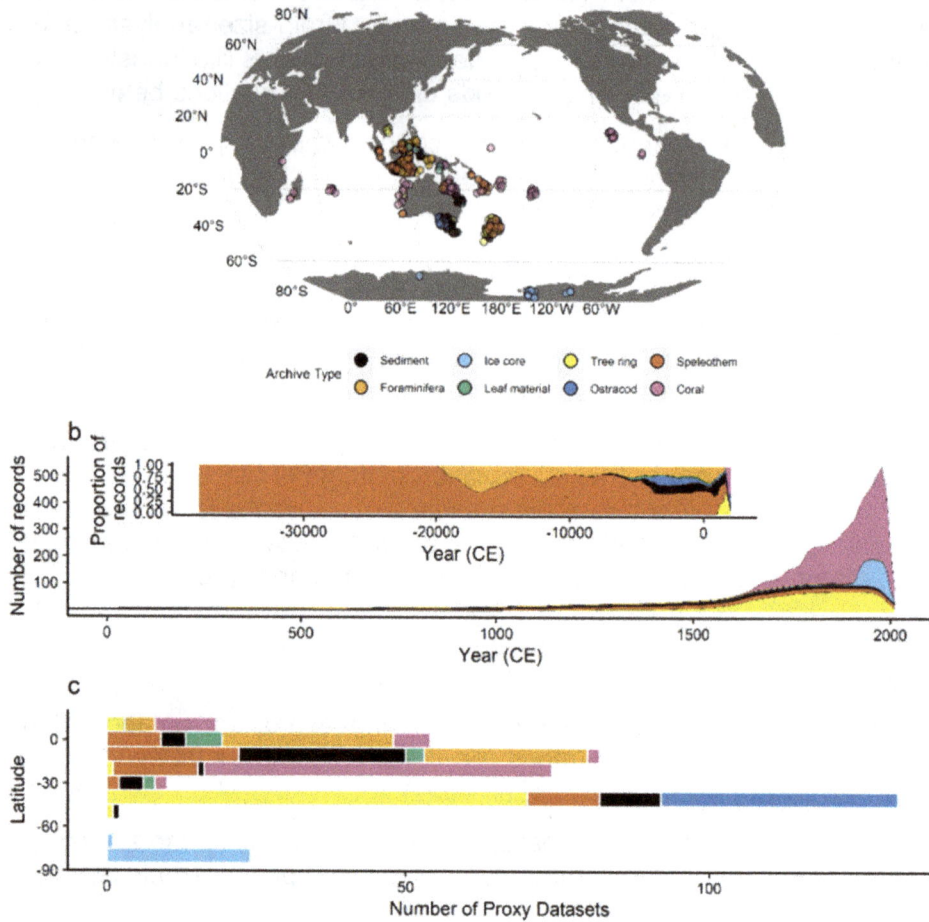

FIG 3 – Overview of the locations and temporal coverage of PalaeoWISE proxy database (Croke et al, 2021a).

In the development of PalaeoWISE, Croke et al (2021a) also assessed the correlation of each proxy and its ability for derived recreations to accurately reflect climate across Queensland. The proxies' reconstruction correlation to measured variables including rainfall, evapotranspiration, temperature assed. A sample of the correlation results across Queensland catchments is shown in Figure 4 below. This work allows for practitioners in Queensland to quickly identify a palaeo proxy for climate reconstruction application.

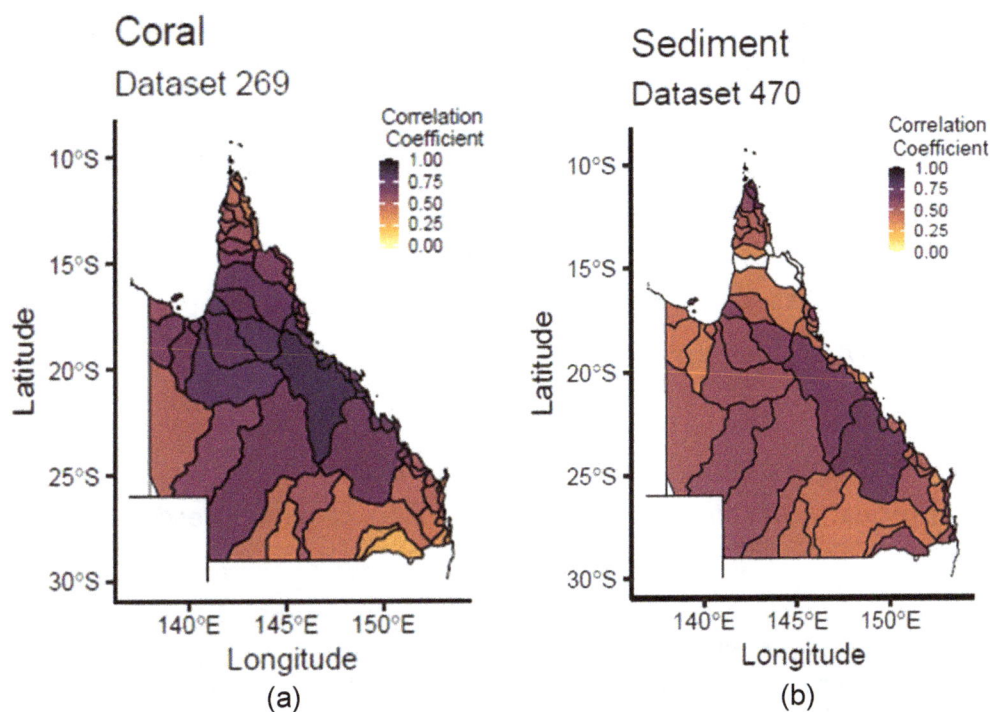

FIG 4 – Examples of proxy-climate correlation maps: (a) coral data set correlation with annual rainfall totals, (b) sediment data set correlation with annual temperature (Croke *et al*, 2021a).

EXAMPLES OF PALAEO-RAINFALL AND PALAEOFLOOD DERIVATION IN AUSTRALIA

Some case studies of palaeo-rainfall and palaeoflood recreations in Australia are presented below. This is not intended as an exhaustive list, rather, a presentation of a cross-section of proxies and findings of palaeohydrology in an Australian context.

East Alligator River palaeofloods, Northern Territory

- **Palaeo proxy:** geomorphic features – slackwater deposits, alluvial stratigraphy, sediment deposits – optically stimulated luminescence dating.

- **Age (Period):** 6380 BCE – 2007 CE.

- **Intensity (Flood):** Palaeofloods identified at East Alligator Gorge were three times the Australian Envelope Curve for Probable Maximum Flood (PMFs) (Mike *et al,* 2020).

Ross River palaeofloods, Central Queensland

- **Palaeo proxy:** geomorphic features – slackwater deposits, alluvial stratigraphy, palaeochannels, sediment deposits – thermoluminescence dating, radiocarbon dating.

- **Age (Period):** 450–1250 CE.

- **Intensity (Flood):** Flooding up to 10 km wide covering entire floodplain; flood heights as great as 12.8 m above the modern floodplain (Paton, Pickup and Price, 1993).

Burnett, Mary, and Brisbane catchment palaeofloods

- **Palaeo proxy:** geomorphic features – slackwater deposits within semi-alluvial channels, sediment deposits – optically stimulated luminescence dating, radiocarbon dating.

- **Barambah Creek (Burnett Basin)**

  o **Age (Period):** 1860 CE ± 90 years, 865 CE ± 20 years.

- **Intensity (Flood):** 9000 m³/s (~0.9 per cent Annual Exceedance Probability (AEP)), 9000 m³/s (~0.9 per cent AEP).

- **Mary River**

  - **Age (Period):** 1765 CE ± 50 years, 1650 CE ± 85 years, and 1425 CE ± 90 year.

  - **Intensity (Flood):** 6750 m³/s (~0.9 per cent AEP), 3800 m³/s (~7 per cent AEP), 3700 m³/s (~7 per cent AEP).

- **Lockyer Creek (Brisbane Basin)**

  - **Age (Period):** 1755 ± 35 years.

  - **Intensity (Flood):** 2700 m³/s (~2 per cent AEP) (Lam *et al*, 2017).

Murray Darling Basin palaeo-rainfall reconstruction

- **Palaeo proxies:** tree rings – Western Australia, coral luminescence – Queensland, speleothem – Wombeyan Caves.

- **Age (Period):** 749 BCE to 1980 CE.

- **Rainfall findings:** Rainfall reconstruction found three periods of wetter wet periods than the instrumental record, suggesting increased flood risk during these periods (Ho, Kiem and Verdon-Kidd, 2015).

OPPORTUNITIES FOR PALAEOCLIMATE APPLICATIONS IN MINE CLOSURE

Mine closure practitioners are faced by many uncertainties when designing for long-term time horizons to achieve safe, stable, and environmentally sustainable post-mining landscapes. There are opportunities to supplement current practices based on instrumental records and climate change projections with palaeoclimate reconstructions to reduce uncertainties about future climate conditions.

There are potentially significant consequences of underestimating of climate risks to mine closure. Structural failures, flood containment/conveyancing failures, loss of containment of contaminated materials can result in loss of lives, impacts to environmental, social, and cultural values.

The current hydrotechnical design practice is presented in following sections, highlighting the opportunities where palaeoclimate and palaeohydrology can improve mine closure outcomes.

CURRENT HYDROTECHNICAL MINE CLOSURE DESIGN PRACTICE

Hydrotechnical mine closure design data

Hydrotechnical closure design criteria and associated design climatic data estimates are required for a wide variety of design and assessment processes, with common assessment processes including:

- Design storm, run-off and flood discharge estimation for hydraulic design and analysis.

- Siting and boundary condition establishment, such as riverine and tidal limits, including consideration of sea level rise, and storm surge.

- Final landform cover or capping layer design and analysis, such as rock mulch, earthen and clay barrier, phyto-capping (vegetative) systems.

- Design, simulation and analysis of water storage inventories, such as residual void pit lake and post-closure water storage studies.

- Hydro-geotechnical and hydro-geological design and analysis.

The aims of closure design ultimately being that mine sites are to be rehabilitated to be 'safe to humans and wildlife, non-polluting (does not cause environmental harm), stable and able to sustain an agreed post-disturbance land use' (Department of Environment, Science and Innovation (DESI), 2024a).

Ultimately, climate data products required for mine closure assessments typically comprise static or transient estimates (over the post-closure period) including:

- **Design rainfall depth data** – design rainfall curves are typically expressed as point rainfall depths (mm) versus probability (Annual exceedance probability) and event duration (hrs) sometimes termed intensity-frequency-duration (IFD) curves. The basis of estimates varies with event probability. A wide variety of post-fact adjustment processes are used with point rainfall estimates depending on use, such as storm temporal pattern ensembling or areal reduction.

  In Australia, design rainfall classes are described per Australian Rainfall and Runoff (ARR) (Ball *et al*, 2019) with the Annual Exceedance Probability (AEP) principally used to refer to:

  o Frequent: AEP ≥ 10 per cent.

  o Infrequent: 10 per cent ≤ AEP < 1 per cent.

  o Rare: 1 per cent ≤ AEP < 0.5 per cent.

  o Extreme: 0.5 per cent ≤ AEP < AEP of the Probable Maximum Precipitation (PMP) event.

  Common to intermediate probability estimates derived from analysis of station data are available through the Bureau of Meteorology (BoM, 2016) IFD program. The technical development of the BoM (2016) IFD is beyond the scope of this paper but produces a gridded result based on statistical fitting models based on observed data with variable geographic input data density (Ball *et al*, 2019; BoM, nd). Alternative estimates are available in some locations through additional council programs, such as the Lockyer, Ipswich and Moreton Bay region (LIMB) program (WMA Water, 2021). Rare to extreme estimates are estimated based on: a) estimation of the PMP depth utilising the Generalised Short Duration Method (GSDM) (BoM, 2003a), Revised Generalised Tropical Storm Method (GTSMR) (BoM, 2003b), Generalised Southeast Australia Method (GSAM) (BoM, 1996) and/or West Coast Tasmania Method (Xuereb, Moore and Taylor, 2001) guidelines; and b) statistical interpolation between the PMP depths and IFD estimates with a common technique applied being the Cooperative Research Centre – Focused Rainfall Growth Estimation (CRC-Forge) growth method (Siriwardena and Weinmann, 1998). Regionally, similar methods are available, however are less developed in a platform offering sense. For example, in New Zealand *A Guide to Probable Maximum Precipitation in New Zealand* (Thompson and Tomlinson, 1993) provides short duration PMP estimates based on a storm maximisation technique, similar to GSDM (BoM, 2003a).

- **Continuous climatic data estimates** – Continuous data refers to calendar-based data which may have resolution typically in the daily, sub-daily or ~minute scale. Continuous data is available in Australia through monitoring gauge data operated by BoM, local and state government agencies.

  Collated database products are typically available via the Bureau of Meteorology for weather station data (for climate observations) and flow gauge network (for water way gauge data) allowing site specific or proxy data use in instances, depending on monitoring network density.

  For climatic data products, the requirements of data completeness have resulted in the development of synthetic, geographically kriged and synthetic infill methods. Commonly used, the Scientific Information for Land Owners (SILO) program (The State of Queensland, 2024) provides point (infill of a weather station) and grid (kriged data) across Australia. Observed, synthetic and derivative estimates are available for rainfall, temperature, vapour pressure, evaporation, solar radiation, humidity, evapotranspiration and pressure.

  Other continuous data sets include the Australian Water Availability Project (AWAP) and Daily Aggregated Weather (DAW) products from the bureau of meteorology, and recent developments for radar rainfall products (Su *et al*, 2022).

Consideration of future climate

The International Council on Mining and Metals (ICMM) (2025) considers that the incorporation of climate change in mine closure designs and risk evaluations is now considered standard practice, and it is the only approach compatible with passive closure or relinquishment of a mine.

The ARR 2024 revision incorporated climate change adjustment of a changing climate to future flood estimates, stating that 'unadjusted historical observations are no longer a suitable basis for design flood estimation: they must be adjusted to reflect the impacts of rising global temperatures'. The revision also acknowledges that a fundamental issue in the characterisation of climate change is separating the influences of naturally occurring climate variability from anthropogenic climate change, based on a relatively short climate record.

Current practice for consideration of climate change in key closure data is as follows:

- **Design rainfall depth data** – Adjustment procedures are outlined in the ARR 2024 revision (Ball *et al*, 2024). The procedure aligns with the framework outlined in Climate Change 2021: The Physical Science Basis (IPCC, 2021a). The approach combines regional hydrological meta-analysis (Wasko *et al*, 2024) and duration adjustment methods (Visser *et al*, 2021).

  An apparent scaling approach based on projected temperature change from the Coupled Model Intercomparison Projects (CMIP) 6 generation of global climate models, is applied to a baseline for the 2016 IFD. The method produces a series of uncertainty estimates, based primarily on the: (a) selected Shared Socio-economic Pathway (SSP); (b) global surface average temperature (GSAT) increase; and c) relationship between temperature increase and rainfall depths. The procedure produces uncertainty estimates for the selected inputs.

  The procedure notionally applies over the range of studied inputs, however due to differences in the development of the PMP depth, the application of the method to the PMP is considered speculative (Ball *et al*, 2024). There is no consensus for adjustment of depths in the rare to extreme design rainfall classes.

- **Continuous climatic data estimates** – There is no industry standard approach or guidance developed for the incorporation of climate change into continuous hydrology modelling. A review found a variety of procedures in use:

  o **Adjusted continuous data series** – Perturbed or continuous data series, such as the Biophysical CCS models (Long Paddock, 2024) are occasionally utilised to inform potential future climates. These series produce adjusted inter-decadal climate data estimates based on CMIP5 (Working Group on Coupled Modelling (WGCM), 2013) and 6 (WGCM, 2019).

  o **Annual, seasonal and monthly factoring processes** – Adjustment of climatic data often occurs by factor adjustment of key climate variables (such as rainfall and temperature). Commonly utilised factors are available from various data portals, including Queensland Future Climate Dashboard (Long Paddock, 2025b), Climate Change in Australia (DCCEEW, 2025) and proprietary sources (ClimSystems Ltd, 2025).

  o **Design rainfall embedding** – Incorporating selective design rainfall adjustments per (Ball *et al*, 2024) for high rainfall depth periods as an embedding procedure, in addition to scaling methods is occasionally completed.

  o **Non-stationarity analysis and trend inclusion for specific gauge data** – Occasional site-specific non-stationarity analysis and trend adjustment is completed however is rare and has limited long-term predictive power.

  The factoring processes utilised represent data portals and dashboard summaries which are based primarily on CMIP5, and occasionally CMIP6. Regardless of the process followed, most processes inherently rely on the CMIP5 (WGCM, 2013) and CMIP6 (WGCM, 2019) project model outcomes.

  These climate models generate a timeseries of rainfall and other hydrometric parameters. The current generation of climate models do not perform well in the generation of rainfall time series, especially in an Australian setting (Nishant *et al*, 2022). Extreme events such as flooding and drought are not well replicated by climate models (Nishant *et al*, 2022). However, the performance of climate models to simulate temperature is well proven, and temperature can be used as a proxy to generate rainfall time series (Kiem *et al*, 2021).

It is noted the development of localised model data is typically completed as an adjustment to historical baseline data. Additionally, most uses of CMIP data group model sets with limited approaches distinguishing between models with varying assumed future conditions.

The underlying CMIP5 study focus is limited to near-term (2035) with 2006–2100 sequences produced as a subsequent extrapolation exercise (Taylor, Stouffer and Meehl, 2009), which limits the robustness of the estimates. Similarly, CMIP6 focuses on estimations prior to 2100, with extended simulations for 2300 included as an extension (Eyring *et al*, 2016).

Accordingly, methods of continuous adjustment of data rely on substantial model projects by WGCM then local or proprietary bodies. Typically, the available adjustments data currency lag behind the studied design rainfall adjustments methods available. Most model methods aggregate the individual model outcomes from the CMIP series, with limited approaches considering the difference in assumed future conditions of sub-categories of CMIP models.

Notably, at the time of writing, the CMIP6 estimates have not gained widespread adoption, however the Seventh Assessment Report (AR7) and CMIP7 (WGCM) projects have commenced (IPCC, 2025). The CMIP7 project objectives (Dunne *et al*, 2024) specifically include the review of extreme weather statistics, including rainfall. Additionally, the study aims to answer how human responses may affect the carbon cycle, and the risks of tipping points for future climate outcomes.

Post-closure period

For the purpose of mine closure, consideration of climate change adjustments in design and planning usually occurs through the designation of a closure project timeline particularly the post-closure period.

Most approaches do not specifically require the application of climate change adjustments:

- Climate change impacts for closure were a key consideration, including related closure asset costs for ICMM members (ICMM, 2019) with opportunities to build resilience by considering climate change impacts identified.

- The Mining Closure Framework (Towards Sustainable Mining (TSM), 2008) requires life cycle consideration for closure planning.

- The Ranger Uranium Mine closure requirements specify a 10 000-year post-closure period for tailings containment in technical assessments (Energy Resources of Australia (ERA), 2023).

- Queensland EPA 1994 policy and guideline framework (Queensland Government, 1994) is transitioning to a progressive rehabilitation framework, which involves public interest, application and departmental decision stages. Progressive rehabilitation and closure requirements are outlined in ESR2019/4964 (DESI, 2024b). A variety of technical assessments interact with long-term climate conditions, prompting climate change positions, such as:

  o Voids in flood plains – floodplain assessments.

  o Flooding – flood levels and flood risk profile.

  o Landform design – hydrological, hydraulic and erosion assessments.

  o Water management – long-term management requirements.

  o Revegetation – seed mix, growth media and topsoil requirements.

  o Tailings and Voids – Water balance, water storage, hydrogeological assessments.

  o Underground mining – subsidence and hydrogeological model.

- Global Industry Standard on Tailings Management (GISTM) (Global Tailings Review (GTR), 2020) requirement 3.1 encourages climate change knowledge to be used in adaptive management of tailings storage facilities in order to reduce facility risks to As-Low-As-

Reasonably-Practicable (ALARP). Water balance modelling including climate change is suggested. Flood design criteria for closure of tailings dams is 1:10 000 AEP.

- Guidelines on Tailings Dams (Australian National Committee on Large Dams Incorporated (ANCOLD Inc), 2019) specifies a design period of 1000 years for post-closure consideration for tailings storage facilities.

Future probability

Mine closure design and analyses utilises a variety of event probabilities depending on application and model. An overview of typical design ranges is shown below in Table 2.

TABLE 2

Summary of hydrologic data required for mine closure assessments.

Mine closure assessment type	Data type	Probability classes				
		Frequent	Infrequent	Rare	Extreme	Probable Maximum
Flood assessment	Design rainfall depth data			▓	▓	▓
Hydraulic structure design	Design rainfall depth data			▓	▓	▓
Drain design	Design rainfall depth data		▓	▓	▓	
Run-off volume assessments	Continuous data	▓	▓	▓	▓	
Vegetation assessment	Continuous data	▓	▓	▓	▓	
Cover performance assessment	Continuous data	▓	▓	▓	▓	
Erosional modelling	Continuous data	▓	▓	▓	▓	
Groundwater modelling	Continuous data	▓	▓	▓		

Choosing a path

Once agreement that climate change adjustments are required, such as for mine closure design and analysis, a key challenge is in making appropriate assumptions for potential future climates.

Within most climate change adjustment processes, the underpinning framework for climate change adjustments is outlined in Climate Change 2021: The Physical Science Basis (IPCC, 2021a).

A key challenge is that the IPCC (2021a) framework requires the assumption of a Shared Socio-economic Pathway (SSP), which are used to describe potential global climate change responses. In IPCC (2021b) these are described as shared pathways and net radiative increases (W/m^2).

In example, the SSP3.0–7.0 pathway represents Pathway 3 ('regional rivalry') or 'rocky-road' which describes global responses and broad economic and environmental pathways, and 7.0 W/m^2 of effective radiative forcing at 2100, which associates to very high level of greenhouse gas emissions and negligible mitigation.

Review of adopted SSPs in recent public studies suggests:

- Adoption of upper bound SSP5–8.5 IPCC trajectory is common, however recent guidance suggest that this scenario is very unlikely to be realised (Long Paddock, 2025a) and represents a worst case, whereas a 'business as usual' trajectory of Representative Carbon Pathway 6.0 (equivalent to SSP4–6.0). Notably, SSP4–6.0 trajectory pathway estimates are not typically available through the estimation and adjustment methods, which may lead to default adoption of SSP5–8.5.

- A specific reference to an SSP is frequently not made, with temperature adjustments referred to directly.

- Consideration is given to one selected SSP rather than a range of possible outcomes.

Limitations of current practice – forming a consensus

Looking to an uncertain future

Based on the review completed, a matrix of typical post-closure periods aligned to post-closure assessment data needs and available adjustment methods (by underlying data) is shown in Figure 5. The key gap identified relates to climate change impacts for rare to extreme probability design rainfall depths. Whilst ARR methods (Ball *et al*, 2024) can be applied, these are speculative and generally unsupported by the literature due to differences in derivation methods.

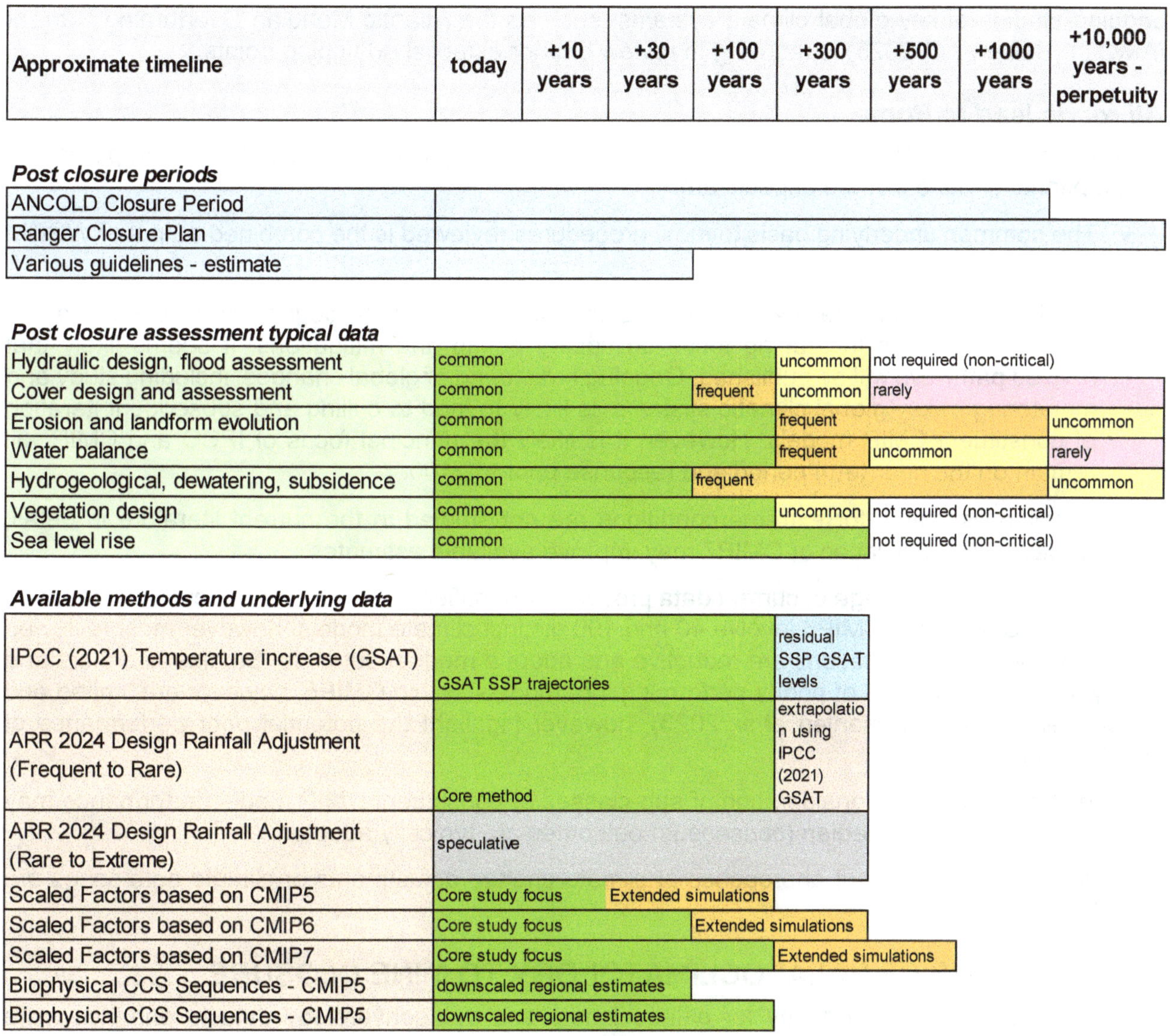

Approximate timeline	today	+10 years	+30 years	+100 years	+300 years	+500 years	+1000 years	+10,000 years - perpetuity

Post closure periods

	today	+10	+30	+100	+300	+500	+1000	+10,000
ANCOLD Closure Period								
Ranger Closure Plan								
Various guidelines - estimate								

Post closure assessment typical data

	today	+10	+30	+100	+300	+500	+1000	+10,000
Hydraulic design, flood assessment	common				uncommon	not required (non-critical)		
Cover design and assessment	common		frequent	uncommon	rarely			
Erosion and landform evolution	common			frequent			uncommon	
Water balance	common			frequent	uncommon		rarely	
Hydrogeological, dewatering, subsidence	common		frequent				uncommon	
Vegetation design	common				uncommon	not required (non-critical)		
Sea level rise	common				not required (non-critical)			

Available methods and underlying data

	today	+10	+30	+100	+300	+500	+1000	+10,000
IPCC (2021) Temperature increase (GSAT)	GSAT SSP trajectories							residual SSP GSAT levels
ARR 2024 Design Rainfall Adjustment (Frequent to Rare)	Core method							extrapolation using IPCC (2021) GSAT
ARR 2024 Design Rainfall Adjustment (Rare to Extreme)	speculative							
Scaled Factors based on CMIP5	Core study focus		Extended simulations					
Scaled Factors based on CMIP6	Core study focus			Extended simulations				
Scaled Factors based on CMIP7	Core study focus				Extended simulations			
Biophysical CCS Sequences - CMIP5	downscaled regional estimates							
Biophysical CCS Sequences - CMIP5	downscaled regional estimates							

FIG 5 – Post-closure periods and assessments.

The review also highlights a key gap in the establishment of climate change estimates in the far-term, with typical post closure periods unlikely to be effectively serviced by available data products. IPCC AR7 and CMIP7 projects are ongoing. The publication of project outcomes is likely to lead to additional SSP developments, and published time series (such as GSAT) to a horizon of 2500.

Design rainfall adjustments may simply be updated by reference to new GSAT increases at this time, however the potential addition of new pathways may complicate adoption and industry understanding.

Including subsequent downscaling projects to produce local continuous climate estimates is likely to take time. Based on the overall review, the propagation of design rainfall climate change adjustment methods has been reasonably prompt, relative to incorporating climate change impacts in continuous data due to delayed localisation of CMIP model outcomes.

Overall, there is a strong possibility of climate change adjustment process revision, and incorporating climate change impacts in closure design and analysis represents a moving target.

Based on current IPCC (2021a) trajectories, design processes which primarily consider temperature increase, such as design rainfall adjustments, might reasonably be enveloped within current estimates, however residual GSAT increases currently estimated for 2200–2300 may not provide reliable or robust estimates for the majority of the closure period.

Notably, the focus of AR5, AR6, CMIP5 and CMIP6 was on system response with 2100–2300 estimates provided as extended simulations. The focus of IPPC AR7 and CMIP7 projects, and ongoing studies of key global climate systems, such as the Atlantic Meridian Overturning Current (AMOC) (Baker *et al*, 2025) is resulting in review of prior established tipping points.

All roads lead to Rome

Based on the current IPCC framework, the following issues in establishing long-term, or in-perpetuity assumptions for mine closure aspects are:

- The common underlying basis to most procedures reviewed is the combined output of WGCM and IPCC.

- The focus of the climate change scientific community is on anthropogenic caused impacts and responses with an increasing focus on tipping points and mitigations. It seems likely that revised pathways will be published. Ongoing monitoring of global changes, including study and monitoring of key global climatic systems, is likely to lead to culling and subsequent iteration of constituent CMIP models. However, it is likely the principal focus of IPCC and CMIP will remain on the near-term period and response prior to 2100.

- Detailed estimates for extreme conditions are not studied in the current literature in detail, however the publication of CMIP7 may improve available estimates.

Additionally, a key challenge of current data processes identified relates to the structure of the CMIP model sets. CMIP5 and CMIP6 involve 40 and 100 distinct climate models, however most reviewed methodologies and applications are reductive and adopt a median outcome. Studies to understand the hydrological influence of poorly performing 'hot models' within CMIP6, however are limited and impacts are unclear (Assenjén *et al*, 2023), however highlight the potential poor performance of constituent models.

For the closure context, consideration of sub-classes or constituent CMIP model performance may be warranted, however median (consensus) outcomes are typically adopted.

Overall, limited independent approaches of climate change adjustments or climate data series are available.

APPLICATIONS OF PALAEOCLIMATOLOGY TO MINE CLOSURE

There is a significant opportunity for palaeoclimate and palaeohydrology data to enhance mine closure design by providing long-term climate and hydrological insights that are not captured in the current approach.

Whilst climate change estimates under the IPCC framework are improving and expanding in scope, there are substantial gaps in rare to extreme design conditions. Pathways and CMIP model sets and outcomes will change. Additionally, improvements to continuous data lag behind the documented studies, and have not been widely adopted.

Palaeoclimatology offers an alternative approach that is independent to the established IPCC framework and is less prone to impact from subsequent revision and refinement. Additionally, the consideration of data beyond the modern monitoring is possible, with consideration of conditions across geological time possible. This process can augment the existing approaches and provide static reference points.

Palaeoclimatology can envelope the natural variability of conditions over the post-closure period, particularly when consideration from 1000 years to perpetuity is required, enabling closure practitioners to design for climate risks with greater confidence. Whilst events occurring over the geological epoch may seem uncertain, adopting possible future conditions subsequent to the climate change crisis resolution requires an imaginative approach.

Palaeoclimatology can provide new and independent inputs to adaptive design processes for post-closure period. The process can also assist in communicating the credibility of estimated events with rare probabilities (such as the PMF).

A simple framework is proposed when considering the applicability of palaeoclimate to a mine closure project for design hydrology and continuous hydrology applications, see Figure 6.

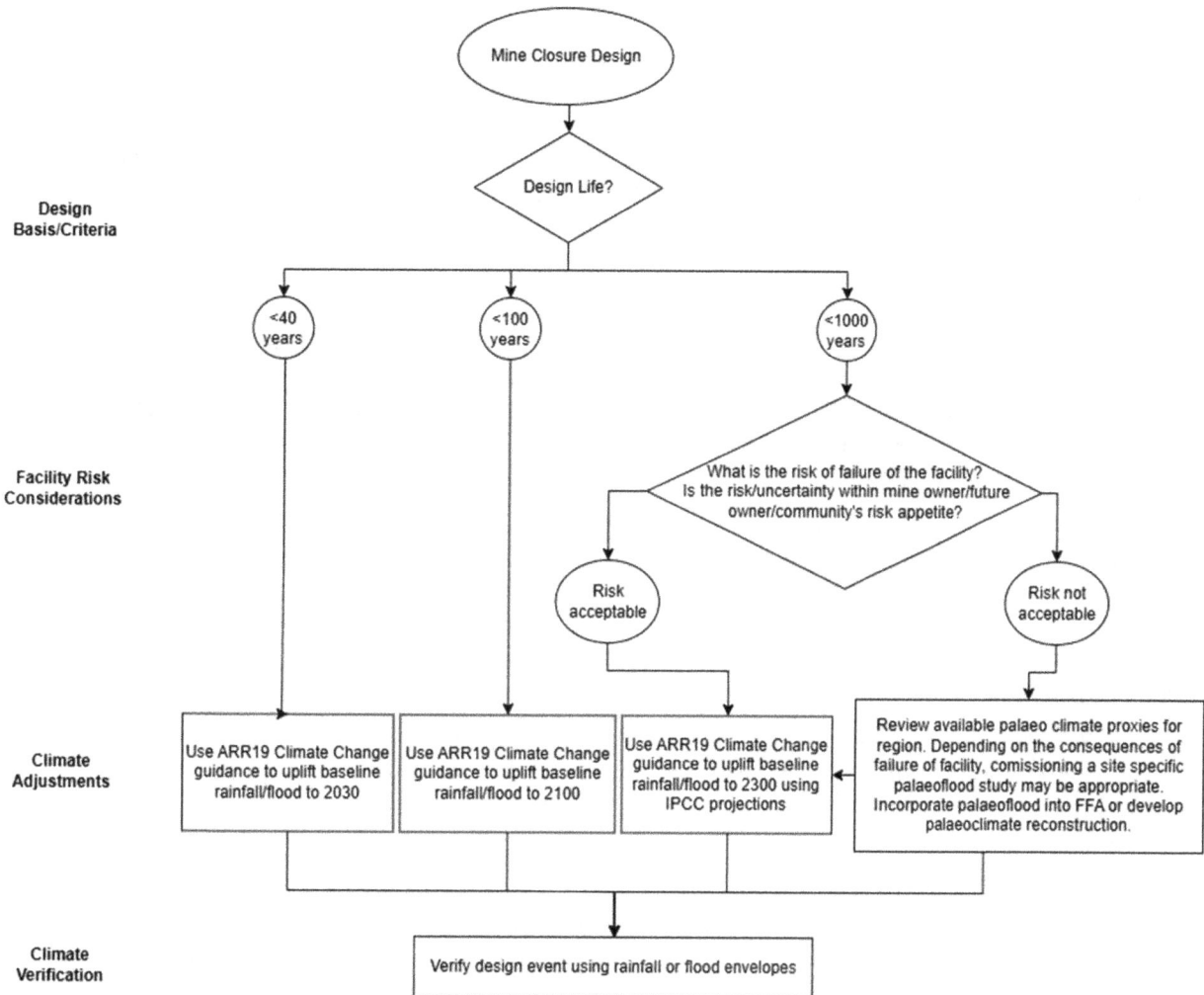

FIG 6 – Proposed framework for the application of palaeoclimate and palaeohydrology to mine closure design.

Key identified palaeoclimate processes for application include:

- Palaeo-flood estimates provide alternative discharge estimation for extreme flood events. Envelopes for large and rare flood events can be provided, providing alternative events for consideration discretely or as inputs to existing hydrological processes (such as Flood Frequency Analysis (FFA)).

- Design of stable landforms and drainage systems can be informed by palaeoclimate-derived time series. Landform evolution modelling using palaeorainfall inputs could provide better confidence in ensuring that a closed mine site remains resilient to climatic changes through perpetuity.

- Water balance, groundwater and climate-interfacing models considering post-closure periods benefit from incorporating palaeoclimate-derived time series, providing alternative adjustment procedures for long-term consideration.

Palaeoclimate informed design hydrology for extreme flood events

Mine closure designs require hydraulic features such as final landform, spillways, and other drainage features to withstand an agreed upon design event – often the PMF or a suitably rare design event. There are many uncertainties associated with the calculation of the PMP and resulting PMF, as described in previous sections.

The United States Geological Survey (USGS; Harden *et al*, 2021) recommends palaeofloods to be incorporated into FFAs based on instrumental flood records using the Expected Moments Algorithm (EMA). This allows palaeoflood data to be included as interval estimates rather than point estimates to account for uncertainty inherent in the palaeoflood reconstruction (Harden *et al*, 2021). In Australia, the ARR 2019 indirectly proposes the incorporation of ungauged flood records into FFAs using the Bayesian estimation framework, which also allows for interval censoring to input flood events as a range (Ball *et al*, 2024). The ARR 2019 cautions studies related to the value of palaeoflood estimates in FFAs in Australia are limited, however, its potential has been indicated by studies which have identified large floods in Australia.

Inclusion of palaeoflood events in FFAs can refine the upper tail of the Flood Frequency (FF) distribution, resulting in significantly reduced uncertainty of rare, extreme flood estimates, especially when available instrumental records are short (Reinders and Muñoz, 2021; Lam *et al*, 2017; Harden *et al*, 2021). Reinders and Muñoz (2021) found that the inclusion of more than six palaeoflood events is beneficial for an FFA, with diminishing returns beyond ten palaeo events. Location and scale parameters of a FFA are less influenced by palaeoflood data, and better estimated using long instrumental records (Reinders and Muñoz, 2021).

Though uncertainties in flood characterisation can be reduced by use of palaeoflood estimates, palaeoflood estimates themselves can introduce uncertainty. Large errors in palaeoflood discharge estimates can reduce the effectiveness of flood frequency analyses (Reinders and Muñoz, 2021). ARR 2019 suggests that although high accuracy is not possible with palaeoflood estimates, they may only be marginally less accurate than other estimates that require extrapolation of rating curves (Ball *et al*, 2024).

The USGS suggests analysis should account for non-stationarity in the flood record. Changes to flooding regimes can occur due to changes to climate, land-use, flood geometries. If non-stationarity is found, alternative FFA methods may require incorporation (Harden *et al*, 2021). Though, Lam *et al* (2017) found in his study that the FFA results were not greatly sensitive to potential changes in channel capacity over time.

There are also questions as to whether climate change may increase the magnitude and frequency of extreme floods, however, similar atmospheric processes consistent with predicted future climates have occurred in the past. Climate change is unlikely to create entirely new kinds of flood-generating weather and climate that have never occurred on Earth (Baker *et al*, 2022). Identifying floods or climate patterns which occurred during pre-instrumental times may be indicative of the possible futures under climate change (Baker *et al*, 2022).

The concept of integrating palaeoflood hydrology into flood assessments have been generally accepted in the USA, where the guidance is provided in the USGS flood flow frequency guidelines (Harden *et al*, 2021). In the USA, it has been used to inform probabilistic flood risk assessments for dam and nuclear powerplants (Baker *et al*, 2022). In China, the use of pre-instrumental records to inform design flood calculations has been incorporated into regulation to ensure water resources and hydropower infrastructure account for extreme flood events (Baker *et al*, 2022). In Australia, though methodologies for the inclusion of pre-instrumental flood events into FFA are provided in ARR 2019, incorporation of palaeoflood in design hydrology is not widely adopted by industry.

Palaeoclimate generated continuous hydrology

Many mine closure features like the final landform including mine voids require a continuous hydrologic assessment to ensure risks to the closure designs have been materially reduced to as low as possible.

There are examples of work undertaken for the incorporation of palaeoclimate to generate inputs for a continuous hydrology assessment globally (Tierney, deMenocal and Rosenthal, 2017; Cook *et al*, 2015) and in Australia (Tozer *et al*, 2018; Ho, Kiem and Verdon-Kidd, 2015; Croke *et al*, 2021b). Applications of the climate time-series reconstructions have ranged from water security/supply modelling, to developing insights into non-stationarity of rainfall.

The reconstruction of a climate time series using a palaeoclimate proxy can be undertaken if a statistically significant correlation between the proxy and climate parameter can be determined. For example, summer sea salt levels in Antarctic ice cores are linked to rainfall in Eastern Australia (Tozer *et al*, 2016).

Once a correlation is determined, calibration can be undertaken with periods where instrumental records exist, and a regression model can be developed to match the statistical distribution of the instrumental record to local settings (Tozer *et al*, 2018; Ho, Kiem and Verdon-Kidd, 2015). The regression can be applied in a number of ways, including to develop scaling factors to apply to instrumental records, or to develop a synthetic time series.

The reconstructed time series can then be applied to a continuous hydrology assessment. Palaeoclimate reconstructions have been applied to assessments so far for simulation of dam inflows, water security assessments including stress testing against drought, and flood risk assessments (Croke *et al*, 2021b).

Rainfall recreations from palaeoclimate data provides a greater understanding of the range of possible hydroclimate variability, without human induced climate change. The length of the continuous hydrology time series plausible to be reconstructed is dependent upon the specific proxy on which the reconstruction is based as indicated in Table 1 and Figure 3. If multiple proxies are available that have statistically significant correlations to a study area, methodologies exist to incorporate multiple proxies to best leverage the palaeo data available (Croke *et al*, 2021b).

Continuous hydrology assessments using palaeo data has found that reliance on only instrumental records can underestimate hydrologic and hydraulic risks (Tozer *et al*, 2018). Wetter wet epochs and drier dry epochs are common findings when recreating Australian palaeoclimates (Tozer *et al*, 2018; Ho, Kiem and Verdon-Kidd, 2015). Such epochs would not be captured in the current hydrology practice due to assumption of stationarity of climate. In relation to mine closure, misunderstanding of plausible climates have potential to result in structural failures of facilities with consequences to people, environment, and social values.

Given the availability of publicly available palaeoclimate proxy databases, and in Queensland, the availability of correlation coefficients associated with proxies (Croke *et al*, 2021a), it seems an untapped opportunity to not consider the use of palaeoclimate insights in engineering design. As highlighted in earlier discussions, extended continuous hydrology time series would have applications to inform mine closure design such as of a final landform through landform evolution modelling, or final mine void water balance modelling, exactly relevant for design horizons at temporal scales current hydrological modelling practices are lacking.

Verifying design events against envelopes

Design rainfall or flood events should be verified against published envelope values. These can provide context for palaeoclimate derived or informed design events. Published envelopes have been provided below.

Observed rainfall depth envelopes

Notable and maximum rainfall depths (BoM, 2016; WMO, 2009; Griffiths and McKerchar, 2014) have been used to generate envelope curves for rainfall depth and duration as shown in Figure 7. Design rainfalls, PMP estimates, and palaeo reconstructed rainfall depths should be benchmarked against this envelope for contextualisation and verification.

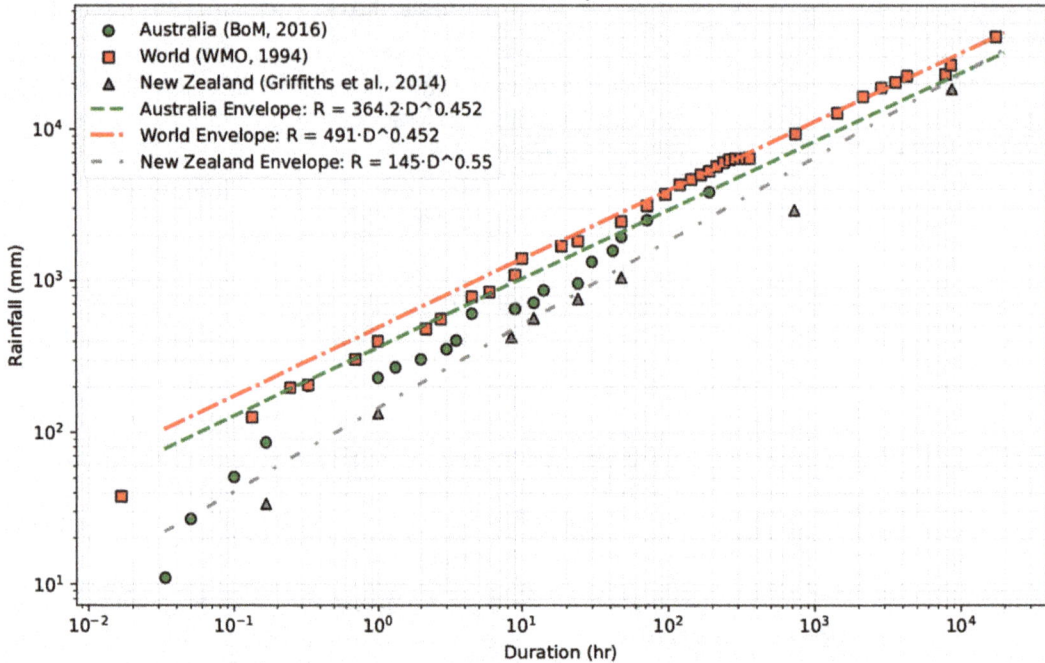

FIG 7 – Rainfall depth envelope (Ladson, 2016).

Flood peak discharge envelopes

Envelopes for flood peak discharge has been undertaken based on gauged flood records around the world. A combined plot of these envelopes has been provided as Figure 8. Design flood, PMF, and palaeoflood estimates should be benchmarked against the envelope for contextualisation and verification of results.

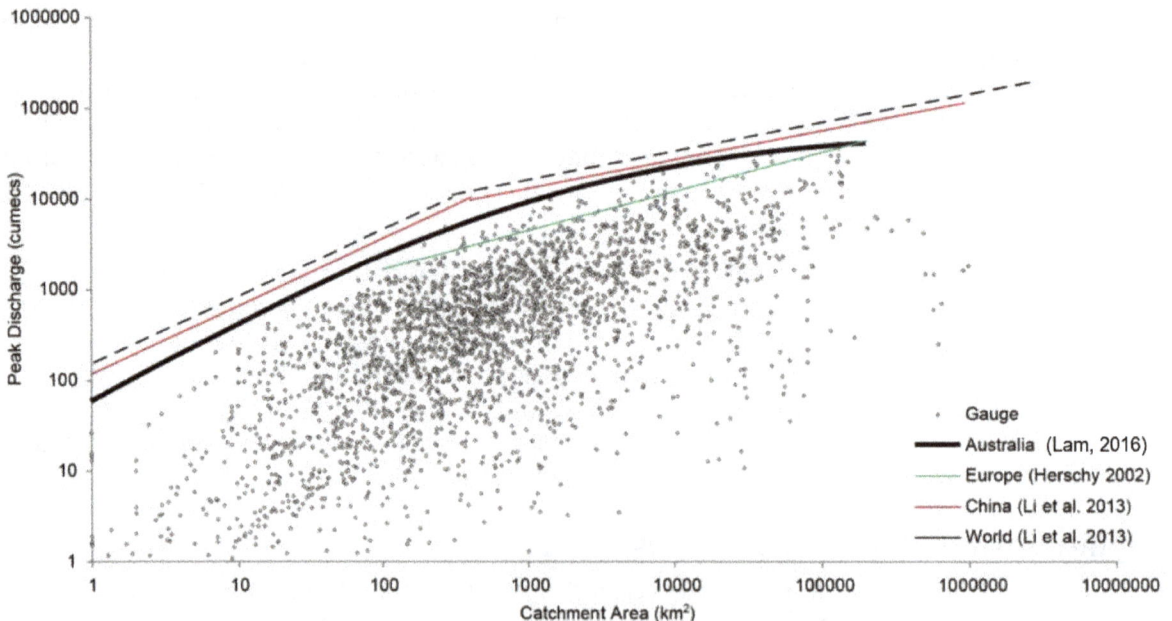

FIG 8 – Peak discharge envelopes (Lam, 2021).

Improving perceptions around uncertainty

The uncertainties associated with current hydrotechnical assessments is an ongoing challenge for mine closure practitioners. Mine owners rightfully question the feasibility of the magnitude and occurrence of extreme events that are often perceived as 'unprecedented'. Designing for the occurrence of a hypothetical extreme events could result in overengineering of facilities, or perceived so. In a very simple way, the documented occurrence of extreme events such as palaeofloods is a practical way to counter claims that the occurrence of the PMFs and climate change impacted design events are unprecedented or improbable (Baker, 2008).

Examples of palaeoclimate applications

Published applications of palaeoclimate related to mine closure in Australia is limited to the Ranger Mine tailings storage facility (TSF) rehabilitation example provided below (Mike *et al*, 2020). A published international example was not found at the time of writing of this paper. It is plausible that there have been unpublished examples of palaeoclimate applications to mine closure designs within the mining industry.

An example of palaeoclimate informed continuous hydrology assessment applied for mine closure could not be found. Instead, a water balance example from the hydropower industry has been presented.

Rehabilitation of in-pit tailings storage facilities at Ranger Mine, Northern Territory (Mike et al, 2020)

The designs of final landforms for the rehabilitation of two in-pit TSFs at Ranger Mine in the Northern Territory were informed by site-specific palaeoflood investigations. At this site, the containment of tailings was critical – the environment downstream of the TSFs consists of Magela and Gulungul creeks that drain to World Heritage listed Magela Creek wetlands in Kakadu National Park. Release of tailings or impacted water to the environment has potential for lasting environmental damage to the heritage sites. Legislated requirements also required the rehabilitated mine area must establish an environment similar to adjacent areas of Kakadu National Park, including the landform's erosion characteristics.

The site specific palaeo investigation consisted of pits excavated at five strategic locations exhibiting evidence of palaeoflood slackwater deposits. Samples were taken for optically simulated luminescence dating. Figure 9 shows the location of the palaeoflood investigation, including pit locations.

FIG 9 – Plan view of the East Alligator site showing the main features of the upstream slackwater site, the survey line for palaeo-discharge calculations, and the locations of the pits in the alluvial deposits. Flow is from left to right (Mike *et al*, 2020).

A summary of the findings from each pit have been presented in Table 3.

TABLE 3

Summary of key findings by dated pits from East Alligator site.

Pit	Number of palaeoflood units	OSL-dated flood events	Observations
1A	5	~1678 CE ~1535 CE ~895 BCE ~1120 BCE ~6380 BCE	Located at the highest elevation of the sedimentary deposits, ~20 m above the low-flow channel, minimising the risk of contamination from slopewash or recent sediment. Units compacted with sharp boundaries inclined toward the river. Contain millimetre-scale parallel laminations and occasional charcoal fragments. No angular erosional unconformities were found, but erosion cannot be ruled out.
1	4	2007 CE ~1535 CE ~950 CE ~80 CE	0.5 m lower than Pit 1A. Uppermost unit is loose sand, attributed to 2007 CE flood. Oldest unit lack laminations, suggesting rapid deposition from high sediment concentration flow.

The 2007 CE flood was used as a benchmarking and calibration event. An estimate of the total rainfall across the catchment for the 2007 CE flood was made based on radar data. River flow data for the event was extracted from a single gauge. This event was used to inform the design flood for landform stability modelling of the closure design required at Ranger Mine.

The peak discharge associated with each identified flood event was estimated using the modified Chezy equation and modified Manning equation, which both rely on the dimensions of the channel and channel slope (Bjerklie, Bolster and Vazquez, 2005). The nine palaeofloods identified were estimated across 8400 years, to be of similar peak magnitude, with one flood that may have been larger.

These floods were assessed by Mike *et al* (2020) to be equivalent to the PMF at the site, based on comparison to published PMF calculations for two nearby catchments, as well as the position of the East Alligator floods relative to Lam's Australian Envelope Curve (Lam, Thompson and Croke, 2017) (Figure 10).

The East Alligator floods were also estimated to have AEPs estimated at 0.3 per cent or 1:390, based on the most recent five floods forming a stationary series. Given that 0.3 per cent AEP for the PMF is orders of magnitudes higher than the notional AEPs assumed for PMFs (10^{-6}), the study does raise questions about the reliability of PMP-based PMF estimates. The region is characterised by bare rock and minimal soil cover, high run-off efficiency, and exposure to intense tropical rainfall from cyclones and monsoon lows – conditions, that make the region highly responsive to extreme rainfall events, increasing the likelihood of PMF-scale floods.

Similarities between palaeoflood discharge rates and PMF estimates (based on current methodology) suggests that PMFs may occur more frequently than previously assumed. The study helped define realistic upper bounds for PMF estimation which informed the design of the closure landform. Using PMP to estimate PMF in this instance would have resulted in overengineering of the closure landform by an order of magnitude.

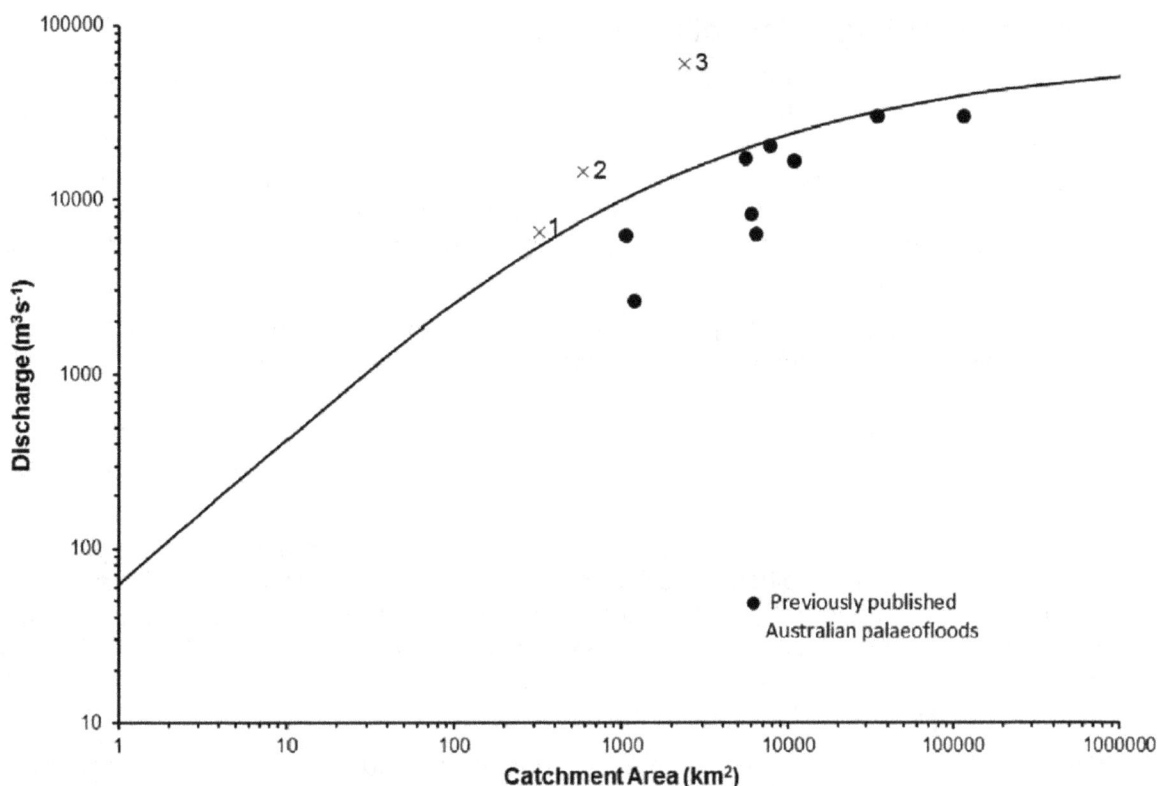

FIG 10 – Palaeofloods and PMFs in relation to the Australian Envelope Curve (Lam, Thompson and Croke, 2017) from which were also taken the 'Previously Published Australian palaeofloods'. Crosses denote the following sites: 1, Koolpin gorge PMF; 2, Magela Creek PMF; 3, East Alligator gorge palaeofloods (Mike *et al*, 2020).

Lake Burbury hydroelectric system water balance, Western Tasmania (Verdon-Kidd et al, 2025)

This case study integrated seasonal palaeoclimate reconstructions into a water balance model for a hydroelectric system in Tasmania. Two seasonal streamflow reconstructions (Austral summer ~1000 years, winter ~400 years) were derived from tree-ring chronologies. The reconstructions were downscaled to daily resolution using a bootstrapping method for use in a Source water balance model. The palaeoclimate data was coupled with a Long Short-Term Memory Artificial Neural Network model which simulated hydroelectric water extractions. This allowed for stress-testing the current hydroelectric system under a broader range of climatic conditions than the instrumental record alone provides.

The water balance model using palaeoclimate reconstruction resulted in a number of findings not reflected in the instrumental record. 18th century showed prolonged low storage levels and reduced extraction potential. The 11–13th centuries showed wet conditions, with high spill volumes.

For mine closure design, this study offers several parallels:

- Method for palaeoclimate reconstruction: the method of converting seasonal palaeoclimate data into daily inputs can be applied for water balance modelling, limited equilibrium modelling, and or contaminant transport modelling in post-closure landscapes.

- Stress-testing designs: mine closure designs for cover systems, water balance models for final voids, mine site flood resilience can be evaluated under a broader range of historical scenarios to assess erosion and water quality risks.

- Use of machine learning: where dynamic water balance inputs were required but are sparse or unavailable for pre-instrumental period for pairing with palaeoclimate time-series a machine learning model can be used to supplement water balance model inputs. In a mine closure context, this could be water extractions for community and industrial use, and varying contamination loads.

Limitations of adopting palaeoclimate reconstructions

As mentioned as a limitation of the design hydrology approaches in ARR 19, non-stationarity is a feature of natural climate variability and must also be considered in palaeoclimate research. Palaeoclimate reconstructions should be grounded by data assimilation and integration with physically based frameworks or climate modelling (Emile-Geay *et al*, 2024; Gallant, Phipps and Karoly, 2013). It is important to note that it is not suggested that palaeoclimate informed hydrology be adopted as a single source of truth, rather, results should be considered in the context of instrumental records, as well as climate change projections.

There are many studies comparing flood and rainfall estimates based on palaeoclimate reconstructions versus instrumental records. However, studies comparing flood and rainfall estimates based on palaeoclimate reconstructions and climate change projections are limited. Differences should become apparent as closure practitioners incorporate palaeoclimatology and climate change into closure design. It is encouraged that these findings be shared with industry as the studies become available.

In areas of Australia, there will likely be spatial gaps for where palaeoclimate proxies, or proxies with correlation are available. As additional studies and proxy data are collected, it is the hope that existing databases continue to get updated, and studies like PalaeoWISE (Croke *et al*, 2021a) are undertaken for other states in Australia. Awareness and adoption of these methodologies in growing industries like mine closure would likely drive an appetite for further studies. However, again, it does rely on sharing of proxy data and findings in the public sphere.

Given there is a current lack of specific language or guidance in Australian legislation regarding the methodologies to consider future climate in mine closure designs, there is an opportunity for industry to drive best practice. However, given the unavoidable uncertainties in climate science and hydrology, there may be resistance from mine owners to invest in studies beyond what is legislatively required and or considered current best practice. However, in the pursuit of the reduction of risks to as-low-as-reasonably-practicable, risk reduction measures may include studies to reduce uncertainties in hydrological risks, and palaeoclimate assessments have potential to provide benefit.

A CALL TO ACTION

To advance the application of palaeoclimate insights in mine closure, we recommend the following actions:

- **Practitioners** should consider incorporation of palaeoclimate-informed flood and rainfall reconstructions into design hydrology and continuous hydrology assessments, particularly for features with long design lives as is typical for mine closure infrastructure and systems.

- **Regulators and Industry Bodies** should consider updating closure planning guidelines to explicitly allow or encourage the use of palaeoclimate data as a complementary line of evidence alongside instrumental records and climate change projections.

- **Researchers** should focus on refining the translation of palaeoclimate proxies into hydrologically relevant parameters, improving spatial coverage of proxy data sets, and developing tools for integrating these insights into existing modelling frameworks.

- **Industry and academia** should collaborate to share case studies and methodologies, helping to build a body of evidence that supports broader adoption and regulatory acceptance.

CONCLUSIONS

This paper demonstrates that palaeoclimate and palaeohydrology reconstructions offer a valuable, independent lens through which mine closure designs can be strengthened—particularly for long-term and in-perpetuity planning horizons. By extending the climatic record beyond the instrumental era, these reconstructions provide a more comprehensive understanding of natural climate variability and extreme hydrological events. This is especially critical in the context of rare to extreme design conditions, where current climate models and adjustment methods remain speculative or incomplete.

The integration of palaeoclimate data into mine closure planning can:

- Reduce uncertainty in design rainfall and flood estimates, particularly for rare and extreme events.

- Support the development of adaptive, resilient landforms and water management systems.

- Enhance stakeholder confidence by grounding design assumptions in observed historical extremes.

- Provide a robust benchmark for validating climate change projections.

Ultimately, embracing palaeoclimate reconstructions as part of a multi-line-of-evidence approach will not only improve the robustness of mine closure designs but also contribute to a more resilient and scientifically grounded approach.

In conclusion, the palaeoclimate data not only strengthen the scientific foundation of the mine closure design but also contribute to broader environmental stewardship goals, aligning with regulatory expectations and community values.

ACKNOWLEDGEMENTS

Big thanks to Alex Karrasch for being a sounding board and helping pull together this paper.

This paper heavily leverages works completed by others such as Jacky Croke and Daryl Lam.

REFERENCES

Assenjén, N, Martel, R, Arsenault, R and Brissette, F, 2023. Understanding the influence of 'hot' models in climate impact studies: a hydrological perspective, *Hydrology and Earth System Sciences*, 27:1011–1032.

Australian National Committee on Large Dams Incorporated (ANCOLD Inc), 2019. Guidelines On Tailings Dams - Planning, Design, Construction, Operation And Closure, revision 1, ANCOLD Inc. Available from: <https://ancold.org.au/product/guidelines-on-tailings-dams-planning-design-construction-operation-and-closure-revision-1-july-2019/>

Baker, J A, Bell, J A, Jackson, L C, Vallis, G K, Watson, A J and Wood, R A, 2025. Continued Atlantic overturning circulation even under climate extremes, *Nature*, 638:987–994.

Baker, V R, 2008. Palaeoflood hydrology: origin, progress, prospects, *Geomorphology*, 101(1–2):1–13.

Baker, V R, Benito, G, Brown, A, Khan, N S, Kale, V S, Costa, J E, Zielhofer, C and Wohl, E E, 2022. Fluvial palaeohydrology in the 21st century and beyond, *Earth Surface Processes and Landforms*, 47(1):58–81.

Baker, V R, Burr, D M, Carling, P A, Komatsu, G and Parish, K, 2009. Channeled Scabland morphology, in *Megaflooding on Earth and Mars* (eds: D M Burr, P A Carling and V R Baker), pp 65–77 (Cambridge: Cambridge University Press).

Ball, J, Babister, M, Nathan, R, Weeks, W and Weinmann, E, 2019. *Australian Rainfall and Runoff: A Guide to Flood Estimation*, Version 4.2, Commonwealth of Australia.

Ball, J, Babister, M, Nathan, R, Weeks, W and Weinmann, E, 2024. *Australian Rainfall and Runoff: A Guide to Flood Estimation*, Version 4.3, Commonwealth of Australia.

Bjerklie, D M, Bolster, C H and Vázquez, C, 2005. Comparison of constitutive flow resistance equations based on the Manning and Chezy equations applied to natural rivers, *Water Resources Research*, 41(11):W11415.

Bureau of Meteorology (BoM), 1996. Development of the Generalised Southeast Australia Method for Estimating Probable Maximum Precipitation, HRS Report No. 4, Hydrology Report Series, Bureau of Meteorology, Melbourne.

Bureau of Meteorology (BoM), 2003a. The Estimation of Probable Maximum Precipitation in Australia: Generalised Short-Duration Method, Bureau of Meteorology, Melbourne.

Bureau of Meteorology (BoM), 2003b. Revision of the Generalised Tropical Storm Method for Estimating Probable Maximum Precipitation, Hydrology Report Series, Report No. 8, Bureau of Meteorology, Melbourne.

Bureau of Meteorology (BoM), 2016. Australia's Record Rainfall [online]. Available from: <http://www.bom.gov.au/water/designRainfalls/rainfallEvents/ausRecordRainfall.shtml> [Accessed: 18 June 2025].

Bureau of Meteorology (BoM), nd. How were the 2016 design rainfalls estimated? [online]. Available from: <http://www.bom.gov.au/water/designRainfalls/ifd/ifd-estimation.shtml> [Accessed: 10 April 2025].

ClimSystems Ltd, 2025. SimClim AR5 AR6 [online]. Available from: <https://climsystems.com/simclim/> [Accessed: 1 June 2025].

Cook, B I, Smerdon, J E, Seager, R and Cook, E R, 2015. A 1,200-year perspective of 21st century drought in southwestern North America, *Science Advances*, 1(1):e1400082.

Croke, J, Lam, D, Thompson, C and Price, P, 2021a. A palaeoclimate proxy database for water security planning in Queensland, Australia, *Scientific Data*, 8(292).

Croke, J, Lam, D, Thompson, C and Price, P, 2021b. Palaeoclimate analysis for water resource assessments: Final report, Department of Environment and Science (DES) and Seqwater.

Department of Climate Change, Energy, the Environment and Water (DCCEEW), 2025. Climate Change in Australia [online]. Available from: <https://www.climatechangeinaustralia.gov.au/en/obtain-data/> [Accessed: 25 June 2025].

Department of Environment, Science and Innovation (DESI), 2024a. Information Sheet – Voids in flood plains, ESR/2019/4966. [Online] Queensland Government.

Department of Environment, Science and Innovation (DESI), 2024b. Progressive rehabilitation and closure plans (PRC plans). ESR2019/4964 Version 3.01, Queensland Government Department of Environment, Science and Innovation.

Dunne, J P, Eyring, V, Sillmann, J, Krinner, G, Ruane, A C, Gutowski, W J, Boucher, O and Stouffer, R J, 2024. An evolving Coupled Model Intercomparison Project phase 7 (CMIP7) and Fast Track in support of future climate assessment, EGUsphere [preprint].

Emile-Geay, J, Evans, M N, Dee, S G, Feng, R and Tierney, J E, 2024. Temporal Comparisons Involving Palaeoclimate Data Assimilation: Challenges and Remedies, *Journal of Climate*.

Energy Resources of Australia (ERA), 2023. Ranger Mine Closure Plan – Executive Summary, Version 1.23.0, Energy Resources of Australia.

Eyring, V, Bony, S, Meehl, G A, Senior, C A, Stevens, B, Stouffer, R J and Taylor, K E, 2016. Overview of the Coupled Model Intercomparison Project Phase 6 (CMIP6) experimental design and organization, *Geoscientific Model Development*, 9:1937–1958.

Gallant, A J E, Phipps, S J and Karoly, D J, 2013. Nonstationary Australasian Teleconnections and Implications for Palaeoclimate Reconstructions, *Journal of Climate*, 26(22):8827–8849.

Global Tailings Review (GTR), 2020. Global Industry Standard on Tailings Management (GISTM) [online], Global Tailings Review. Available from: <https://globaltailingsreview.org/global-industry-standard/>

Griffiths, G A and McKerchar, A I, 2014. Towards prediction of extreme rainfalls in New Zealand, *Journal of Hydrology (New Zealand)*, 53(1):41–52.

Harden, T M, Koenig, R K, O'Connell, J E, Fabian, J M and Koenig, K J E, 2021. Historical and palaeoflood analyses for probabilistic flood-hazard assessments—Approaches and review guidelines, US Geological Survey, Scientific Investigations Report 2020-5089, 94 p.

Ho, M, Kiem, A S and Verdon-Kidd, D C, 2015. A Palaeoclimate Rainfall Reconstruction in the Murray-Darling Basin (MDB), Australia, Assessing Hydroclimatic Risk Using Palaeoclimate Records of Wet and Dry Epochs, *Water Resources Research*, 51(10):8380–8396.

Intergovernmental Panel on Climate Change (IPCC), 2021a. Climate Change 2021: The Physical Science Basis, Contribution of Working Group I to the Sixth Assessment Report of the Intergovernmental Panel on Climate Change (Cambridge University Press).

Intergovernmental Panel on Climate Change (IPCC), 2021b. Summary for Policymakers, in Climate Change 2021: The Physical Science Basis (Cambridge University Press).

Intergovernmental Panel on Climate Change (IPCC), 2025. Seventh Assessment Report [online]. Available from: <https://www.ipcc.ch/assessment-report/ar7/> [Accessed: 16 June 2025].

International Council on Mining and Metals (ICMM), 2019. Adapting to a Changing Climate – Building resilience in the mining and metals industry, International Council on Mining and Metals.

International Council on Mining and Metals (ICMM), 2025. Integrated Mine Closure – Good Practice Guide, 3rd edn, International Council on Mining and Metals.

Kiem, A S, Henley, B J, Baker, P J and Johnson, F, 2021. Stochastic Generation of Future Hydroclimate Using Temperature as a Climate Covariate, *Water Resources Research*, 57(12).

Ladson, T, 2016. Envelope curve for Australian Rainfall [online]. Available from: <https://tonyladson.wordpress.com/2016/02/08/envelop-curve-for-record-australian-rainfall/> [Accessed: 18 June 2025].

Lam, D, 2021. What the PMF? [online]. Available from: <https://awschool.com.au/content/uploads/PMF-Webinar-Envelope-Curves-Daryl-Lam-1.pdf> [Accessed: 16 June 2025].

Lam, D, Thompson, C and Croke, J, 2017. Improving at-site flood frequency analysis with additional spatial information: a probabilistic regional envelope curve approach, *Stochastic Environmental Research and Risk Assessment*, 31(8).

Lam, D, Thompson, C, Croke, J, Choy, S, Wasson, R, Erskine, W and Mike, S, 2017. Reducing Uncertainty with Flood Frequency Analysis: The Contribution of Palaeoflood and Historical Flood Information, *Water Resources Research*, 53(3):2312–2327.

Long Paddock, 2024. Biophysical Modelling (CCS) [online]. Available from: <https://www.longpaddock.qld.gov.au/seasonal-climate-outlook/CCS/>

Long Paddock, 2025a. Factsheets – CMIP5 and CMIP6 [online]. Available from: <https://www.longpaddock.qld.gov.au/qld-future-climate/factsheets/6/>

Long Paddock, 2025b. Queensland Future Climate [online]. Available from: <https://longpaddock.qld.gov.au/qld-future-climate/dashboard/> [Accessed: 2025].

Mike, S, Wasson, R, Erskine, W and Lam, D, 2020. Holocene palaeohydrology of the East Alligator River, for application to mine site rehabilitation, Northern Australia, *Quaternary Science Reviews*, 249.

NASA, nd. The raw truth on global temperature records [online]. Available from: <https://science.nasa.gov/earth/climate-change/the-raw-truth-on-global-temperature-records/> [Accessed: 12 June 2025].

National Climate Centre, 2000. Australian Data Archive for Meteorology, Bureau of Meteorology.

NCEI, nd. Palaeo Data Search [online]. Available from: <https://www.ncei.noaa.gov/access/palaeo-search/> [Accessed: 11 June 2025].

Nishant, N, Evans, J P, Westra, S, Remedio, A R C, Coppola, E and Di Luca, A, 2022. Evaluation of Present-Day CMIP6 Model Simulations of Extreme Precipitation and Temperature over the Australian Continent, *Atmosphere*, 13(9).

PAGES, nd. Past Global Changes [online]. Available from: <https://pastglobalchanges.org/> [Accessed: 9 June 2025].

Paton, P C, Pickup, G and Price, D M, 1993. Holocene Palaeofloods of the Ross River, Central Australia, *Quaternary Research*, 40(2):201–212.

Queensland Government, 1994. Environmental Protection Act 1994, Office of the Queensland Parliamentary Counsel.

Queensland Government, 2018. Palaeoclimatology for the future [online]. Available from: <https://palaeoclimate.com.au/palaeoclimatology/> [Accessed: 20 April 2025].

Reinders, J B and Muñoz, S E, 2021. Improvements to Flood Frequency Analysis on Alluvial Rivers Using Palaeoflood Data, *Water Resources Research*, 57(4).

Siriwardena, L and Weinmann, P E, 1998. A technique to interpolate frequency curves between frequent events and probable maximum events, CRCCH Research Report 98/9, Cooperative Research Centre for Catchment Hydrology.

Su, C, Jakob, D, Xu, H, Xie, Y and Zhu, H, 2022. BARRA2: Development of the next-generation Australian regional atmospheric reanalysis, Bureau of Meteorology.

Taylor, K E, Stouffer, R J and Meehl, G A, 2009. A Summary of the CMIP5 Experiment Design [online]. Available from: <https://pcmdi.llnl.gov/mips/cmip5/docs/Taylor_CMIP5_design.pdf> [Accessed: 10 June 2025].

The State of Queensland, 2024. SILO – Australian climate data from 1889 to yesterday [online]. Available from: <https://www.longpaddock.qld.gov.au/silo/> [Accessed: June 2025].

Thompson, F and Tomlinson, A I, 1993. A Guide to Probable Maximum Precipitation in New Zealand, *Journal of Hydrology (NZ)*.

Tierney, J E, deMenocal, P B and Rosenthal, Y, 2017. A 1.3-million-year record of East African hydroclimate, *Science*, 3566340:93–97.

Towards Sustainable Mining (TSM), 2008. Mine Closure Framework [online], Available from: <https://mining.ca/wp-content/uploads/dlm_uploads/2021/08/TSM_Mine_Closure_Framework.pdf> [Accessed: 16 June 2025].

Tozer, C R, Kiem, A S, Vance, T R, Roberts, J, Lorrey, A, Gallant, A J E, Sisson, S A, Curran, M A J and Moy, A D, 2018. Reconstructing pre-instrumental streamflow in eastern Australia using a water balance approach, *Journal of Hydrology*, 558:632–646.

Tozer, C R, Vance, T R, Roberts, J, Lorrey, A, Sime, L C, Curran, M A J and Moy, A D, 2016. An ice core derived 1013-year catchment-scale annual rainfall reconstruction in subtropical eastern Australia, *Hydrology and Earth System Sciences*, 20(5):1703–1717.

US Army Corps of Engineers, nd. Palaeoflood Examples [online]. Available from: <https://www.hec.usace.army.mil/confluence/sspdocs/ssptutorialsguides/palaeoflood-examples> [Accessed: 10 June 2025].

Vance, T R, Roberts, J L, Plummer, C T, Kable, S H and van Ommen, T D, 2013. A millennial proxy record of ENSO and eastern Australian rainfall from the Law Dome ice core, East Antarctica, *Journal of Climate*, 26(4):710–725.

Verdon-Kidd, D C, Kiem, A S, Johnson, F and Westra, S, 2025. Using seasonal palaeo-flow reconstructions and artificial neural networks for daily water balance modelling: A case study from Tasmania, Australia, *Global and Planetary Change*, 246.

Visser, A, Wasko, C, Sharma, A and Nathan, R, 2021. Eliminating the 'hook' in Precipitation-Temperature Scaling, American Meteorological Society.

Wasko, C, Visser, A, Sharma, A and Nathan, R, 2024. A systematic review of climate change science relevant to Australian design flood estimation, *Hydrology and Earth System Sciences*.

WMA Water, 2021. Updated local design rainfalls for Brisbane, Ipswich, Lockyer Valley and Moreton Bay – Final Report, WMA Water.

Working Group on Coupled Modelling (WGCM), 2013. Working Group on Coupled Modelling CMIP Phase 5 (CMIP5) [online]. Available from: <https://wcrp-cmip.org/cmip-phases/cmip5/>

Working Group on Coupled Modelling (WGCM), 2019. Working Group on Coupled Modelling CMIP Phase 6 (CMIP6) [online]. Available from: <https://pcmdi.llnl.gov/CMIP6/>

World Meteorological Organization (WMO), 2009. Manual on estimation of Probable Maximum Precipitation (PMP), WMO-No. 1045, World Meteorological Organization.

Xuereb, K, Moore, G and Taylor, J, 2001. Development of the Method of Storm Transposition and Maximisation for the West Coast of Tasmania, HRS Report No.7, Hydrology Report Series, Bureau of Meteorology.

Neutralisation effectiveness of an alkaline brine treatment for a copper heap leach facility

M E J Hilton[1], A M Robertson[2], P M Long[3] and I Sheppard[4]

1. Senior Consultant, RGS Environmental Consultants Pty Ltd, Brisbane Qld 4108.
   Email: melinda@rgsenv.com
2. Director, RGS Environmental Consultants Pty Ltd, Brisbane Qld 4108. Email: alan@rgsenv.com
3. Senior Consultant, RGS Environmental Consultants Pty Ltd, Brisbane Qld 4108.
   Email: pete@rgsenv.com
4. Executive Consultant, Aeris Resources, Brisbane Qld 4000.
   Email: isheppard@aerisresources.com.au

ABSTRACT

Heap leach residues from processing of copper ore pose challenges for remediation due to their inherent acidity and both short- and long-term potential to produce acid, saline, and metalliferous drainage (AMD). As a potential improvement to a conventional cover system design, liquid neutralants can be employed to reduce the dissolution of pH-dependent major and trace metals/metalloids and improve seepage water quality. Produced as a by-product from coal seam gas extraction, sodium carbonate alkaline brine is potentially beneficial for neutralising acidic materials due to the availability of high concentrations of CO_3^{2-} and HCO_3^- anions.

A field-scale heap leach facility containing spent residues was constructed under controlled conditions and monitored over 16 months as alkaline brine was added. Despite challenges with permeability, site access due to Covid-19 travel restrictions, and weather conditions, neutralisation of the acidic heap leach residues was achieved. The pH behaviour of seepage water was strongly influenced by the slow release of retained acidity from jarosite. Neutralisation of the combined actual and retained acidity, from both sulfide and sulfate minerals, was required to achieve circumneutral seepage quality pH conditions.

The mobility of major and minor trace metals/metalloids was dependent on the pH of the heap leach residues. The initial pH 4 conditions produced elevated concentrations of dissolved arsenic, copper, nickel, and zinc, which diminished to low or negligible concentrations at pH 8. Conversely, concentrations of sodium and chloride reached a high plateau as the heap leach residues became saturated with brine. Saline drainage is an expected consequence of this neutralisation technique.

This study presents the benefits, challenges, and limitations of using an alkaline brine by-product to achieve improved seepage water quality from copper heap leach materials.

INTRODUCTION

Mining and mineral processing industries can generate environmentally challenging waste and by-product materials, which have the potential to generate acidic leachate, as well as elevated concentrations of dissolved metals, metalloids and salts. Hereafter, these characteristics are collectively termed AMD (Commonwealth of Australia, 2016; INAP, 2025). At legacy sites, AMD treatment can include ongoing active neutralisation of AMD and precipitation of metals using alkaline reagents such as hydrated lime ($Ca(OH)_2$). The resulting neutralised waters can generally be recycled, or depending on the receiving environment, may require expensive secondary treatment. The mining and processing of limestone to produce hydrated lime produces carbon dioxide (CO_2) in the decomposition of limestone to produce lime and the combustion of fossil fuels used to mine and facilitate the decomposition of limestone.

A potential alternative source of an alkaline material for neutralisation of AMD results from coal seam gas (CSG) production and is contained in co-produced water, which commonly has high concentrations of dissolved salts (Davies, Gore and Khan, 2015). The co-produced water requires Reverse Osmosis (RO) treatment to reduce the salt concentration and reclaim water for beneficial reuse. Subsequently, a much smaller volume of highly concentrated brine containing substantial alkalinity from sodium carbonate [Na_2CO_3] (64 per cent by weight). The brine can be dried and go to landfill although this overall process is expensive, due to the energy required to generate the dry salt as well as transport and landfill costs.

Due to its high solubility with carbonate ions yielding high acid neutralising potential, and metal precipitation per unit mass, sodium carbonate offers some advantages over calcium-based neutralants. Sodium carbonate will not foul RO membranes through precipitation of gypsum ($CaSO_4$). Carbonate-rich brines and salts generated from the CSG industry are therefore an alternative source of neutralant material, with the advantage that the output of one resource extraction industry has the potential to be reused as a valuable input to benefit another industry.

Over the past few years, an Australian CSG producer has commissioned several studies to investigate the potential for reuse of this alkaline brine/salt by-product in the mining industry for the remediation of acidic mine waste materials and AMD (Chen *et al*, 2017; Cohen *et al*, 2017; Robertson *et al*, 2018). These studies demonstrated the potential for the alkaline brine/salt by-product to be successfully used in these activities.

The results of the studies cited above were used to inform the design of a heap leach pad neutralisation field trial, which operated for from late 2021 to mid-2023. The objective of the project was to assess whether outcomes of the laboratory tests would be applicable at a field scale and provide some advantages with respect to planned closure of the heap leach facility.

This paper summarises the findings from the third and final phase (field scale trials). A brief description of the precursory small scale laboratory testing and the heap leach pad design is also included here (Robertson *et al*, 2022). The key focus of this paper is the operation and effectiveness of a heap leach neutralisation field trial using the alkaline brine as part of rehabilitation plans for a copper heap leach facility in New South Wales (NSW), Australia.

METHODOLOGY

Phase 1 – laboratory scale trials

In 2017, 40 kg samples of spent heap leach material from the Murrawombie Copper Mine in NSW were collected at selected site locations from 0 to 5 m depth. The samples were classified as either High Sulfur (HS) with ~4 per cent total sulfur (TS) or Low Sulfur (LS): with <1 per cent TS. The materials were subjected to Acid Base Account (ABA) tests and were classified as Acid Forming (AF) (INAP, 2025). Soil quality, mineralogy, and physical testing was also performed on these samples.

Seventy-five (75) kg composite samples of either HS or LS materials underwent kinetic leaching in 60 L tanks with sodium carbonate alkaline brine, under various unsaturated and saturated conditions (Figure 1). After three months of leaching, the laboratory results verified that the alkaline brine was effective in neutralising the heap leach materials from acidic (pH 2.9 to 3.8) to neutral (pH 7.5) or alkaline (pH 9.0) (Table 1). The concentrations of dissolved metals/metalloids (eg Cu, Ni, Mn, Pb, Se, U, and Zn) contained in leachate were also reduced (RGS, 2020). A second stage of testing over three months verified that the neutralised pH could be maintained under wet and dry conditions, when the materials were leached with de-ionised (DI) water.

FIG 1 – Side and top view of 60 L KLC neutralisation tanks.

TABLE 1

Laboratory KLC set-up and results.

KLC #	Description	Brine (x / ✔)	Leaching regime	Start pH	End pH
KLC 1	Control test for LS heap leach material	x	Progressive leaching with 20 L deionised water (weekly × 4 cycles, monthly × 2 cycles)	2.99	3.07
KLC 2	Control test for HS heap leach material	x	Progressive leaching with 20 L deionised water (weekly × 4 cycles, monthly × 2 cycles)	2.76	2.54
KLC 3	Saturated immersion in concentrated liquid brine LS heap leach material	✔	20 L of concentrated brine in first cycle followed by leaching with 20 L DI water (weekly × 4 cycles, monthly × 2 cycles)	8.16	4.24
KLC 4	Saturated immersion in concentrated liquid brine HS heap leach material	✔	20 L of concentrated brine in first cycle followed by leaching with 20 L DI water (weekly × 4 cycles, monthly × 2 cycles)	8.29	4.05
KLC 5	Sprinkled with diluted liquid brine HS heap leach material	✔	11 L of diluted brine (1:2 v:v) in first cycle, followed by leaching with 3 L of diluted brine (weekly × 4 cycles, monthly × 2 cycles)	3.80	7.46
KLC 6	Sprinkled with diluted liquid brine HS heap leach material	✔	11 L of diluted brine (1:2 v:v) in first cycle, followed by leaching with 3 L of diluted brine (weekly × 4 cycles, monthly × 2 cycles)	2.93	9.03

Phase 2 – field scale heap leach pad set-up

In November 2021, a field scale heap leach pad remediation trial (HLPRT) was constructed as planned in a fully lined, high density polyethylene (HDPE) working area on the Murrawombie mine site. Embankments were designed and constructed to support and contain the working area, such as berms, benches, underdrains and drains, and were comprised of benign waste rock that had been geochemically tested. The dimensions of the HLPRT were 9 m × 9 m × 3 m (h). HS and LS heap leach materials to be treated with the brine were placed in a 1 m grid arrangement to ensure even representation of these material types. Each 1 m³ cell of HS or LS material was carefully placed to minimise mixing prior to HLPRT operation. A schematic flow diagram of the HLPRT and its components is shown in Figure 2.

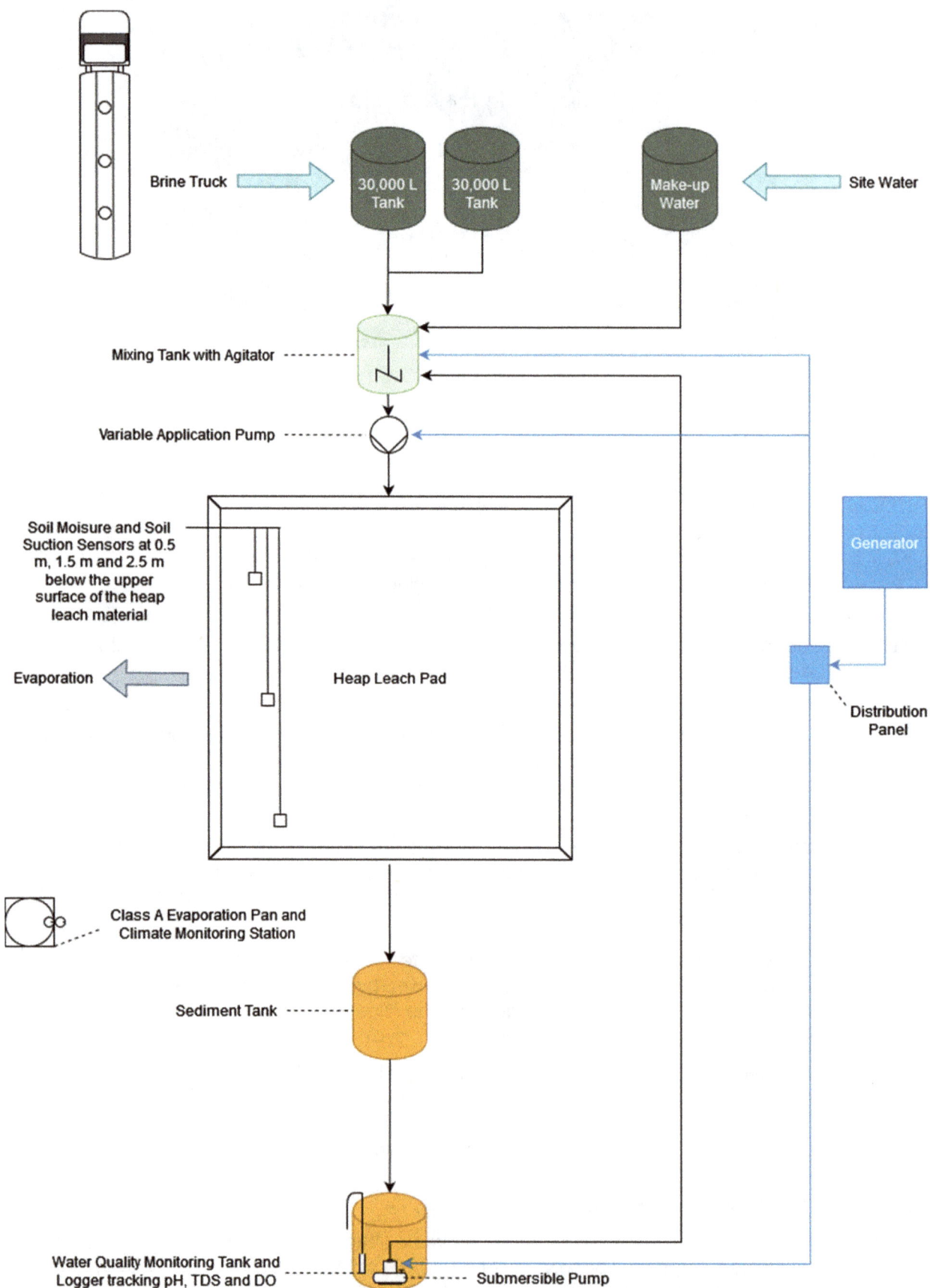

FIG 2 – HLPRT flow diagram.

By design, Soda Ash Brine (SAB) would be delivered by truck (44 000 L) and pumped into two 30 000 L poly tanks in bunded areas on the HLPRT area. Brine would then be gravity fed into the mixing tank. The SAB return line was used to pump the brine from the SAB mixing tank onto the HLPRT area. The brine percolated through the HLPRT material to the base of the HLPRT and was then filtered through the drainage layer into the SAB sediment tank to remove particulate matter. Sampling and analysis of the resultant seepage liquor was undertaken from the SAB recovery tank (RGS, 2021).

Phase 3 – field scale operation

Based on the laboratory scale testing, it was planned that nine loads of brine (total ~396 000 L) would need to be applied to the HLPRT over the first six months of operation. The HLPRT seepage was then to be monitored for a further 12 months as the neutralisation reactions occurred. Due to extreme wet weather in early 2022, this could not be achieved, and instead four loads of brine (total ~176 000 L) were applied over a 13 month period.

Difficulties were also encountered with reduced material permeability, which was attributed to chemical reactions and the high rate of brine delivery through the 'wobbler' sprinklers applying the brine. Initially, the brine was applied at a rate of 50 mm/hr for eight hours. Under this moisture regime, the surface of the heap leach materials eventually became impermeable. To remedy the issue, an excavator was used to rip the surface of HLPRT. The sprinklers applying the brine were then resumed at a reduced rate of 10 mm/hr.

While the mechanical mixing resolved the reduced permeability issues, it also disrupted some of the internal sensors that had been installed in the HLPRT. Maintenance work was required to remedy the sensor operation. Approximately six months of sensor data was lost between Dec 2021 and June 2022. Despite the disruption, sensors were still able to demonstrate that the HLPRT had met the test objectives. Twelve months of uninterrupted monitoring data was collected once the sensors were repaired.

It was intended that a water balance monitoring program would be part of the HLPRT. Quantifying brine and rainfall inputs and outputs to and from the HLPRT had the potential to verify if there were any aqueous losses due to leaks in the HDPE liner or gains due to incident rainfall. However, the La Nina climatic condition in 2021 and 2022 with above average and more frequent rainfall made this difficult. Some error would have been introduced when it was not possible to place a protective tarpaulin over the trial for every single rain event. As the water balance quantification could not be achieved, it is assumed that some dilution of the leachate has occurred due to rainfall inputs.

From December 2021 to March 2023, 34 samples of leachate were collected from the seepage outlet. The frequency was daily for the first two weeks, weekly for the following two months, and then fortnightly or monthly thereafter. The seepage samples were subjected to water quality analyses for unfiltered and 0.45 μm filtered metals/metalloids as well as nitrogen and phosphorous compounds, and hydrocarbons/BTEXN (RGS, 2023).

RESULTS AND DISCUSSION

Climate and sensor data

Rainfall data was collected daily at a climate monitoring station and is summarised in Figure 3. Trends show that Aug 2021 to May 2022 was an unusually wet season and was followed by a much lower rainfall period from June 2022 to June 2023. The annual rainfall was 758 mm for July 2021 to June 2022, and 196 mm for July 2022 to June 2023. The average annual rainfall for the area is 450 mm. The first two brine applications occurred during the wet period and the next two were during the drier year that followed.

Pan evaporation was also measured. Lower rates of evaporation correlated to wet, cooler periods, as well as seasonal conditions through winter to summer. The climate at the mine is described as mild and temperate.

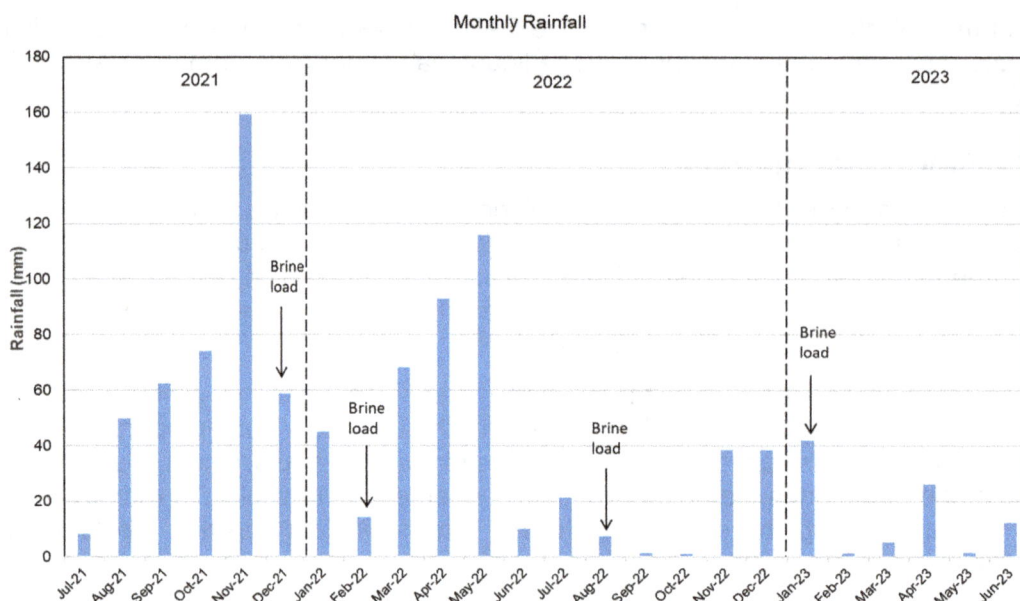

FIG 3 – Monthly rainfall at the HLPRT site.

Soil moisture, soil temperature, and soil water potential were measured in the HLPRT using installed sensors. Each parameter had three nodes spatially distributed and placed at variable depths within the materials at approximately 0.8, 1.2, and 2.2 m. All three nodes for each measurement reported similar results, despite disruptions to some of the sensors.

At the beginning of the trial period, soil moisture in the HLPRT was minimal. Soil moisture rose to ~80 per cent in June 2022 and then remained relatively steady between 70 to 90 per cent for the remainder of the trial period.

Soil temperature was measured at ~25°C in December 2021. The soil temperature then gradually decreased to ~13°C in July 2022 and rose again to ~27°C in Feb 2023. The results inferred that weather patterns of summer and winter had the strongest influence on soil temperature, regardless of the geochemical reactions occurring inside the HLPRT.

The water potential in the HLPRT decreased in an inverse pattern to soil moisture. The soil water potential began at ~1 pF in Dec 2021 and then remained steady between 0 to 2 pF for the remainder of the trial period.

Brine and seepage quality

The laboratory testing was conducted by ALS, Stafford, Brisbane. Analyses were NATA-accredited and performed using Australian standard methods (eg APHA 4500: American Public Health Association (APHA), 1998; Ahern, McElnea and Sullivan, 2004).

The initial brine sample was characterised as alkaline (pH 10.2) and saline (EC 195 mS/cm). Major cations in the brine were dominated by Na^+ (47 400 mg/L), and K^+ (1080 mg/L). Major anions were CO_3^{2-} (67 000 mg/L), Cl^- (15 000 mg/L), SO_4^{2-} (924 mg/L), Br^- (61 mg/L), and F^- (56 mg/L). Dissolved trace metals/metalloids were generally below the laboratory limit of reporting (LoR), with the exceptions of Al, B, Cu, Li, Mo, Rb, Sr, and Zr. Only Al and Cu had higher concentrations in the heap leach material seepage than in the brine.

In the HLPRT seepage, the initial pH was slightly acidic (3.7 to 4.3) but rose steeply from pH 4.2 to pH 8.0 at Week 40, which indicates that an inflexion point had been reached (Figure 4). The production of actual acidity can be attributed to iron sulfide minerals, such as pyrite. It is hypothesised that the stability at ~pH 4 from Week 1 to Week 40 is due to the slow release of retained acidity by the iron sulfate mineral jarosite, which is unstable at higher pH. At pH 4.7, jarosite becomes relatively soluble and transforms into goethite (FeO(OH)) (Ryu and Kim, 2022). This decomposition reaction occurs more readily when excess water is available. The inflexion point, therefore, marks a transition whereby most of the pyrite and jarosite has been neutralised, and alkalinity from the brine becomes the predominant influence over pH.

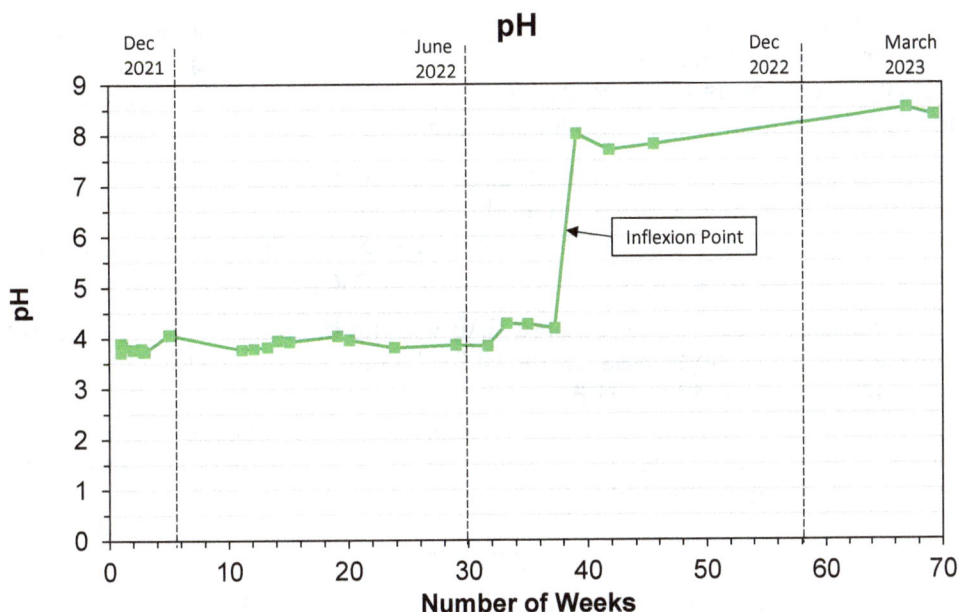

FIG 4 – pH measurements in HLPRT seepage water.

Electrical conductivity (EC) in the HLPRT seepage increased initially from 3.73 to 141 mS/cm, then decreased and appeared to stabilise at approximately 65 mS/cm at Week 40. Acidity (as $CaCO_3$) also increased from 1320 to 17 700 mg/L and then reduced and remained at <500 mg/L, indicating a very low rate of acid production by the end of the trial period. Total alkalinity (as $CaCO_3$) was below the laboratory LoR until Week 40 and then rose to over 13 000 mg/L, confirming that alkalinity dominated over acidity from this point onwards.

Comparable results were observed for concentrations of most dissolved trace metals/metalloids of environmental concern. For example, Cu followed a dome-shaped trend, rising to 1080 mg/L in Week 10, and eventually diminished after Week 40 to relatively low levels of <50 mg/L (Figure 5). Analogous behaviour was recorded for concentrations of Al, As, Ba, Cd, Co, Fe, Mn, Ni, Pb, Se, Sr, Th, U, and Zn, most of which are known to have increased mobility under acidic conditions (eg Pitre, Boullemant and Fortin, 2014; Hezarkhani, Williams-Jones and Gammons, 1999).

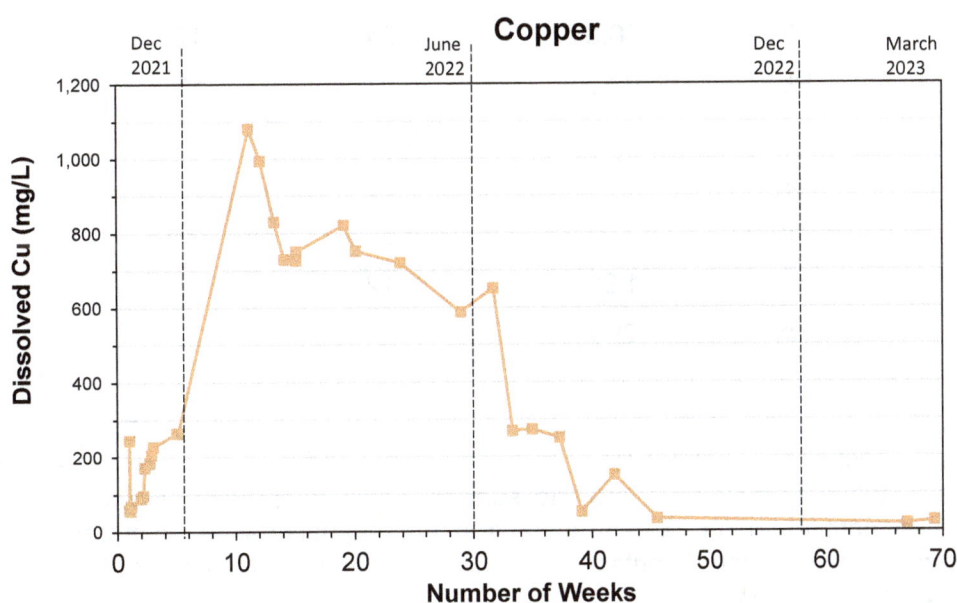

FIG 5 – Dissolved cooper concentrations in HLPRT seepage water.

The release of some dissolved major and trace metals/metalloids, such as B, Cl, Mo, Na, and Rb, showed a different trend. For example, the starting concentration of Cl was 100 mg/L, which rose to 7800 mg/L by Week 30 as brine was continually added to the HLPRT. This concentration then sharply decreased to 4090 mg/L at Week 40, presumably when the acid and base concentrations were equalising. The concentration of Cl then increased up to ~8800 mg/L and remained steady to Week 70. This pattern was generally repeated for the majority of metals/metalloids that had a higher concentration in the brine than in the heap leach materials.

Sulfate (SO_4^{2-}) was the only parameter that appeared to have a composite graph, which was attributed to that fact that it is present in both the brine and the heap leach materials. The concentration of SO_4^2 at the end of the trial remained steady at ~30 000 mg/L, which is higher than the original brine concentration of 924 mg/L, but lower than the peak concentration of ~70 000 mg/L.

Despite less than half of the estimated required SAB volume being applied, the results discussed in this paper verify that the key objectives of the HLPRT were successfully achieved. The AF heap leach materials were neutralised to ~pH 8, and the concentration of metals/metalloids in seepage were typically reduced by a factor of 10 to 20 (Table 2). The neutralisation was most effective for metals/metalloids whose solubility is pH-sensitive (eg Al, Cu, Th, and Zn) and less effective for elements that can dissolve over a variable pH range (eg Ba, Mn, and Se).

TABLE 2
Seepage concentration results for the HLPRT.

Element	Peak concentration (mg/L)	Final concentration (mg/L)	Factor of difference
Al	3880	0.17	>20 000
As	0.26	0.01	26
Ba	0.04	0.005	8
Cd	3.29	0.19	17
Co	51.6	3.67	14
Cu	1 080	32.7	33
Fe	16.3	0.94	17
Mn	191	22.1	9
Ni	16.3	0.94	17
Pb	0.13	0.01	13
Se	0.81	0.11	7
Sr	1.08	0.09	12
Th	1.6	0.01	160
U	1.98	0.12	17
Zn	290	7.5	39

The dissolved water quality of the final leachate is above the applied ANZG (2018) criteria for livestock drinking water for metals/metalloids and, coupled with the final salinity of ~40 000 mS/cm, would not be suitable for discharge into site surface water catchments. However, the treatment shows a marked improvement in concentrations from the original heap leach materials and, coupled to a well-designed cover system, merits further study and investigation as part of an overall closure and final rehabilitation strategy.

Approval for the HLPRT was prematurely rescinded by the NSW Environmental Protection Authority (EPA) due to a clerical error on their part and a retrospective attempt to apply Acid Sulfate Soil guideline criteria to a small scale rehabilitation trial on mining operation. Hence, the brine addition to the HLPRT could not be continued as planned. A 12-month monitoring period for the HLPRT was also scheduled to commence, followed by deconstruction of the HLPRT and geochemical characterisation of the neutralised heap leach materials. It is hoped that the available results from this study will inspire future projects utilising CSG brine by-products to mitigate AMD in accordance with the NSW Waste and Sustainable Minerals Strategy 2041 (DPIE, 2021).

CONCLUSIONS

The field scale trial has demonstrated the potential to utilise the alkalinity from bicarbonate and carbonate species in a CSG by-product, to neutralise the acidity in the heap leach materials from a copper heap leach operation. Decreased metal/metalloid concentrations measured after the inflexion point indicates the immobilisation of dissolved metal species through precipitation. The release of actual acidity from pyrite and retained acidity from jarosite is responsible for the acid forming properties of the heap leach materials, which buffers the seepage at ~pH 4 until this acidity reservoir is overcome by the brine. Thereafter, the neutralised heap leach materials produce seepage that is circumneutral.

The quality of the neutralised seepage increases the salinity and some oxyanion concentrations that are associated with the brine or the dissolution of iron sulfate minerals. These properties are outweighed by the significantly reduced concentration of potentially hazardous metals/metalloids present in the neutralised seepage.

Despite the early cessation of the trial, it is hoped that the successful application of the alkaline CSG waste product at this mine site will lead to further research in the reuse of CSG by-products in other industries, and a positive outcome for the industry stakeholders, regulators, and the environment.

ACKNOWLEDGEMENTS

The authors would like to thank Aeris Resources Ltd and the Murrawombie Copper Mine for the opportunity to participate in the HLPRT and the permission to present outcomes of the study at the Life-of-mine | Mine Waste, and Tailings Conference 2025.

REFERENCES

Ahern, C R, McElnea, A E and Sullivan, L A, 2004. Acid Sulfate Soils Laboratory Methods Guidelines, Queensland Department of Natural Resources, Mines and Energy.

American Public Health Association (APHA), 1998. APHA 4500-F Standard methods for the Examination of Water and Wastewater, American Public Health Association.

Australian and New Zealand Guidelines (ANZG), 2018. Australian and New Zealand Guidelines for Fresh and Marine Water Quality, Australian and New Zealand Environment Conservation Council and Agriculture and Resource Management Council of Australia and New Zealand, Canberra, ACT.

Chen, Q, Cohen, D R, Andersen, M S, Robertson, A M, Jones, D R and Kelly, B, 2017. Application of alkaline coal seam gas waters to remediate AMD from historical sulfide ore mining operation, in *Mine Water and Circular Economy* (eds: C Wolkersdorfer, L Sartz, M Sillanpää and A Häkkinen), vol II, pp 923–931 (Lappeenranta University of Technology).

Cohen, D R, Chen, Q, Andersen, M S, Robertson, A M, Jones, D R and Kelly, B, 2017. Chemical and economic feasibility study of alkaline CSG waters in AMD Remediation, in *Proceedings of the Ninth Australian Workshop on Acid and Metalliferous Drainage* (eds: L C Bell, M Edraki and C Gerbo), pp 231–238 (The University of Queensland: Brisbane).

Commonwealth of Australia, 2016. Leading Practice Sustainable Development Program for the Mining Industry, Preventing Acid and Metalliferous Drainage, Canberra ACT.

Davies, P J, Gore, D B and Khan, S J, 2015. Managing produced water from coal seam gas projects: Implications for an emerging industry in Australia, *Environ Sci and Pollut Res*, 22:10981–11000.

Department of Planning, Industry and Environment (DPIE), 2021. NSW Waste and Sustainable Materials Strategy 2041. Stage 1 2021–2027, Published by the NSW Department of Planning, Industry and Environment, June.

Hezarkhani, A, Williams-Jones, A E and Gammons, C H, 1999. Factors controlling copper solubility and chalcopyrite deposition in the Sungun porphyry copper deposit, Iran, *Mineralium deposita*, 34:770–783.

International Network for Acid Prevention (INAP), 2025. The Global Acid Rock Drainage Guide, International Network for Acid Prevention. Available from: <https://www.gardguide.com>

Pitre, D, Boullemant, A and Fortin, C, 2014. Uptake and sorption of aluminium and fluoride by four green algal species, *Chemistry Central Journal*, 8:1–7.

RGS, 2020. Murrawombie Laboratory Scale Heap Leach Neutralisation Trial, Report prepared by RGS Environmental Pty Ltd for Aeris Resources Ltd and Santos Ltd, 26 February 2020.

RGS, 2021. Heap Leach Pad Remediation Trial: Field Sampling Plan, Report prepared by RGS Environmental Consultants Pty Ltd for Tritton Resources, 15 October 2021.

RGS, 2023. Heap Leach Pad Remediation Trial: Interim Technical Report, Report prepared by RGS Environmental Pty Ltd for Aeris Resources Ltd, 30 June 2023.

Robertson, A, Maddocks, G, Kelly, B and Sheppard, I, 2018. Application of a coal seam gas waste product as part of the rehabilitation program for a copper heap leach operation, in *11th International Conference on Acid Rock Drainage* (International Mine Water Association).

Robertson, A, Maddocks, G, Long, P and Woods, D, 2022. Alkaline brine treatment of a heap leach facility as part of the mine closure planning process, in 12th International Conference on Acid Rock Drainage.

Ryu, J G and Kim, Y, 2022. Mineral transformation and dissolution of jarosite coprecipitated with hazardous oxyanions and their mobility changes, *Journal of Hazardous Materials*, 427:128283.

Re-imagined futures – Indigenous workforce leadership

Z Hughes[1]

1. Academic Lead and Course Coordinator – Mine Closure and Transition, School of Molecular Life Sciences, Faculty of Science and Engineering, Curtin University, Bentley WA 6102. Email: zane.hughes@curtin.edu.au

ABSTRACT

This paper presents a transformative workforce design model that positions First Nation Australians as Rights Holders rather than stakeholders in the extractive industry. Grounded in the United Nations Declaration on the Rights of Indigenous Peoples (UNDRIP; United Nations General Assembly, 2007) and aligned with the Life of Asset framework, this research proposes embedding Indigenous workforce leadership across six key stages: Exploration, Feasibility, Design and Construction, Operations, Pre-Closure, and Closure and Transition. The model integrates Free, Prior and Informed Consent (FPIC), cultural governance, and benefit sharing to create just, inclusive, and regenerative extractive sector practices. By shifting from traditional 'pay-out' models to meaningful participation frameworks, this approach strengthens Social License to Operate while advancing Indigenous self-determination and economic sovereignty.

INTRODUCTION

The extractive industry stands at a critical juncture where traditional models of Indigenous engagement must evolve to meet contemporary standards of justice, sustainability, and human rights. As custodians of Country with sovereign rights recognised under Australian and international law, First Nation Australians possess inherent authority over lands that contain significant mineral resources. Yet current workforce practices often position Indigenous communities as peripheral stakeholders rather than central decision-makers in projects that fundamentally affect their territories and futures.

Legal framework for First Nations land rights in Australia

Understanding the complex legal landscape of First Nations land recognition is essential for developing appropriate workforce strategies across extractive asset life cycles. Australia's system of Indigenous land recognition operates through multiple, sometimes overlapping frameworks that have evolved over nearly five decades of legal and political reform (Table 1).

This legal complexity creates both opportunities and challenges for extractive industry workforce development. While Native Title provides the broadest recognition framework, its requirements for continuous connection and extensive proof can limit its applicability. Conversely, statutory land rights schemes offer more certain tenure but vary significantly in scope and application across jurisdictions.

For mining companies, this means that workforce strategies must be tailored to the specific land tenure context of each project. Operations on Native Title land may require different consultation and agreement-making processes compared to those on statutory Aboriginal land, while projects crossing multiple tenure types require sophisticated coordination across different legal frameworks.

TABLE 1

First Nations land recognition frameworks in Australia.

Framework	Legislation/basis	Jurisdiction	Key features	Land coverage
Native Title	Native Title Act 1993 (Cth) Mabo v Queensland (No. 2) (1992)	National	Recognition of pre-existing rights under traditional law and custom; requires continuous connection since sovereignty; co-exists with some other interests	~32% of continental Australia recognised
Northern Territory Land Rights	Aboriginal Land Rights (Northern Territory) Act 1976 (Cth)	Northern Territory	First legislative recognition of Aboriginal land ownership; inalienable freehold title; requires traditional ownership proof	~50% of NT (~85% of coastline)
NSW Land Rights	Aboriginal Land Rights Act 1983 (NSW)	New South Wales	Crown land claims through Local Aboriginal Land Councils; compensation for historic dispossession	Specific claims-based coverage
SA Land Rights	Aboriginal Land Trust Act 2013 (SA) Pitjantjatjara Land Rights Act 1981 (SA) Maralinga Tjarutja Land Rights Act 1984 (SA)	South Australia	Mixed system: general trust lands plus specific Traditional Owner grants	Variable coverage by act
Queensland Schemes	Aboriginal Land Act 1991 (Qld) Torres Strait Islander Land Act 1991 (Qld) Deed of Grant in Trust (DOGIT)	Queensland	Dual system for Aboriginal and Torres Strait Islander peoples; some former reserves converted to freehold	~5% state coverage
Victorian Recognition	Aboriginal Lands Act 1970–1991 (Vic) Traditional Owner Settlement Act 2010 (Vic)	Victoria	Specific land transfers; alternative settlement framework for native title resolution	Site-specific grants
WA Trust System	Aboriginal Affairs Planning Authority Act 1972 (WA)	Western Australia	Crown land reserves managed by Aboriginal Lands Trust; leasing arrangements but no freehold transfer	~24 million hectares (~10% of state)
Tasmanian Arrangements	Aboriginal Lands Act 1995 (Tas)	Tasmania	Limited land council ownership in trust for Aboriginal people	Specific areas only

Implications for workforce development

This paper argues for a paradigm shift that recognises First Nation Australians as Rights Holders throughout the extractive asset life cycle. Drawing on the United Nations Declaration on the Rights of Indigenous Peoples (UNDRIP; United Nations General Assembly, 2007) and the Life of Asset framework developed through CRC TiME's Songlines Project (2021–2024), we propose a comprehensive workforce design model that embeds Indigenous leadership from exploration through post-closure transition.

The urgency of this transformation extends beyond moral imperatives to encompass practical business considerations. Companies increasingly recognise that sustainable operations require genuine Indigenous partnership, not merely compliance with minimum regulatory requirements. Environmental, Social and Governance (ESG) frameworks demand demonstrable commitment to Indigenous rights, while investors scrutinise companies' relationships with Traditional Owners as indicators of long-term viability.

THEORETICAL FRAMEWORK – UNDRIP AND THE LIFE OF ASSET MODEL

UNDRIP as foundation for Indigenous workforce rights

The United Nations Declaration on the Rights of Indigenous Peoples (United Nations General Assembly, 2007) establishes fundamental principles that must underpin any legitimate approach to Indigenous workforce development in the extractive sector. Article 18 affirms Indigenous peoples' rights to participate in decision-making processes affecting them, while Article 19 mandates that states obtain Free, Prior and Informed Consent before adopting legislative or administrative measures that may affect Indigenous peoples.

Article 20 specifically addresses economic participation, stating that Indigenous peoples have the right to maintain and develop their political, economic and social institutions. In the mining context, this translates to Indigenous communities having agency over workforce structures, training programs, and employment policies rather than being passive recipients of externally designed initiatives.

Article 32 provides perhaps the most direct application to extractive industries, requiring states to consult and cooperate with Indigenous peoples to obtain their free and informed consent prior to the approval of any project affecting their lands. This article establishes that Indigenous consent is not merely consultative but constitutes a fundamental prerequisite for legitimate resource development.

The life of asset framework

The Life of Asset framework, as articulated through CRC TiME's Songlines Project (2021–2024), provides an operational structure for embedding Indigenous rights across mining project phases. This framework recognises that Indigenous engagement requirements and opportunities vary significantly across the asset life cycle, from initial exploration through post-closure transition.

The framework's strength lies in its recognition that workforce engagement must be phase-specific, culturally informed, and regionally grounded. Rather than applying uniform approaches across all project stages, the model enables targeted strategies that align with both operational requirements and Indigenous community priorities at each phase.

WORKFORCE DESIGN ACROSS THE ASSET LIFE CYCLE

Exploration and permitting phase

The exploration phase presents the first critical opportunity to establish Indigenous workforce leadership patterns that will influence all subsequent project phases. During this stage, workforce design should prioritise Indigenous roles in cultural heritage assessments, land and water monitoring, and environmental surveying. These positions leverage Indigenous Ecological Knowledge (IEK) while fulfilling UNDRIP Article 31 requirements to maintain and protect cultural heritage and traditional knowledge systems.

Early Indigenous workforce engagement serves multiple strategic purposes. First, it establishes trust and credibility with Traditional Owner communities, creating foundation relationships that support subsequent project phases. Second, it enables the embedding of Free, Prior and Informed Consent protocols within operational structures rather than treating FPIC as a separate consultative process. Third, it creates pathways for Indigenous workers to develop technical skills and project familiarity that prepare them for expanded roles in later phases.

Case study – BHP's Tjiwarl Partnership

BHP's Comprehensive Agreement with the Tjiwarl people in Western Australia, signed in 2018, demonstrates effective early-phase engagement. The wide-ranging native title agreement was influenced heavily by the community's ambitions and priorities in areas including culture, land management, business development, education and employment, laying the foundation for generational change and future collaboration. More than 50 Tjiwarl Traditional Owners have completed an eight-week Work Readiness program, securing a Certificate II Resource Infrastructure and Work Preparation and a pathway into the resources industry.

The exploration phase workforce should include Indigenous cultural heritage officers, environmental monitors, and community liaison personnel. These roles require specialised training that combines traditional knowledge with contemporary monitoring techniques, creating professional development opportunities that strengthen both individual capacity and community cultural continuity.

Feasibility, design and construction phase

The feasibility and construction phases present expanded opportunities for Indigenous workforce development across technical, administrative, and operational domains. During these phases, workforce design should incorporate comprehensive training pathways that prepare Indigenous workers for long-term operational roles while meeting immediate construction labour demands.

Vocational Education and Training (VET) pathways must be designed to address the specific challenges identified by Pearson and Daff (2011) regarding linguistic barriers and alternative assessment models. Traditional recruitment processes often exclude Indigenous candidates due to literacy requirements that do not reflect actual job competencies. Alternative assessment approaches focusing on disposition, community standing, and adaptive capacity have demonstrated success in identifying Indigenous candidates with high potential for technical roles.

Case study – Fortescue's VTEC Program

Fortescue Metals Group's Vocational Training and Employment Centre (VTEC) provides a comprehensive model for construction phase workforce development. Established in 2006, VTEC has graduated over 1500 Aboriginal people across its South Hedland and Roebourne facilities. The program is built on the concept that following completion of training, participants are guaranteed employment. In FY21, the program had 85 participants with 20 participants from earlier intakes successfully gaining permanent employment during that year. This approach directly addresses the trust deficit that often undermines Indigenous training programs by providing concrete employment pathways rather than training without guaranteed outcomes.

Micro-credentialing programs aligned with Resource and Infrastructure Industry (RII) Training Package requirements can provide flexible pathways for Indigenous workers to develop specific competencies without requiring extended absence from community and family obligations. These programs should integrate Indigenous learning preferences, including storytelling, land-based learning, and visual pedagogies that enhance retention and engagement.

The construction phase also presents opportunities for Indigenous enterprise development through supply chain integration. The Indigenous Procurement Policy (IPP) framework provides mechanisms for embedding Indigenous businesses within project supply chains, creating multiplier effects that extend employment benefits beyond direct workforce participation.

Operations and production phase

During operational phases, Indigenous workforce participation should extend across all operational domains, from site operations and maintenance to administration and strategic planning. This phase represents the peak employment opportunity and requires sophisticated workforce development strategies that support both individual career progression and community economic development.

Eva et al (2023) demonstrate that Indigenous-owned businesses consistently employ Indigenous people at rates far exceeding national benchmarks. This finding suggests that Indigenous enterprise development should be integrated with workforce strategies to maximise Indigenous employment outcomes. Rather than focusing solely on individual job placement, operational phase workforce design should support Indigenous business development that creates sustainable employment ecosystems.

Case study – Rio Tinto's Indigenous employment model evolution

Rio Tinto's experience demonstrates both the challenges and potential of operational phase workforce development. In the 1990s, Indigenous workers comprised less than 0.5 per cent of Rio Tinto's total Australian workforce. By 2010, this figure had increased to 7 per cent, representing a fourteen-fold improvement. This transformation was achieved through a comprehensive model incorporating cultural mentorship, leadership development, and family-friendly working contexts.

However, research by Parmenter *et al* (2023) on Rio Tinto's Argyle Diamond Mine reveals complex outcomes regarding whether Aboriginal employees achieved 'good careers at the same time as following their culture' as articulated in the Indigenous Land Use Agreement. The study found that while employment provided economic benefits, the intersection of workplace demands and cultural obligations remained challenging for many Indigenous employees. This highlights the importance of designing operational workforce systems that genuinely accommodate Indigenous cultural practices rather than merely tolerating them.

Success story – BHP's Indigenous leadership development

BHP's Indigenous Development Program (IDP), running since FY2015, demonstrates effective operational phase leadership development. The program has achieved remarkable outcomes: 49 per cent of participants have moved into new roles, and 20 per cent have been promoted into leadership positions. The success led to creation of the Indigenous Leadership Program (ILP) in FY2019, supporting Indigenous leaders to develop capabilities for higher-level leadership roles. As of 2022, BHP exceeded its Indigenous employment target of 3 per cent three years ahead of schedule, with over 1500 Indigenous employees and contractors across Australian operations.

The operational phase also requires embedding Indigenous governance structures within site management systems. These structures should align with UNDRIP Articles 20 and 23, which affirm economic self-determination and program participation rights. Indigenous governance groups can provide ongoing oversight of employment practices, training program effectiveness, and community benefit realisation.

Procurement strategies during operations should prioritise Indigenous suppliers and service providers, creating secondary employment effects that strengthen regional Indigenous economic capacity. This approach requires proactive supplier development programs that support Indigenous businesses to meet technical and commercial requirements for mining industry contracts.

Pre-closure and closure phase

As extraction activities wind down, workforce strategies must shift toward skills diversification, enterprise transition, and land rehabilitation training. This phase presents unique challenges and opportunities that require careful planning to avoid the welfare dependency trap identified by Holcombe and Kemp (2020).

Case study – Transitioning beyond mining: Nhulunbuy experience

Research by O'Neil and Smith (2025) on East Arnhem Land's Nhulunbuy provides insights into closure phase challenges. The Yolŋu Aboriginal people, who have inhabited the region for over 60 000 years, face complex decisions about post-mining economic development. The study demonstrates that while Indigenous tourism is viewed as a potential alternative, the transition requires significant infrastructure maintenance costs and time lags between mine closure and asset relinquishment that can create 'shock to the system' effects for Indigenous communities.

Pre-closure workforce planning should begin early in the operational phase to ensure adequate time for skills transfer and enterprise development. Indigenous workers should be supported to develop capabilities in environmental monitoring, land rehabilitation, and cultural site management that will be required during closure and post-closure phases.

The closure phase also presents opportunities for Indigenous communities to define post-mining land uses that align with cultural values and economic aspirations. Workforce development during this phase should support Traditional Owners to develop capabilities in sustainable land management, cultural tourism, and environmental stewardship that can provide long-term employment alternatives.

Training programs during pre-closure should emphasise transferable skills that can support diverse post-mining economic activities. These might include project management, environmental monitoring, cultural heritage management, and small business development capabilities that can be applied across multiple sectors.

Post-closure and transition phase

The post-closure phase completes the transformation from extractive employment to restorative economy models. During this phase, employment should focus on long-term cultural stewardship, sustainable land enterprises, and ongoing environmental monitoring.

Indigenous ranger programs provide an established framework for post-closure employment that combines cultural maintenance with environmental stewardship. These programs can be expanded to include mine site rehabilitation monitoring, cultural site protection, and traditional ecological management practices that support landscape restoration.

Article 32 of UNDRIP affirms Indigenous rights to determine land use and development priorities, and post-closure employment arrangements must reflect this principle. Rather than imposing external definitions of appropriate post-mining activities, workforce strategies should support Traditional Owner communities to implement their own visions for country management and economic development.

CULTURAL COMPETENCE AND TRAINING INNOVATION

Culturally responsive training models

Successful Indigenous workforce development requires training models that integrate Indigenous knowledge systems and learning preferences with technical skill development. The Songlines Project recommends curriculum frameworks that blend Western technical knowledge with Indigenous pedagogical approaches, including storytelling, land-based learning, and visual communication methods.

These approaches recognise that Indigenous learners often bring sophisticated knowledge systems that can enhance rather than compete with technical training content. For example, Indigenous ecological knowledge provides deep understanding of environmental systems that can strengthen environmental monitoring and land management capabilities.

Training programs should also address the cultural dimensions of workplace participation, including cross-cultural communication, conflict resolution, and leadership development that prepares Indigenous workers for supervisory and management roles. These capabilities are essential for Indigenous workforce leaders to effectively navigate complex organisational environments while maintaining cultural integrity.

Alternative assessment and recruitment

The recruitment innovations trialled by Rio Tinto and documented by Pearson and Daff (2011) demonstrate the potential for alternative assessment methods that identify Indigenous candidates with high potential for technical roles. These approaches focus on disposition, community standing, and adaptive capacity rather than formal qualifications or literacy levels.

Alternative assessment methods should be integrated throughout the workforce development process, not only at initial recruitment. Performance evaluation systems should recognise different learning styles and communication preferences while maintaining technical competency standards. This approach enables Indigenous workers to demonstrate their capabilities through multiple modalities rather than being constrained by narrow assessment criteria.

FROM PAY-OUT TO PARTICIPATION – REIMAGINING BENEFIT SHARING

The traditional model of mining compensation through royalty payments has failed to generate sustainable long-term benefits for Indigenous communities. Holcombe and Kemp (2020) argue for a fundamental shift from 'pay-out to participation' that recognises true benefit sharing through co-created economic opportunities, equitable employment, and community-defined development pathways.

Case study – Fortescue's Billion Opportunities program

Fortescue Metals Group's Billion Opportunities program exemplifies innovative benefit sharing through economic participation. Launched in 2011 with the goal of awarding one billion dollars in contracts to Aboriginal businesses within two years, the program has exceeded expectations, awarding over $3 billion to Aboriginal businesses and joint ventures by 2022. The program includes practical measures such as a $50 million ANZ Bank funding scheme where Fortescue acts as guarantor, eliminating security requirements for Aboriginal businesses while ensuring they own assets at contract expiry.

A key success story is Balyku woman Elsa Derschow's Brindabella Resources, a plant hire business servicing Fortescue's Cloudbreak mine alongside four other female Pilbara Traditional Owners. The business was awarded a three-year contract in 2022, demonstrating how Indigenous enterprise development creates both direct employment and community leadership opportunities. Fortescue reports that Aboriginal people comprise more than 11.7 per cent of their direct workforce, significantly exceeding national benchmarks.

This transformation requires embedding employment and training commitments within Community Development Agreements (CDAs) and Indigenous Land Use Agreements (ILUAs) as core components rather than peripheral additions. Employment provisions should specify not only job numbers but also career progression pathways, training investments, and Indigenous enterprise development support.

The participation model also requires shared governance structures that give Indigenous communities ongoing influence over workforce policies and practices. These structures should align with CRC TiME's emphasis on changing the discourse from 'stakeholder' to 'Rights Holder'—a shift that realigns mining practices with Indigenous law, sovereignty, and aspirations.

IMPLEMENTATION CHALLENGES AND SOLUTIONS

Organisational culture change

Implementing Indigenous-centred workforce models requires fundamental organisational culture change that extends beyond policy statements to operational practices. This transformation requires leadership commitment, staff training, and performance measurement systems that reward Indigenous engagement outcomes.

Cultural competence training for non-Indigenous staff is essential but insufficient. Organisations must also examine their recruitment, promotion, and decision-making processes to identify and address systemic barriers that prevent Indigenous advancement. This includes reviewing communication styles, meeting structures, and performance evaluation criteria that may inadvertently exclude Indigenous perspectives.

Regional capacity building

Effective Indigenous workforce development requires regional capacity building that extends beyond individual companies to encompass educational institutions, government agencies, and community organisations. This systems approach recognises that sustainable Indigenous employment outcomes require coordinated investment across multiple sectors and stakeholders.

Regional approaches also enable economies of scale in training delivery and reduce duplication of effort across multiple projects. Shared training facilities, equipment, and expertise can provide higher quality programs at lower per-participant costs while building regional Indigenous institutional capacity.

Measuring success

Traditional employment metrics focused on job numbers and duration provide insufficient insight into Indigenous workforce development effectiveness. Success measurement should include indicators of Indigenous advancement to leadership roles, enterprise development outcomes, and community-defined benefit realisation.

Evidence from research – Indigenous employment impacts

National census data analysis by Biddle and Hunter (2016) provides empirical evidence of mining's positive employment effects for Indigenous Australians. Indigenous employment is higher in mining areas than non-mining areas in both remote areas (4 percentage points higher) and non-remote areas (7 percentage points higher). Average incomes are also higher in mining areas, and there are fewer low-income Indigenous households in mining areas compared with non-mining areas. However, the research also identifies challenges, noting that housing stress for low-income households has increased as a result of the mining boom despite overall income improvements.

Research on Indigenous employee retention in the Pilbara region reveals critical success factors for sustainable workforce outcomes. The findings suggest that a culturally competent non-Indigenous workforce, culturally appropriate support mechanisms and access to professional development opportunities are key retention factors. This research emphasises that successful Indigenous workforce development requires systemic cultural competence throughout organisations, not merely Indigenous-specific programs.

Cultural indicators should also be incorporated to assess whether workforce participation strengthens or undermines Indigenous cultural continuity. These might include measures of traditional knowledge transfer, cultural site protection, and community governance capacity development. The Argyle Diamond Mine case study demonstrates the complexity of measuring cultural outcomes, revealing tension between economic advancement and cultural maintenance that requires ongoing negotiation rather than simple metrics.

CONCLUSION

The extractive sector must evolve toward a model where Indigenous workforce participation is embedded as a structural imperative of project design, execution, and legacy rather than a compliance obligation. The Life of Asset framework, underpinned by UNDRIP principles, provides a roadmap for co-developing workforce systems with First Nations Australians that reflect cultural governance, regional autonomy, and intergenerational equity.

This transformation requires more than policy reform—it demands fundamental reconceptualisation of the relationship between Indigenous peoples and extractive industries. By recognising First Nation Australians as Rights Holders with inherent authority over their territories, the industry can move beyond paternalistic approaches toward genuine partnership models that strengthen both Indigenous communities and business outcomes.

The workforce design model presented in this paper demonstrates that Indigenous-centred approaches enhance rather than constrain operational effectiveness. By embedding Indigenous knowledge, governance, and enterprise throughout the asset life cycle, mining companies can improve their Social License to Operate, strengthen ESG performance, and contribute to the reconciliation agenda that is essential for Australia's social and economic future.

The choice facing the extractive industry is clear: continue with incremental adjustments to fundamentally flawed engagement models or embrace the transformative potential of Indigenous-centred workforce design. The latter path requires courage, commitment, and cultural humility, but it offers the possibility of redefining the relationship between Country and industry for the benefit of all Australians.

REFERENCES

Biddle, N and Hunter, B, 2016. The economic impact of the mining boom on Indigenous and non-Indigenous Australians, *Asia Pacific Economic Papers*, 386:1–29. https://doi.org/10.1002/app5.99

CRC TiME, 2021–2024. *Songlines project: First Nations short form guides on the life of mine.* Available from: <http://www.crctime.org.au/>

Eva, C, Bodle, K, Foley, D, Harris, J and Hunter, B, 2023. The importance of understanding Indigenous employment in the Indigenous business sector, *Australian Journal of Social Issues*, 58(1):1–29. https://doi.org/10.1002/ajs4.12345

Holcombe, S and Kemp, D, 2020. From pay-out to participation: Indigenous mining employment as local development?, *Sustainable Development*, 28(5):1122–1135. https://doi.org/10.1002/sd.2067

O'Neil, S and Smith, R, 2025. Beyond mining: Transitioning indigenous communities to sustainable tourism in Australia's Northern Territory, *Journal of Sustainable Tourism*, 33(1):45–67. https://doi.org/10.1080/09669582.2025.2494046

Parmenter, J, Dowell, K, Alexander, R and Holcombe, S, 2023. Aboriginal employment outcomes at Argyle Diamond Mine: What constitutes success and for whom?, *The Extractive Industries and Society*, 14:101243. https://doi.org/10.1016/j.exis.2023.101243

Pearson, C A L and Daff, S, 2011. Extending boundaries of human resource concepts and practices: An innovative recruitment method for Indigenous Australians in remote regions, *Asia Pacific Journal of Human Resources*, 49(3):325–343. https://doi.org/10.1177/1038411111413216

United Nations General Assembly, 2007. United Nations Declaration on the Rights of Indigenous Peoples, A/RES/61/295. Available from: <https://www.un.org/development/desa/indigenouspeoples/wp-content/uploads/sites/19/2018/11/UNDRIP_E_web.pdf>

APPENDIX 1

Further reading on various subjects can be found here:

PRINCIPAL LEGISLATIVE SOURCES

Commonwealth legislation

- Aboriginal Land Rights (Northern Territory) Act 1976 (Cth)
- Native Title Act 1993 (Cth)
- Native Title Legislation Amendment Act 2021 (Cth)
- Corporations (Aboriginal and Torres Strait Islander) Act 2006 (Cth)

State and Territory land rights legislation

- Aboriginal Land Rights Act 1983 (NSW)
- Aboriginal Land Trust Act 2013 (SA)
- Pitjantjatjara Land Rights Act 1981 (SA) [renamed Anangu Pitjantjatjara Yankunytjatjara Land Rights Act 1981 (SA)]
- Maralinga Tjarutja Land Rights Act 1984 (SA)
- Aboriginal Land Act 1991 (Qld)
- Torres Strait Islander Land Act 1991 (Qld)
- Traditional Owner Settlement Act 2010 (Vic)
- Aboriginal Affairs Planning Authority Act 1972 (WA)
- Aboriginal Lands Act 1995 (Tas)

Landmark legal cases

- Mabo v Queensland (No. 2) [1992] HCA 23; (1992) 175 CLR 1
- Wik Peoples v Queensland [1996] HCA 40
- Western Australia v Ward (2002) 213 CLR 1
- Milirrpum v Nabalco Pty Ltd (1971) 17 FLR 141 (Gove Land Rights Case)
- McGlade v Native Title Registrar & Ors [2017] FCAFC 10

Government policy documents

- National Guiding Principles for Native Title Compensation Agreement Making (2021)
- Indigenous Procurement Policy (IPP)
- National Indigenous Australians Agency. (2024). *About native title.* https://aiatsis.gov.au/about-native-title

MAJOR CASE STUDIES

BHP Indigenous development programs

- Indigenous Development Program (IDP) – Running since FY2015, achieving 49% participant progression to new roles and 20% promotion to leadership positions
- Indigenous Leadership Program (ILP) – Established FY2019 to support Indigenous leaders in higher-level leadership development

- Tjiwarl Comprehensive Agreement (2018) – Native title agreement resulting in 50+ Traditional Owners completing Certificate II Resource Infrastructure and Work Preparation
- Indigenous employment target achievement – Exceeded 3% target three years ahead of schedule with 1500+ Indigenous employees and contractors

Fortescue Metals Group (FMG) programs

- Vocational Training and Employment Centre (VTEC) – Established 2006, graduated 1500+ Aboriginal people with guaranteed employment pathway
- Billion Opportunities Program – Launched 2011, awarded $3+ billion to Aboriginal businesses and joint ventures
- Brindabella Resources Success Story – Female Pilbara Traditional Owners' plant hire business servicing Cloudbreak mine
- Indigenous workforce representation – 11.7% of direct workforce, significantly exceeding national benchmarks

Rio Tinto Indigenous employment evolution

- Historical progression from <0.5% (1990s) to 7% (2010) Indigenous workforce representation
- Argyle Diamond Mine case study – 40-year operation examining cultural and economic outcomes for Aboriginal employees
- Cultural Onboarding program and Indigenous employee engagement initiatives
- Cultural heritage management and FPIC implementation following Juukan Gorge lessons

Regional impact studies

- Pilbara Region Indigenous Employment – Mining accounts for two-thirds of Indigenous male employment and one-third of Indigenous female employment
- East Arnhem Land Transition Study – Nhulunbuy post-mining economic development challenges and opportunities for Yolŋu Aboriginal people
- Bowen Basin Social Impact Assessment – Six-community study examining mining boom effects on Indigenous and non-Indigenous communities

Innovative recruitment and training models

- Alternative assessment methods (Rio Tinto) – Non-literacy based assessments focusing on disposition, community standing, and adaptive capacity
- Cross-cultural competence programs – Industry-wide initiatives to improve Indigenous employee retention through culturally safe workplaces
- Indigenous ranger program development – Post-closure employment in environmental stewardship and cultural site management

Algae-based technologies for mine site rehabilitation and closure

A H Kaksonen[1], J Ayre[2], K Schipper[3], M Ginige[4], M Edraki[5], K Y Cheng[6], L Trevaskis[7], D Purcell[8], N Moheimani[9], J Wolf[10], P A Bahri[11] and B Hankamer[12]

1. Senior Principal Research Scientist, CSIRO Environment, Waterford WA 6152. Email: anna.kaksonen@csiro.au
2. Postdoctoral Fellow, Algae Innovation Hub, School of Environmental and Conservation Sciences, Murdoch University, Murdoch WA 6150. Email: jeremy.ayre@murdoch.edu.au
3. Casual Senior Principal Researcher, Institute for Molecular Bioscience, The University of Queensland, St Lucia Qld 4072. Email: kira.schipper@uq.edu.au
4. Senior Research Scientist, CSIRO Environment, Waterford WA 6152. Email: maneesha.ginige@csiro.au
5. Associate Professor, Sustainable Minerals Institute, The University of Queensland, St Lucia Qld 4072. Email: m.edkari@uq.edu.au
6. Senior Research Scientist, CSIRO Environment, Waterford WA 6152. Email: kayu.cheng@csiro.au
7. Managing Director, Valarion Pty Ltd, Brisbane Qld 4000. Email: l.trevaskis@valarion.com.au
8. Savage River Rehabilitation Project Manager, EPA Tasmania, Hobart Tas 7001. Email: diane.purcell@epa.tas.gov.au
9. Professor, Algae Innovation Hub, School of Environmental and Conservation Sciences, Murdoch University, Murdoch WA 6150. Email: n.moheimani@murdoch.edu.au
10. Research Fellow, Institute for Molecular Bioscience, The University of Queensland, St Lucia Qld 4072. Email: j.wolf@imb.uq.edu.au
11. Pro Vice Chancellor, College of Science, Technology, Engineering and Mathematics, Murdoch University, Murdoch WA 6150. Email: p.bahri@murdoch.edu.au
12. Professor, Institute for Molecular Bioscience, The University of Queensland, St Lucia Qld 4072. Email: b.hankamer@imb.uq.edu.au

EXTENDED ABSTRACT

Mining is a key driver for Australia's economic prosperity; however, it can also create long-term environmental legacies, such as land disturbance, atmospheric emissions, waste, and mine drainage that threaten valuable land and water resources, and local communities. Sustainable mine closure is critical to ensure alignment with expectations of net-zero emissions, circular economic transitions, and delivery of nature-positive outcomes. Integrating on-site algal culture and the application of algal biomass into mining offers emerging nature-based solutions with strong alignment to circular economy and decarbonisation principles. Examples of algal integration within mining context could be the use of algae for mine water treatment, CO_2 sequestration, dust suppression, mine waste stabilisation, and the facilitation of ecological rehabilitation through its use as a fertiliser and biostimulant. Furthermore, algal biomass offers significant commercial potential, serving as a feedstock for producing various commodity products, such as bioplastics, biofuels, pigments, and animal feed. Thus, algal applications have the potential to accelerate ecological recovery and reduce the environmental impacts of mining, whilst creating business opportunities, turning post-mining land into economically productive ecosystems. This dual approach not only enhances rehabilitation outcomes but also ensures long-term sustainability for mine sites through the development of business opportunities for ongoing value generation beyond mine closure. Additionally, algae cultivation at mine sites may create jobs and offer new opportunities to engage Traditional Owners and other local communities in remote locations in Australia and facilitate community skill development and economic empowerment.

The selection of suitable algal species and cultivation systems for implementation at mine sites can include: 1) reviewing mine site characteristics and objectives for algal applications, 2) selecting algal species based on their growth rate and suitability to the mine site conditions and application, and 3) identifying necessary mine water treatment or amendments. This can be followed by the selection of 4) algal cultivation systems and 5) algal harvesting and processing systems. Other steps include 6) the verification of algal end products in terms of yield and quality, and 7) the optimisation of the process.

Large-scale microalgae cultivation can be conducted in open raceway ponds, biofilm-based algal turf scrubbers or closed photobioreactors that are supplied with water from eg mine pits if there is sufficient water of good enough quality available. Macroalgae cultivation systems can use free-floating algae, or algal biomass adhered to the bottom of the water body or solid substratum, such as rope or raft systems. Commonly used microalgal harvesting methods include centrifugation, filtration, and flocculation. Macroalgae can be harvested manually with cutting tools or through mechanical harvesting. Mechanical harvesting methods include the use of amphibious vehicles, boats, land-based long-armed vehicles equipped with suction apparatus, rotating mowers, cutters, rotating blades, or dredgers. Whole algae biomass can be directly used in multiple applications, such as animal feeds, supplements, and fertilisers. However, in some scenarios cell disruption is required for extracting valuable components, such as pigments, proteins, or lipids. This can be conducted through chemical, enzymatic, mechanical, thermal, electrical, or other emerging technologies.

Regulatory requirements need to be considered if planning algal cultivation and use at mine sites. These include land use agreements (eg Indigenous land use agreements and mining lease commercial conditions) and environmental permits and compliance (eg water use license, waste management and rehabilitation requirements). Moreover, health and safety standards for occupational, algal process and product safety must be followed, and biodiversity and ecosystem protection (eg non-invasive species and ecosystem monitoring) must also be ensured.

The application of algal technologies at mine sites may offer notable economic, social, and environmental benefits. Estimates of algal production cost vary with various cultivation systems and geographic locations. Typical costs are 10–100 AU$ kg^{-1} of dry algal biomass, however not all applications require drying of algae. The algae markets are rapidly increasing with a global value of US$1.9–5.3 billion in 2023, estimated annual growth of approximately 5–6 per cent, and projection to reach over US$2.5–7.5 billion by 2030. The value of algal products and market size vary depending on the application. Multi-product biorefineries help offset production costs, making bulk algae products more cost-competitive and profitable. Algae production costs can also be reduced through technical refinement, policy optimisation and economy of scale. Moreover, algal cultivation can foster industrial ecology or industrial symbiosis, where for example, CO_2 emissions and nutrient-rich wastewater streams from nearby industries support algal growth, while the algae purify the exhaust gases and wastewater.

The creation of a business case for the establishment of algae production and application within the context of Australian mine sites will require the consideration of many economic, environmental, and social factors, which are dependent on algal strains, cultivation and harvesting technology, end-product(s), mine site characteristics and location. The implementation of algae technologies within a mining context will involve various phases, beginning with site selection and definition of the site-specific objective. These will inform the selection of the most feasible technologies that could be implemented. Due diligence is required to determine economic, social, and environmental opportunities as well as risks. To successfully commercialise algae-based products from a mining context, a comprehensive path to market is required, incorporating market research and analysis, product development and testing, commercial partnerships, regulatory compliance, and marketing efforts. Key considerations include: 1) defining project scope and objectives, 2) site and resource availability, 3) technologies, 4) market analysis, 5) financial projections, 6) regulatory and compliance considerations, 7) risks and mitigation strategies, 8) social and community impacts, 9) sustainability and environmental impacts, 10) strategic partnerships and collaboration, 11) pathways to market, and 12) long-term vision and scalability.

This presentation provides an overview of the findings of the Cooperative Research Centre for Transformations in Mining Economies (CRC TiME) project 3.15, which explores algae-based technologies for improved environmental outcomes and sustainable post-mining futures, and some examples of case studies from literature. The project is supported by CRC TiME via the Australian Government Department of Industry, Science and Resources through the Cooperative Research Centres Program; CSIRO, The University of Queensland, Murdoch University, South 32, Fortescue Metals Group, Rio Tinto, Heidelberg Materials, Mineral Research Institute of Western Australia, Queensland Mine Rehabilitation Commissioner, and Energy Australia.

Advancing post-mining rehabilitation – utilising rainfall simulators to enhance erosional stability and landform design

A M Khalifa[1], H B So[2] and G Maddocks[3]

1. Principal Consultant, SLR Consulting Australia Pty Ltd, Brisbane Qld 4000.
   Email: akhalifa@slrconsulting.com
2. Professor of Soil Science, Griffith University, Brisbane Qld 4000. Email: b.so@griffith.edu.au
3. Technical Director – SLR Consulting Australia Pty Ltd, Brisbane Qld 4000.
   Email: gmaddocks@slrconsulting.com

ABSTRACT

The mining industry plays a vital role in economic development, yet it faces challenges in designing stable post-mining landforms. Rainfall simulators have emerged as essential tools for understanding and mitigating soil erosion, enabling the design of effective landform cover systems to ensure long-term stability. This study explores the development, calibration, and application of rainfall simulators to assess soil erodibility and erosion rates in rehabilitated mining landscapes. Through experiments conducted with portable rainfall simulators, calibrated using Christiansen's uniformity coefficient and raindrop kinetic energy, the findings demonstrate their effectiveness in sourcing data for more accurately modelling erosion processes. Data generated from these simulators have been integrated with numerical models, such as MINErosion, and WEPP to predict and optimise landform stability. Case studies in Queensland illustrate the role of rainfall simulators in meeting regulatory requirements and designing sustainable post-mining landforms. The results highlight the potential of combining field measurements and modelling to guide rehabilitation practices and achieve long-term environmental sustainability.

INTRODUCTION

Rehabilitating post-mining landscapes requires the integration of soil science, hydrology, and geomorphology. Soil erosion, particularly under intense rainfall and steep gradients, presents a major challenge to the integrity of landform cover systems. In Queensland, the Progressive Rehabilitation and Closure (PRC) framework mandates demonstrable long-term stability of rehabilitated landforms prior to regulatory sign-off.

Erosion modelling is vital for predicting and managing the long-term performance of these landforms. Tools such as WEPP (Flanagan and Livingston, 1995), MINErosion (Khalifa, 2010), and landscape evolution models like SIBERIA (Willgoose, Bras and Rodriguez-Iturbe, 1991) and CAESAR-Lisflood (Coulthard et al, 2013) are widely adopted. However, the accuracy of these models depends heavily on the input parameters, especially soil erodibility. Standard parameter estimation using pedotransfer functions (PTFs) often fails to account for site-specific spoil variability, particularly in dispersive or saline overburdens.

The predictive accuracy of erosion models such as WEPP is heavily influenced by the sensitivity of key input parameters—particularly effective hydraulic conductivity (Keff), rill erodibility (Kr), and critical shear stress (τ_c). These parameters not only govern run-off generation and sediment detachment but also dictate how the model responds to variations in slope gradient, slope length, and vegetation cover. Studies such as Zheng et al (2020) have demonstrated that small changes in Keff and Kr can significantly alter run-off and soil loss predictions, especially on steep gradients. Moreover, WEPP's sensitivity to rill spacing and its effect on shear stress becomes critical when modelling constructed landforms with high spatial variability, such as those found in post-mining settings. This highlights the importance of site-specific calibration, as model performance deteriorates when relying on default or literature-derived inputs. For reliable simulation of erosion processes in mining rehabilitation, empirical parameterisation using rainfall simulation data is essential to constrain model sensitivity and reduce predictive uncertainty.

THE ROLE OF RAINFALL SIMULATORS IN EROSION MODELLING

Rainfall simulators provide a controlled and repeatable means to generate run-off and sediment yield data for quantifying rill and interrill erodibility (Kr and Ki), critical shear stress (τ_c), and other hydrological parameters. Unlike *pedotransfer function (PTF)-derived* parameters, which are based on general soil properties such as texture and organic matter, rainfall simulation tests generate *site-specific values* that capture the actual erosional response of the material under varying slope and rainfall conditions (Aksoy *et al*, 2012). In WEPP and MINErosion applications, the use of rainfall simulator-derived parameters has been shown to reduce uncertainty, improve regulatory confidence, and enable more precise landform design, thereby reducing over-engineering and unnecessary earthworks. Accurate parameterisation remains a critical factor in improving the predictive performance of erosion models, particularly in complex post-mining environments where soil characteristics often deviate significantly from those in natural or agricultural settings. One of the most influential advancements in recent years has been the use of rainfall simulators to obtain direct, site-specific measurements of soil erodibility parameters. These include interrill and rill erodibility (Ki and Kr), critical shear stress (τ_c), and infiltration-related hydrological properties—factors that govern the detachment and transport of sediment under varying slope and rainfall intensities.

A growing body of research has demonstrated that erosion models calibrated using rainfall simulator-derived parameters consistently outperform those using estimated values generated from pedotransfer functions (PTFs), especially when applied to highly dispersive or sodic spoil materials. For instance, Mahmoodabadi and Cerdà (2013) evaluated the WEPP model under semi-arid conditions and reported that default or PTF-derived inputs led to a substantial underestimation of interrill erosion rates—by a factor of 14.5 on average. However, when calibrated with measured values obtained through rainfall simulation, the model's prediction efficiency improved dramatically (from -2.46 to 0.90). Similarly, long-term modelling using the SIBERIA landscape evolution model has shown that erosion predictions closely match field-measured values (eg from ^{137}Cs tracer studies) only when calibrated with measured soil erodibility parameters using a flume experiment. Hancock *et al* (2011) confirmed that uncalibrated runs using inferred or database-derived inputs produced unrealistic channel forms and erosion patterns.

These findings collectively underscore the importance of empirical calibration using site-specific data. Measured values derived from rainfall simulation not only capture the true erosional behaviour of local materials but also account for variations in surface conditions, such as rock fragment cover and sediment concentration, which standard PTFs fail to represent—particularly under the intense storm conditions common in Central Queensland (So *et al*, 1998, 2018; Sheridan *et al*, 2000; Khalifa, 2010).

Beyond improving model accuracy, rainfall simulators also play a vital role in addressing a long-standing challenge in erosion assessment: the determination of acceptable erosion thresholds. Despite early efforts to define 'tolerable' erosion rates (eg Browning, Parish and Glass, 1947), there remains no universal consensus on what constitutes an acceptable rate of erosion for rehabilitated landforms. Commonly cited tolerable values—such as <11.2 t/ha/annum for deep fertile soils and <4.5 t/ha/annum for shallow soils—are derived from U.S. agricultural contexts and are not directly applicable to Australian mine rehabilitation settings. In recognition of this, the Queensland Department of Mines and Energy previously recommended a broader, more pragmatic range of 12–40 t/ha/annum (Welsh *et al*, 1994; Williams, 2001), reflecting the erosivity of local conditions and the operational requirement to maintain long-term landform stability.

In recent years, however, the Department of the Environment, Tourism, Science and Innovation (DETSI, formerly DESI) in Queensland has revised its stance and adopted a more restrictive interpretation of acceptable erosion thresholds. It now asserts that erosion rates on rehabilitated landforms must not exceed the 'natural' background erosion rate estimated for Queensland landscapes. This shift has been informed largely by the findings of Lu *et al* (2003), who used the RUSLE model, coupled with remotely sensed data, to predict sheetwash and rill erosion across the Australian continent. Based on this study, DETSI has adopted a generalised average erosion rate of 5–6 t/ha/annum, with a maximum tolerable value of 10 t/ha/annum for worst-case scenarios.

While this approach aims to enforce conservative benchmarks, it raises several concerns. First, it assumes that erosion rates on newly constructed, disturbed landforms—often composed of spoil or dispersive materials—should match those of stable natural soils that have evolved over thousands of years. This equivalence is scientifically questionable given the fundamental differences in profile development, material properties, and hydrological behaviour. Second, and more critically, the very study on which DESI relies (Lu *et al*, 2003) derived its erosion estimates using RUSLE—a model which DESI itself considers inadequate for PRCP-level assessments due to its empirical structure and inability to simulate key erosion processes such as deposition, channel development, or landform evolution.

This inconsistency highlights the need for a more robust, evidence-based framework that integrates site-specific data derived from rainfall simulation, validated process-based modelling, and an understanding of long-term geomorphic function. Rainfall simulators provide a defensible empirical basis for determining such thresholds. By generating run-off and erosion data under controlled conditions that reflect local slope gradients, soil properties, and rainfall intensities, they allow erosion rates to be benchmarked against adjacent undisturbed or successfully rehabilitated areas. This approach ensures that thresholds are not only scientifically valid but also tailored to the geomorphic, climatic, and operational context of each site. When aligned with the objectives of the Progressive Rehabilitation and Closure (PRC) Plan, such thresholds support landform designs that are both technically robust and environmentally and socially responsible.

SLR'S CAPABILITY AND THE RAESIM SIMULATOR

SLR Consulting has developed and widely implemented the RaESIM (Rainfall and Erosion Simulator) as a versatile tool across numerous mine rehabilitation projects. RaESIM is specifically designed to generate high-quality, site-specific data essential for calibrating and validating erosion models such as WEPP, MINErosion, and SIBERIA. It enables precise measurement of key parameters including rill and interrill erodibility, critical shear stress, run-off rates, infiltration, and sediment yield—under both laboratory and field conditions.

Beyond model calibration, RaESIM plays a critical role in establishing baseline erosion and run-off rates directly at the site, which allows for the empirical determination of acceptable erosion thresholds tailored to each mine's environmental and operational context. This approach ensures that the thresholds used in rehabilitation design are both realistic and defensible, supporting compliance with regulatory frameworks such as the Progressive Rehabilitation and Closure (PRC) Plan. The integration of RaESIM data enhances the scientific rigour of erosion assessments while enabling more efficient, cost-effective, and sustainable landform designs.

The earliest rainfall simulators were classified as non-pressurised systems, where raindrops were formed by allowing water to pass through devices such as perforated pipes, hanging threads, or arrays of syringe needles. These droplets were then released from a minimum height of 9.1 m to ensure they achieved a terminal velocity like that of natural rainfall. However, this design was highly susceptible to wind interference due to the long drop distance, necessitating large wind shields. The size and structural complexity of these simulators made them unsuitable for field use, significantly limiting their portability. Subsequently, pressurised rainfall simulators were developed, using nozzles to deliver water under controlled pressure. These systems gained widespread adoption thanks to their compact design (typically 2–3 m in height), portability, lower construction cost, and suitability for both field and laboratory applications (Khalifa and So, 2024).

For these reasons, SLR RaESIM was purposefully designed as a pressurised nozzle-type simulator. Its development prioritised key features to ensure practicality and performance in mining rehabilitation contexts (Figure 1), including:

- **High portability**: Lightweight, modular components that allow easy transport and rapid assembly across different sites.

- **Water efficiency**: Optimised to operate with limited volumes of quality water, which is often a constraint in remote locations.

- **Field reliability**: Engineered for durability and ease of on-site maintenance, minimising downtime during campaigns.

- **Operational simplicity**: Designed to be managed by a small field team (2–3 people), supporting efficient deployment and data collection.

FIG 1 – SLR RaESIM in action.

As Pressurised Rainfall Simulators (Nozzle's type), The RaESIM consists of the following main parts: the structural frame, the rain drops generator system, the water supply unit (tank), and the flume/specimen container. The structural frame was constructed from aluminium tubing of 38 mm outside diameter (OD) and a 3 mm wall thickness. The bases of the upright tubes are bent to form detachable legs, this design make it possible for the frame to set on 3 m hydraulic-tipper trailer when it used in laboratory/warehouse so it can benefit from the slope making mechanism of trailer to make a slope up to 65 per cent (Figure 2); as well as make it possible to be pegged to the ground, if the apparatus taken to the open fields.

FIG 2 – The structure of RaESIM and its ability to make the slopes.

To ensure the RaESIM is easily transportable, its structural frame was designed as a modular system comprising 12 aluminium tubes that can be quickly assembled and disassembled. These tubes are connected using internally fitted nylon joiners, each approximately 180 mm in length and precision-machined from solid nylon stock, providing both durability and ease of use during field deployment.

The nozzle boom is constructed from a 3 m length of aluminium tubing with a 38 mm outer diameter and a wall thickness of 1.6 mm. It is mounted on two SB210ZZ-Kart bearings that not only facilitate smooth rotation but also minimise lateral movement for stability during operation. Welded at 1 m intervals along the boom are three male-threaded ½' BSPT unions, onto which check valves and selected nozzles are fitted. These check valves prevent leakage or dripping when the system is idle. The water is supplied via a 1¼' BSPT tee fitting located opposite the nozzle row, with an additional tapping to accommodate two pressure gauges for real-time monitoring and adjustment of operating pressure.

The raindrop generation system includes a selection of VeeJet nozzles (models 8010, 8015, 8020, and 80100), chosen based on the rainfall intensity and drop characteristics required. These nozzles are spaced 1 m apart along the boom to ensure complete coverage of the 3 m × 1 m flume area. The decision to use VeeJet nozzles was informed by the authors' extensive field experience and supported by previous research, including studies by Bubenzer (1979), Loch *et al* (2001), and Blanquies, Scharff and Hallock (2003). These nozzles have proven effective in simulating the drop size distribution and kinetic energy profiles of intense natural storms, achieving the terminal velocities needed for realistic erosion modelling scenarios (Horne, 2017).

The water delivery system comprises a Matrix 10-5 VFD pump unit, built around the Ward 10-5 horizontal multistage stainless-steel pump (Figure 4). It is powered by a 2.2 kW single-phase motor drawing 13 amps and features a 40 mm BSP female inlet and a 32 mm BSP female outlet. Controlled by a SteadyPress variable speed drive, the system can deliver up to 200 L/min. During operation, the simulator draws water from a 100 L tank, which is connected to a mains water tap for replenishment. For field deployments, a 0.5 m^3 auxiliary tank supplies the main reservoir, ensuring consistent operation even in remote areas.

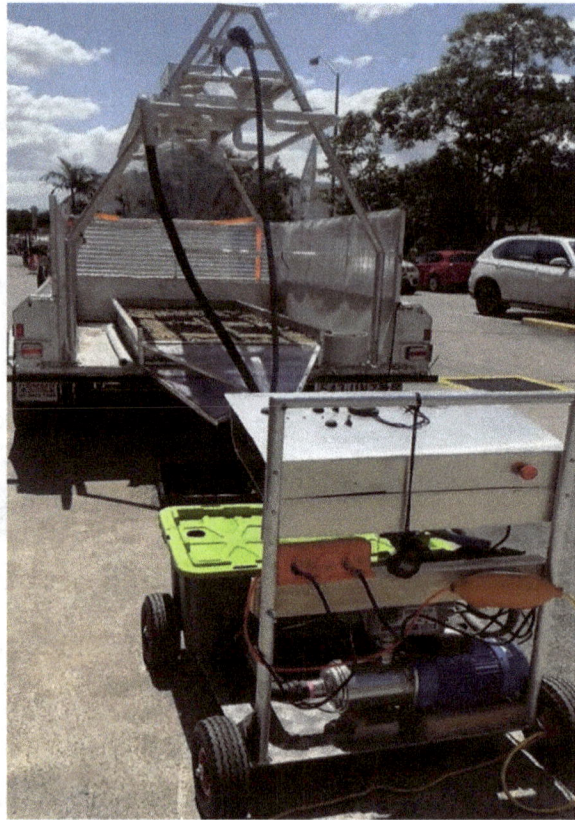
FIG 4 – RaESIM water supply unit.

The RaESIM operates using a continuous flow of water through its nozzles, delivering rainfall across the target plot. Any water that falls outside the designated flume area is collected and recycled through a system of catch trays made from galvanised aluminium. These trays are positioned on either side of each oscillating nozzle to capture excess spray (Figure 3), redirecting it back into the main water tank to conserve water and maintain consistent system performance.

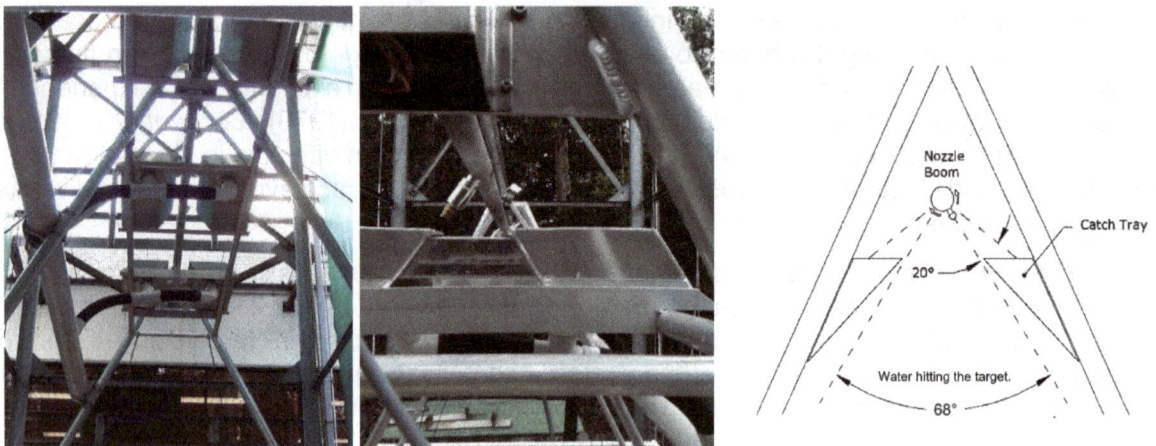
FIG 3 – Excess water collecting trays.

During operation, the nozzles oscillate across an arc of approximately 107°. The central 67° of this sweep is dedicated to applying rainfall directly onto the test flume, while the remaining 20° at each end ensures overlapping coverage to prevent gaps in rainfall distribution. The excess water from these peripheral sweeps is effectively captured by the catch trays.

Rainfall intensity is controlled by adjusting the oscillation frequency of the nozzles using an integrated stepper motor control system. This system includes a DVP-14SS211T2 programmable logic controller (PLC), an EM806 stepper driver, a four-button digital control interface, and a set of transistor switching circuits. The microcontroller operates in a standalone configuration using internal

RAM and EEPROM for program and data storage. The digital interface allows users to precisely set the delay between nozzle sweeps in 0.1 sec increments, providing fine control over rainfall intensity and event duration. The transistor switching circuit ensures proper voltage and current delivery to the stepper driver for stable motor performance.

RaESIM is equipped with a dedicated flume/specimen container that mounts onto the hydraulic ramp of a trailer when used for indoor experiments in a laboratory or warehouse setting. For field applications, the testing plot is defined using 3 mm thick metal edging plates, which are hammered into the soil to a depth of 7–10 cm to contain the soil and run-off. A gutter is installed at the downslope edge to collect and channel run-off for sampling and measurement, replicating real erosion scenarios under controlled yet field-representative conditions.

RaESIM performance and calibration

The RaESIM was comprehensively evaluated to ensure its performance and reliability under simulated storm conditions. The assessment focused on four key parameters:

1. **Rainfall intensity and the consistency of spatial distribution** of raindrops across the flume,

2. **Uniformity of rainfall** over the flume area, assessed through spatial analysis of water collection to determine evenness of coverage,

3. **Raindrop size distribution**, to verify that the drop sizes generated align with those typically observed in natural erosive storms, and

4. **Raindrop terminal velocity and kinetic energy**, ensuring that the droplets possess sufficient impact force to replicate the erosive power of real rainfall events.

Together, these evaluations confirmed that the RaESIM can deliver rainfall events that closely mimic natural conditions, making it a dependable tool for erosion studies and landform stability assessments.

Rainfall intensity

RaESIM uses nozzles to generate the rainfall drops, it manages and controls the rainfall intensity by adjusting both water discharge\pressure from the water pump and the movement and shape of its nozzles, using the stepper motor controller and driver, to change the sweep and waiting time pattern. The RaESIM will successfully produce rainfall events with intensities of 30 mm/hr to 150 mm/hr (±10 per cent mm/hr); that was verified by considering the average amount of water falling on the flume area measured by the pan method, as well as by measuring the water flow rate (discharge) from each nozzle separately by sticking PVC tube around the nozzle and collecting effluent from the tube for 5 mins (multiply by 12 to get the flow mm/hr).

Spatial Uniformity of rainfall over the flume area

The uniformity of rainfall distribution across the plot surface was evaluated by collecting simulated rainfall at varying intensities (each tested in triplicate) using 18 identical graduated beakers. These beakers were systematically arranged in a 6-row by 3-column grid to evenly cover the plot area. Rainfall was collected over a standard duration of ten mins for each run. The volume of water captured in each beaker was then used to calculate the rainfall distribution uniformity using Christiansen's Coefficient of Uniformity (CU), as defined by Equation 1 (Christiansen, 1942). This method provided a quantitative measure of the spatial consistency of rainfall application delivered by the RaESIM system.

$$CU = 100 \left[1 - \left(\frac{\sum (|D_i - D_m|)}{n * D_m} \right) \right] \tag{1}$$

where

CU	is the coefficient of uniformity (%)
D_i	is the depth of water in the graduated beakers (cm)
D_m	is the mean depth of water in graduated beakers (cm)

n   is the number of graduated beakers

Depending on the results out of Equation 1, the RaESIM is performing uniformly over the tested area with a coefficient of uniformity above 86.55 per cent for all intensities and replicates (Table 1), According to Sousa, Mendes and Siqueira (2017), several researchers consider CU any values above 80.0 per cent are acceptable for the uniformity of the rainfall distribution.

TABLE 1

The values of coefficient of uniformity measured from different rainfall intensities generated by RaESIM.

Pan No.	30 mm/hr Designed Event			60 mm/hr Designed Event			90 mm/hr Designed Event			100 mm/hr Designed Event		
	Measured Rainfall Intensity (mm/hr)	Di	|Di-Dm|	Measured Rainfall Intensity (mm/hr)	Di	|Di-Dm|	Measured Rainfall Intensity (mm/hr)	Di	|Di-Dm|	Measured Rainfall Intensity (mm/hr)	Di	|Di-Dm|
1	28	13.83	1.61	56	28.00	2.29	84	42.17	3.88	91	45.55	6.34
2	37	18.56	3.12	62	31.04	0.75	86	43.18	2.87	88	43.86	8.03
3	34	17.21	1.77	57	28.34	1.95	81	40.49	5.57	88	43.86	8.03
4	38	19.23	3.79	62	31.04	0.75	92	45.88	0.17	101	50.61	1.28
5	31	15.59	0.15	65	32.35	2.06	98	48.82	2.77	106	52.94	1.05
6	35	17.54	2.10	61	30.36	0.07	86	43.18	2.87	101	50.61	1.28
7	31	15.29	0.15	61	30.59	0.29	92	45.88	0.17	112	55.88	3.99
8	38	18.82	3.38	66	32.94	2.65	98	48.82	2.77	118	58.82	6.93
9	29	14.71	0.73	61	30.29	0.00	96	48.24	2.18	118	58.82	6.93
10	27	13.53	1.91	62	31.18	0.88	95	47.65	1.59	118	58.82	6.93
11	29	14.41	1.03	67	33.53	3.24	93	46.47	0.41	118	58.82	6.93
12	21	10.59	4.85	58	28.82	1.47	94	47.06	1.00	106	52.94	1.05
13	22	11.23	4.21	64	31.91	1.62	99	49.65	3.59	106	53.19	1.30
14	35	17.71	2.27	61	30.70	0.41	101	50.61	4.55	103	51.28	0.61
15	35	17.54	2.10	59	29.35	0.94	88	43.86	2.20	88	43.86	8.03
16	26	13.16	2.28	55	27.33	2.97	88	43.86	2.20	101	50.61	1.28
17	28	14.12	1.32	60	30.00	0.29	100	50.00	3.94	106	52.94	1.05
18	30	14.84	0.60	55	27.50	2.80	86	43.18	2.87	101	50.61	1.28
	Dm	15.44		Dm	30.29		Dm	46.06		Dm	51.89	
	n	18		n	18		n	18		n	18	
	Sum |Di-Dm|	37.37		Sum |Di-Dm|	25.42		Sum |Di-Dm|	45.61		Sum |Di-Dm|	72.35	
	CU	86.55		CU	95.34		CU	94.50		CU	92.25	

The Drop Size Distribution (DSD) and Kinetic energy

The ability of any rainfall simulator to generate raindrops that approximate the volumetric size distribution of the droplets that occur during rainstorms in nature is highly influential in our judgment of the efficiency and quality of the rainfall simulator design. This is because the distribution of grain size over the different classes of drop size (volume in mm^3) affects the total kinetic energy generated from the simulated rainstorm, whereas the kinetic energy of a single drop is a function of a grain's mass, which is related to its size (volume) as well as its terminal velocity when it hits the ground (Serio, Carollo and Ferro, 2019; Ngezahayo, Michael and Ghataora, 2019). In general, the sizes of raindrops in nature range from 0.5 mm in diameter to the large drops associated with heavy rainfall reaching up to 6 mm in diameter, with median droplet diameter varying depending upon the storm intensity but usually ranging from 2 mm to 3 mm (Mhaske, Pathak and Basak, 2019; Horne, 2017). There are several methods to measure the drop size distribution, the Flour Method (Bentley, 1904) was used to calibrate RaESIM as it is widely accepted, standardised test method in the research field (Eigel and Moore, 1983; Horne, 2017; Ngezahayo, Michael and Ghataora, 2019). The mean diameter of the raindrops out of each examined rain event (30 mm/hr, 45 mm/hr, 60 mm/hr, 90 mm/hr, and 100 mm/hr rainfall intensities) was calculated using Equation 2.

$$D_r = \sqrt[3]{\left(\frac{6}{\pi}\right) W m_R}$$

(2)

Where

D_r is the mean raindrop diameter (mm)

W   is the mean weight of the raindrops (mg)

m_R is the ratio of the mass of the raindrop to the mass of the pellet which is obtained using the flour-calibration line (Laws and Parson, 1943).

According to the calculations, the median size distribution (D_{50}) of RaESIM range from 3.36 to 2.32 mm in diameter, the kinetic energy from individual raindrops was then using the following equation:

$$KE = \tfrac{1}{2}\,mv^2$$

Where:

KE   is the kinetic energy (J)

m   is the mass (kg) of the raindrop (calculated from the relation between the volume, density, and mass)

v   is the terminal velocity (m/s) at which the drop hits the soil surface where the values for examined rainfall intensities by RaESIM were obtained from ASTM (2015) chart that correlates the fall velocity, fall height and raindrop diameter.

The calibration results for the raindrops size distribution and kinetic energy are shown in Table 2.

TABLE 2

The Drop Size Distribution (DSD) and Kinetic energy produced by RaESIM during the calibration process.

Rainfall intensity (mm/hr)	Number of pellets	Weights of dry pellets (g)	Average weight of dry pallets (mg)	The mean raindrops' diameter D_r (mm)	Kinetic energy, KE (μ J) of the rain drop
100	35	0.298	8.52	3.36	300.10
90	39	0.253	6.54	3.02	217.44
60	44	0.229	5.26	2.77	168.98
45	52	0.198	3.80	2.47	118.90
30	57	0.179	3.16	2.32	98.84

Salles, Poesen and Govers (2000) examined and detailed the minimum kinetic energy levels required to initiate soil particle detachment due to raindrop impact. Their findings indicated that the threshold energy needed to trigger detachment decreases as the soil's median particle diameter (D_{50}) increases from very fine sizes (around 0.001 mm) up to approximately 0.1–0.2 mm. Beyond this range, as D_{50} continues to increase, the required kinetic energy begins to rise again in proportion to the particle size.

Based on calibration results, the raindrops produced by the RaESIM—both in terms of size and kinetic energy were well within the effective range for initiating erosion in soils with D_{50} values between 0.001 mm and 2.5 mm. This range matches the surface textures of the soil samples tested, confirming RaESIM's capability to simulate erosive conditions suitable for a wide spectrum of soil types typically encountered in post-mining environments.

PRACTICAL APPLICATIONS OF RAINFALL SIMULATORS – PREVIOUS PROJECTS

The accurate calibration of erosion models such as WEPP (Water Erosion Prediction Project) and SIBERIA relies heavily on precise measurement of key input parameters. Rainfall simulators like RaESIM provide high-resolution, site-specific inputs that are essential for model calibration and validation. These inputs fall into distinct categories related to hydrology, soil erodibility, and surface conditions. By operating under known rainfall intensity, drop size, and kinetic energy conditions, the

RaESIM creates a controlled environment that enables the direct measurement of sediment yield and run-off under various slope gradients and soil treatments.

WEPP inputs from rainfall simulator data:

- Interrill erodibility (Ki) – measured from sediment yield during overland sheet flow; essential for simulating splash and sheet erosion.

- Rill erodibility (Kr) – derived from sediment concentration in concentrated flow paths or induced rills during the simulation.

- Critical shear stress (τ_c) – calculated by correlating flow energy and sediment transport initiation during high-intensity simulations.

- Effective hydraulic conductivity (K_{eff}) – estimated from steady-state infiltration measurements and surface run-off volumes.

- Run-off rate and timing – measured directly to define hydrologic response curves under controlled rainfall events.

- Surface cover effects – simulated with varied mulch, rock fragments, or vegetation to quantify their effect on erosion resistance.

SIBERIA inputs from rainfall simulator data:

- Erosion-deposition parameters (β, ε) – derived indirectly by calibrating SIBERIA simulations against observed erosion under known rainfall intensity, slope, and soil conditions from simulator tests.

- Transport threshold (eg critical shear or detachment threshold) – estimated from the point at which sediment yield initiates during simulation events.

- Sediment transport capacity functions – informed by sediment yield versus slope/run-off relationships observed during rainfall experiments.

- Rainfall erosivity proxies – simulator-derived rainfall intensity, drop size distribution, and kinetic energy data help estimate the erosivity factor (R) adapted for process-based input.

- Initial surface microtopography – if measured before and after simulator events (eg via photogrammetry or laser scanning), can be used to assess fine-scale surface evolution trends for calibration purposes.

These measured parameters replace uncertain estimates or generalised pedotransfer function values, resulting in more realistic simulations and landform design outcomes tailored to each site's material behaviour and hydrological conditions.

During rainfall simulation experiments, sediment concentration in collected run-off is analysed to determine detachment rates. The steady-state run-off rate, combined with rainfall intensity and slope, is used to estimate hydraulic parameters such as conductivity and τ_c. These measurements are taken under different slope gradients, moisture contents, and soil surface conditions (eg bare, crusted, vegetated), allowing a wide range of model input conditions to be captured. Drop size distribution and rainfall uniformity are also monitored to ensure consistency with natural erosive storms, thereby improving the reliability of derived parameters for input into WEPP and SIBERIA simulations.

Our team members have accumulated substantial hands-on experience in rainfall simulation and erosion modelling, dating back to the landmark ACARP-funded multi-institutional project titled Post-mining Landscape Parameters for Erosion and Water Quality Control (PLPEWC), which ran from 1992 to 1998. This project established one of the most comprehensive soil and spoil erodibility data sets in Australia and pioneered the integration of laboratory and field rainfall simulation into post-mining rehabilitation assessment. The methodologies and expertise developed during this initiative have since underpinned numerous research and consultancy efforts (Figure 5).

FIG 5 – Soil erosion experiment and measurement of erodibility factors using RaESIM, conducted at the SLR Laboratory in Newcastle.

CONCLUSION

Rainfall simulators are indispensable tools in post-mining rehabilitation. They bridge the gap between empirical observation and numerical modelling, enabling the design of stable landforms that meet both regulatory criteria and long-term environmental objectives. The RaESIM has demonstrated substantial advantages over traditional rainfall simulation systems, particularly in terms of portability, precision, water-use efficiency, and ease of deployment in remote field conditions. These benefits enhance its applicability for large-scale erosion data collection campaigns across diverse mining regions.

Beyond their use in calibrating 2D erosion models such as WEPP and MINErosion, rainfall simulators can provide critical input for spatially distributed erosion models like GeoWEPP and SIBERIA. These spatial models allow erosion predictions over heterogeneous terrain and enable the identification of erosion hotspots across complex mine landscapes, thus supporting proactive rehabilitation design. Additionally, rainfall simulator-derived inputs can improve the reliability of long-term simulations using landscape evolution models (eg CAESAR-Lisflood), which incorporate hydrological and geomorphic feedbacks over decadal timescales.

Future research should focus on establishing a harmonised erosion parameter database derived from rainfall simulation across key climatic zones. Integrating these data with both spatial and temporal erosion modelling platforms will enhance the predictive capacity and scientific rigour of erosion assessments for mine rehabilitation.

REFERENCES

Aksoy, H, Unal, N E, Cokgor, S, Gedikli, A, Yoon, J, Koca, K, Inci, S B and Eris, E, 2012. A rainfall simulator for laboratory-scale assessment of rainfall-runoff-sediment transport processes over a two-dimensional flume, *CATENA*, 98:63–72.

American Standards for Testing Materials (ASTM), 2015. Standard Test Method for Determination of Rolled Erosion Control Product Performance in Protecting Hillslopes from Rainfall-Induced Erosion, American Standards for Testing Materials: West Conshohocken.

Bentley, W A, 1904. Studies of raindrops and raindrop phenomena, *Mon Wea Rev*, 32(10):450–456.

Blanquies, J, Scharff, M and Hallock, B, 2003. The Design and Construction Of A Rainfall Simulator, in *International Erosion Control Association (IECA), 34th Annual Conference and Expo*, pp 24–28.

Browning, G M, Parish, C L and Glass, J, 1947. A Method for Determining the Use and Limitations of Rotation and Conservation Practices in the Control of Soil Erosion in Iowa, *Agronomy Journal*, 39(1):65–73. https://doi.org/10.2134/agronj1947.00021962003900010008x

Bubenzer, G D, 1979. Rainfall characteristics important for simulation, in *Proceedings of the Rainfall Simulator Workshop*, pp 22–34, Administration Agricultural Reviews and Manuals.

Christiansen, J E, 1942. Hydraulics of Sprinkling Systems for Irrigation, *Transactions of the American Society of Civil Engineers*, 107(1):221–239. https://doi.org/10.1061/Taceat.0005460

Coulthard, T J, Neal, J C, Bates, P D, Ramirez, J, de Almeida, G A M and Hancock, G R, 2013. Integrating the LISFLOOD-F P, 2D hydrodynamic model with the CAESAR model: implications for modelling landscape evolution, *Earth Surface Processes and Landforms*, 38(15):1897–1906. https://doi.org/10.1002/esp.3478

Eigel, J D and Moore, I D, 1983. A Simplified Technique for Measuring Raindrop Size and Distribution, *Transactions of the ASAE*, 26(4):1079–1084. https://doi.org/10.13031/2013.34080

Flanagan, D C and Livingston, S J, 1995. WEPP User Summary, Flanagan, NSERL Report #11, USDA-ARS National Soil Erosion Research.

Hancock, G R, Coulthard, T J, Martinez, C and Kalma, J D, 2011. An evaluation of landscape evolution models to simulate decadal and centennial scale soil erosion in grassland catchments, *Journal of Hydrology*, 398(3–4):171–183. https://doi.org/10.1016/j.jhydrol.2010.12.002

Horne, M A, 2017. Design and Construction Of A Rainfall Simulator For Large-Scale Testing Of Erosion Control Practices and Products, in *Graduate Faculty of Auburn University: vol Master of Science*, p 166, Auburn University.

Khalifa, A and Bing So, H, 2024. Using Rainfall Simulators to Design and Assess the Post-Mining Erosional Stability, in *Soil Erosion – Risk Modeling and Management*, IntechOpen. https://doi.org/10.5772/intechopen.112240

Khalifa, A M, 2010. MINErosion 4: A user-friendly catchment/landscape erosion prediction model for post mining sites in Central Queensland, PhD thesis, in Griffith School of Engineering: Vol, Griffith University. https://doi.org/10.25904/1912/44

Laws, J O and Parson, D A, 1943. The relation of raindrop-size to intensity, *Trans of the Am Geographical Union Papers Hydrology*, pp 453–460.

Loch, R J, Robotham, B G, Zeller, L, Masterman, N, Orange, D N, Bridge, B J, Sheridan, G and Bourke, J J, 2001. A multi-purpose rainfall simulator for field infiltration and erosion studies, *Australian Journal of Soil Research*, 39(3):599–610. https://doi.org/10.1071/SR00039

Lu, H, Prosser, I P, Moran, C J, Gallant, J C, Priestley, G and Stevenson, J G, 2003. Predicting sheetwash and rill erosion over the Australian continent, *Australian Journal of Soil Research*, 41(6):1037–1062. https://doi.org/10.1071/sr02157

Mahmoodabadi, M and Cerdà, A, 2013. WEPP calibration for improved predictions of interrill erosion in semi-arid to arid environments, *Geoderma*, 204–205:75–83. https://doi.org/10.1016/j.geoderma.2013.04.013

Mhaske, S N, Pathak, K and Basak, A, 2019. A comprehensive design of rainfall simulator for the assessment of soil erosion in the laboratory, *Catena*, 172:408–420. https://doi.org/10.1016/j.catena.2018.08.039

Ngezahayo, E, Michael, P N B and Ghataora, G S, 2019. Rainfall induced erosion of soils used in earth roads, in *7th International Symposium on Deformation Characteristics of Geomaterials*, 92:17006. https://doi.org/10.1051/e3sconf/20199217006

Salles, C, Poesen, J and Govers, G, 2000. Statistical and physical analysis of soil detachment by raindrop impact: Rain erosivity indices and threshold energy, *Water Resources Research*, 36(9):2721–2729. https://doi.org/10.1029/2000wr900024

Serio, M A, Carollo, F G and Ferro, V, 2019. Raindrop size distribution and terminal velocity for rainfall erosivity studies, A review, *Journal of Hydrology*, 576:210–228. https://doi.org/10.1016/j.jhydrol.2019.06.040

Sheridan, G J, So, H B, Loch, R J, Pocknee, C and Walker, C M, 2000. Use of laboratory-scale rill and interill erodibility measurements for the prediction of hillslope-scale erosion on rehabilitated coal mine soils and overburdens (statistical data included), *Australian Journal of Soil Research*, 38:285.

So, H B, Khalifa, A M, Yu, B, Caroll, C, Burger, P and Mulligan, D, 2018. MINErosion 3: Using measurements on a tilting flume-rainfall simulator facility to predict erosion rates from post-mining landscapes in Central Queensland, Australia, *PLoS One*, 13(3):e0194230–e0194230. https://doi.org/10.1371/journal.pone.0194230

So, H B, Sheridan, G J, Loch, R J, Carroll, C, Willgoose, G, Short, M and Grabski, A, 1998. Post-mining Landscape Parameters for Erosion and Water Quality Control, ACARP report. Available from: <https://www.acarp.com.au/abstracts.aspx?repId=C4011>

Sousa, S, Mendes, T and Siqueira, E, 2017. Development and calibration of a rainfall simulator for hydrological studies, *RBRH Scientific/Technical Article*, 22. https://doi.org/10.1590/2318-0331.0217170015

Welsh, D, Hinz, R, Garlipp, D and Gillespie, N, 1994. Coal mines on target with environmental planning, Queensland *Government Mining Journal (Australia)*, 95(1107).

Willgoose, G, Bras, R L and Rodriguez-Iturbe, I, 1991. A coupled channel network growth and hillslope evolution model, Theory, *Water Resources Research*, 27(7):1671–1684. https://doi.org/10.1029/91wr00935

Williams, D J, 2001. Prediction of erosion from steep mine waste slopes, *Environmental Management and Health*, 12(1):35–50. https://doi.org/10.1108/09566160110381913

Zheng, F, Zhang, X-C, Wang, J and Flanagan, D C, 2020. Assessing applicability of the WEPP hillslope model to steep landscapes in the northern Loess Plateau of China, *Soil and Tillage Research*, 197:104492. https://doi.org/10.1016/j.still.2019.104492

Transforming mine closure with GIS and climate data for nature-positive outcomes

E Littlewood[1], A Hutton[2] and J O'Kane[3]

1. Principal Consultant – Sustainability Lead, Integrated Environmental Management Australia, Newcastle NSW 2300. Email: erin.littlewood@iema.com.au
2. Managing and Technical Director, Integrated Environmental Management Australia, Newcastle NSW 2300. Email: andrew.hutton@iema.com.au
3. GIS Associate, Integrated Environmental Management Australia, Newcastle NSW 2300. Email: james.okane@iema.com.au

ABSTRACT

The mining sector stands at a pivotal moment for climate- and nature-positive innovation. With Australia's legislative and policy framework evolving to align with global climate and sustainability goals, the industry faces unprecedented pressure to achieve measurable emissions reductions and deliver nature-positive outcomes. Despite this, the integration of Geographic Information Systems (GIS) with climate data to inform mine closure planning remains an underutilised approach.

This paper highlights the innovative potential of EcoScenario, a spatial and climate data science tool, to bridge the gap between climate science and actionable decision-making in mine closure and post-mining land-use planning. Using a hypothetical example, this paper demonstrates how EcoScenario enables mining companies to model site-specific climate risks—such as bushfires, flooding, and extreme weather—using globally recognised scenarios developed by the Intergovernmental Panel on Climate Change (IPCC). Incorporating these scenarios ensures scientifically robust and comparable outputs across sites, companies, and reporting periods, enhancing decision-making credibility while supporting compliance with the Australian Sustainability Reporting Standards.

By combining spatial analysis with climate science, EcoScenario facilitates the creation of resilient rehabilitation plans and optimised land-use strategies. This innovative approach empowers mining companies to make informed decisions, mitigate liabilities, attract investment, and manage long-term climate risks, while contributing to broader environmental and societal goals. EcoScenario positions mining operations as leaders in sustainable resource management and climate adaptation, equipping them to address present and future challenges in a rapidly changing world.

INTRODUCTION

Climate change is introducing new risks and uncertainties into mine closure planning. Traditional mine closure designs, which often assumed historical climate patterns would continue, may be ill-equipped to endure the increasingly extreme weather events and shifting conditions projected for coming decades (Zakharia, 2021). For example, mine sites globally are facing more intense rainfall events that can overwhelm tailings dams or cause erosion of rehabilitated landforms, as well as longer droughts and hotter temperatures that stress revegetation efforts (Zakharia, 2021). Closure strategies, especially for facilities like waste rock dumps and tailings storage, must remain effective for centuries – even millennia – and faster climate change has ramifications for mine remediation and closure strategies that need to last thousands of years (Zakharia, 2021). This makes it imperative to integrate robust climate risk considerations into life-of-mine closure plans.

Geographic Information Systems (GIS) have emerged as powerful tools to tackle these challenges by enabling spatial integration of climate data, environmental data sets, and engineering designs. EcoScenario is a complex proprietary tool which integrates GIS with climate modelling and ecological data. EcoScenario can collate and analyse vast geospatial data sets – from topography and hydrology to land use and biodiversity – and overlay them with the IPCC's Representative Concentration Pathways (RCPs) (eg RCP 4.5, equivalent to 2.5°C to 3°C of global warming, or RCP 8.5, equivalent to over 4°C of global warming) to visualise potential future scenarios that are specific to a site, rather than relying on regional climate projections, for example the East Coast North sub-cluster in New South Wales (NSW) or Monsoonal North East sub-cluster in Queensland (Qld) (Poole, 2023a). In the context of mine closure, as long as site-specific GIS data is available, EcoScenario can be used to map flood-prone zones on specific mine sites under extreme rainfall regimes, identify slopes at risk of erosion, and locate specific areas where rising temperatures may inhibit vegetation growth across the site. This information can be used to inform mine closure planning and execution, as well as supporting corporate reporting under regimes such as the Australian Sustainability Reporting Standard (ASRS). By using EcoScenario to track risk and performance, identify areas of concern, and track management of closure activities (Poole, 2023a), mining professionals can develop more robust closure designs that remain safe, stable and non-polluting under future climate conditions and use data-driven insights to inform disclosures under mandatory reporting.

Another key driver in modern closure planning is the push for nature-positive outcomes. This concept goes beyond minimising environmental harm – it aims to leave ecosystems in a net better state post-closure, enhancing biodiversity and ecological function. The mining industry, led by organisations like the International Council on Mining and Metals (ICMM), is increasingly committed to outcomes such as *no net loss of biodiversity by the time of closure* and eventually net gains in ecosystem value (Harrison, 2024a). For instance, ICMM member companies have pledged to achieve at least no net loss of biodiversity at all mine sites by closure (against a 2020 baseline) and to work towards net positive impact in support of global biodiversity goals (Harrison, 2024a). Achieving these ambitious targets requires careful planning of rehabilitation and offset programs to restore habitats, create wildlife corridors, and improve ecosystem services. Here again, EcoScenario is instrumental – it helps design and evaluate biodiversity mitigation hierarchies (avoid, minimise, rehabilitate, offset) spatially, ensuring that reclaimed land and any off-site offsets together contribute to landscape-level connectivity and species conservation (University of Queensland, 2022). By combining climate resilience with nature-positive principles, mine closure can be transformed from a conventional compliance exercise into an opportunity for environmental regeneration that endures in a changing climate.

This paper explores how using EcoScenario to integrate GIS and climate data into mine closure planning can yield climate- and nature-positive outcomes, using hypothetical case studies in NSW and Qld to illustrate applications. The NSW case study examines a Hunter Valley coalmine incorporating EcoScenario's GIS-based climate modelling for rehabilitation and biodiversity offset strategies. The Queensland case study focuses on a northern Qld metalliferous mine using EcoScenario to mitigate extreme weather impacts through adaptive landform design. Through these examples, we discuss methodologies, potential results, and challenges of climate-driven closure planning. The goal is to provide mining professionals and GIS specialists with a technical framework to future-proof mine closures – protecting both the environment and the long-term stability of closure works – while delivering positive outcomes for nature and supporting data-driven disclosures for mandatory climate reporting.

METHODOLOGY

Climate change risk assessments in mine closure are often conducted using a broad-brush approach, relying on regional-scale climate projections with limited spatial and temporal resolution. These assessments typically draw on publicly available data sets from national science agencies (such as CSIRO and the Bureau of Meteorology in Australia), which provide projections based on IPCC scenarios (eg RCPs). However, key challenges persist: the selected scenarios are not always clearly aligned with IPCC pathways; time horizons may not match the long-term risks relevant to closure infrastructure; spatial resolution is often too coarse to inform site-specific decisions; and the methodologies behind impact assessments are frequently opaque or undocumented. This can make it difficult for mine planners to translate high-level climate projections into practical, defensible closure designs. By contrast, integrating GIS with robust climate modelling tools—such as EcoScenario—enables a more structured and transparent methodology. This approach combines high-resolution data sets (eg soil maps, hydrology, and ecological layers) with site-specific, downscaled climate projections for any IPCC-aligned scenario, over relevant future periods. This allows closure teams to run targeted scenario analyses and generate spatial decision-support outputs that are tailored to the site, greatly improving the accuracy and usefulness of climate risk assessments in closure planning (Trotta and Ridgway, 2022b). Using these inputs, planners perform a climate risk assessment within the GIS environment that is not only specific to a region, but specific to their closure site.

The authors' recommended approach to scenario analysis entails evaluating mine closure outcomes under multiple climate scenarios; for instance, an RCP 2.6 scenario, an RCP 4.5 scenario, and an RCP 8.5 scenario to ensure alignment with ASRS requirements and visibility of the site's resilience under 1.5°C of global warming as well as more extreme scenarios. The EcoScenario tool can be used in this context to overlay ecological and climate scenario data in GIS to predict how rehabilitation efforts will perform over time. Such projections are incorporated into the GIS as data layers (eg raster grids of future annual rainfall or heatwave days) to facilitate detailed spatial analysis of climate impacts on the closure plan. In practice, this involves simulating vegetation growth under projected rainfall and temperature regimes and mapping which areas of a rehabilitated landscape might be prone to wildfire in the future. The climate risk assessment typically follows frameworks such as ISO 31000 (ISO, 2018) or national guidelines for climate adaptation. For instance, a climate risk scan might be done in early closure planning to identify vulnerabilities (extreme rainfall, drought, heat, bushfire) and trigger more detailed analysis where needed (Trotta and Ridgway, 2022a). This was the approach taken at Ballarat Gold Mine in Victoria, where a screening of climate threats (looking at design records, historical weather extremes, and projection data) informed the closure risk register and design decisions (Trotta and Ridgway, 2022a).

A range of GIS-based tools and models support this analysis. For water-related risks, hydrological modelling can be conducted on the mine's digital terrain model to delineate catchments and predict run-off. By inputting increased design storm intensities (as predicted under different climate change scenarios), GIS can map flood extents and drainage paths, highlighting infrastructure or landforms at risk.

For geotechnical stability, landform evolution modelling (LEM) can be integrated with GIS. Tools like CAESAR-Lisflood or SIBERIA use input topography (eg a dump or tailings landform design from GIS) and simulate long-term erosion under sequences of rain events. By feeding in more intense or frequent extreme events (for example, a 1-in-100-year storm that drops 600 mm in a day), these models can predict whether gullies might form over decades to centuries (Lowry et al, 2024). If simulations show that certain slopes would erode severely or that protective covers could be compromised, the design can be iteratively adjusted (slope angles, addition of contour banks, surface armouring etc) and re-tested in the model. In parallel, soil moisture and drought scenarios can be analysed by GIS-based soil-vegetation-atmosphere models to ensure that the planned cover system (growth medium and vegetation) will survive in different future climate conditions. This is not to recommend introducing non-local species—which could disrupt local biodiversity and ecosystem dynamics—but rather to inform the selection of native species likely to be most resilient under future site conditions. The aim is to enhance the site's resilience to a warming climate while maintaining its core biodiversity values and supporting the continuity of natural ecosystem processes.

From an ecological perspective, the EcoScenario tool contributes to biodiversity-focused planning by informing species selection and habitat restoration strategies. Planners begin by mapping existing habitats and species on and around the mine site, then apply climate niche models to forecast changes in species' suitable ranges under future climate scenarios. Where target species for rehabilitation are projected to lose climate suitability by 2070, alternative species or more drought-tolerant provenances can be considered. However, priority is given to species native to the local ecosystem to preserve biodiversity values and maintain ecological function. Habitat connectivity modelling is also performed: the post-closure landscape (including rehabilitated areas and any offset reserves) is evaluated for how well it connects to surrounding habitat networks. Techniques like least-cost path or circuit theory in GIS help identify corridors through or around the mine that are important for wildlife movement. This is critical for designing biodiversity offsets that truly add ecological value. If an offset is required (eg to compensate for an endangered ecosystem impacted by mining), GIS analysis can locate and prioritise offset sites that enhance existing conservation areas or bridge gaps between habitat 'islands' (University of Queensland, 2024). Prioritising the creation of nature corridors is increasingly important for biodiversity conservation, as global warming continues and species move pole-wards and upwards in search of cooler climates. In practice, multi-criteria GIS models could rank potential offset lands based on factors like climate resilience (areas less likely to be adversely affected by climate change), proximity to the mine (for practical management), and contribution to regional conservation plans.

Throughout this methodology, the planning process is iterative. Results from GIS and climate simulations feed back into mine closure plan updates. For example, if flood mapping reveals a certain haul road will be inundated under future rainfall extremes, the closure design might be revised to reroute or raise that road, and then the model is run again. This iteration continues until the closure plan meets safety and environmental criteria under the range of modelled scenarios. Stakeholder consultation (eg community, regulators, Traditional Owners) is integrated by using GIS visualisations (maps, 3D perspectives) to communicate how the closure will perform and to demonstrate consideration of both climate adaptation and environmental enhancement. By the end of the process, the mine closure plan is not a static document but a scenario-tested strategy: it has been 'stress-tested' against droughts, floods, storms, and heatwaves in silico (Zakharia, 2021), according to robust and reputable IPCC RCP scenarios, and incorporates features to ensure long-term stability and nature-positive outcomes even as conditions change.

Figure 1 illustrates a sample output from the EcoScenario climate risk modelling tool, which integrates terrain analysis, hydrological modelling, and future climate projections to assess flood hazard exposure across mine landscapes. The modelled scenario represents an extreme 1-in-500-year rainfall event, simulating potential impacts under intensifying climatic conditions projected for the region. The figure depicts the mine site boundary (black outline) in relation to areas of differentiated flood risk. Red zones indicate areas subject to high flood hazard, driven by topographic depressions, flow accumulation zones, and existing drainage pathways. In contrast, green areas denote low-risk zones where surface water flow is minimal or naturally diverted. The utility of this tool lies in its ability to spatially visualise how future extreme rainfall events may interact with post-mining landforms and infrastructure. By identifying zones of potential inundation, the model supports the strategic design of protective features—such as levees, armoured spillways, or diversion drains—and informs the placement of sensitive infrastructure away from high-risk areas.

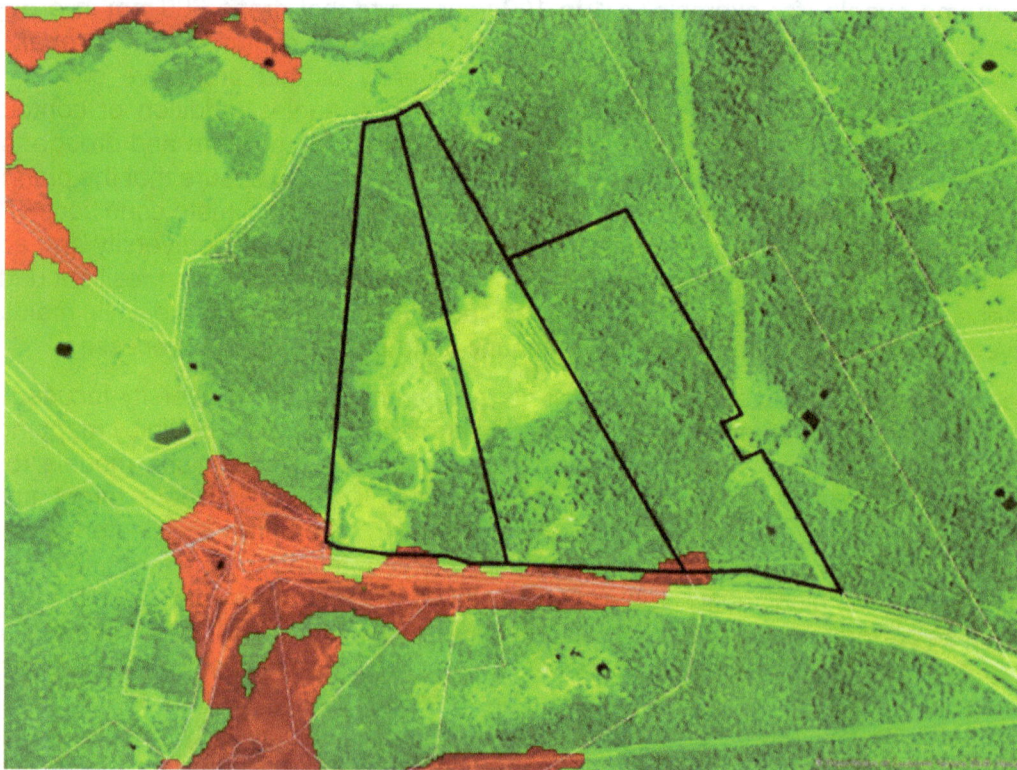

FIG 1 – Example output from a EcoScenario climate risk tool. Model shows terrain and drainage factors leading to high flood hazard under extreme rainfall in a 1-in-500-year flood event. Red areas denote flood affected areas around a mine site (black outline). Green areas are low-risk zones. Such maps enable engineers to identify where protective measures or design changes are needed to adapt the mine closure landform to future flood events.

This approach exemplifies how GIS-based climate risk assessments can be used proactively during mine closure planning to enhance long-term landform stability and resilience. As climate extremes become more frequent and severe, such predictive modelling is essential to reduce residual risk, protect post-closure assets, and meet evolving regulatory and community expectations for sustainable mine relinquishment.

Hypothetical case studies

NSW case study – climate-adaptive rehabilitation in the Hunter Valley

Site context

The hypothetical site is an open cut coalmine in the Upper Hunter Valley of New South Wales, an area already experiencing a warming trend and shifting rainfall patterns. According to the NSW Department of Climate Change, Energy, the Environment and Water (DCCEEW, 2024), by 2090 under a high-emission scenario, the Hunter region is projected to have nearly four times as many hot days above 35°C compared to the late 20th century. Meanwhile, average winter rainfall may decrease by up to 20 per cent or more, even as heavy downpours and bushfire weather become more frequent (DCCEEW, 2024). These projections pose risks to mine rehabilitation – extreme heat and periodic drought could impair revegetation success, and intense rainfall events could trigger erosion or flooding on the reshaped landforms. The mine lease is adjacent to remnant woodlands of high biodiversity value, including habitat for several threatened species. As part of closure, the company has committed to a biodiversity offset program aiming for a net gain in habitat extent for those species.

GIS-based climate modelling application

The mine's closure team uses the EcoScenario tool to integrate climate projections into the rehabilitation plan. First, they divide the large disturbance footprint (which includes backfilled pits and recontoured waste dumps) into soil-landscape units and model future soil moisture regimes. GIS maps of projected rainfall and evaporation for 2050 and 2070 are overlayed on these units to identify where dryness could limit plant growth. The modelling shows, for example, that north-facing slopes on the waste dump could become significantly drier – by 2050, soil moisture during summer is expected to drop below thresholds for the current grass and woodland mix. In response, the rehabilitation prescription in GIS is adjusted: drought-tolerant native species (such as silver-top stringybark and burrowa bush) replace some of the original species, and additional water-harvesting features (swales and berms) are added to those slopes to retain wet-season run-off for use in dry periods. The GIS is used to design and place these features precisely, tying into the natural drainage lines. Risk of bushfire to the site is also mapped using data from IPCC's RCP scenarios – areas projected to face more high fire-danger days are planned to be rehabilitated with native vegetation communities that exhibit greater fire resilience and include strategic fire breaks (eg low-fuel zones or access tracks delineated in GIS).

Next, the closure planners evaluate flood and erosion risk under future extreme rainfall using high-resolution elevation models. Figure 2 represents the closure planners' simulation of a 1-in-100-year storm in 2070, which might be, for example, 30 per cent heavier than historical events. This simulation identifies that this tailings dam can successfully withstand the extreme rain event. Had the tailings dam breached or collapsed under the simulated storm event, the closure planners could consider options for mitigating the risk, such as redesigning the spillway channel through an iterative GIS process or altering slope profiles to a more concave, natural landform shape and increasing rock armouring in vulnerable swales. This example demonstrates how adaptive landform design can be guided by combining GIS and climate data outputs.

FIG 2 – Example output from an EcoScenario climate risk tool. Model represents a 1-in-100-year rain event over 24 hrs (Bureau of Meteorology, 2025). This output demonstrates that this tailings dam can withstand the extreme rain event successfully.

Biodiversity rehabilitation and offsets

The Hunter Valley mine's closure goal is to restore a native woodland ecosystem over the mine site, contributing to a larger habitat network. Before finalising species and planting zones, the team conducts a habitat suitability analysis under climate change, considering RCP 2.6, RCP 4.5 and RCP 8.5 warming scenarios. For a focal threatened species (eg the Regent Honeyeater bird or a rare tree species) that the rehabilitation aims to support, they use species distribution models in GIS to map its potential range in the 2040s and 2070s. Unfortunately, the model indicates that climate suitability for this species in the lower Hunter Valley may contract under warmer scenarios, shifting toward cooler uplands to the north. Acknowledging this, the rehabilitation still includes the species in parts of the site (especially riparian zones that may retain moisture), but the biodiversity offset strategy takes on greater importance to ensure long-term species conservation. Using EcoScenario, the team identifies an offset site in the nearby ranges (~15 km from the mine) that is projected to remain suitable habitat for the species under RCP 4.5 and RCP 8.5 warming scenarios. The chosen offset is a 570 ha area of mostly intact forest that lies between two protected areas. By securing and managing this property, the offset will establish connectivity between a State Conservation Area and a Nature Reserve, as was similarly achieved in another NSW offset plan (GM3 Consulting, 2024). In this case study, this means the offset creates a crucial corridor in the Hunter region, facilitating species movement in response to climate shifts.

The GIS mapping in Figure 3 shows that with the offset in place, there will be an interconnected habitat network spanning from the mine rehabilitation area to high-quality forests in the ranges. The planners quantify anticipated connectivity gains: analysis indicates a ~20 per cent increase in connected core habitat for local fauna when comparing the scenario of full native ecosystem restoration versus a baseline of pre-mining condition. Indeed, a recent study in Queensland's Fitzroy Basin found that restoring mines to native ecosystems (versus merely rehabilitating to pasture) yielded significant connectivity improvements across the landscape (Hernandez-Santin, Gagen and Erskine, 2024). Applying that insight, the Hunter Valley mine chooses restoration as the post-mining land use, even though pasture might have been cheaper – the GIS-aided analysis demonstrates the nature-positive benefit of this decision in terms of biodiversity corridors and resilience. Furthermore, by integrating climate data, the restoration is tailored so that species assemblages planted on-site are those likely to thrive in future conditions, thereby increasing the long-term success of the offset and rehabilitated areas in providing habitat.

FIG 3 – Map of chosen offset area being an area of ~570 ha between two reserves connecting a State Conservation Area and a Nature Reserve.

Throughout this case study, stakeholder and regulator engagement is maintained with the help of GIS visualisations. Regulators are shown maps of the predicted 2050 landscape: a reclaimed mosaic of woodlands and grasslands, with overlay layers illustrating climate-proofing measures (like water catchments) and offset links. This not only satisfies regulatory requirements for climate adaptation in the closure plan (which some jurisdictions, including NSW, now expect under updated mine closure guidelines), but also builds community confidence. Local communities and Traditional Owners can see how the plan mitigates future flood or fire impacts on neighbouring lands and how cultural sites are protected through considered rehabilitation planning. In sum, the Hunter Valley case demonstrates the use of GIS and climate modelling to guide rehabilitation design and offset selection – ensuring the closed mine land evolves into a climate-resilient ecosystem that delivers net positive outcomes for nature.

Queensland case study – extreme weather resilient design in North Qld

Site context

The hypothetical Queensland site is a large metalliferous (eg copper-gold) mine in the seasonally wet-dry tropics of northern Qld. The region endures a monsoonal climate – very intense rainfall and cyclones in the wet season, contrasted with prolonged dry seasons. Climate change projections for the site suggest an increase in the intensity of extreme rain events (even if total rainfall stays similar or slightly decreases) and more erratic cyclone behaviour (fewer cyclones overall but a higher chance of very severe Category 4–5 storms making landfall). The mine's closure scenario includes a large waste rock dump re-shaped into a final landform, a tailings storage facility (TSF) that will be capped, and a pit lake. A key concern is landform stability: the dump and TSF must withstand severe rainfalls without excessive erosion or failure, and the pit lake's spillway must safely pass extreme floods. Additionally, there is a need for the site to not become a source of pollution during extreme weather – for instance, avoiding tailings liquefaction or uncontrolled discharge in floods.

GIS for extreme weather event mitigation

From the outset, the closure design is developed using a GIS-based landform design platform. Designers employ digital terrain modelling to create what is known as a 'landform that mimics natural analogues' – gentle slopes, sinuous drainage lines, and retention basins – aimed at sustainable erosion control. To refine this design, they integrate climate data: specifically, rainfall intensity-duration-frequency (IDF) curves adjusted for a future climate scenario with, for example, 10 per cent heavier extreme precipitation. Using GIS, they derive revised design storm run-off values for each catchment on the landform. The largest catchment is on the leeward side of the waste rock dump; modelling shows a probable maximum precipitation event could generate peak flows exceeding the current channel capacities. Therefore, adaptive engineering is applied: wider channels and armoured spillways are incorporated into the dump's design at strategic points (selected by tracing flow accumulation paths in the GIS model).

Figure 4 presents the results of a hydrological simulation conducted on the hypothetical Queensland mine site, focusing specifically on the response of tailings storage infrastructure to a high-intensity rainfall event. The modelled scenario applied an 800 mm rainfall event—consistent with cyclone-scale precipitation referenced in comparable studies (eg Lowry *et al,* 2024)—across the site's surface infrastructure. The simulation demonstrated that the tailings dams remained structurally resilient, with no evidence of overtopping or structural compromise under the simulated conditions. This suggests that the current design specifications of these containment structures provide a satisfactory level of risk mitigation against extreme weather events of this magnitude.

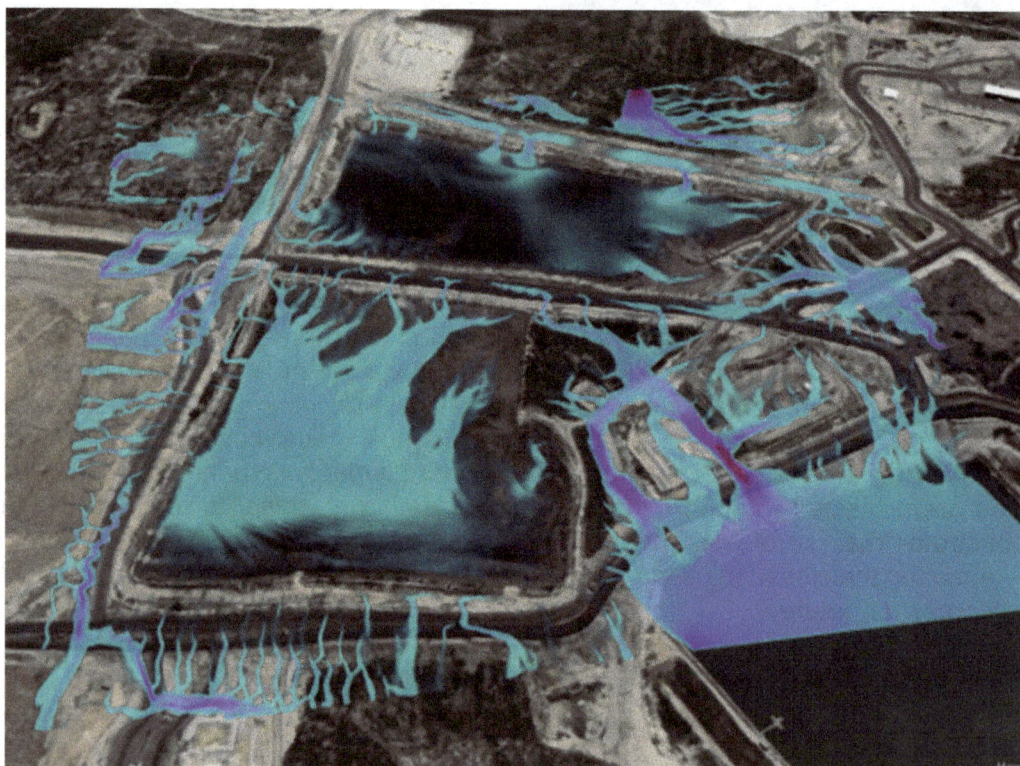

FIG 4 – Simulated overland flow following an 800 mm cyclone event at a Queensland mine site: tailings dams resilient, but high flow velocity observed through heavy vehicle area.

However, the modelling also revealed a significant concentration of overland flow-through a heavy vehicle parking and manoeuvring area. Flow velocities in this area were high enough to present a potential safety hazard, including the possible mobilisation of heavy equipment towards adjacent water storage infrastructure, specifically the site's clear water dam. This outcome underscores the value of scenario-based modelling in mine closure and risk management planning, as it can identify critical vulnerabilities in site layout or stormwater pathways that may not be evident under normal operating conditions.

These findings support the broader argument for incorporating GIS-based hydrological simulations into mine closure planning using the IPCC's RCP scenarios to assess the performance of legacy and operational infrastructure under future climate scenarios. Such tools enable proactive identification and remediation of high-risk zones, contributing to both operational safety and long-term environmental stewardship. Using the IPCC's RCP scenarios also supports broader understanding of the scenarios considered, and the opportunity for management, investors, and other stakeholders to compare the resilience of different sites and prioritise and allocate capital for closure planning accordingly.

Ecological and land use considerations

Although this Qld case focuses on physical stability, nature-positive opportunities are not neglected. The closure plan includes establishing a series of connected wetlands at the toe of the waste rock landform to act as both sediment traps and new aquatic habitats. GIS mapping of pre-mining wetlands and streams guides where to locate these features so that post-closure water flows reconnect with natural watercourses. During cyclonic rains, these wetlands will slow and store run-off (reducing flood peaks), and in normal times they provide breeding grounds for birds and amphibians. This approach aligns with ecosystem-based adaptation – using nature (wetlands, in this case) to buffer climate extremes while also creating ecological value. The mine's surroundings in north Qld include savanna woodland used for cattle grazing. After closure, about half of the disturbed area will be returned to grazing, using drought-tolerant pasture species, and the remainder rehabilitated to native savanna. A GIS-based land capability analysis, incorporating future rainfall and evaporation projections, informs the allocation of land uses. Moist, lower slopes are designated for grazing, while steeper upper slopes—prone to intense run-off—are revegetated with dense native species to enhance erosion control, restore structural complexity, and support ecosystem processes. This climate-adapted zoning approach aims to balance future productivity with the preservation of ecosystem dynamics.

The extreme weather focus of this case also drives innovation in monitoring. The plan includes a network of environmental sensors (rain gauges, soil moisture probes etc) across the site, all georeferenced in GIS. This effectively creates a 'digital twin' of the closed mine where incoming data can be visualised spatially. If an anomalous extreme event occurs (for example, a rainstorm larger than the scenarios used in design), the GIS will help technicians pinpoint any areas of damage or erosion (for example, via drone imagery analysis) for repair. Such adaptive management is crucial in the decades after closure, as climate patterns continue to evolve.

Through the Queensland case, we see EcoScenario enabling a proactive design that is specific to the probability of extreme events occurring on that particular site, rather than relying on more general predictions from climate data associated with a larger climate zone, such as the Monsoonal North East sub-cluster of Australia. The outputs – flood risk maps, erosion simulations, and land-use suitability maps – lead to concrete design measures like armoured channels, regraded slopes, and constructed wetlands. These interventions are all mapped and documented, creating an evidence-based mine closure plan. The final landform is not only physically robust under future extreme climate stresses on the site, but it also contributes positively to the environment (stable landforms, functional ecosystems and water features) and neighbouring communities (reducing downstream flood risk, providing grazing and habitat). It exemplifies how, with the aid of EcoScenario – GIS and climate data – mines can be closed in a way that is future-proofed, leaving a resilient landscape aligned with engineering standards, nature-positive aspirations, and stakeholder expectations.

RESULTS AND DISCUSSION

Both hypothetical case studies demonstrate tangible benefits from integrating GIS-based climate modelling into mine closure planning. Key findings from these simulations and analyses include:

Identification of vulnerabilities

By overlaying climate projections specific to the mine sites, the GIS analyses pinpointed specific vulnerabilities that traditional closure plans might overlook. In the NSW case, spatial modelling highlighted that certain rehabilitated slopes would likely become too arid or fire-prone for the planned vegetation mix by mid-century, prompting a change in species selection that would be more resilient in, but still appropriate for, the local ecosystem. In Queensland, flood routing maps clearly showed where extreme rainfall would concentrate flows, revealing weak points in the draft landform design. Early identification of these problem areas allowed timely design modifications – avoiding potential failures such as washouts, gullying, or die-off of vegetation. These findings underscore that spatial scenario analysis combined with site-specific climate data adds a layer of rigour to risk management, minimising residual risk and increasing confidence that closure criteria (stability, no pollution, self-sustaining ecosystems) will be met under future conditions (Poole, 2023a).

Improved design interventions

The case studies translated modelling results into practical design interventions which were iteratively tested in GIS. For example, the Hunter Valley flood risk maps reinforced confidence in the tailings dam design, while the extreme weather event predictions in Queensland drove changes in site layout. The data-driven insights ensured that these engineering measures were precisely targeted – neither over-designed (which can waste resources) nor under-designed (which leaves residual risk). The result is more resilient closure landforms that are tailored to endure the changes in climate predicted to occur at the site under specific RCP scenarios. Notably, by simulating events up to worst-case scenarios (eg RCP 8.5), the designs have built-in buffers. In essence, the process 'climate-proofed' the mine closures. This aligns with industry observations that mines must be stress-tested against drought and flood scenarios to adapt properly (Zakharia, 2021). A collateral benefit is that many of these adaptations (like gentler slopes, larger water features) also yield environmental co-benefits; for instance, gentler slopes are easier to revegetate and larger spillways can double as ephemeral streams or wetlands.

Nature-positive outcomes quantified

EcoScenario in the NSW case allowed the team to quantify how a nature-focused rehabilitation scenario compares to a conventional one. The connectivity analysis showed measurable increases in habitat linkage by opting for native ecosystem restoration and strategic offset placement. According to the simulation, such a strategy could significantly improve species dispersal potential across the landscape, as evidenced by analogous real-world findings of connectivity gains with restorative rehabilitation (Hernandez-Santin, Gagen and Erskine, 2024). Furthermore, biodiversity habitat models indicated that, by including climate-hardy species, the rehabilitated ecosystem is more likely to persist and continue developing positively, rather than suffering setbacks such as vegetation collapse in a hotter, drier 2060. These anticipated nature-positive benefits support the business case for investing in higher-quality rehabilitation and robust offsets to lower residual risk, meet nature-positive aspirations, and ensure successful relinquishment. They also give substance to corporate sustainability goals; the plan can demonstrate, via GIS metrics, that the site will achieve a net gain in certain ecosystem services or habitat indices over time. This is crucial for meeting commitments such as ICMM's 'no net loss by closure' target and for responding to emerging frameworks like the Taskforce on Nature-related Financial Disclosures (TNFD) (Harrison, 2024b), which will expect data-backed reporting on biodiversity outcomes.

Adaptive management and monitoring

One interesting outcome is how these approaches set the stage for adaptive management. The models produce expectations (eg vegetation cover will reach X per cent by year ten if rainfall follows scenario Y). These become hypotheses to be monitored. With EcoScenario and remote sensing, the actual performance (via satellite-derived vegetation indices, drone surveys of landform integrity etc) can be compared against the modelled expectations. Large deviations might indicate either an unforeseen climate stress or an over- or underperformance of the design, triggering management actions or even model recalibration. The case studies presume such a feedback loop. Thus, a climate-informed closure plan is inherently a living plan; something regulators and miners may revisit periodically as new climate data emerges. This is a cultural shift from the traditional approach to closure. It does, however, raise questions about long-term custodianship: Who will be responsible for adaptive interventions decades post-closure if conditions diverge from projections? Regulatory frameworks may need to adapt to enable or enforce this adaptive capacity; perhaps through requirement of a bond or a post-closure management fund that can finance adaptive measures if needed.

Despite these encouraging findings, several challenges were noted in the climate-driven closure planning process:

Data uncertainty

Climate projections themselves carry uncertainty, especially regarding the timing and magnitude of extreme events. Planning for a range of scenarios is necessary, but it does not guarantee that reality will mirror any single one. This means designs might still be surprised by an outlier event. The case studies tried to mitigate this by using climate data specific to the site's latitude and longitude in pessimistic scenarios (like RCP 8.5 or worst-case historical analogues) to err on the side of caution. Nonetheless, uncertainty remains a factor, possibly requiring conservative design factors or more flexible closure strategies (eg designing landforms that can be adaptively modified if future monitoring indicates the need).

Technical complexity and skills

EcoScenario is a technically complex tool, requiring the integration of GIS, climate data, and ecological data. It demands a multidisciplinary team with skills in climate science, hydrology, geotechnical engineering, GIS analysis, and ecology. Many mine sites or local regulators may lack some of this expertise. There is a learning curve to interpret climate model outputs and to run specialised GIS tools. As noted by industry experts, a general scarcity of expertise in measuring and managing biodiversity and climate impacts can hinder setting and achieving realistic goals (Wood, 2024). Upskilling staff or engaging specialists (eg sustainability experts, GIS modellers) will be essential, which has cost and coordination implications.

Computational and data challenges

Working with high-resolution spatial data and running hundred-year simulations can be computationally intensive. In practice, some sites might simplify analyses (eg using coarser resolution or shorter time frames) to reduce run times, which could sacrifice detail. Data management is also non-trivial – keeping track of multiple climate data sets, versions of terrain models, and model outputs requires good data governance. The case studies assumed these challenges were managed by using cloud-based GIS processing for heavy tasks and by implementing strict data version control for the closure models.

Regulatory and closure plan integration

Another challenge is fitting these advanced analyses into existing regulatory requirements. Many regulators are only beginning to ask for climate change considerations in mine closure plans. In some jurisdictions, there might not be a clear mandate or guidance on how to include such information. Proponents might be wary that adding scenario analyses could complicate approval if, say, it reveals worst-case outcomes that regulators then expect to be fully mitigated. There is a delicate balance in communicating that a design is robust yet not overcommitting to handle literally every conceivable scenario (which could be prohibitively costly). The case studies showed that transparency about assumptions and including safety margins can help – for instance, explaining that 'even under a 20 per cent increased rainfall scenario, the closure criteria are met' provides assurance.

Evolving guidance, for example—Western Australia's updated Mine Closure Plan guidelines, CSIRO guidance, and the national Leading Practice handbooks (Department of Industry, Innovation and Science, 2016; Australian Government, 2016; CSIRO, 2024a, 2024b) are increasingly referencing climate adaptation, which will hopefully standardise expectations and encourage all operations to follow suit.

Long-term financing and responsibility

Designing for extreme scenarios often means higher upfront costs – moving more soil to flatten slopes, establishing intensive rehabilitation, etc. The pay-off is reduced long-term risk, but those benefits are sometimes outside the typical accounting frame of a mine (especially if the company aims to relinquish the site shortly after closure). Justifying these investments requires forward-thinking corporate governance and possibly external pressure (investors or insurers concerned about liabilities). Additionally, truly achieving nature-positive outcomes often means going above minimum compliance (eg creating additional habitat beyond what is required). It can be challenging to secure budget for these voluntary enhancements unless they are tied to offsets or license conditions. New mechanisms like biodiversity credits or carbon sequestration credits (for restoring forests) might emerge to co-fund nature-positive closure actions.

In discussion, it is clear that climate-adaptive, GIS-informed closure planning represents a paradigm shift in how we think about the end of mine life. Historically, mine closure was often about meeting compliance and immediate rehabilitation requirements. Now, the case studies illustrate a future where closure is viewed through the lens of resilience and stewardship: we design a landscape that should thrive and remain safe, stable and non-polluting in an environment our great-grandchildren will live in, not just based on past environments our grandparents knew. This approach yields more sustainable outcomes – socially (protecting communities from residual risks like dam failures or dust storms), environmentally (improving ecosystems), and financially (reducing the risk of costly remedial work later). It aligns mining with the broader global efforts on climate adaptation and biodiversity conservation, turning a legacy liability into a potential community and environmental asset.

CONCLUSION

The incorporation of GIS and climate data tools – such as EcoScenario – into mine closure planning is becoming not just best practice, but a necessary evolution for responsible mining in the 21st century. As demonstrated, a climate-informed approach can significantly bolster the durability of closure designs and help deliver nature-positive outcomes that extend well beyond a mine's operational life. To successfully integrate these practices, the following best practices and recommendations are offered:

Plan early and iteratively

Climate and environmental scenario analysis should be embedded in the early stages of closure planning – ideally from pre-feasibility through to final closure plan revisions. Early identification of climate risks allows mines to adjust mine plans (eg location/geometry of waste dumps, progressive rehabilitation scheduling) long before closure, when changes are cheaper and easier. The closure plan should remain a living document, updated as new climate projections or monitoring data become available. This aligns with the concept of Progressive Rehabilitation and Closure (PRC) planning now required in jurisdictions like Queensland (Queensland Government, 2023; Marshall et al, 2024), ensuring that climate adaptation measures can be incorporated progressively during operations.

Leverage GIS for decision support

GIS should be treated as a core tool for closure teams, not an afterthought. Mining companies are encouraged to invest in building strong GIS capabilities or partnerships, so that they can fully utilise functionalities such as high-resolution terrain analysis, habitat modelling, and spatial risk mapping. The full extent of site data (historical maps, monitoring data, remote sensing) should be fed into a centralised GIS database (Hernandez-Santin, Gagen and Erskine, 2024). This ensures institutional knowledge is captured spatially – an advantage when staff turn-over or redundancies occur or consultants change. The result is more technically defensible closure plans, with GIS maps and figures that clearly communicate the design intent and risk mitigation to stakeholders.

Adopt a multi-scenario climate modelling strategy

Use the IPCC's RCP scenarios to promote transparency and allow for comparative analysis between sites and use multiple climate scenarios – such as RCP 2.6, RCP 4.5 and RCP 8.5 – to cover best-to worst-case scenarios in analyses to bracket the range of possible futures (Trotta and Ridgway, 2022a). This will help in designing robust measures that are effective across scenarios and in identifying trigger points where different actions might be needed whilst also supporting the company to comply with mandatory climate reporting requirements. For instance, a closure plan could include contingency 'Trigger Action Response Plans' (TARPs) – if monitoring shows climate trends moving towards an extreme scenario, then certain additional interventions would be implemented (such as augmenting water storage, extra planting of drought-hardy native species etc). Regulators should be amenable to such adaptive pathways, as they demonstrate proactive planning without unnecessarily over-engineering for highly uncertain far-future events.

Prioritise nature-based solutions

Wherever feasible, incorporate nature-based solutions in closure design as they often provide resilience more cost-effectively and align with nature-positive goals. Examples include using vegetation buffers to reduce erosion and filter water, creating wetlands to manage floodwaters and water quality, or establishing conservation land uses (like a wildlife reserve) that can adapt naturally over time. These approaches tend to be self-sustaining and can improve over time (a forest can grow denser, a wetland can expand), in contrast to engineered structures that may degrade. Moreover, they contribute to corporate Environmental, Social, and Governance (ESG) objectives. Companies can highlight these nature-based closure elements in sustainability reports as contributions to climate adaptation and biodiversity targets, strengthening their social license to operate.

Strengthen regulatory frameworks

It is recommended that regulators and industry bodies (like AusIMM or ICMM) update closure planning guidelines to explicitly include climate change and biodiversity outcome considerations. The ICMM Integrated Mine Closure Good Practice Guide (2019) already emphasises that closure plans should be integrated with overall life-of-mine planning (Poole, 2023b); building on this, future revisions could provide specific guidance on climate risk assessment and nature-positive indicators for closure in view of climate reporting standards such as the ASRS. Government regulators might also consider incentives for companies that exceed baseline requirements (for example, reduced financial assurance if a company can demonstrate via GIS modelling that their closure design is low-risk under extreme climate scenarios). Additionally, guidelines for long-term monitoring and maintenance in the context of climate change would be helpful – eg recommending climate resilience audits at regular intervals post-closure.

Monitor and share outcomes

As the mining industry gains experience with climate-adapted, GIS-driven closure projects, it is crucial to monitor their performance and share lessons learned. Key performance metrics might include rates of erosion observed versus predicted, vegetation establishment success versus predicted under climate stresses, or biodiversity return (species counts) over time. Where outcomes are positive, they validate the modelling approach and can justify its wider use. Where discrepancies occur, they provide valuable data to refine models (for instance, how a habitat model handles sudden ecosystem shifts). An open database of case studies, perhaps through forums like the AusIMM Life-of-Mine Conference, can accelerate learning across the sector. Mining companies, consulting firms, and research institutions (eg CSIRO, universities) should collaborate, as was done in projects like the Fitzroy Basin rehabilitation connectivity study (Centre for Mined Land Rehabilitation, 2023; Hernandez-Santin, Gagen and Erskine, 2024), to advance the state of practice.

In conclusion, transforming mine closure in the face of climate change is both a challenge and an opportunity. By harnessing GIS and robust climate and ecological data in tools like EcoScenario, mine closures can be designed to not only withstand the test of time but to leave a positive legacy – landscapes that are safe, stable, and biologically rich. The two hypothetical case studies illustrate that with thoughtful application of such tools, nature-positive mine closure is an attainable goal, even under the pressures of a changing climate. The mining industry stands at the cusp of this transformation. Implementing the recommendations above will require commitment and possibly additional upfront investment, but the returns come in the form of reduced long-term liability, improved stakeholder trust, and the knowledge that our generation of miners has acted as a responsible custodian of the environment. In essence, embracing GIS and climate data in closure planning allows the industry to 'future-proof' its rehabilitation efforts and ensure that closed mine sites can become resilient new ecosystems that benefit nature and society for generations to come.

REFERENCES

Australian Government, 2016. Leading Practice Sustainable Development Program for the Mining Industry [online]. Available from: <https://www.industry.gov.au/sites/default/files/2019-05/lpsdp-mine-closure-handbook-english.pdf> [Accessed: 16 Jun 2025].

Bureau of Meteorology, 2025. Design Rainfall Data System (2016) [online]. Available from: <http://www.bom.gov.au/water/designRainfalls/revised-rfd> [Accessed: 31 Mar 2025].

Centre for Mined Land Rehabilitation, 2023. Mapping biodiversity corridors and mine rehabilitation opportunities [online]. Available from: <https://smi.uq.edu.au/cmlr-research/ecosystem-assessment-restoration-resilience/project/mapping-biodiversity-corridors-mine-rehabilitation-opportunities> [Accessed: 24 Mar 2025].

CSIRO, 2024a. Climate adaptation in mining [online]. Available from: <https://www.csiro.au/en/work-with-us/industries/mining-resources/Social-and-environmental-performance/Climate-adaptation> [Accessed: 20 Mar 2025].

CSIRO, 2024b. COBRA: Coal burst risk assessment and CRATER: Climate adaptation in mining [online]. Available from: <https://research.csiro.au/msci/projects-mining-coal-burst-risk-assessment-cobra-climate-adaptation-mining-crater> [Accessed: 20 Mar 2025].

Department of Industry, Innovation and Science, 2016. Mine Closure: Leading Practice Sustainable Development Program for the Mining Industry [online]. Available from: <https://www.industry.gov.au/sites/default/files/2019-05/lpsdp-mine-closure-handbook-english.pdf> [Accessed: 24 Mar 2025].

GM3 Consulting, 2024. Dendrobium Mine Strategic Biodiversity Offset [online]. Available from: <https://gm-3.com.au/wp-content/uploads/2024/06/dendrobium-mine-strategic-biodiversity-offset.pdf> [Accessed: 20 Mar 2025].

Harrison, C, 2024a. Mining: nature's most unexpected guardian [online]. Available from: <https://theintelligentminer.com/2024/07/31/mining-natures-most-unexpected-guardian/> [Accessed: 20 Mar 2025].

Harrison, J, 2024b. Understanding nature-positive frameworks: Implications for mining and biodiversity reporting, *Environmental Sustainability Journal*, 12(1):34–47.

Hernandez-Santin, L, Gagen, E J and Erskine, P D, 2024. Setting restorative goals with a regional outlook: Mine-rehabilitation outcomes influence landscape connectivity, *Journal of Environmental Management*, 357:120778.

International Council on Mining and Metals (ICMM), 2019. Integrated Mine Closure Good Practice Guide, 3rd edition, International Council on Mining and Metals.

International Organisation for Standardisation (ISO), 2018. ISO 31000:2018, Risk management – Guidelines.

Lowry, J, Saynor, M, Hancock, G R and Coulthard, T, 2024. Using a landform evolution model to model the effect of extreme rainfall events on the geomorphic stability of a rehabilitated landform, in *Proceedings of the 18th International Conference on Mine Closure* (eds: A B Fourie, M Tibbett and G Boggs), pp 59–70 (Australian Centre for Geomechanics: Perth).

Marshall, D J, Preece, N D, Setterfield, S A, Speldewinde, P C, Wardle, G M and Andersen, A N, 2024. Setting restorative goals with a social license to operate: How expert knowledge can be used as a tool for mine site rehabilitation, *Journal of Environmental Management*, 350:120094.

NSW Department of Climate Change, Energy, the Environment and Water (DCCEEW), 2024. Hunter region climate change projections [online]. Available from: <https://www.climatechange.environment.nsw.gov.au/my-region/hunter> [Accessed: 20 Mar 2025].

Poole, A A, 2023a. Good GIS functionality and practices in mine closure planning, in *Proceedings of the 16th International Conference on Mine Closure* (eds: B Abbasi, J Parshley, A B Fourie and M Tibbett), pp 1–14 (Australian Centre for Geomechanics: Perth).

Poole, B, 2023b. Integrated mine closure: An evolving approach to climate resilience, paper presented to 16th International Conference on Mine Closure (Australian Centre for Geomechanics: Perth).

Queensland Government, 2023. Progressive Rehabilitation and Closure Planning Framework [online]. Available from: <https://environment.des.qld.gov.au/__data/assets/pdf_file/0035/87659/mined-land-rehabilitation-policy.pdf> [Accessed: 20 Mar 2025].

Trotta, L and Ridgway, J, 2022a. Climate scenario planning for mining: A practical guide, *Resources Policy*, 78:102827.

Trotta, L M and Ridgway, T H, 2022b. Embedding climate change risk into mine closure planning: A case study of tailings closure design at Ballarat Gold Mine in *Proceedings of the 15th International Conference on Mine Closure* (eds: M Tibbett, A B Fourie and G Boggs), pp 259–266 (Australian Centre for Geomechanics: Perth).

University of Queensland, 2022. Spatial analysis for mine closure planning, Sustainable Minerals Institute, UQ Brisbane.

University of Queensland, 2024. Mapping biodiversity corridors for mine rehabilitation opportunities [online]. Available from: <https://smi.uq.edu.au/cmlr-research/ecosystem-assessment-restoration-resilience/project/mapping-biodiversity-corridors-mine-rehabilitation-opportunities> [Accessed: 20 Mar 2025].

Wood, C, 2024. Mining: nature's most unexpected guardian [online]. Available from: <https://theintelligentminer.com/2024/07/31/mining-natures-most-unexpected-guardian/> [Accessed: 24 Mar 2025].

Zakharia, N, 2021. Reducing mining risk through climate adaptation [online]. Available from: <https://www.australianresourcesandinvestment.com.au/2021/08/26/reducing-mining-risk-through-climate-adaptation/> [Accessed: 20 Mar 2025].

Integrating circular economy principles into mine closure planning

E Littlewood[1], A Hutton[2], A Jiang[3] and W Rifkin[4]

1. Principal Consultant – Sustainability Lead, Integrated Environmental Management Australia, Newcastle NSW 2300. Email: erin.littlewood@iema.com.au
2. Managing and Technical Director, Integrated Environmental Management Australia, Newcastle NSW 2300. Email: andrew.hutton@iema.com.au
3. Chief Executive Officer, Go Circular, Newcastle NSW 2300. Email: anniej@gocircular.org.au
4. Emeritus Professor, University of Newcastle, Newcastle NSW 2300. Email: will.rifkin@newcastle.edu.au

ABSTRACT

Traditionally focused on compliance and rehabilitation, mine closure strategies can be enhanced by embedding circular economy (CE) concepts and systems to deliver long-term environmental, social, and economic benefits. The integration of CE principles into mine closure planning provides the mining sector with a transformative opportunity to align with global sustainability standards, such as the Global Reporting Initiative (GRI) and the Task Force on Climate-related Financial Disclosures (TCFD).

This paper explores how innovative CE business models and systems thinking – from material reuse to landscape repurposing to regenerating nature – can reshape mine closure planning where practical and legal parameters allow. Reimagining waste streams as resources can – with appropriate research, planning and investment – provide mining companies with opportunities to reduce closure costs, generate new economic and employment opportunities, enhance sustainability and secure the challenging 'social licence to close' in resource regions. For example, repurposing mine tailings for construction or soil rehabilitation can support local economies while promoting biodiversity restoration, aligning with GRI's environmental impact and resource efficiency objectives.

The potential of CE integration is illustrated through examples from construction, manufacturing, and agriculture. These sectors demonstrate the effectiveness of circular principles in driving resource efficiency and waste reduction. Lessons from the reuse of by-products in construction can inform approaches to mine waste reprocessing, while regenerative agriculture offers guidance on restoring post-mining landscapes to improve soil health and biodiversity.

Case studies from mining regions – from Lusatia to the Hunter Valley – highlight the emerging potential of CE approaches to foster sustainable post-mining livelihoods, reduce environmental risks, and achieve community benefit-sharing outcomes, whilst the primary objectives of safe stable and non-polluting landforms are prioritised. Collaborative efforts – which are integral to many CE strategies – can involve mining companies, regulators, local communities and other partners in designing closure plans that transform environmental, economic and social licence liabilities into assets. By considering CE principles and learning from other industries, the mining sector can create resilient, regenerative post-mining landscapes that deliver lasting benefits for people, ecosystems, and the industry.

INTRODUCTION

Mine closure is among the most pressing challenges facing the global mining sector. Once extraction ceases, companies must manage long-term impacts on land, water, and communities. Historically, many mines were abandoned or minimally rehabilitated, causing environmental damage and socio-economic decline (World Economic Forum, 2019; TJ Ryan Foundation, 2023). In Queensland, for instance, there are around 120 complex abandoned mine sites—most predating 2000, when legislative reforms strengthened environmental and rehabilitation standards (Queensland Government, 2025). These legacy sites reflect closure liabilities that have fallen to governments and communities (McCabe and Morris, 2016), with issues such as acid mine drainage, safety hazards, and lost post-mining value highlighting the need for more innovative closure strategies. Without meaningful change, the industry risks greater regulatory pressure and financial burdens driven by public and community expectations.

Integrating CE principles into closure planning offers a potential paradigm shift. The International Organisation for Standardisation (ISO) defines a CE as 'an economic system that uses a systemic approach to maintain a circular flow of resources, by recovering, retaining or adding to their value, while contributing to sustainable development' (ISO, 2024). In practice, this involves eliminating waste, recirculating resources, and regenerating natural systems (Acaroglu, 2017). When applied to closure, CE encourages rethinking waste, land, and infrastructure as assets for reuse or repurposing—contrasting with the traditional 'take-make-dispose' model. This shift can reduce environmental liabilities and unlock long-term social and economic opportunities (Whitbread-Abrutat and Lowe, 2024).

Today's closure challenges span environmental rehabilitation, socio-economic transition, compliance, and stakeholder engagement. In Australia, mining operations must prepare Mine Closure Plans and provide financial assurances. Globally, however, regulatory consistency is lacking—nearly 25 per cent of jurisdictions do not require closure plans at all (Centre for Strategic and International Studies, 2022). Even where planning is mandated, the quality varies. Best practice promotes integrated closure from the outset of mine design (Global Reporting Initiative, 2025), but many plans still focus narrowly on returning land to pre-mining conditions, missing opportunities to leave enduring community value. The emerging concept of a 'social licence to close' reflects heightened expectations that companies contribute positively beyond compliance.

CE principles align with these expectations, framing closure as a transition to new value cycles. Infrastructure, land, and materials can be repurposed to support new industries or community uses—creating regenerative, resilient outcomes. This enables mine closure to serve as a catalyst for ecological recovery and economic diversification.

However, CE solutions may not be feasible or permissible at all sites, due to site-specific constraints and regulatory frameworks. Further analysis is needed to determine where CE approaches are technically, economically, and politically viable. This paper advocates for a circular lens in closure planning, while emphasising the importance of research, long-term investment, and, where necessary, legislative reform to enable tailored implementation across varied contexts.

METHODOLOGY

This paper employs a qualitative, case study-based methodology grounded in a review of industry data, policy frameworks, and sustainability literature. The aim is to analyse how CE principles can be concretely integrated into mine closure planning and to evaluate outcomes from real-world examples. The methodological approach consists of several key components.

Literature review and framework analysis

An extensive review of academic literature, industry reports, and guidance documents informed the evaluation framework. Key references included the GRI Mining Sector Standards, TCFD recommendations, Australia's Mine Closure Handbook, and ICMM's Integrated Mine Closure guide—all of which emphasise early, iterative, and stakeholder-informed planning. Policy documents from Australian jurisdictions and the EU were analysed to understand regulatory support for circular approaches, such as land use planning and incentives for progressive rehabilitation. These insights shaped the criteria used to assess case studies, focusing on early planning, stakeholder engagement, and material reuse.

Data collection – industry statistics and trends

A data-driven approach was used to quantify the scale and opportunity of circular mine closure. This included statistics on mine closures, land affected, and rehabilitation liabilities, eg 17 coalmines across 130 000 hectares in NSW's Hunter Region are expected to close within two decades (Hunter Joint Organisation, 2024). Economic data on circular initiatives, such as job creation and regional diversification, were sourced from reports by governments, consultancies, and NGOs. EY (2022), for instance, modelled economic benefits of repurposing mine lands for renewables in the Hunter. Similar data from Lusatia highlighted job transitions post-closure. Sustainability reports from relevant mining companies were also reviewed.

Case study selection

Two contrasting case studies were selected: Hunter Valley (Australia) and Lusatia (Germany). The Hunter illustrates closure within a regulatory and resource-dependent context, while Lusatia demonstrates state-supported landscape regeneration. Each was assessed using historical context, circular initiatives, and measurable outcomes, drawing on policy documents, academic literature, media, and site-level data. In identifying lessons from the Lusatia region for the Australian context, the authors acknowledge key differences between the two settings that may influence the transferability of CE approaches. These differences include contrasts in geology and mine characteristics—such as Lusatia's deep black coal pits with saline voids and extensive unconsolidated spoil, versus Australia's predominantly shallow soft coal operations, which often involve large-scale extraction and site-specific geotechnical closure requirements. Lusatia also benefits from various governmental incentives and structural support programs aimed at fostering economic diversification and sustainable development in the post-mining era, particularly for regions of the former East Germany. These differences should be noted in interpreting case study outcomes, particularly their potential for application across different coal mining regions.

Cross-sector comparative analysis

To validate circular strategies, comparisons were drawn from other sectors applying CE principles—construction, manufacturing, and agriculture. Examples include the reuse of industrial by-products in cement, product-life extension in manufacturing, and regenerative agriculture on former mine lands. These analogies, while not formal case studies, provide real-world validation of circular strategies proposed for the mining sector.

DATA – STATISTICS, REGULATIONS, AND INDUSTRY TRENDS

Mine closure is an escalating global challenge. The ICMM projects hundreds of large mine closures in the next decade, driven by resource depletion, market transitions, and coal phase-outs. Closure liabilities are substantial, with major firms facing obligations of $1.5–4.5 billion each (van Zyl et al, 2016). Traditional rehabilitation may no longer suffice, prompting interest in circular economy (CE) strategies to manage risks and reduce long-term liabilities. Australia's approximately 60 000 abandoned mine sites exemplify legacy risks (McCabe and Morris, 2016), while the rarity of successful relinquishment underscores the need for productive post-mining land uses. CE approaches that promote reuse and regeneration can improve closure outcomes and community acceptance.

Regulatory frameworks are evolving. In Australia, closure plans are now required at approval and must be regularly updated. Financial provisioning schemes, such as Queensland's pooled fund (holding $397.9 million as of June 2024), aim to mitigate default risks (Queensland Treasury, 2024). However, regulators acknowledge that relinquishment hinges not just on physical outcomes, but also socio-economic viability. National guidelines (Australian Government, 2016) and international standards (ICMM, 2019) emphasise early consultation and pre-agreed land uses, favouring innovative repurposing over default options like grazing.

Globally, gaps remain—nearly a quarter of jurisdictions lack mandatory closure plans or financial sureties (IGF, 2020). Investor expectations are helping bridge this, with institutions like the World Bank and IMF linking funding to robust closure practices. Voluntary frameworks such as the GRI and TCFD are also driving transparency around closure plans, provisions, and long-term risks (Lamm and Johnson, 2022; Global Reporting Initiative, 2025).

Interest in post-mining land use is rising. In the Hunter Valley, reports estimated that repurposing mine lands for renewables, tourism, and conservation could generate $3.7 billion in output and over 13 000 jobs—on par with the current mining workforce (EY, 2022; Murphy and Bernasconi, 2022). Germany's coal exit strategy similarly supports regional regeneration, as seen in the Lusatian lake district (World Economic Forum, 2019). In New South Wales, programs such as the Hunter Expert Panel and Hunter Future Industries Fund show alignment with CE principles (EY, 2022).

Technology is also enabling circular approaches. Sweden's LKAB is recovering critical minerals from tailings (Vitti and Arnold, 2022), while Australian projects are converting legacy sites into sources of construction sand. Former quarries now host solar farms, and underground mines are being adapted for energy storage (Northern Australia Infrastructure Facility (NAIF), 2023). These examples signal a shift toward industrial symbiosis and circular closure, demonstrating how rising expectations are being met with practical, scalable solutions (Hunt *et al,* 2019).

CASE STUDIES

This section focuses on two case studies that embody many of these data points in practice and illustrate the tangible outcomes of applying circular principles to mine closure.

Case study 1 – Hunter Valley, Australia

Circular economy initiatives in closure planning

Over the past decade, the Hunter region has emerged as a leader in applying CE principles to post-mining land use and regional transition planning. The Hunter Joint Organisation (2025) has advocated for regional planning pathways that repurpose former mine sites for new industries aligned with net-zero and CE goals. In response, the NSW Government funded feasibility studies for alternative land uses, supporting a shift from default rehabilitation (eg grassland) toward productive reuse—such as renewable energy, advanced manufacturing, recycling, and high-value agriculture.

Repurposing mine sites for renewable energy is already underway. Former open cut mines offer ideal conditions for solar farms due to flat terrain and existing grid connections. The Maxwell Solar Farm, developed by Malabar Resources, exemplifies the transformation of former mining sites into renewable energy zones. Situated on rehabilitated land from the Drayton open cut coalmine in New South Wales, this 25 MW solar farm is projected to generate over 60 GWh annually, supplying power to approximately 10 000 homes (Malabar Resources, 2024). Another innovative project, the Muswellbrook Pumped Hydro Project, is converting the decommissioned Muswellbrook Coal Mine pit in the Hunter Valley into a pumped hydro energy storage facility (Parkinson, 2024). Similar to Queensland's Kidston Pumped Hydro model (Genex Power/ARENA, 2024), it utilises mine voids for energy storage (ARENA, 2021). This project is a collaboration between AGL Energy and Idemitsu Australia, aiming to provide 400 MW of energy generation with eight hrs of storage capacity (Parkinson, 2024). Sites like these demonstrate how mining infrastructure can support the clean energy transition through circular reuse (Genex Power/ARENA, 2024).

Key data points

The economic and social data emerging from the Hunter Valley's CE initiatives are compelling. As noted above, EY (2022) estimated over 13 000 jobs and A$3.7 billion added to the regional economy from reusing mine lands in the Upper Hunter for sustainable industries. This figure includes jobs in renewables – solar and wind farm construction and operation – land rehabilitation works, ecosystem services, and associated manufacturing and services. For context, approximately 14 000 people currently work directly in Hunter Valley coalmines; so these reuse scenarios have the potential to offset a significant portion of employment losses due to mine closures (Wilkinson, 2023). Another data point is land availability: the analysis found that even dedicating a relatively small fraction of the total mined land to industrial reuse (with the rest undergoing ecological rehabilitation or returning to agriculture) could yield those economic benefits (EY, 2022). This result suggests a high leverage effect; a few thousand hectares used effectively for CE projects can drive large economic outcomes.

From an environmental perspective, ongoing monitoring in the Hunter Valley has started to capture the benefits of circular approaches. For example, trials of using compost and sewage biosolids on mine rehabilitation areas (a circular practice linking urban waste management to mine rehabilitation) have markedly improved soil quality and vegetation growth on mine dumps (Interstate Technology and Regulatory Council (ITRC), 2025). Data from one trial at a coalmine in the Upper Hunter Valley, New South Wales showed that treatments with recycled organics significantly improved tree growth and survival rates compared to traditional methods which used only imported topsoil and fertiliser (NSW Environment Protection Authority, 2008). These results suggest that 'wastes' from cities – green waste and biosolids – can be looped into mine site restoration to accelerate ecosystem recovery; a clear circular opportunity that turns one waste into a resource for another sector (ITRC, 2025). Of course, these results are context-specific; use of organic materials from urban centres may not be feasible for all mining projects, particularly those in remote locations where transport costs and logistical challenges may prevent the use of these materials at scale.

Environmental and social outcomes

The Hunter Valley's early adoption of circular closure strategies is delivering tangible, if early-stage, results across several domains.

Economic diversification

Projects such as the proposed Hunter 'Hydrogen Valley' and new recycled product manufacturing facilities are drawing investor interest and government funding (Parkinson, 2021). These initiatives, often located on or near former mine sites, are diversifying the local economy ahead of mine closures (McGhee, 2024). For example, a disused mine site is being transformed into a motorsport and tourism facility, forecast to deliver $360 million in economic impact, 450 construction jobs, and 230 ongoing roles (NSW Government, 2024).

Community engagement and social licence

Companies such as Glencore and BHP have engaged local communities through government-facilitated forums to discuss post-mining land use (Upper Hunter Mining Dialogue, 2024). Suggestions have included repurposing pits as recreational lakes or industrial heritage parks, drawing inspiration from German models. While pit lakes are not universally viable, in some contexts they offer dual benefits: recreation and flood mitigation. Local advocacy groups are also pushing for stronger circular closure strategies that support environmental regeneration and long-term community resilience (Hunter Renewal, 2023). Community input is increasingly tied to social licence, with support for mine extensions often dependent on credible closure and repurposing plans (BHP, 2024; Glencore, 2025).

Environmental remediation

Circular practices are also improving remediation outcomes. At one site, coal reject piles are being reprocessed to recover fuel for cement production and the remaining material repurposed as construction aggregate (Hunter Valley Operations, 2024). This reduces waste volumes, stabilises landforms, and supports emissions reduction at the cement plant, illustrating industrial symbiosis.

Policy and governance innovations

NSW's draft policy on 'Mine Site Rehabilitation and Futures' embeds CE principles by encouraging post-mining land uses aligned with regional development goals (NSW Government, 2023). It proposes a register of rehabilitated sites for potential reuse, institutionalising circular closure as part of planning and investment frameworks.

In summary, the Hunter Valley demonstrates how CE-aligned mine closure can support economic transition, environmental outcomes, and stronger community engagement. While challenges such as site suitability and funding remain, the region is laying a foundation for closure approaches that offer long-term value beyond compliance.

Case study 2 – Lusatia, Germany

Circular economy applications in post-coal landscapes

Germany's post-mining restoration in Lusatia stands as one of the world's largest land transformation efforts, closely aligned with CE principles (Deshaies, 2020). Rather than backfilling vast pits, planners repurposed them into lakes, forests, and recreational spaces—establishing over 20 artificial lakes interconnected by canals, forming Europe's largest man-made lake district (Connolly, 2014; World Economic Forum, 2019). This strategy added new ecological and economic value through tourism, fisheries, and flood mitigation (Krawczyk and Bartosik, 2020), representing a circular reuse of post-mining landscapes.

Central to this effort was the concept of 'landscape recycling' (Yellishetty et al, 2023). Overburden was reshaped into hills, terraces, and lake shorelines, reusing site materials to stabilise terrain and encapsulate contaminants, while creating a safe and visually appealing environment (World Economic Forum, 2019). In some areas, new vineyards were planted on former spoil piles, diversifying land use and introducing wine tourism.

Lusatia's restoration also included large-scale reforestation, with over 500 million trees and shrubs planted to re-green spoil heaps and regenerate ecosystems (World Economic Forum, 2019). Some forests were designed for future economic use, aligning environmental restoration with long-term productivity. Industrial heritage was preserved as part of the regional transition. Facilities like Brikettfabrik Knappenrode were repurposed into museums and event venues (Energiefabrik Knappenrode, 2020), while decommissioned mining machinery became tourist attractions at sites like Zwölfer Schacht (Internationale Bauausstellungen, 2025). Renewable energy projects, including wind turbines on elevated mine dumps and solar farms on flat terrains, further demonstrate circular reuse of land and energy infrastructure (GP Joule, 2025; Deshaies, 2020).

This circular, integrated, approach demonstrates how large-scale mine closure can create enduring social, ecological, and economic value through circular land use planning.

Key data points

The transformation of Lusatia's former coalmines into a lake district has involved over $17 billion in rehabilitation and water management since 1990 (World Economic Forum, 2019). By the late 2010s, 80 per cent of pit flooding was complete, creating around 14 000 hectares of lakes and an equivalent area of new forests and green spaces. The region attracted approximately one million annual tourists pre-COVID, with Lake Senftenberg alone drawing 200 000 visitors (Connolly, 2014). Socio-economically, Lusatia saw a rise in jobs in tourism, services, and renewables, helping slow youth out-migration (Deshaies, 2020). However, while thousands of tourism-related jobs were created, they did not completely replace the tens of thousands lost in mining—often offering lower pay and seasonal work. Local sentiment reflected pride in environmental recovery but concern over long-term employment (Deshaies, 2020). This outcome underscores that a circular regenerative approach greatly improved environmental and social quality of life, but economic transition needed multiple pathways, which Germany has since been addressing via new industrial investments and government offices relocated to the region (Deshaies, 2020).

Environmental and social outcomes

Lusatia's transformation of coal pits into lakes addressed major environmental challenges, including acidic groundwater (Connolly, 2014). Careful flooding and treatment improved water quality, enabling recreation and biodiversity recovery, with thriving fish populations and returning wetlands (Müller and Morton, 2018). Air and water quality have drastically improved since the 1990s, and former industrial sites now support tourism, fishing, and conservation. Socially, the shift from a desolate mining identity to the 'Lusatian Lakeland' brought renewed community pride, new jobs in tourism and services, and collaborative governance through the Lusatian Lake District Authority to manage ongoing maintenance and water control (Müller and Morton, 2018).

This case offers key lessons in circular mine closure. It shows how viewing closed mines as assets can support regional regeneration, though long-term management (eg water treatment) remains essential. Challenges like acid mine drainage and the limits of seasonal tourism highlight the need for both technical solutions and economic diversification. Lusatia exemplifies circular principles—regenerating nature, keeping assets in use, and designing for future needs—and now serves as a model for similar post-coal transitions in Germany, Poland, and the Czech Republic, supported by EU Just Transition funding.

CROSS-SECTOR COMPARATIVE ANALYSIS

Implementing CE principles in mine closure can draw on successful strategies from other industries. The examples below illustrate how construction, manufacturing, and agriculture have embraced circular approaches that minimise waste, repurpose land, and regenerate ecosystems, offering analogies for post-mining planning.

Material reuse in construction and manufacturing

Other sectors have found innovative ways to treat waste as a resource, extending the life of materials and reducing the need for virgin extraction.

Industrial by-products in construction

The construction industry routinely substitutes conventional materials with industrial by-products. For example, coal fly ash (from powerplants) and blast-furnace slag (from steelmaking) are used to replace a significant portion of cement in concrete (Global Cement and Concrete Association, 2025). Typical replacement rates of 30–50 per cent are common, improving concrete properties while diverting these by-products from landfills. This practice not only reduces the carbon-intensive content in cement but also gives a second life to what would otherwise be waste. Such material reuse shows how mine tailings or waste rock might similarly be processed as construction inputs in a circular model, such as for road base or use as aggregate.

Metal recycling and closed-loop manufacturing

Manufacturers have long recirculated metals and materials. Steelmakers and aluminium producers already rely heavily on recycled scrap. For example, approximately 45 per cent of Germany's steel production feedstock comes from scrap, with approximately 57 per cent for aluminium and 41 per cent for copper (Schmid et al, 2022). These high recycling rates illustrate a circular flow where end-of-life metal from vehicles, buildings, and products is melted down and reused, reducing demand for new mining of ore.

Additionally, companies are extending product life cycles through remanufacturing programs. A notable example is Caterpillar's remanufacturing of heavy equipment components, which recirculates over 68 000 t of iron annually by restoring end-of-life parts to like-new condition (Caterpillar Inc, 2025). This 'reuse/remake' model saves material and energy. The strategy could be analogous to mining firms reprocessing and reusing equipment, infrastructure, or even mineral wastes instead of discarding them.

Land repurposing for productive new uses

Industrial land, once its initial use ends, need not remain derelict. Examples from around the world show how formerly contaminated or developed sites can be repurposed into valuable assets, aligning with CE goals by turning liabilities into opportunities.

Renewable energy

Large-scale initiatives are converting closed industrial sites and brownfields (land previously used for industry or waste disposal) into renewable energy farms. In the United States, for example, over 500 renewable energy projects (solar and wind installations) have already been sited on brownfields, including former mines and landfills (The Nature Conservancy, 2024). These 'brightfields' projects produce clean power on land unsuitable for agriculture or development, effectively transforming idle, polluted lands into green energy assets. Policies encourage this shift by offering incentives for projects in 'energy communities' (often former coal mining areas), illustrating how a mine's footprint can host new industries like solar farms after closure (The Nature Conservancy, 2024).

Adaptive reuse of industrial sites

Many regions have creatively redeveloped old factories, mills, and mine sites for new uses. In Germany's Ruhr Valley – a historic coal and steel region – hundreds of hectares of industrial land have been rehabilitated into parks, housing, and commercial centres. For example, the Duisburg-Nord Landschaftspark preserves an old ironworks facility, now repurposed as a public park and adventure space with gardens, dive pools and climbing walls built into the former plant (Varnon, 2014). Across the Ruhr, former sites of industry have been converted into apartments, shopping centres, offices and other commercial spaces, helping to revitalise the local economy (Varnon, 2014). These developments showcase how a mined land or processing plant site, once closed, could be integrated back into community life – whether as business hubs, recreational parks, or other productive assets – rather than remaining an abandoned wasteland.

Regenerative land use in agriculture

Agriculture offers models for how degraded lands can be actively restored and put back into productive use, aligning with circular principles of regenerating natural systems. Particularly on marginal or previously industrial lands, regenerative practices heal ecosystems while providing economic benefits. In the UK's National Forest, over 9.5 million trees have been planted on former coalmines and quarries, raising forest cover from only 6 per cent in the early 1990s to approximately 25 per cent today (The Telegraph, 2024). This collective effort has transformed a once-barren post-industrial area into a green, productive landscape. New woodlands now support biodiversity, recreation, and a local timber economy, demonstrating how land regeneration can create value. Similarly, regenerative agriculture focuses on restoring soil fertility and ecosystem functions on degraded land, ensuring that land can sustain livelihoods long after its original use.

Simple, low-cost practices can yield dramatic restoration. In Niger, West Africa, subsistence farmers embraced Farmer Managed Natural Regeneration (FMNR), protecting and pruning sprouts from tree stumps and seeds in their fields. Over decades, this approach has 're-greened' about 6 million hectares of degraded semi-arid land (Forest Landscape Restoration, 2020). Tree densities increased tenfold – from 4 to 40 trees/ha – which improved soil structure, reduced erosion, and enhanced moisture retention (Forest Landscape Restoration, 2020). The return of tree cover – after decades of decimation due to increased grazing and collection of firewood – has boosted crop yields and even local water tables. Impressively, communities saw up to 500 000 tonnes per annum (tpa) more cereal production and as much as $1000 extra income per household due to the restored productivity (Forest Landscape Restoration, 2020). What was once impoverished, exhausted land is now fertile again; a living example of how ecological regeneration directly supports economic resilience.

For mine closure, this result suggests that even heavily degraded or cleared lands can be nurtured back to health (eg through planting deep-rooted vegetation, agroforestry, or soil amendments), yielding benefits for local communities, such as agriculture, forestry or grazing opportunities on rehabilitated mine lands. It also indicates that circular strategies are not just for wealthy Western economies; that lower income, agriculturally focused communities in developing economies can benefit.

Each of these cross-sector examples underscores a common theme: waste streams and worn-out lands are opportunities rather than end points. The construction and manufacturing sectors demonstrate that materials like ash, slag, scrap, and old equipment contain value that can be cycled back into the economy rather than disposed. Land reuse examples show that with planning and innovation, a closed industrial site can host new activities – from clean energy generation to community recreation – instead of existing as liabilities. Regenerative agriculture demonstrates the potential to heal environmental damage on degraded soils, restoring natural capital and productivity for future generations. Applying these lessons in a mining context – reprocessing mine waste, repurposing mine lands, and restoring ecosystems – can reshape mine closure from a cost and liability into an opportunity for the mining company and for nearby communities. By looking to these successful implementations of CE strategies, the mining industry can design closure plans that not only remediate, but revitalise, delivering lasting environmental, social, and economic value.

CONCLUSIONS

The integration of CE principles into mine closure planning is emerging as a compelling strategy to address the multi-dimensional challenges of shutting down mining operations. This paper offers a new perspective on mine closure by applying a CE lens to reframe closure not merely as an end-of-life obligation but as an opportunity for regeneration, value creation, and long-term community benefit. While previous literature has explored individual elements of CE or closure, this paper synthesises these concepts to showcase how CE principles could be integrated into closure planning and sets out a series of conclusions and policy recommendations for consideration in strategic planning..

Transformative potential

CE approaches can transform mine closure from a liability-driven process toward being an opportunity-driven process. In both the Hunter Valley and Lusatia, data demonstrates how reimagining 'waste' land and materials as valuable resources can lead to innovative uses – renewable energy hubs, recreational lake districts, and new industries – that deliver lasting benefits. These outcomes stand in stark contrast to traditional closure, which has aimed to contain or cover up the remnants of mining. By designing closure projects that create new economic, social, and environmental value, mining companies can significantly enhance outcomes for communities and shareholders alike. Circular closure planning fundamentally reframes the traditional approach by shifting the focus from the disposal or stabilisation of residual mining elements to the strategic repurposing and regeneration of assets to deliver ongoing environmental, social, and economic value.

Environmental and social opportunities

A circular approach can align mine closure with broader sustainability and community objectives. It inherently supports environmental restoration; by prioritising reuse and regeneration, less waste is left to potentially pollute, and more land is restored to ecological function (Yellishetty *et al*, 2023; Zhang, Wang and Unluer, 2021). In the Hunter Valley, using tailings for construction or compost for rehabilitation improves land and water outcomes; in Lusatia, creating lakes and forests remedied a century of degradation. Socially, these approaches can maintain or create employment and help communities develop a positive post-mining identity (eg the Hunter Valley as a clean energy powerhouse, Lusatia as a lake tourism region).

Importantly, involving communities in envisioning and planning these new uses builds trust and helps to cultivate social licence. Communities are more likely to support mining and mine closures when they see a credible plan for a healthy post-mining use of the land that benefits them. The concept of a 'social licence to close' is achieved when the local stakeholders agree that the closure plan is acceptable and even desirable – something far more attainable under a circular model that emphasises community gains than under a minimalist compliance model.

Economic resilience

Circular mine closure contributes to economic resilience in mining regions. By diversifying the economy and creating new revenue streams (whether through resource recovery, renewable energy generation, or tourism and agriculture on rehabilitated land), regions are less vulnerable to the boom-bust cycle of commodity markets. The data showed that thousands of jobs can be generated through such diversification (EY, 2022). While not a panacea – major transitions still require extensive economic development efforts – circular initiatives are critical building blocks in the broader transition strategy. They can also reduce the financial burden on governments. If companies and new investors take on projects to repurpose mine infrastructure, there is less risk of sites becoming derelict and requiring public funds for clean-up or social support for displaced workers.

Alignment with policy and standards

Integrating CE principles into closure aligns mining practices with evolving regulatory trends and sustainability standards. As discussed, global frameworks like GRI now explicitly query companies on closure planning and outcomes (Global Reporting Initiative, 2025), and climate-focused frameworks like TCFD implicitly encourage forward-looking adaptive reuse of assets (TCFD, 2017). By adopting circular closure strategies, companies can demonstrate compliance with, and even leadership in, these frameworks. They can report reductions in waste, contributions to renewable energy, and positive community impacts as part of their ESG metrics.

Policymakers, for their part, are increasingly recognising and supporting such approaches – whether through funding transition authorities, providing grants for innovative rehabilitation projects, or updating mining laws to require consideration of post-mining economies. The case studies reflected this shift: the Hunter region benefited from State Government initiatives encouraging renewable projects on mine land, and Lusatia's transformation was underpinned by government investment and policy in the form of Germany's *Act to Reduce and End Coal-Fired Power Generation*

('Kohleausstiegsgesetz', commonly known as Germany's coal phase-out law). The authors anticipate that future policies will further incentivise circular approaches, for example by offering faster relinquishment or bond releases if a company delivers certain beneficial reuses, or by integrating mine closure plans with regional land-use plans, breaking the silo between mine regulators and urban planners.

Policy recommendations

Building on these insights, several policy recommendations can be made to foster the integration of CE principles into mine closure on a wider scale.

Mandate early post-mining land-use planning

Governments should require that mine closure plans (submitted at project approval and updated periodically) include an assessment of potential post-mining land uses and resource reuse opportunities. Regulators could stipulate that at least one alternative, value-adding land use scenario be evaluated (such as renewable energy, commercial, or conservation use) in addition to a 'rehabilitate to original condition' scenario. This strategy would push mining companies to engage in circular thinking from the outset and during periodic plan updates. Western Australia, NSW and Queensland are moving in this direction by asking for future land use objectives in closure plans; this sort of regulation should be standard practice globally.

Incentivise circular closure projects

Policy instruments like tax credits, co-funding grants, or expedited permitting can incentivise companies to pursue circular closure projects. For example, a company that commits to install a solar farm on a portion of a mined land or to donate/sell land for public use could receive tax relief or access to government transition funds. Likewise, if a company develops a way to use its mine waste in local industry (reducing the long-term monitoring needs), regulators might allow a reduction in the required financial assurance (since the risk is mitigated). Such incentives align financial interest with circular outcomes.

Facilitate multi-stakeholder partnerships

The best CE projects at closure often involve partnerships – between mining firms, renewable energy companies, local governments, community groups, and sometimes academia (for technical or policy innovation). Policymakers and industry bodies should facilitate platforms for these stakeholders to collaborate. Establish 'Post-Mining Transition Taskforces' in regions with clusters of mine closures, comprising all relevant parties, to jointly plan and attract investment for reuse projects. This collaborative approach was employed in Lusatia through formal commissions, and it can be replicated. Additionally, government agencies – or government supporting establishment of a new class of intermediaries – can help to broker deals where, for example, a mining company might not want to become a renewable energy operator but can prepare the land and hand it over to an energy developer, streamlining the transaction to the benefit of all stakeholders.

Enhance data sharing and guidelines

As more pilot projects and case studies emerge, it is critical to share knowledge of what works and what does not. Industry associations (like ICMM) and research collaborations (like Australia's Cooperative Research Centre for Transformations in Mining Economies (CRC TiME)) should document and disseminate guidelines on circular closure. This effort can encompass technical guidelines (eg how to treat tailings for safe reuse in construction, how to design pit lakes for recreation and ecology) and social guidelines (eg how to conduct community visioning exercises for post-mining land use). Having clear guidance can reduce uncertainty and support wider uptake of these approaches, especially among mid-tier or smaller mining companies that may lack in-house expertise.

Monitoring, metrics, and accountability

Regulators should incorporate specific metrics related to CE in closure into their monitoring frameworks. For example, track the percentage of mine materials reused versus landfilled, the area of land put into productive post-mining use, or the number of community facilities established on former mine sites. By measuring and publicising these outcomes, authorities can hold companies accountable and also celebrate successes, which in turn builds momentum. Companies could be required to report these metrics annually as part of their rehabilitation progress; some already voluntarily do so in sustainability reports. Ultimately, a mine should be judged not just on how well it minimises negative impacts at closure, but on how much positive value it leaves; a mindset shift that metrics can help to enforce.

Implications for the mining industry

For mining companies, embracing CE principles in closure planning has strategic implications. It requires a more integrated, forward-thinking approach to mine life planning; breaking down the silos between the production phase and the closure phase. Progressive companies are already moving toward a 'life cycle' business model where the closure scenario influences decisions during operations (for example, how to segregate waste materials to facilitate later reuse, or investing in community development projects that will outlast the mine). This approach can become a source of competitive advantage. Companies known for leaving behind thriving post-mining sites will enhance their brand in the eyes of investors, reduce legacy liabilities to government, and potentially face lower regulatory hurdles in new projects (due to earned trust).

Furthermore, by aligning closure with a CE, mining companies position themselves as partners in sustainable development, rather than just extractors of resources. This shift in image is increasingly important as society's expectations of corporate responsibility grow. Investors, too, are paying attention: with the rise of ESG investing, how a company handles closure and reclamation is seen as indicative of its overall risk management and stewardship. Demonstrating CE projects (like a successful conversion of a mine to a solar farm or a community agro-forestry project on reclaimed land) can be a powerful narrative in sustainability reporting and investor communications.

The industry must also consider capacity and skill development. Implementing circular closure solutions often means bringing in new expertise (renewable energy specialists, ecologists, circular design experts) that traditionally might not have been part of a mining company's staff. The new expertise brings opportunities for interdisciplinary collaboration and possibly new business units within mining firms that focus on closure project development. Some companies might even find new revenue in the closure phase. That can occur, for example, by processing and selling recovered materials from waste or generating power to sell to the grid from a post-closure renewable project. Mining companies that prefer to focus on their core competencies, though, can partner with consultancies or other organisations to develop these transition enterprises.

Finally, the shift to circular closure underscores the idea that the winding down of a mine can be the beginning of something new. The mining industry, which for centuries was accustomed to a pattern of extraction and abandonment, is now at the cusp of a new era where each mine closure could give rise to a renewable energy facility, a park, a research centre, or a resource recovery plant. This repurposing not only mitigates negative impacts – and extended liability for the mine lease holder or government – but can contribute solutions to other global challenges (like climate change, by enabling more renewables, or resource scarcity, by providing secondary materials). In essence, it is a move toward a more CE at large, in which mining is not an isolated linear process but part of a regenerative system.

The case studies of the Hunter Valley and Lusatia show that while challenges exist, the vision of circular mine closure is actionable and can be beneficial. CE approaches have the potential to transform mine closure practices by creating more sustainable and value-driven outcomes. While their implementation may not be universally applicable due to varying site conditions and regulatory requirements, this diversity underscores the need for targeted, context-specific solutions. With appropriate research into what approaches can work in which settings, strategic investment, and, where necessary, supportive policy reform, CE principles can play a meaningful role in shaping the future of mine rehabilitation.

ACKNOWLEDGEMENTS

The authors would like to acknowledge their respective organisations IEMA, Go Circular, and the University of Newcastle for their support and permission to publish this paper.

REFERENCES

Acaroglu, L, 2017. Tools for Systems Thinkers: Designing Circular Systems [online]. Available from: <https://medium.com/disruptive-design/tools-for-systems-thinkers-designing-circular-systems-2c54cbf9cb43> [Accessed: 3 Apr 2025].

Australian Government, 2016. Mine Closure: Leading Practice Sustainable Development Program for the Mining Industry [online]. Available from: <https://www.industry.gov.au/publications/leading-practice-handbooks-sustainable-mining/mine-closure> [Accessed: 4 Apr 2025].

Australian Renewable Energy Agency (ARENA), 2021. Queensland gold mine lives on as pumped hydro plant [online]. Available from: <https://arena.gov.au/blog/queensland-gold-mine-lives-on-as-pumped-hydro-plant/> [Accessed: 3 Apr 2025].

BHP, 2024. Understanding community perspectives on mine closure – Factsheet [online]. Available from: <https://www.bhp.com/-/media/project/bhp1ip/bhp-com-en/documents/about/our-business/minerals-australia/241209_understandingcommuityperspectivesonmineclosure_factsheet2024.pdf> [Accessed: 3 Apr 2025].

Caterpillar Inc, 2025. The benefits of remanufacturing [online]. Available from: <https://www.caterpillar.com/en/company/sustainability/remanufacturing/benefits.html> [Accessed: 7 Apr 2025].

Centre for Strategic and International Studies (CSIS), 2022. ESG Best Practices in Mining – notes nearly one-quarter of jurisdictions lack mandatory closure plan requirements [online]. Available from: <https://www.csis.org/analysis/environmental-social-and-governance-best-practices-applied-mining-operations> [Accessed: 3 Apr 2025].

Connolly, K, 2014. Life after lignite: how Lusatia has returned to nature [online], *The Guardian*. Available from: <https://www.theguardian.com/environment/2014/sep/10/lusatia-lignite-mining-germany-lake-district#:~:text=LMBV%20floods%20the%20old%20mines%2C,of%20new%20canals%20that%20didn%E2%80%99t> [Accessed: 7 Apr 2025].

Deshaies, M, 2020. Metamorphosis of Mining Landscapes in the Lower Lusatian Lignite Basin (Germany): New Uses and New Image of a Mining Region, *Les Cahiers de la recherche architecturale, urbaine et paysagère*, 7:1–25.

Energiefabrik Knappenrode, 2020. Das Museum – Energiefabrik Knappenrode [online]. Available from: <https://www.energiefabrik-knappenrode.de/museum> [Accessed: 3 Apr 2025].

EY, 2022. Transforming mining land in the Hunter Valley: Economic opportunities for repurposing post-mining land [online]. Available from: <https://assets.nationbuilder.com/lockthegate/pages/7844/attachments/original/1653605597/EY_final_report_Transforming_mining_land_in_the_Hunter_Valley_26_May_2022.pdf> [Accessed: 3 Apr 2025].

Forest Landscape Restoration, 2020. Niger: Farmers taking restoration into their own hands. [online]. Available from: <https://www.forestlandscaperestoration.org/case-studies/niger-farmers-taking-restoration-into-their-own-hands/> [Accessed: 7 Apr 2025].

Genex Power/ARENA, 2024. Genex-Kidston Pumped Hydro Project [online]. Available from: <https://arena.gov.au/assets/2024/04/Genex---Kidston-Pumped-Storage-Hydro-Project---Lessons-Learnt-Report-10.pdf> [Accessed: 3 Apr 2025].

Glencore, 2025. Closure planning [online]. Available from: <https://www.glencore.com/sustainability/esg-a-z/closure-planning> [Accessed: 3 Apr 2025].

Global Cement and Concrete Association (GCCA), 2025. Circular Economy – Essential Concrete [online]. Available from: <https://gccassociation.org/essential-concrete/circular-economy/> [Accessed: 7 Apr 2025].

Global Reporting Initiative (GRI), 2025. Sector Standard for Mining [online]. Available from: <https://www.globalreporting.org/standards/standards-development/sector-standard-for-mining/> [Accessed: 3 Apr 2025].

GP Joule, 2025. Creating regional value with sun and hydrogen - Lusatia Energy Park [online]. Available from: <https://www.gp-joule.com/en/references/solar/lusatia-energy-park/> [Accessed: 3 Apr 2025].

Hunt, J, Byers, E, Riahi, K and Lilliestam, J, 2019. Gravity energy storage with suspended weights for abandoned mine shafts, *Applied Energy*, 239:201–206. Available from: <https://www.sciencedirect.com/science/article/pii/S0306261919302466> [Accessed: 3 Apr 2025].

Hunter Joint Organisation, 2024. Re-use Mining Lands advocacy brief [online]. Available from: <https://www.hunterjo.com.au/wp-content/uploads/2024/08/20240624-Advocacy-Re-use-Mining-Lands-Print-Inhouse.pdf> [Accessed: 3 Apr 2025].

Hunter Joint Organisation, 2025. Advocacy: Mining Land Redevelopment – Post-mining land use [online]. Available from: <https://hunterjo.nsw.gov.au/wp-content/uploads/2025/03/Advocacy-Mining-Land-Redevelopment-Post-mining-land-use-Online.pdf> [Accessed: 3 Apr 2025].

Hunter Renewal, 2023. After the coal rush, the clean up – A community blueprint to restore the Hunter [online]. Available from: <https://assets.nationbuilder.com/lockthegate/pages/8176/attachments/original/1690764718/Blueprint_final_1.pdf?1690764718> [Accessed: 7 Apr 2025].

Hunter Valley Operations (HVO), 2024. Annual Environmental Management Report 2023 [online]. Available from: <https://www.hvo.com.au/wp-content/uploads/2024/04/ARR0001240_Annual-Report_28Mar2024-139pm-1.pdf> [Accessed: 3 Apr 2025].

Intergovernmental Forum on Mining, Minerals, Metals and Sustainable Development (IGF), 2020. Mine Closure Challenges for Government and Industry [online]. Available from: <https://www.igfmining.org/mine-closure-challenges-government-industry/> [Accessed: 3 Apr 2025].

International Council on Mining and Metals (ICMM), 2019. Integrated Mine Closure: Good Practice Guide, 2nd Edition [online], International Council on Mining and Metals. Available from: <https://www.icmm.com/integrated-mine-closure>

International Organisation for Standardisation (ISO), 2024. ISO 59004:2024 – Circular economy — Vocabulary, International Organisation for Standardisation.

Internationale Bauausstellungen, 2025. F60 Visitors' Mine – A 'Horizontal Eiffel Tower' in Lusatia [online]. Available from: <https://www.internationale-bauausstellungen.de/en/history/2000-2010-iba-furst-puckler-land-werkstatt-fur-neue-landschaften/f60-visitors-mine-a-lying-eiffel-tower-in-the-lausitz-region/> [Accessed: 3 Apr 2025].

Interstate Technology and Regulatory Council (ITRC), 2025. Potential applications for reuse of solid mining waste [online]. Available from: <https://mw-1.itrcweb.org/4-potential-applications-for-reuse-of-solid-mining-waste/> [Accessed: 3 Apr 2025].

Krawczyk, J and Bartosik, A, 2020. Post-Mining Lakes: The Status of Planning and Implementation of the Water-Filled Post-Mining Landscape Transformation in Lusatia [online], *Minerals*, 10(2):133–153. [online]. Available from: <https://www.mdpi.com/2075-163X/10/2/133> [Accessed: 7 Apr 2025].

Lamm, K and Johnson, J, 2022. Environmental, social and governance best practices as applied to mining operations, Center for Strategic and International Studies (CSIS) [online]. Available from: <https://www.csis.org/analysis/environmental-social-and-governance-best-practices-applied-mining-operations> [Accessed: 3 Apr 2025].

Malabar Resources, 2024. Quarterly Report for the Period Ending 31 March 2024 [online]. Available from: <https://malabarresources.com.au/wp-content/uploads/2024/04/Malabar-Quarterly-ending-31-March-2024-1.pdf> [Accessed: 8 Apr 2025].

McCabe, M and Morris, T, 2016. Dark side of the boom: What we do and don't know about mine closure and rehabilitation in Australia [online] *The Australia Institute* p1–67. Available from: <https://australiainstitute.org.au/wp-content/uploads/2020/12/P192-Dark-side-of-the-boom-web.pdf> [Accessed: 3 Apr 2025].

McGhee, A, 2024. Malcolm Turnbull's pumped hydro company eyes NSW coal country for renewable energy bid [online]. Available from: <https://www.abc.net.au/news/2024-02-26/turnbulls-pumped-hydro-upper-hunter-renewable-energy-bid/103510232> [Accessed: 7 Apr 2025].

Müller, K and Morton, T, 2018. Lusatia and the coal conundrum: The lived experience of the German Energiewende, *Energy Policy*, 99:277–287. Available from: <https://www.sciencedirect.com/science/article/pii/S0301421516302543> [Accessed: 3 Apr 2025].

Murphy, B and Bernasconi, A, 2022. Hunter Valley mining land could create 13 000 jobs if reused sustainably, report finds [online]. Available from: <https://www.abc.net.au/news/2022-06-15/hunter-valley-mining-land-reuse-could-boost-economy-report-finds/101149714> [Accessed: 3 Apr 2025].

Nature Conservancy, The, 2024. Mining the Sun: A Vision for Renewable Energy and Mine Land Transformation [online]. Available from: <https://www.nature.org/content/dam/tnc/nature/en/documents/Mining_the_Sun_Report_Final_5.23.24.pdf> [Accessed: 7 Apr 2025].

Northern Australia Infrastructure Facility (NAIF), 2023. Heritage Minerals Mount Morgan Tailings Processing and Rehabilitation Project [online]. Available from: <https://www.naif.gov.au/our-projects/heritage-minerals-mount-morgan-tailings-processing-and-rehabilitation-project/> [Accessed: 7 Apr 2025].

NSW Environment Protection Authority, 2008. Application of recycled organics in mine site rehabilitation [online]. Available from: <https://www.epa.nsw.gov.au/sites/default/files/080371-mine-site-rehab.pdf> [Accessed: 7 Apr 2025].

NSW Government, 2023. Draft policy: Mine Site Rehabilitation and Futures, NSW Government Department of Planning and Environment.

NSW Government, 2024. 450 jobs for Hunter as coal mine transforms into motor park and tourist resort [online]. Available from: <https://www.nsw.gov.au/media-releases/450-jobs-for-hunter-as-coal-mine-transforms-into-motor-park-and-tourist-resort> [Accessed: 7 Apr 2025].

Parkinson, G, 2021. Hydrogen Valley plan unveiled to turn Hunter into a green hydrogen hub [online]. Available from: <https://reneweconomy.com.au/hydrogen-valley-plan-unveiled-to-turn-hunter-into-a-green-hydrogen-hub/> [Accessed: 3 Apr 2025].

Parkinson, G, 2024. AGL adds pumped hydro project based around old coal mine to EPBC pipeline [online]. Available from: <https://reneweconomy.com.au/agl-adds-pumped-hydro-project-based-around-old-coal-mine-to-epbc-pipeline/> [Accessed: 8 Apr 2025].

Queensland Government, 2025. Management of abandoned mines [online]. Available from: <https://www.qld.gov.au/environment/land/management/abandoned-mines/management> [Accessed: 27 May 2025].

Queensland Treasury, 2024. Financial Provisioning Scheme Annual Report 2023–24 [online]. Available from: <https://s3.treasury.qld.gov.au/files/2023-24-FPS-Annual-Report.pdf> [Accessed: 7 Apr 2025].

Schmid, S, Angerer, G, Fichter, K and Ziegler, A, 2022. Towards a circular economy: conditions and strategic approaches for resource efficiency in the value chain, *Mineral Economics*, 35:45–59.

Task Force on Climate-related Financial Disclosures (TCFD), 2017. Final Report – Recommendations of the Task Force on Climate-related Financial Disclosures, 2017.

Telegraph, The, 2024. The glorious forest Britain built from scratch, Travel [online]. Available from: <https://www.telegraph.co.uk/travel/news/the-glorious-forest-britain-built-from-scratch/> [Accessed: 7 Apr 2025].

TJ Ryan Foundation, The, 2023. Mining report finds 60,000 abandoned sites, lack of rehabilitation and unreliable data [online]. Available from: <https://tjryanfoundation.org.au/environment-energy/mining-report-finds-60000-abandoned-sites-lack-of-rehabilitation-and-unreliable-data/> [Accessed: 3 Apr 2025].

Upper Hunter Mining Dialogue, 2024. 2024 Community Forum [online]. Available from: <https://miningdialogue.com.au/engagement/2024-community-forum> [Accessed: 7 Apr 2025].

van Zyl, D, Sassoon, M, Fleury, A M and Kyeyune, S, 2016. Guidelines for mine closure planning, *Journal of Cleaner Production*, 129:308–316.

Varnon, R, 2014. Decaying factories become vital tourist destinations [online]. Available from: <https://www.ctpost.com/local/article/Decaying-factories-become-vital-tourist-5925249.php> [Accessed: 7 Apr 2025].

Vitti, C and Arnold, B J, 2022. The reprocessing and revalorisation of critical minerals in mine tailings, *Mining, Metallurgy and Exploration*, 39:49–54.

Whitbread-Abrutat, P and Lowe, J, 2024. 102 Things to Do with a Hole in the Ground [online]. Available from: <https://www.icmm.com/en-gb/case-studies/2024/102-things> [Accessed: 7 Apr 2025].

Wilkinson, M, 2023. Mining legacy: a void the size of Sydney Harbour [online]. Available from: <https://www.inkl.com/news/mining-legacy-a-void-the-size-of-sydney-harbour> [Accessed: 3 Apr 2025].

World Economic Forum, 2019. Germany is turning its old mines into a tourist hotspot [online]. Available from: <https://www.weforum.org/stories/2019/06/germany-is-turning-its-old-mines-into-a-tourist-hotspot/> [Accessed: 3 Apr 2025].

Yellishetty, M, Mudd, G M, Shukla, R and Graham, P, 2023. Recycling and reuse of mine tailings: A sustainable pathway for a circular mining economy, *Remediation Journal,* 33(2):5–16.

Zhang, L, Wang, Y and Unluer, C, 2021. Recycling and reuse of mine tailings in building materials: A review, *Journal of Cleaner Production*, 314:127933.

A novel and practical approach to identify durable rock for landform rehabilitation at the Mt Arthur Coal Mine

S Mackenzie[1], A Sampaklis[2], G Wesley[3], T Trickey[4], N Stackman[5] and W Ringland[6]

1. Founder, Mine Earth, O'Connor WA 6163. Email: shannon@mineearth.com.au
2. Specialist Closure Planning, BHP NSW Energy Coal, Muswellbrook NSW 2333.
   Email: andrew.sampaklis@bhp.com
3. Principal Environmental Geologist, Mine Earth, O'Connor WA 6163.
   Email: glendon@mineearth.com.au
4. Senior Geologist, BHP NSW Energy Coal, Muswellbrook NSW 2333.
   Email: teresa.trickey1@bhp.com
5. Project Manager Engineering Projects, BHP NSW Energy Coal, Muswellbrook NSW 2333.
   Email: nigel.stackman@bhp.com
6. Manager Technical Services, BHP NSW Energy Coal, Muswellbrook NSW 2333.
   Email: william.ringland@bhp.com

INTRODUCTION

The Mt Arthur Coal Mine is located approximately 5 km south-west of Muswellbrook, in the Upper Hunter Valley region of New South Wales (Figure 1). The project is scheduled for closure by 2030.

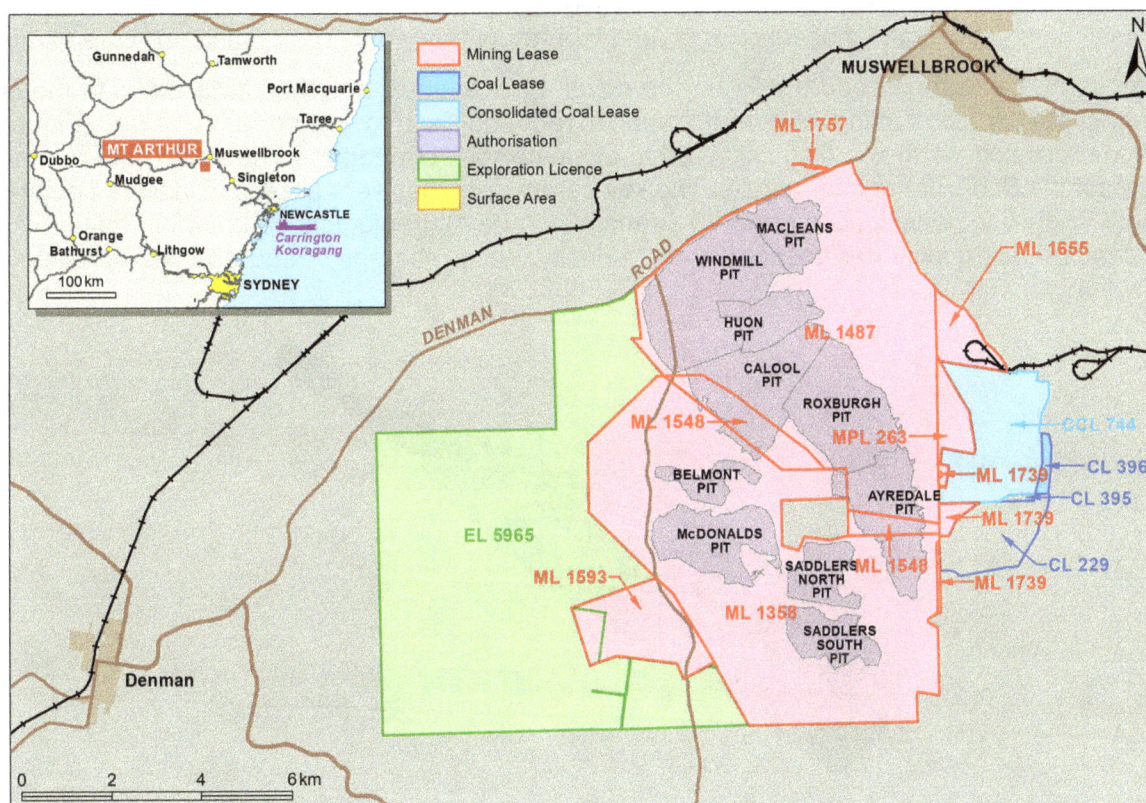

FIG 1 – Location of the Mt Arthur Coal Mine.

Disturbed land at the project will be rehabilitated using innovative geomorphological designs specifically developed for mining landforms. A large volume of durable rock will be required for the construction of the drainage channels that will be incorporated into the final slopes of the rehabilitated landforms.

A preliminary assessment did not identify a suitable source of durable rock on-site for constructing the drainage channels. To address rehabilitation requirements, alternative options were considered including importing rock from off-site sources and fabricating specialised concrete matting off-site. Following the evaluation of these options, a comprehensive and collaborative reassessment of on-site rock resources was conducted to identify a viable source of durable rock suitable for drain construction and other rehabilitation applications.

The reassessment resulted in the discovery of an abundant source of durable rock and the development of a novel and practical approach to its characterisation and identification in the field.

Target rock types

The most abundant and accessible source of durable rock identified on-site was the lower, massive portion of the Woodlands Hill sandstone unit (WHMS). This material comprises massive sandstone that was observed to be consistently hard and durable from all pits and stockpile exposures that were inspected. The upper portion (approximately the upper half) of the Woodlands Hill sandstone unit was observed to be finely laminated and interbedded and not suitable as a source of durable rock.

Whilst less abundant and accessible, a secondary source of potentially durable rock identified on-site was the Warkworth sandstone unit (WWS).

The WHMS will be the focus of this paper given it is the most abundant and accessible source of durable rock on-site.

APPROACH

A practical field procedure was developed for implementation by operational staff. This led to the collection of data over an intensive six-month period. Qualitative and quantitative data were collected from field observations and measurements, and laboratory test work on grab and drill core samples.

The focus of the field procedure was on the novel application of a 'Schmidt hammer' to measure rock strength by conducting a non-destructive rebound test to derive a 'Q-value' (Figure 2). Additional physical data was also collected to evaluate rock strength and durability and to provide context for the Schmidt hammer results. The key parameters that were assessed included point load strength (Figure 3), uniaxial compressive strength (Figure 4), geological-hammer hit, mineralogical composition, rock quality designation (RQD), density, water absorption and as-mined particle size distribution.

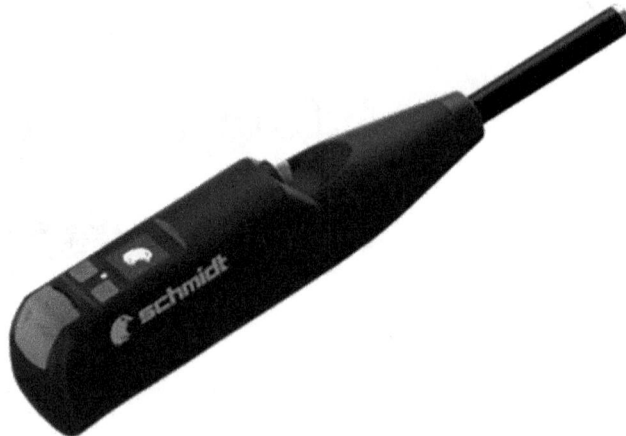

FIG 2 – Schmidt hammer.

FIG 3 – Point load test.

FIG 4 – Uniaxial compressive strength test.

A threshold Q-value was ultimately developed for operational staff to distinguish suitable durable rock from non-durable rock. The threshold Q-value was developed from a comprehensive review of the data, collected over the six-month campaign and from collaborative verification by geological, engineering and landform design personnel.

Interpretation

Rock strength classification systems utilising Schmidt hammer, point load strength and uniaxial compressive strength (UCS) results have been developed by several authors including Goudie (2006), Hawley and Cunning (2017), Hoek and Brown (1997), Berkman (1995) and CIRIA, CUR and CETMEF (2007). The classification system used by Hawley and Cunning (2017) in Guidelines for mine waste dump and stockpile design is applicable for categorising waste rock on the slopes of mining waste rock landforms in Australia.

The classification system presented in Hawley and Cunning (2017) utilises the rating system and UCS thresholds listed by Hoek and Brown (1997). This has been adapted further to include point load index (Hoek and Brown, 1997) and Schmidt hammer results (Goudie, 2006).

The modified rock strength classification system presented in Table 1 was applied to the results to provide context and categorise the waste rock sources.

TABLE 1

Rock strength classification system (modified from Hawley and Cunning, 2017).

Type	UCS	Point Load (Is50)	Adjusted Q-Value
5	>250 MPa	>10 MPa	>70
	100-250	4-10	60-70
4	50-100	2-4	50-60
3	25-50	1-2	45-50
2	5-25	0.1-1	30-45
1	1-5	<0.1	20-30
	<1		<20

FINDINGS

The most prospective target rock type on-site was the WHMS. There are two key sources of the WHMS unit: (i) Roxborough pit and (ii) Saddlers pit. The WHMS from both locations was classified as suitable for drain construction by the landform design team.

Weathered WHMS is exposed along the crest of the Roxborough pit (Figure 5). It was drilled and blasted in preparation for recovery and use as durable rock for drain construction. Fresh WHMS will be mined from the Roxborough pit in the coming years. The blasted rock is comprised of large, massive sandstone fragments up to ~2 m in diameter with a relatively low proportion of fines. These larger fragment sizes are reflective of the way the rock on the pit crest was blasted, with more energy being released at surface than in a pit setting.

FIG 5 – Woodland Hill Massive Sandstone from Roxborough pit.

Fresh WHMS was drilled, blasted, loaded, hauled and dumped in a large stockpile at the Saddlers pit (Figure 6). Whilst a smaller fragment size was observed, this is predominantly due to the rock being blasted within a confined pit setting and then undergoing mechanical handling. The blasted rock is comprised of massive sandstone fragments up to ~1.5 m in diameter with a low proportion of fines.

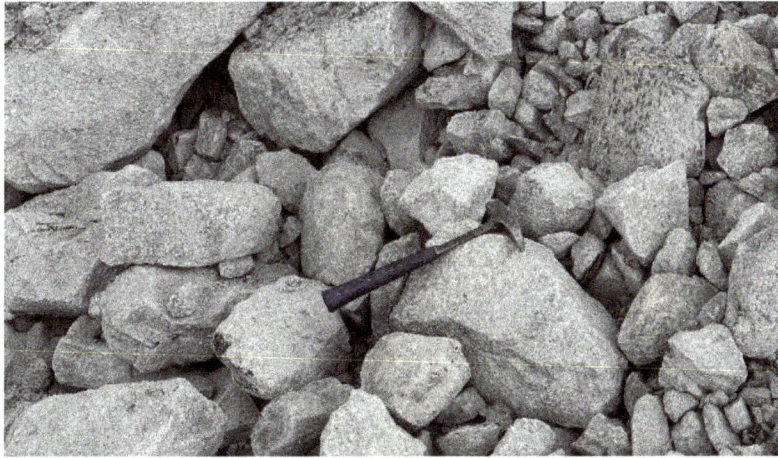

FIG 6 – Woodland Hill Massive Sandstone from Saddlers pit.

Only the results from the key material parameters are presented in this paper including Schmidt hammer, point load, UCS and mineralogy results.

Schmidt hammer results

Schmidt hammer mean Q-values were obtained for transitional to fresh WHMS from the Roxborough pit (113 results) and Saddlers pit (24 results).

Histograms of the mean Q-values for Roxborough and Saddlers pits are presented in Figures 7 and 8 respectively. The histograms show a bimodal distribution of the mean Q-values, with the highest numbers of results within the ranges of 40–50 (Types 2–3) and 65–70 (Type 5).

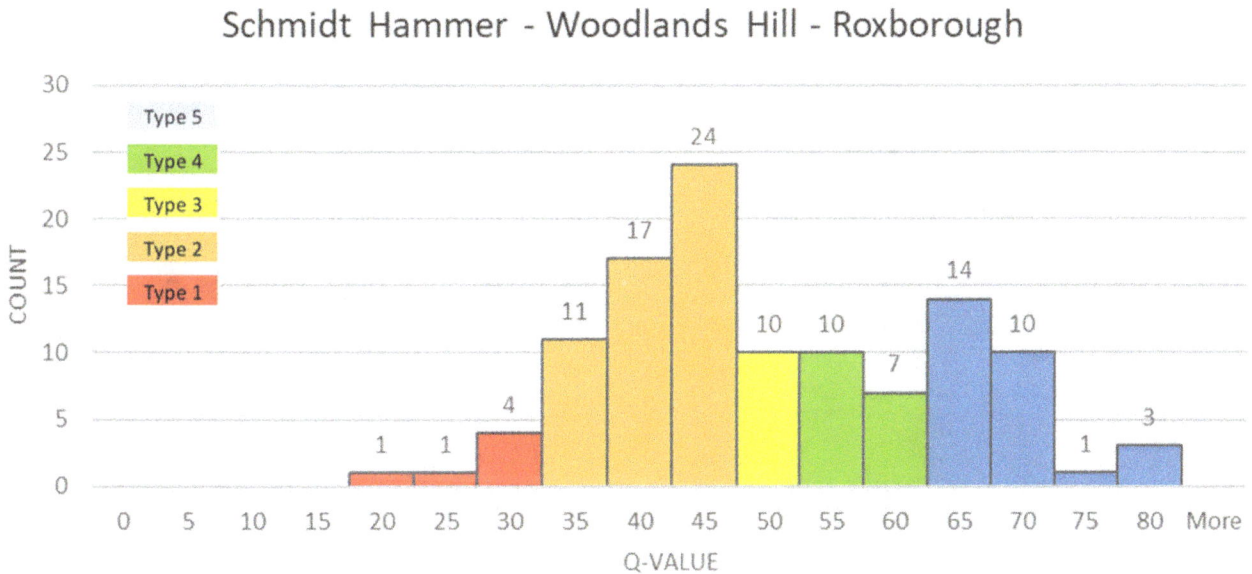

FIG 7 – WHMS mean Q-values for Roxborough pit (colours align with the Hawley and Cunning (2017) classification).

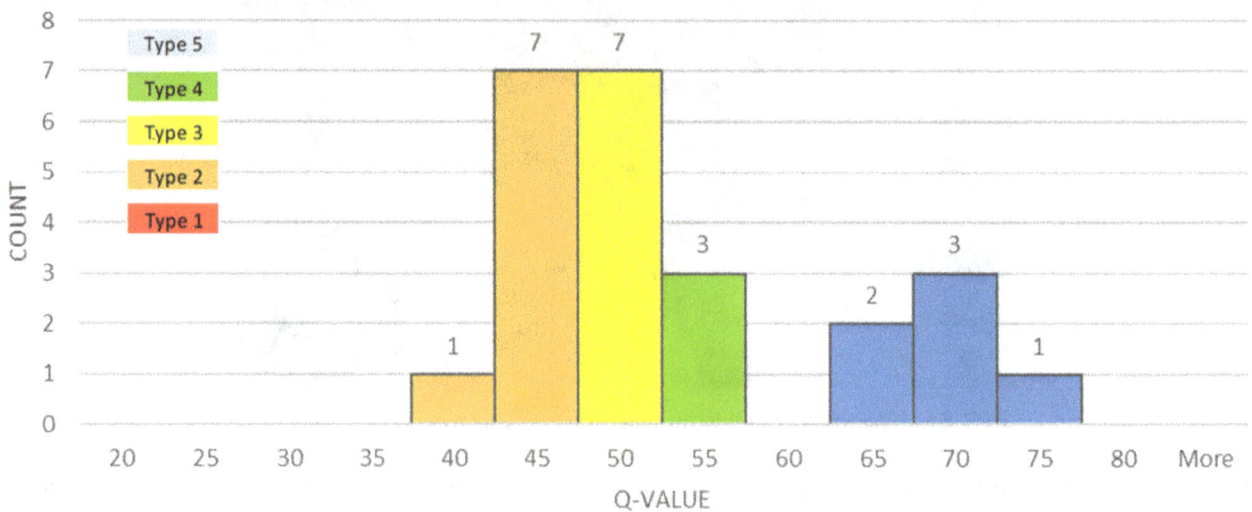

FIG 8 – WHMS mean Q-values for Saddlers pit (colours align with the Hawley and Cunning (2017) classification).

A review of the correlation between the WHMS Schmidt hammer data and the hardness rating assigned by geological hammer hits is presented in Figure 9. The subjective hardness classifications of low, moderate and high, as rated by site personnel, correlate consistently to Q-values of <30 (low), 30–50 (moderate) and >50 (high), and to rock strength Types 1–2 (low), Type 3 (moderate) and Types 4–5 (high). This adds confidence to the use of a geological hammer to quickly and easily rate rock hardness in the field.

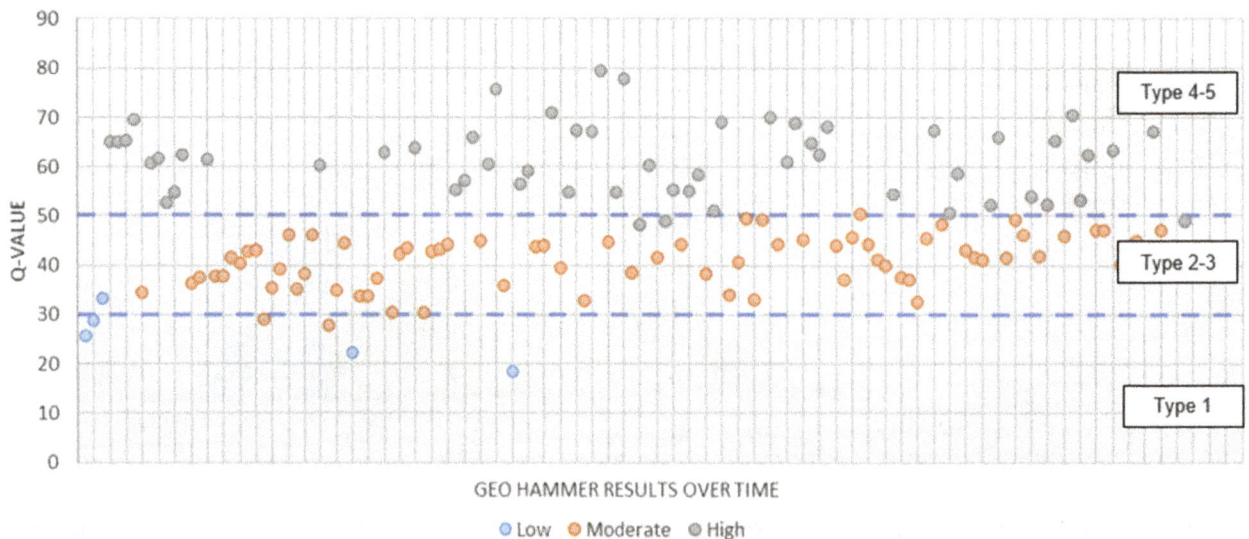

FIG 9 – Correlation of geological hammer hardness ratings with Q-values (Types align with the Hawley and Cunning (2017) classification).

Point load results

From the analysis of the field and drill core samples, point load results were obtained for transitional to fresh WHMS from the Roxborough pit (32 field/5 core) and Saddlers pit (9 field/3 core).

Histograms of the point load results for Roxborough and Saddlers pits are presented in Figures 10 and 11 respectively. Rock strength Types 2–4 are the most prevalent within the point load results, however the results within the Type 5 class can show significantly high point load values (>6 MPa).

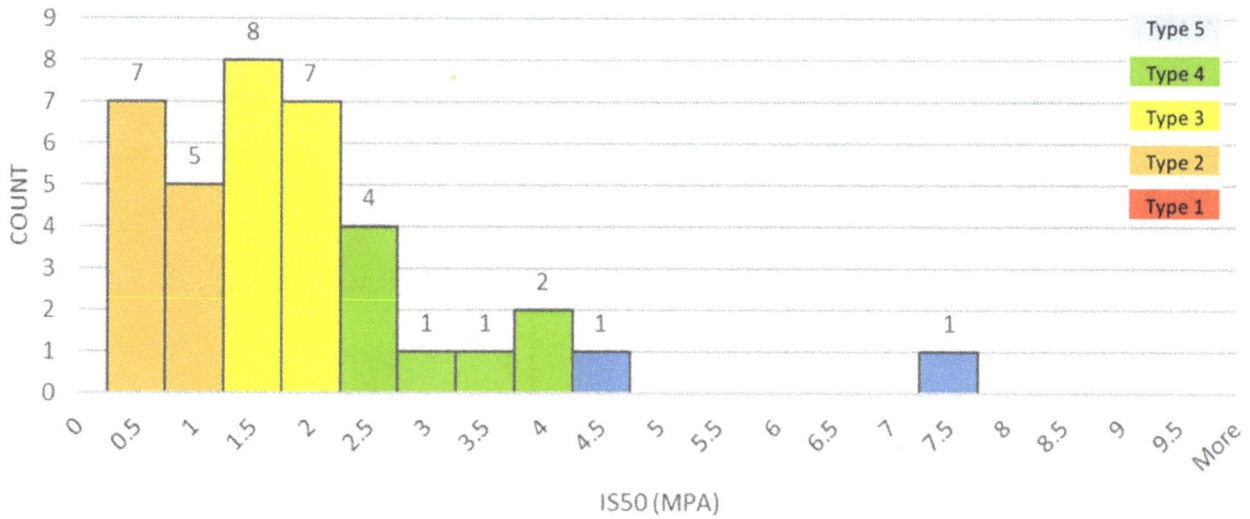

FIG 10 – WHMS point load results for Roxborough Pit.

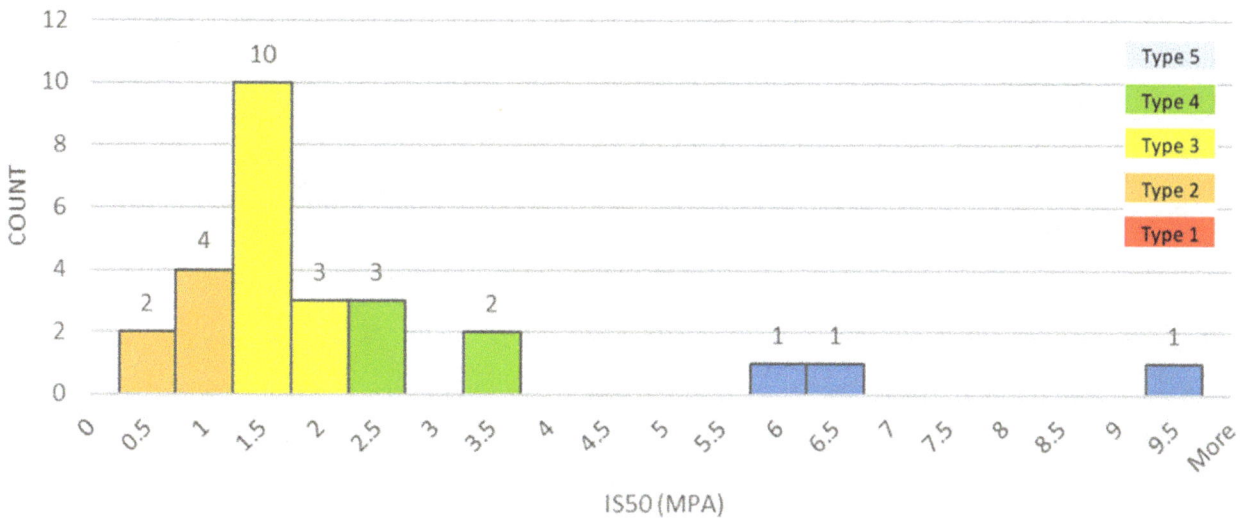

FIG 11 – WHMS point load results for Saddlers Pit.

UCS results

The UCS data was obtained for transitional to fresh WHMS from the Roxborough pit and Saddlers pit by testing drill core samples.

Histograms of the UCS results for Roxborough and Saddlers pits are presented in Figures 12 and 13 respectively. All UCS results for the drill core samples fall within the Type 3–4 rock strength classes.

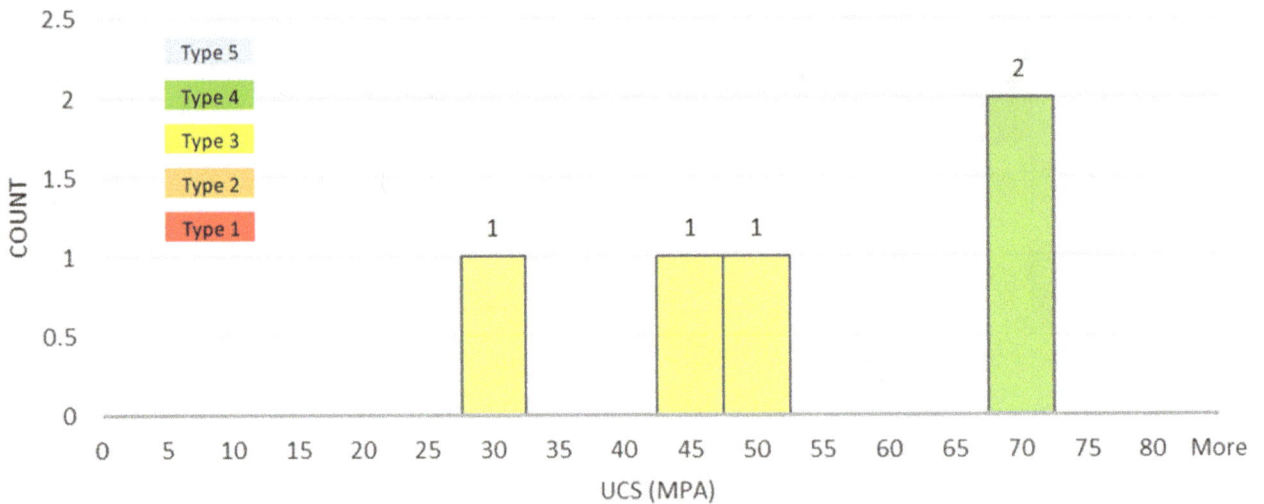

FIG 12 – WHMS UCS results for Roxborough Pit.

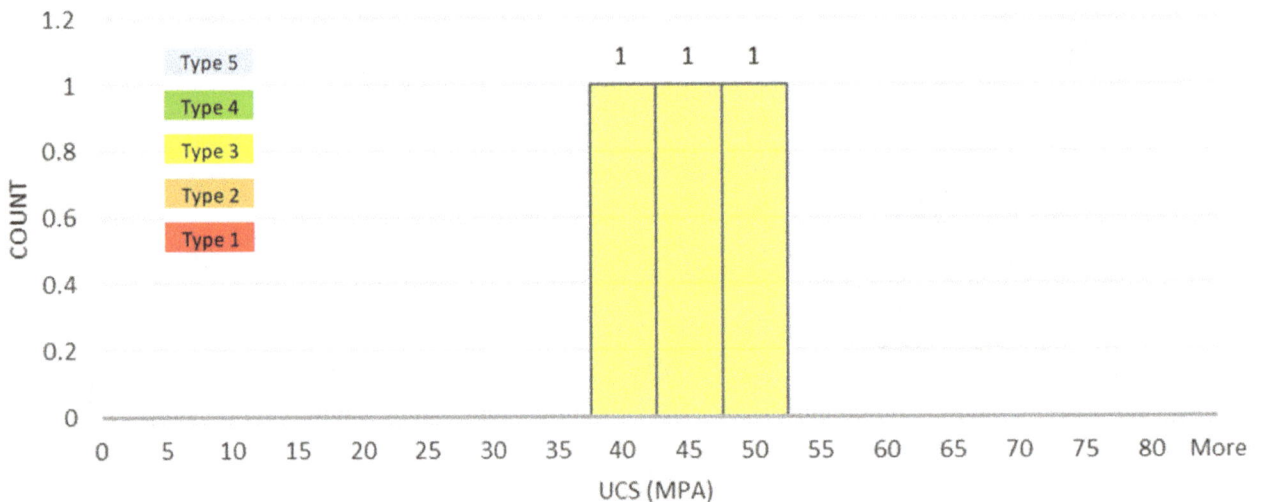

FIG 13 – WHMS UCS results for Saddlers Pit.

Mineralogy results

Mineralogical test work was undertaken on six samples from Roxborough pit and eight samples from Saddlers pit. The samples typically contained dominant quartz grains and a variety of lithic fragments within a matrix of silica, clays and carbonate. All samples were grain-supported, containing >60 per cent sedimentary grains and 6–37 per cent matrix. Photomicrographs of a WHMS sample from Saddlers pit are presented in Figure 14.

Mineral composition provides an indication of rock strength, susceptibility to dissolution and overall erodibility. The dominance of quartz and igneous rock fragments within the tested samples indicates that the granular component of the rocks is both physically and chemically durable. The amount of matrix and the ratio of silica (high durability) to carbonate and clays (low durability) cementing the grains within the sandstones can affect the overall ability of the rocks to resist erosion from weathering, and fragmentation during blasting and mechanical disturbance. When carbonates and clays form a higher proportion of the rock matrix, the durability of the rock structure can be more susceptible to physical and chemical weathering over time, increasing the porosity.

FIG 14 – Photomicrograph of WHMS sample from Saddlers pit.

Inventory

It has been estimated that approximately 1 M BCM of durable rock will be required for rehabilitation purposes on-site. Even with a material recovery factor of 50 per cent, the volume of WHMS available from Roxborough and Saddlers pits (3.3 M BCM) greatly exceeds estimated demand.

CONCLUSION

The WHMS unit is the most abundant, accessible and homogeneous source of durable waste rock available on-site and has been classified as suitable for drain construction by the landform design team.

Results for Schmidt hammer, point load index, UCS, RQD, density, water absorption, porosity and mineralogy were assessed to provide an overall indication of rock strength and durability. A rock strength classification system, modified from Hawley and Cunning (2017), was applied to the results to provide context and help categorise the key waste rock sources.

Importantly, the results showed Schmidt hammer measurements to be a quick and reliable indicator of rock strength in the field.

The Schmidt hammer, point load index, UCS and geological hammer hit results showed that the WHMS from the Roxborough and Saddlers pits was similar in strength:

- The Schmidt hammer results showed a bimodal distribution of mean Q-values with the highest numbers of results between material Types 2–3 (Q-value 40–50) and Type 5 (Q-values 65–70).

- The point load results fell within material Types 2–4 and some within Type 5.

- All UCS results were within the range for material Types 3–4.

From a comprehensive review of the data and from collaborative verification by geological, engineering and landform design personnel, threshold Q-values were developed to distinguish suitable durable rock from non-durable rock in the field:

- Rock strength Type 1 (Q-value <30) material is not suitable for use as durable rock.

- Rock strength Type 2 (Q-value 30–45) material is suitable for general applications where durable rock is required.

- Rock strength Type 3 and above (Q-value >45) material is suitable for applications where the best durable rock is required.

To further develop the data set and improve confidence for defining material thresholds, ongoing data collection will continue.

ACKNOWLEDGEMENTS

The authors would like to acknowledge the contribution of all on-site geology, mine planning, technical services, engineering projects, environmental and rehabilitation personnel and the contribution of the consulting landform design team.

REFERENCES

Berkman, D A (ed), 1995. *Field Geologists' Manual*, third revised edition, Monograph 9, 382 p (The Australasian Institute of Mining and Metallurgy: Melbourne).

CIRIA, CUR and CETMEF, 2007. *The Rock Manual: The use of rock in hydraulic engineering*, second edn, Construction Industry Research and Information Association (CIRIA), the Netherlands Centre for Civil Engineering Research and Codes (CUR), Centre d'Etudes Techniques Maritimes et Fluviales (CETMEF) (London: CIRIA).

Goudie, A S, 2006. The Schmidt Hammer in geomorphological research, *Progress in Physical Geography*, 30(6):703–718.

Hawley, M and Cunning, J, 2017. *Guidelines for Mine Waste Dump and Stockpile Design* (CSIRO: Clayton South).

Hoek, E and Brown, E T, 1997. Practical estimates of rock mass strength, *International Journal of Rock Mechanics and Mining Sciences*, 34(8):1165–1186.

Integrating environmental, rehabilitation, and closure workflows using an intelligence system framework

W Mitry[1], B Baker[2] and C Cooper[3]

1. Associate Consultant, IEMA, Wollongong NSW 2500. Email: will.mitry@iema.com.au
2. Manager Environment, Narrabri Coal Operations, Baan Baa NSW 2390.
   Email: brentbaker@whitehavencoal.com.au
3. Principal Consultant, IEMA, Brisbane Qld 2291. Email: chris.cooper@iema.com.au

ABSTRACT

Mining companies must navigate increasingly strict rehabilitation and closure obligations and regulatory scrutiny through the life-of-mine, requiring robust systems to track obligations status, rehabilitation planning and record-keeping.

At an organisational level, mining companies are seeking to invest in technologies and tools to manage closure planning and implementation and develop governance frameworks to provide adequate oversight of sites' closure planning status. In response to these challenges, Integrated Environmental Management Australia Pty Ltd (IEMA) has collaborated with mining companies to develop the Integrated Mining, Environmental, Rehabilitation, and Closure System (IMERCS). This advanced cloud-based platform integrates management plan obligations seamlessly with execution records, consolidating workflows and enhancing regulatory compliance. Leveraging spatial intelligence, automation, and artificial intelligence, IMERCS facilitates data-driven decision-making and operational transparency.

In collaboration with Whitehaven Coal's Narrabri Coal Operations (NCO), located near Gunnedah, NSW, IEMA has developed thirteen specialised modules built in the IMERCS intelligence framework to streamline data management and operational workflows. Centralised Cloud Data Stores provide organisation-wide access to structured and secure data, supporting seamless collaboration across the organisation.

IMERCS adheres to the 'Plan – Do – Check – Act' continuous improvement process. Tailored modules address diverse mine site obligations, embedding task planning, execution, record management, and quality assurance processes to ensure comprehensive compliance management. Customisable dashboards provide users within the organisation live information regarding the status of rehabilitation and closure planning metrics.

This paper describes the NCO IMERCS case study to demonstrate the value a structured and centralised spatial data store can deliver for mine site management. By combining innovative technologies into a mine site intelligence framework, including automations and AI, IMERCS can deliver enhanced rehabilitation and closure planning knowledge retention and verification evidence. Environmental, rehabilitation, and closure workflows significantly improve rehabilitation implementation practices, quality assurance and support compliance with regulatory obligations, as well as provide internal governance tools for the organisation.

INTRODUCTION

In the contemporary mining landscape, achieving and maintaining regulatory compliance in environmental management is increasingly challenging. Rising stakeholder expectations, evolving legislation, and the complexity of environmental rehabilitation require mining companies to re-evaluate traditional systems and embrace more innovative, data-driven solutions. These challenges brought a team within IEMA together from diverse backgrounds in mining, government and consulting to find a better way to achieve success for the industry. What began as a discovery process to understand the industry's key problems, rapidly turned into full steam development. The need for change within the industry was real, and IEMA formed relationships with industry partners to develop and implement the IMERCS solution.

IMERCS is redefining how environmental, rehabilitation and closure management is achieved, embedding an intelligence system framework to build integrated solutions with an industry leading technology stack.

THE IMERCS SOLUTION – DESIGN AND CORE ARCHITECTURE

Developed using a technology stack that includes AWS Cloud, ESRI's ArcGIS Online and Field Map technologies, FME automation software and Open AI's API's, IMERCS is a cloud-based platform designed to house environmental and operational workflows and intelligence capabilities. IMERCS consolidates complex environmental and rehabilitation requirements using the IMERCS Intelligence framework (Figure 1).

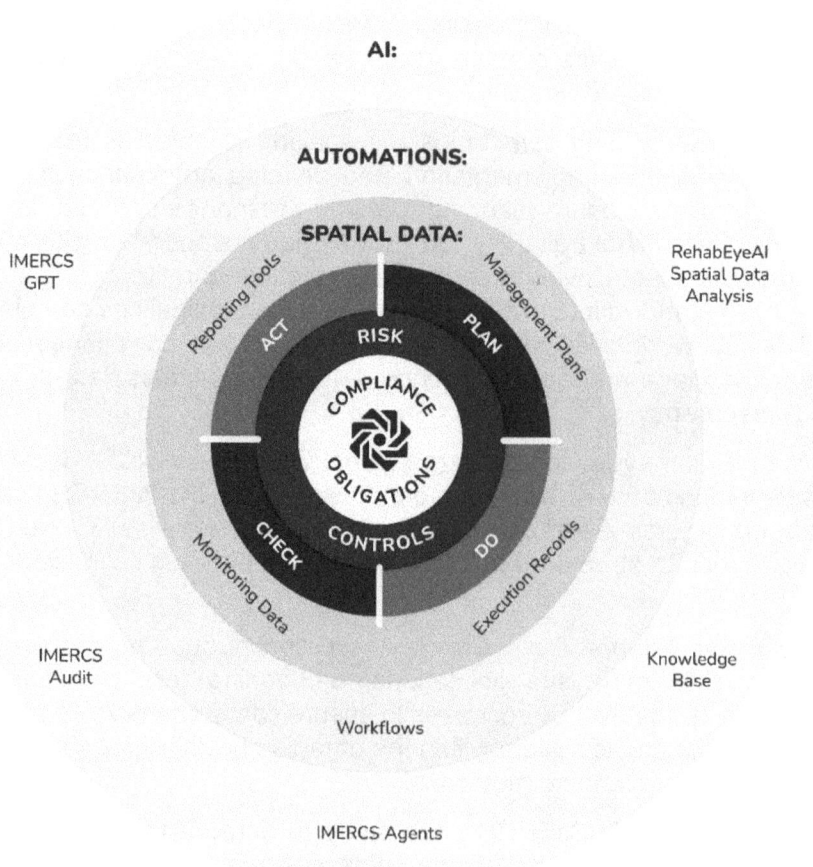

FIG 1 – IMERCS Intelligence framework.

The intelligence system framework is made up of the following components.

1. Centralised Cloud Data Stores:

    o Provides organisation-wide access to structured and secure data.

    o Supports seamless collaboration across departments and stakeholders.

2. Structured Data Framework:

    o Obligation focus: system designed around compliance requirements.

    o Risk Management: Embedded understanding of risk and control.

    o QA/QC Integration: Embedded quality assurance and control into workflows.

    o Continual Improvement: Implements Plan-Do-Check-Act cycles to achieve site management goals.

3. Workflow Modules:

- Approvals Management: Facilitates permit-to-work processes and approval workflows.
- Operations Monitoring: Tracks inspections, borehole activities, and cultural heritage assessments.
- Rehabilitation Planning: Manages execution, monitoring, and reporting of rehabilitation activities.
- Closure Reporting: Streamlines final reporting and compliance documentation.

4. Technology Stack:

- **Spatial Intelligence:** Utilises ESRI-supported spatial databases for enhanced location intelligence and field-to-desktop data synchronisation.
- **Automations:** Enables streamlined integrations with business systems, automated notifications, data updates, and reporting triggers.
- **Artificial Intelligence:** AI capabilities built into the system so all data and records within the knowledge base (trained on management plans, execution records, regulations, guidelines, and industry best practice documents etc) can be analysed and actioned by AI agents. The IMERCS GPT tool can provide domain-specific insights based on a company specific knowledge base. This AI integration is enabling future development of AI agents such as the embedded Audit AI tool. This tool can perform audits of documentation streamlining processes for audit preparation and preparedness.

This comprehensive digital system enhances environmental management, compliance tracking, and operational safety.

The following section breaks down the various components of the IMERCS system as used at NCO.

NARRABRI COAL OPERATIONS CASE STUDY

NCO partnered with IEMA to design and implement a comprehensive intelligence system—IMERCS—that would transform how mining operations manage environmental compliance, rehabilitation, and closure planning. This paper presents the journey of IMERCS from concept to implementation and highlights its profound impact on operational transparency, data integrity, and regulatory compliance.

Identifying the challenge

The trigger for change at NCO was a desire to develop effective spatial data controls to manage surface disturbance activities in accordance with environmental approvals. These activities, governed by the Permit to Work (PTW) system, depended on a cumbersome paper-based process that provided limited visibility, was prone to human error, and often lacked critical real-time data required for proactive environmental management.

NCO recognised that the PTW process had to evolve into a more intelligent, automated, and integrated system to ensure that all works were carried out in compliance with approval conditions, while also enabling streamlined internal governance and reporting.

Alternative mine planning solutions

Geospatial tools with AI integrations are an emerging powerful force to drive efficiencies for managing a range of operational functions including mine planning, developing geological models and intelligent remote sensing applications such as erosion monitoring or ecological monitoring tools.

However, the authors are unaware of any comparable suite of tools developed to assist mines specifically manage mine rehabilitation planning and execution processes and record management, that is aligned with Australian regulatory requirements.

WORKFLOW APPLICATIONS

A core functionality of the IMERCS system are the embedded workflow capabilities. The workflow capabilities follow a Plan-Do-Check-Act cycle to achieve site management goals and continual improvement.

NCO have developed 13 specialised workflow modules including the following:

1. Drill hole rehab tracker
2. ePTW
3. Aboriginal cultural heritage
4. Dam inspections
5. Subsidence management
6. Weed management
7. Rehabilitation planning
8. Rehabilitation monitoring
9. Rehabilitation reporting
10. Enviro monitoring locations
11. General inspections
12. GDE monitoring
13. RCE.

Each one of these workflows have spatial databases as the primary information storage using ESRI's WebGIS and field maps capabilities. This enables spatial intelligence to be embedded through the workflow cycle. Embedded automations enable streamlined process flow leading to improved operational efficiency, and compliance with obligations. The following sections introduce the Workflow Modules at NCO.

ePTW workflow web application

A key IMERCS module, the ePTW Workflow Web Application (ePTW App) replaces paper forms with an online workflow application to manage the permit to work process for all on surface activities for new and existing disturbance areas (Figure 2). The ePTW App integrates spatial layers, such as environmental approvals, mine plans, and concealed infrastructure, directly into the permit workflow. The intent was to provide greater visibility and control of the disturbance workflow, ensuring that all necessary risk assessments and approvals are complete before any physical work begins.

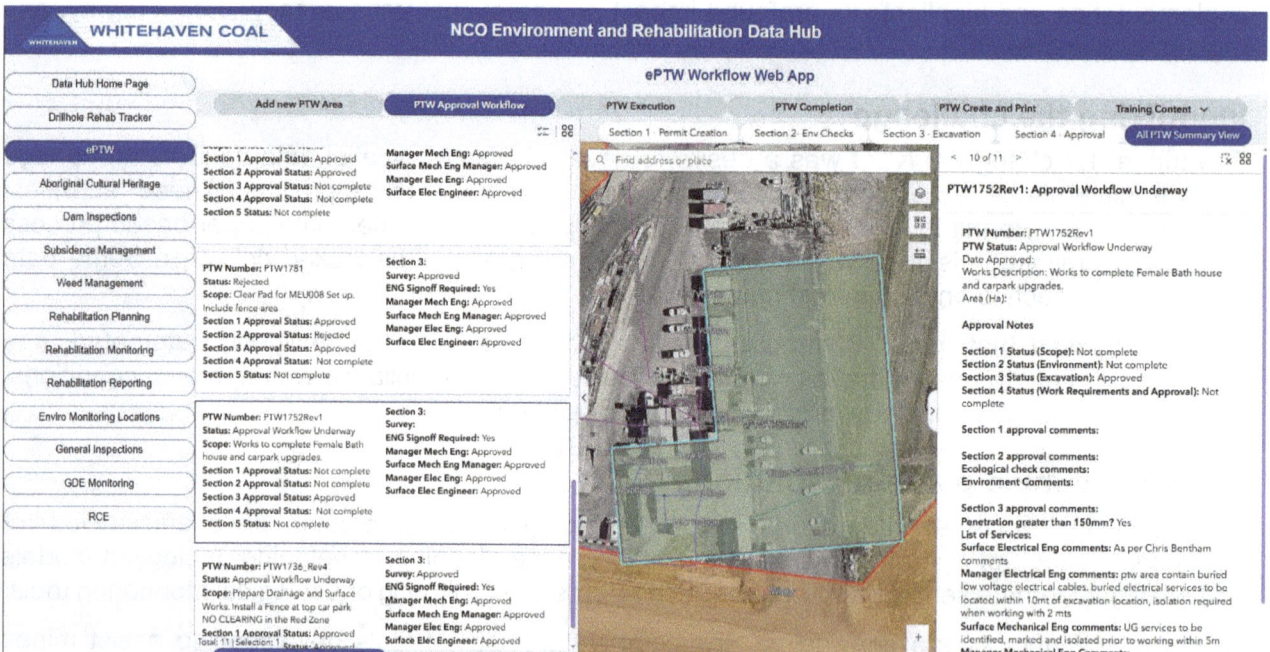

FIG 2 – ePTW Workflow Web Application example.

Key functionalities include:

- Digital, cloud-hosted and field enabled workflows replacing manual processes.
- Automated integration with Survey team (mine plans, services) and IMERCS apps.

- Automated validation for uploaded PTW areas. DXF files are uploaded by Survey team, with the automation validating and submitting to ePTW workflow (ArcGIS feature service).

- Electronic permit approvals with mandated risk-owner signoffs.

- Automated email notifications triggered by workflow progress.

- Dashboards for tracking permit approval status and execution progress.

- In-field surveys for environmental and cultural heritage assessments as well as clearing records and daily sign on.

- Accessed from both desktop and mobile devices.

Email notifications are automated throughout the ePTW App approval workflow across multiple operational departments. The ePTW App system also enables the collection of execution records, including ecological surveys, daily sign on and records of the clearing works with embedded QA/QC steps. Automations support steps required for ePTW close out once all works have been completed.

This system integrates systems, syncing data between the Survey teams CAD records and the ePTW App ensuring real time data updates within the system.

Drill hole rehabilitation tracker

With over 3000 drill holes installed since operations began, tracking rehabilitation status was previously a manual process managed through spreadsheets, paper forms and contractor systems. IMERCS now enables end-to-end life cycle tracking, from planning through decommissioning to rehabilitation, using a spatial interface linked to survey-controlled data. The module enables collection of all records throughout the borehole life cycle including, grouting, cementing, decommissioning, rehabilitation and follow up inspections. Borehole records are synced nightly through a scheduled automation with the Survey Teams data set ensuring seamless integration across the organisation.

This tool has significantly improved record keeping and risk management and ensures compliance with rehabilitation requirements.

Dam safety inspections

A tailored application enables trained personnel to conduct and record dam inspections using mobile devices in the field. Photos, corrective actions, and compliance records are automatically synchronised with the central database, ensuring visibility and audit readiness for internal and external stakeholders.

Subsidence management

The Subsidence Management Workflow enables for the collection of subsidence records locations in the field, rehabilitation and follow up inspection records. Smart symbology and data update automations are used to show what features exceed time frames for rehabilitation. This helps maintain compliance with the rehabilitation time frames and performance.

Aboriginal cultural heritage inspections

The Aboriginal Cultural Heritage module enables the capture of annual audits for each ACH features as required by the management plan. Collection of fence locations is also completed within the field app and this data integrated with the PTW process to ensure compliance is maintained. Data is synced with the ACH dashboard which enable real-time updates of inspections and required actions.

Rehabilitation planning and reporting module

Building on the ePTW App workflow, these modules allow planners to identify areas available for rehabilitation. It assigns tasks to civil teams and enables verification by environmental personnel using real-time dashboards. Progress is automatically tracked against regulatory rehabilitation targets, improving accountability and reporting quality.

Automations are embedded within the workflow to update Field capture records following a verification step by the Environment team. Once verified, the rehabilitation boundary is moved into the Master data set by the Automation. This is master data set is used for submission to the Regulator for annual reporting compliance.

REHABILITATION COST ESTIMATE

A rehabilitation cost estimate module has been created to store all spatial data for the RCE submission. Domain spatial data is sourced from the master data set mentioned above and is displayed in the RCE domain dashboard. The dashboard has been designed to directly inform the RCE submission tool for areas of rehabilitation.

Infrastructure spatial layers have been created that match the RCE tool cost schedule table and these layers are feed into the RCE Infrastructure dashboard. These values directly populate the RCE tool for submission to the Regulator.

Automations have been developed for creating a RCE comparison dashboard, whereby the past submission data is compared with the new submission data, giving a real-time update/reporting of how the RCE data changes from one submission to another.

Module development

Additional modules have been built for Groundwater Dependent Ecosystem Inspections, Weed Management and Rehabilitation Monitoring however, are not discussed in this paper. As management plans differ across sites, regions and jurisdictions, module development is always completed to meet the sites specific requirements within the IMERCS intelligence framework.

IMPLEMENTATION JOURNEY AND CHANGE MANAGEMENT

Implementation of IMERCS followed a phased, risk-managed approach over 2022–2024 (Figure 3). The project began with a discovery phase involving a scoping workshop with key stakeholders across operations, including surveyors, planners, and environmental teams. This collaborative approach ensured system requirements were defined clearly, and user needs were understood.

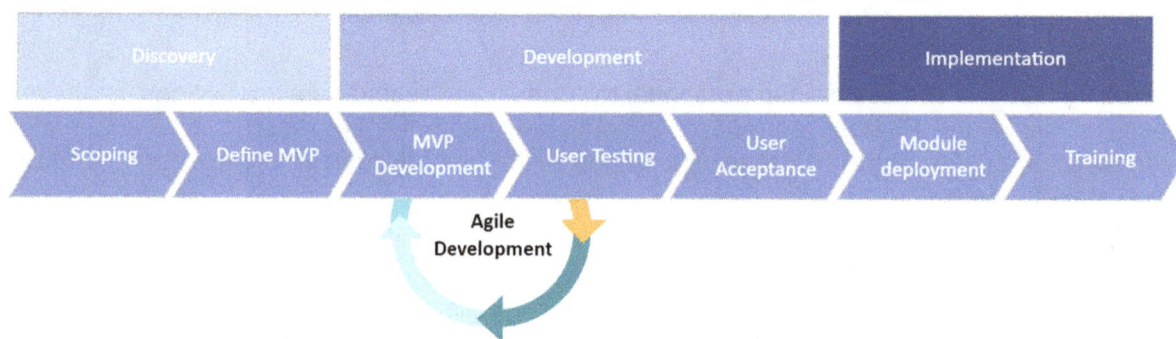

FIG 3 – Phases of project development for each workflow module.

A pilot/minimum viable product (MVP) version of the ePTW App was evaluated alongside existing paper workflows, allowing for bug fixes, feedback, and data validation before full-scale deployment. Training and ongoing support were essential components of the rollout, especially given the initial hesitancy among some users unfamiliar with digital systems.

A comprehensive risk assessment identified potential risks in shifting to digital records and data workflows. Mitigation measures were implemented, including user access controls, automated data backups, and procedural changes aligned with Whitehaven Coal's risk management framework.

BENEFITS, OUTCOMES, AND MEASURABLE IMPACT

The implementation of IMERCS has delivered wide-ranging benefits to NCO's operational, environmental, and compliance performance:

- **Operational Efficiency:** Over 100 ePTWs have been processed without incident, showcasing the system's reliability. The digital interface has saved personnel significant time and reduced errors, while also streamlining interdepartmental approvals.

- **Compliance and Audit Readiness:** During a 2024 audit by the NSW Resources Regulator, IMERCS was recognised as a 'comprehensive and robust system for managing inspections and monitoring data.' The audit team praised its transparency and ease of access to historical and real-time data.

- **Enhanced Rehabilitation Performance:** Greater than 95 per cent of available rehabilitation areas in the northern mining zone were completed successfully. The system enabled prioritisation of rehabilitation works and efficient allocation of civil resources, leading to significant environmental improvements.

- **Real-Time Environmental Intelligence:** Dashboards displaying live rehabilitation progress, disturbance status, and PTW approval metrics empower managers and supervisors with actionable insights. This facilitates proactive decision-making and timely intervention (Figure 4).

- **Cultural Shift and Workforce Engagement:** Through targeted training and ongoing support, IMERCS fostered a culture of accountability and digital literacy. Teams now rely on live data for planning and decision-making, enhancing internal governance.

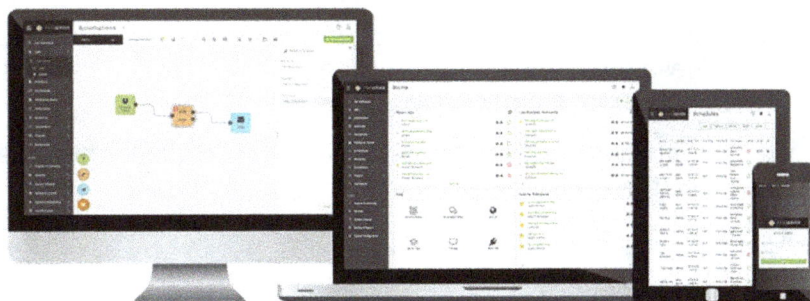

FIG 4 – The IMERCS system operates within a central cloud hosted portal, with one source of data accessed across all platforms (desktop to handheld mobile).

End-user testimonials

Feedback from NCO staff reinforces the system's value:

- **Surface Operations Manager:** 'The Data Hub has brought about a remarkable transformation in how we manage and track environmental and rehabilitation data. We've exceeded compliance requirements and operational goals.'

- **Registered Surveyor:** 'We used to lose hours chasing signatures or replacing lost paperwork. Now, with ePTW, everything is digital and integrated. It's increased both speed and accuracy.'

- **Surface Drilling Coordinator:** 'The single access point for all drilling and rehab data has saved us considerable time and confusion. Crews can now see exactly what's required, and where.'

Leading practice and transferability

The IMERCS architecture is designed with adaptability and scalability in mind. While developed for NCO, its foundation on a robust industry leading technology stack makes it transferable to other mining operations with appropriate technical support. The system's modular design allows companies to tailor workflows, spatial layers, and dashboards to their specific regulatory and operational contexts. While IEMA manages the development and bug fixes for NCO, the system allows for users to create their own maps and apps with the data within the ESRI AGOL environment.

Its success at NCO serves as a benchmark for the broader mining sector, demonstrating how digital transformation can unlock efficiency, accountability, and compliance excellence in environmental management.

AI CAPABILITIES

With the intelligence revolution upon us, the IMERCS intelligence model has recently got a significant update. Using the Open AI API, a vector database within the IMERCS cloud environment and the combination of structured and unstructured data, a paradigm shift in how mine site intelligence and is managed is unfolding. The IMERCS system now has embed AI capabilities including:

- Knowledge Base – Vector store for site records, documents, workflow execution records, management plans etc.

- IMERCS GPT – AI interface to seek intelligence from the knowledge base.

- Regulator GPT – AI tool trained on all Regulatory information.

- Audit AI – An AI agent that compares a compliance requirement document (eg NSW Resources Regulator Targeted Assessment Program Guidelines or Project Approvals) with site-based documentation and provides a report showing where gaps exist.

- Rehab Eye AI – this tool currently in development uses AI to support aerial image processing and analysis for erosion, rehab performance etc (watch this space).

- AI Agents – AI Agents have been developed for weed management reporting and RCE analysis reporting. These agents deliver powerful workflows with embedded reasoning to get more out of the data and support decision-making.

The AI revolution is a new frontier of possibility and the IMERCS platform is ready to deliver the operational improvements that this technology promises.

CONCLUSION

IMERCS represents a step-change in how mining operations manage environmental compliance, rehabilitation, and closure obligations. By combining innovative technologies into a mine site intelligence framework – including automations and AI, this structured, centralised system is delivering enhanced rehabilitation and closure planning knowledge retention and verification evidence. Surface disturbance and rehabilitation workflows significantly improve rehabilitation implementation practices, quality assurance and compliance with regulatory obligations, while providing internal governance tools for the organisation.

Tailored modules address diverse mine site obligations, embedding task planning, execution, record management, and quality assurance processes to ensure comprehensive compliance management. Customisable dashboards provide users with live information on the status of rehabilitation and closure planning metrics.

The transformation achieved at NCO underscores the immense value of structured, intelligent data systems in modern mining. As the industry continues to evolve, solutions like IMERCS will be central to maintaining social license, achieving sustainability goals, and ensuring operational excellence in an increasingly complex regulatory landscape.

ACKNOWLEDGEMENTS

We thank Whitehaven Coal and the Narrabri Coal Operations Environment team for the ongoing support in the development of this technology.

The development of a Queensland approach for field-scale trials of mine waste cover system designs

L Nicolson[1], J Dunlop[2] and J Purtill[3]

1. Lead Technical Advisor, Office of the Queensland Mine Rehabilitation Commissioner, Brisbane Queensland 4000. Email: louisa.nicolson@qmrc.qld.gov.au
2. Principal Technical Advisor, Office of the Queensland Mine Rehabilitation Commissioner, Brisbane Queensland 4000. Email: jason.dunlop@qmrc.qld.gov.au
3. Queensland Mine Rehabilitation Commissioner, Office of the Queensland Mine Rehabilitation Commissioner, Brisbane Queensland 4000. Email: james.purtill@qmrc.qld.gov.au

ABSTRACT

Impacts from AMD seepage from mine wastes are commonly observed. The properties of mine waste, availability of suitable cover materials and climatic factors are unique on every site. Field-scale trials are often used to identify an appropriate cover system design to manage AMD.

To facilitate a consistent approach, we developed advice for the design and delivery of cover system trials. We sought to understand how companies have historically approached trials in Queensland and other jurisdictions, developed a series of case studies, and interrogated international guidance. We then applied these learnings to describe an approach to undertake mine waste cover trials that is applicable in Queensland and similar geographic settings elsewhere.

Our review of the literature and international guidance identified factors that should be considered when planning a field-scale trial and the activities that should be completed prior to undertaking such a trial.

A comparative review of the case studies on cover system trials and cover system performance was also undertaken. We found there is wide variation between the methods used to plan, implement and report results. The case study trials were challenged by remote operating environments, equipment failure, construction deviating from design and other barriers to successful demonstration of performance.

Considering that most mines in Queensland are located where evaporation exceeds rainfall, the work focused on store-and-release cover design with net percolation used as the key performance measure. A field-scale trial was defined as being at a scale sufficient to test the constructability of the design, using full size machinery and built using materials that meet design specifications.

We developed a five-step approach for undertaking a field-scale trial. We present a method to compare cover system design options, identify the most appropriate study design for a field-scale trial to test its constructability and importantly, performance in reducing net percolation.

INTRODUCTION

Rehabilitating mine waste, whether held in tailings storage facilities, waste rock dumps or heap leach piles, is a critical challenge for the mining industry in Queensland. Impacts of acid and metalliferous drainage (AMD) from mine wastes are commonly observed. Cover systems are an integral part of controlling this interaction and therefore a key component of effective mine rehabilitation. The properties of mine waste, availability of suitable cover materials and climatic factors are unique on every site and field-scale trials are often used to identify an appropriate cover system design.

During early stakeholder interactions, leading practice advice regarding cover systems for mine wastes in Queensland with the potential to generate AMD was identified as a priority area for research by the Queensland Mine Rehabilitation Commissioner (QMRC). In Queensland, most mines with this risk profile are in north and central Queensland, with the largest concentration of mines in the North-west Minerals Province near Mt Isa. Regulatory and industry stakeholders have both requested guidance around trials that companies undertake when comparing various cover systems.

In conjunction with technical experts the authors have developed a five-step approach for undertaking a field-scale trial of the preferred cover system design, tailored specifically to site conditions. This is significant as the authors are not aware of any other such approach targeted at Queensland mine sites. The approach was developed following a significant review and evaluation of published literature, Progressive Rehabilitation and Closure plan (PRC plan) submissions, publicly available monitoring reports and conference proceedings and a comparative review of case studies describing field-scale trials of mine waste covers. The approach applies store-and-release cover design principles and is focused on the primary AMD transport mechanism of net percolation with the intent of presenting a pragmatic and cost-effective method to undertake a field-scale trial.

This paper describes the process taken to develop the five-step approach and demonstrate its relevance for application to cover system trials at field-scale in Queensland.

METHODS

Literature review

To support the development of an approach to field-scale trials, the first stage of this work involved a review of the literature and international guidance. This provided an insight into what factors should be considered when planning a field-scale trial, and the necessary preliminary steps leading up to a field-scale trial such as establishing objectives of the cover design, collecting baseline information, and developing cover alternatives.

A broad review of academic literature on mine waste cover trials identified 199 peer reviewed articles and conference proceedings published between 1985 and 2024. Of the 199 documents examined, 70 were related to cover trials or monitoring. These 70 documents spanned 13 countries, with documents pertaining to Australia, Canada, and the USA being the most numerous.

To further understand important facets and recommended elements of cover trial design, leading practice technical documents and guidelines were also reviewed. These included US EPA (1994), MEND (2004), Reid (2005), Price (2009); (Australian Government, 2016a, 2016b), INAP (2017) and Department of Environment, Science and Innovation (2024). Fundamental texts on experimental design were also included.

Identification and comparative analysis of case studies

The authors next considered other information from a range of sources describing mine waste cover performance monitoring and trials. This information was drawn from publicly available conference proceedings, peer reviewed papers, Environmental Authorities and PRC Plans. The information was reviewed to identify cover trials that were described in sufficient detail to include as case studies in a comparative analysis. The focus was on case studies in Queensland, but other studies outside of Queensland were also included where these provided results and information relevant to the climate and environment in Queensland. Each type of document was prepared for a different purpose, and this variation made it difficult to compare directly between studies.

In total, 27 case studies were developed. These include studies from 13 mineral mines in Queensland, two mineral mines each from the Northern Territory and New South Wales and one from Western Australia. One study was from a Queensland site of unknown commodity. An international case study was also included from a mineral mine in South Africa. Most of the studies identified were undertaken on metals mines. Case studies were identified for three coalmines in Queensland where field-scale trials were undertaken. Where information was not publicly available and it was necessary for these to remain confidential, studies were deidentified and are referred to by a case study number.

Meetings were also held with State and Territory government regulatory agencies across Australia to identify case studies and learn how each jurisdiction approaches issues related to management of AMD. In addition, a request was made to the Queensland Department of Environment, Tourism, Science and Innovation (DETSI) for information describing waste cover field-scale trials and monitoring. The information collated for this phase of work represents the best available information on mine waste cover system trials and cover performance, however, it is recognised that there may be information that was not discoverable.

Categorisation of case studies and review criteria

The case studies reviewed vary in terms of their purpose and complexity. Three key study types were identified:

1. Trials on existing structures.
2. Trials in test cells.
3. Environmental monitoring of an existing waste structure.

Case studies were classified as falling into one of these three categories. Each category represents a different approach to test waste cover designs, undertaken during the operation stages of mine life or nearing closure. The case studies reviewed were reported in a range of documents including scientific journals, conference papers or reports prepared by consultants or mining companies as part of a regulatory process.

Case studies were reviewed using a set of criteria. The criteria are unique to this study, however they are consistent with the key activities described in the International Network for Acid Prevention (INAP) framework for the design and testing of mine waste covers (INAP, 2017). Criteria used to review case studies are presented in Table 1.

TABLE 1

Criteria used to review case studies of mine waste cover system trials and monitoring programs.

Criteria	Aspects of case studies considered
Description of study	Study design (eg trials on existing structures, trials in test cells, environmental monitoring of an existing waste structure).
	Description of experimental design (eg controls, replication, plot size, duration of study).
Waste cover design rationale and modelling	Description of the type of cover(s) and layers being evaluated (eg store-and-release, water-shedding, capillary break, reduced permeability layers, capillary breaks, membranes or other layers).
	Justification for the waste cover designs (site-specific climate and AMD risk).
	Modelling supporting design of waste cover(s) (eg erosion, geochemistry, seepage and geohydrological risk models).
	Feasibility of construction (price/material availability/technical difficulty).
Materials	Description of AMD potential, physical and chemical properties of waste material and material used to construct the cover system.
Objectives of the cover design and trial	Description of the criteria used to assess against study objectives (eg Net Infiltration, NP as % of total, contaminant loads or metal concentrations).
Construction and QA/QC	Information describing the techniques used to construct the waste covers being evaluated including quality assurance to check cover construction/installation was as specified.
Monitoring of waste covers	Description of monitoring undertaken including locations, indicators monitored (eg rainfall, run-off, evaporation, soil moisture, surface run-off, groundwater) and the frequency and duration of monitoring.
Limitations to achieving study objectives	Description of factors that may have influenced the success of the study (incomplete, failure of monitoring equipment, lack of maintenance, failure of the waste covers due to damage from animal or extreme weather, change of staff or management).
Reporting of results and recommendations	Description and interpretation of monitoring data against objectives.
	Reports on cover performance and identification of best waste cover option.

Development of a leading practice approach to undertake field-scale trials

Consultation with technical experts identified a store-and-release style cover system is appropriate for most locations in Queensland where evaporation exceeds rainfall. These systems are designed to store infiltrated water within the root zone, allowing the water to be released later by evapotranspiration thereby minimising deep percolation of water and preventing transport of contaminated seepage to the receiving environment.

The scope of the leading practice approach was developed following further discussions with technical experts. The scope considered that:

- a field-scale trial (for the purpose of this leading practice approach) is a series of test cells constructed in the same way and with the same materials that would be used to construct a full-scale cover system.

- net percolation (NP) is regarded as the key measure of performance for a store-and-release design.

- prior to commencement of a field-scale trial, there are several early steps critical to underpin the tested design.

- water flux modelling is required to model cover performance and identify a design or series of cover designs most suitable to trial at field-scale.

- learnings from earlier phases of this work provide a clear understanding of the different approaches taken by mine operators, researchers and regulators as to how trials can be conducted and should inform development of the approach.

- conclusions from case studies regarding the performance of tested cover systems present important learnings and must be incorporated into the approach.

It is recognised that other elements of cover system performance such as arresting capillary rise of salts, are important considerations. However, in store-and-release systems, net percolation is the primary transport mechanism for impacts from seepage on receiving environments. The resultant approach to undertaking field-scale trials is focused on addressing other variables early in the cover design process with the field component dedicated to measuring the key performance objective of reducing net percolation.

The five-step approach focuses on the requirements to meet key design objectives under conditions relevant to Queensland. It also draws from INAP (2017) and the Global Acid Rock Drainage (GARD) Guide (INAP, 2009). The approach is presented in a simplified process that includes clear recommendations for its practical implementation and documentation. This process serves as a useful resource for industry and researchers to plan and undertake cover system trials for mine sites across Queensland and more broadly.

Following its initial development, the approach was presented to several peer reviewers and their feedback incorporated.

RESULTS

The authors' intent was to develop a pragmatic, cost-effective approach to undertake field-scale trials that avoids prescribing excessive requirements. Each phase of work provided a number of interesting perspectives and outcomes, useful to inform the development of an approach for undertaking trials.

Literature review

Key findings from the initial literature review include:

- Developing a thorough understanding of the site where a cover is to be located in the field is crucial to ensure the correct cover system type and designs are chosen.

- Correct characterisation and sampling of waste materials and potential cover materials is critical to understand their acid generation and neutralisation potential and geotechnical properties. This will inform cover design.

- Modelling plays an important role during cover system design and may be used in an iterative manner to test and compare designs ahead of field studies. It can also be a useful tool to investigate aspects of cover performance that are not practical to test during a field-scale trial, such as different time frames, comparing multiple cover designs and undertaking a robust experimental design. Data collected from field trials may also be used to validate model results and improve predictions. Small/medium scale testing can also provide an intermediate step between modelling and a field-scale trial.

- Several examples of incorrect construction of covers for field-scale trials were identified in the literature. Undertaking quality control checks during construction and potentially trialling the construction methodology may be necessary.

Limiting NP was a common objective for field-scale trials identified in the literature. In some cases, no clear objective of the trial was identified, and it was difficult to determine whether the trial was successful. Often, capturing significant weather events was a key factor in determining the duration of a trial, but several cases were identified in the literature where the desired weather events were not captured during the trial. To determine whether the objective of the trial has been met monitoring of key variables, such as NP, is necessary. However, monitoring was identified as a challenge during field-scale trials.

Case studies

Comparative review of case studies

The comparative review of case studies identified several important learnings. Our observations are not a criticism of trials as presented, as each was published in a different setting with different objectives and levels of detail. The comparison was undertaken to better understand trial methodologies employed and their relationship to trial results.

Studies on existing structures provided a means to look back at performance of an as-constructed cover. Other studies were designed to test the performance of alternative mine waste covers to inform the design process prior to construction.

Experimental designs used in trials varied widely. The size of field-scale trial plots reviewed ranged from 30 m × 30 m up to 100 m × 100 m. Field-scale trials studied rarely included controls. When controls were included, they were of natural soil or bare tailings. Few field-scale trials used replication.

The duration of a field-scale trial is a critical consideration for store-and-release covers where rainfall and evaporation are key drivers for moisture storage and NP. Settlement of materials after construction may occur. Results were reported at varying time frames after cover construction or commencement of a field-scale trial. The trials reviewed operated between 6 months and up to 15 years for the longest reported trial at one site. These time frames are indicative only as some of the reviewed trials did not report an end date and may be ongoing once covers are established and monitoring is continued.

Modelling provides a useful way to predict the permeability of cover materials and assess whether cover designs can meet a set objective such as a threshold of NP. Although modelling was undertaken in some studies it was not routinely reported in the case studies.

Twenty of the 27 case studies provided information describing the geotechnical and/or geochemical properties of the cover material.

Objectives varied between studies. The two most stated objectives were to control oxygen ingress to mine waste material and control infiltration of rainfall. Various NP targets were used to benchmark success or performance of a cover. Examples of NP targets used in case studies were 10 per cent or 15 per cent. A target of 10 per cent NP was referred to as an 'industry standard' by some but was not universally adopted. The purpose of some studies was to validate modelling as a step in between final selection of a cover design.

The methods used to construct trial plots were also either not described in full or not described. Many studies did not present information to verify that covers had been built according to their design specifications making it difficult to interpret the results.

A common cause of failure was the disconnect between cover design and construction. It is assumed this occurred where planning and design of the cover system was not integrated with construction methods in the field. This is potentially related to poor communication between the various consultants involved in the process, site staff and construction teams involved, lack of suitable materials, and a failure to ensure rigorous quality assurance and quality control processes. A lack of quality assurance and quality control on material properties and to verify the construction process may also lead to uncertainty when interpreting the results.

In several case studies, monitoring was undertaken to meet regulatory requirements or to inform the regulator and other stakeholders of the performance of the cover system.

The analysis of case studies demonstrated a need for more focused and practical guidance on how to plan and undertake a trial. Existing guidance in the INAP framework (INAP, 2017) requires both a pilot and a prototype field-scale trial prior to full-scale cover construction. This would require significant time and effort to implement in practice. Accordingly, this appears to have limited its adoption to date by industry. A hybrid type of field-scale trial that can trial one preferred design or several alternate designs (similar to a pilot trial as described in INAP) but is constructed at a scale large enough to test construction methods (as required for a prototype trials) is more likely to be applied at mine sites. This hybrid approach would allow covers to be tested in a focused way over shorter time frames and was instrumental in informing the leading practice approach described here.

Cover performance considerations

For those case studies where information regarding cover performance was presented, several failure modes were observed. Broadly, they were associated with:

- a lack of proper materials characterisation.
- poor cover design or selection of the wrong type of cover system.
- the failure to construct the cover system as per the design (driven by poor quality assurance processes).
- increased infiltration due to surface ponding and preferential flow.
- ongoing consolidation of underlying tailings materials, generating low points on the surface of the cover.

Leading practice approach to undertake field-scale trials

The approach presents a method to develop a series of cover system design options, identify the most appropriate design and progress that design to a field-scale trial to test its constructability and importantly, performance in reducing net percolation.

A five-step process is proposed (Figure 1):

Step 1 – Determine cover system objectives and design options (identify cover objectives, collect baseline information, model cover performance and identify a suitable cover design for a field-scale trial, prepare construction specifications).

Step 2 – Plan the field-scale trial (identify trial objectives, determine the trial design, choose a suitable location, determine trial location and size of test cells, set study duration, and design the monitoring system).

Step 3 – Undertake the field-scale trial (construct and monitor).

Step 4 – Refine the design based on trial outcomes.

Step 5 – Report on the trial.

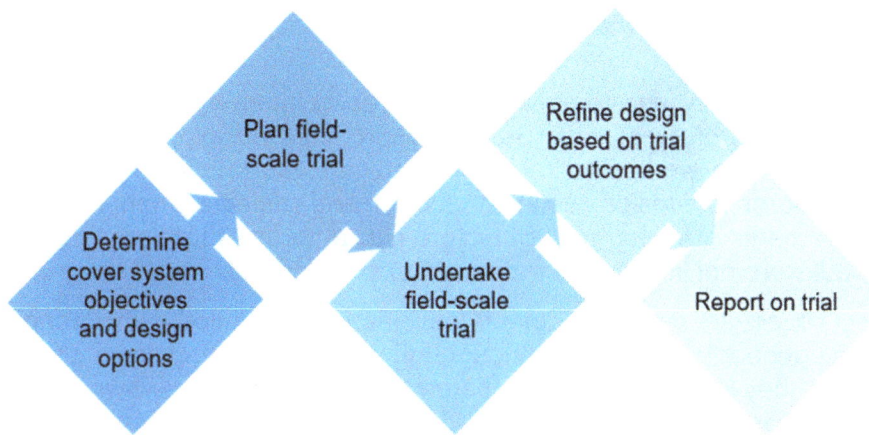

FIG 1 – The five-step process to undertake a leading practice field-scale trial of a cover system.

The first step in the approach begins by considering the site-specific closure scenario in order to determine the design objectives for the cover system. The design objectives for the cover system should then be stated upfront and include a quantitative performance measure related to limiting net percolation. Baseline studies and information gathering activities are key to inform which cover system design(s) will advance for testing at field-scale. Information gathered at this stage should describe the geochemical and geotechnical characteristics of both the mine waste and cover system materials, the climate, topography and hydrogeological settings. A site-specific water flux model and cover system performance assessment should be prepared. Modelling should predict the long-term performance of the cover system under representative climate conditions. Large laboratory column tests and smaller field tests can be undertaken to test material properties and check modelling assumptions. Prior to construction of the field-scale trial, design specifications for each layer in the cover system must be prepared and a materials inventory undertaken to locate and quantify the volumes of materials required to construct the cover system.

Planning the field-scale trial requires that clear objectives are developed for the trial and used to inform decisions regarding trial design, location and size of the test cells, trial duration and monitoring. As controlling NP is the primary objective of a store-and-release cover system, lysimeters (or other seepage collection instrument) must be installed to directly measure NP and water balance estimates of NP established. A fully automated meteorological station in close proximity is required to develop potential evapotranspiration rates and identify site-specific weather conditions and patterns.

Undertaking the field-scale trial involves constructing the test cells and monitoring performance of the cover system for a defined period. Diligence is required to ensure the cover system is constructed according to the design, as failure to follow design specifications has previously been identified as one reason why cover systems fail to meet their performance objectives. A QA/QC system must be developed and implemented during the construction phase of the trial. Monitoring data collected during the trial must be stored, including any calibration activities, failure of equipment, repairs and other disrupting events.

Once the trial has been completed, monitoring data will be collated and interpreted to inform the final cover system design. The water balance must be determined and presented along with an analysis to show how water moved through the cover system during the trial (ie reporting on NP). Monitoring data must also be used to calibrate and refine earlier water flux modelling simulations to improve accuracy and reliability regarding the predicted performance of the cover system over time.

The fifth and final step of the approach is to present information describing the trial and its outcomes. A template outline of the structure and content of the report has been prepared. Raw data should be provided as an appendix to the report or be available for scrutiny.

The field-scale trial must be supported by a record management system to ensure each step in the leading practice approach is documented and records maintained for future use.

DISCUSSION

The broader, multi-disciplinary and temporal aspects associated with mine waste cover systems required the authors to carefully consider how to approach the topic and prioritise which aspect should be the focus of the work. The role of vegetation upon cover system performance, the resistance of a landform to erosion, solute transport and impacts upon receiving surface and groundwater environments are examples of factors critical to the successful performance of mine waste structures but were not included in the scope of this work.

Where mine waste covers are properly planned, designed and installed early during operations they can help to limit the generation of AMD. Although mine waste cover systems provide a useful way to reduce AMD risks, they should not be considered as a solution on their own. Instead, mine waste covers should be used as part of a broader strategy to manage AMD across the life of a mine where potentially acid forming (PAF) material is identified and handled selectively to reduce the footprint of PAF materials.

The authors' perception is that there are barriers to undertaking robust, field-scale trials of cover systems, and a lack of transparent data on the performance of these systems. This led to the establishment of the work described in this paper. Trials represent an important step in mine closure planning as it relates to managing AMD generating material. However, while they are used to develop, compare and refine options for mine waste cover designs, they require significant resourcing and are often undertaken in remote and logistically challenging locations. These challenges were seriously considered throughout development of the five-step approach.

Feedback from industry and government members was crucial to informing the priority and scope of this work. During early engagement with stakeholders, questions were raised regarding appropriate time frames for demonstrating cover performance, minimum modelling requirements and parameters, and how water balance influences the geotechnical and geochemical stability of final landforms.

Additional feedback from technical experts in Queensland identified several themes related to issues about mine waste cover systems. They included the following:

- A continuum of practice is required to limit or prevent ongoing production of AMD. Cover systems are one strategy, but other management actions such as material placement and source control must be deployed throughout the life-of-mine.

- While internationally available guidance reflects standard concepts and approaches, do these remain fit for purpose and are they leading the Queensland industry towards a situation where mining leases or mine domains can be surrendered and with an acceptable level of residual risk?

- Quantifying the success of covers should be a focal point. It is important to quantify how successful cover systems are in achieving good rehabilitation outcomes and this should be informed by a discussion around the processes associated with cover failure in Queensland.

- The design, construction and management of cover systems is multi-disciplinary and must consider landform stability, water management and vegetation establishment.

- Materials characterisation and handling remains critical to managing AMD, and waste management should be informed by ongoing sampling and characterisation campaigns throughout the life of a mine.

- Quality assurance and quality control is crucial to ensure cover systems are constructed and operated in accordance with designs. This is particularly relevant with increasing complexity associated with some cover system designs and represents a major risk to failure of the system.

A Queensland-specific approach

Internationally, there are a number of authoritative documents presenting best practice management for AMD producing mine wastes (INAP, 2020, 2017, 2009; MEND, 2004). These are wide-ranging, written for a global audience and comprehensive in the information presented.

Generating a Queensland-centric process narrowly focused on how to undertake a field-scale trial of a preferred cover system design requires an understanding of several technical disciplines (eg geochemistry, geotechnical, soil and erosional processes, solute transport) that feed important information into the cover design development process. The recently published Tool for Acid Rock Drainage and Metal Leaching Prevention and Management (INAP, 2025) provides a method to connect guidance in the GARD guide (INAP, 2009) with site specific actions that require advice from a subject matter expert. However, the work presented here achieves a different outcome in that is presents an approach, broken into a sequence of logical steps, focused on the store-and-release types of cover systems most likely suited to the majority of sites in Queensland.

Each step is supported by specific recommendations, aimed at focusing the reader's attention to a list of tasks that should be completed. To encourage consistent reporting of field-scale trials and transparency of information, a standard reporting template is provided.

Learnings taken from the comparative review of case studies highlight the critical importance quality assurance and quality control procedures play. This applies to the selection of materials used to construct a cover system (at trial or full-scale), availability of the required cover system materials and construction methods. It also applies to ensuring monitoring equipment is routinely checked and maintained. The five-step approach requires investment of time and effort to ensure document management, process control and standard operating procedures are developed and implemented.

A key aspiration of the authors in presenting the five-step approach is to remove barriers regarding how to develop a cover system design to the point of testing at field-scale. It is also hoped that the approach will provide a consistent methodology for trials, encourage more routine application of trials on mine sites throughout Queensland and allow industry and government a more transparent appreciation of cover system performance.

CONCLUSIONS

Field-scale trials are an important part of the broader process necessary to develop and implement strategies to control AMD and potential impacts on the receiving environment.

The five-step approach presents a strategy to develop a series of cover system design options, identify the most appropriate design and progress that design to a field-scale trial to test its constructability and importantly, performance in reducing net percolation. The authors sought to determine how robust field-scale trials may be undertaken in a manner particular to Queensland conditions and for trials to be cost-effective and avoid prescribing excessive or impractical requirements.

Industry and government stakeholders are cognisant of the importance of ensuring cover systems perform effectively to achieve the best possible environmental outcomes and meet regulatory requirements. To match performance and outcomes to Queensland conditions, the sequence and logical inclusions for each step were developed by the authors and then validated against the globally recognised guidelines from INAP (2017).

Key to ensuring the rigour and practical application of the five-step approach were the early phases of work which interrogated academic literature, conference proceedings, PRC plan submissions, monitoring reports and other publicly available information. Also crucial were the experiences and examples presented to the authors during discussions with other jurisdictions and key stakeholders, and in feedback from peer reviewers.

In conclusion, while field-scale trials are the focus of this work, it is important to note they are one component of the continuum of practice necessary to develop, implement and monitor a cover system. The authors look forward to learning of future case studies where the five-step approach has been implemented and practically applied to demonstrate the performance of a cover system design.

This work is presented as three technical papers, available for download at <qmrc.qld.gov.au>.

ACKNOWLEDGEMENTS

This work was funded by a grant from the Financial Provisioning Scheme (Queensland Treasury) which supports research into mine rehabilitation.

The authors wish to acknowledge the technical expertise of Dr Warwick Stewart and Antony Volcich from Environmental Geochemistry International, and the contributions of Dr Megan Clay and Dr Bhasker Rathi.

Thanks also to Professor Mansour Edraki, Gilles Tremblay and officers from the Department of Environment, Tourism, Science and Innovation for their reviews and Professor David Williams and Professor G. Ward Wilson for their input.

REFERENCES

Australian Government, 2016a. Leading practice sustainable development program for the mining industry: Preventing acid and metalliferous drainage, Australian Government.

Australian Government, 2016b. Leading practice sustainable development program for the mining industry: Evaluating performance – monitoring and auditing.

Department of Environment, Science and Innovation (DESI), 2024. Guideline: Estimated rehabilitation cost under the Environmental Protection Act 1994.

International Network for Acid Prevention (INAP), 2009. The global acid rock drainage guide (GARD Guide), Mine Water and the Environment [online], INAP. Available from: <https://www.gardguide.com/index.php?title=Main_Page> [Accessed: 26 June 2025].

International Network for Acid Prevention (INAP), 2017. Global cover system design, The International Network for Acid Prevention.

International Network for Acid Prevention (INAP), 2020. Rock placement strategies to enhance operational and closure performance of mine rock stockpiles phase 1 work program – Review, assessment and summary.

International Network for Acid Prevention (INAP), 2025. Tool for acid rock drainage and metal leaching prevention and management.

Mine Environment Neutral Drainage (MEND), 2004. MEND 2.21.4a Design, construction and performance monitoring of cover systems for Volume 5 Case Studies.

Price, W, 2009. Prediction manual for drainage chemistry from sulphidic geologic materials, MEND Report 1.20.1, Mine Environment Neutral Drainage (MEND), Smithers, British Columbia, Canada.

Reid, G, 2005. How to conduct your own field trials, New South Wales (NSW) Government. Available from: <https://www.dpi.nsw.gov.au/__data/assets/pdf_file/0020/41636/Field_trials.pdf>

US Environmental Protection Agency (US EPA), 1994. Technical document: Acid mine drainage prediction.

Advancing progressive rehabilitation through research and innovation

F Saavedra-Mella[1], G Mullins[2], C Blackburn[3], J Barker[4] and L Commander[5,6]

1. Environmental Research Scientist – Rehabilitation, Alcoa of Australia, Pinjarra WA 6208. Email: felipe.saavedra@alcoa.com
2. Plant Production Superintendent, Alcoa of Australia, Pinjarra WA 6208.
3. Environmental Research Scientist – Flora, Alcoa of Australia, Pinjarra WA 6208.
4. Environmental Research Scientist – Fauna, Alcoa of Australia, Pinjarra WA 6208.
5. Research Manager, Alcoa of Australia, Perth WA 6000.
6. Adjunct Senior Research Fellow, The University of Western Australia, Perth WA 6000.

ABSTRACT

Through decades of collaborative research and site-based experience, Alcoa has developed many innovative approaches to progressive mine rehabilitation. These advancements have contributed to improved ecosystem rehabilitation, informed future practices, and inspired global best practices for creating sustainable and resilient post-mining landscapes. This long-term commitment to research and innovation has enabled the refinement of techniques and significant progress in restoring diverse and self-sustaining ecosystems.

These successes have been achieved through a combination of pioneering rehabilitation techniques including optimised deep ripping, topsoil management, tailored seed mixes incorporating ecologically significant plant species, and the development of innovative tissue culture and germination pre-treatment techniques. Combined with optimised seeding, enhanced plant survival, and lower tree density stocking rates, Alcoa's approach has enabled successful re-establishment of diverse flora communities and fauna habitat.

More recent innovations include using remote sensing tools for rehabilitation integrity assessments and erosion detection, enabling early intervention if necessary. The return of key amphibian, invertebrate, mammalian, avian, and reptilian species, restored soil microbial function, and the reassembly of insect pollinators reveals a positive ecosystem recovery trajectory. Ongoing research is focused on the re-establishment of ancient grasstrees in rehabilitated areas in recognition of their cultural significance to Noongar People and understanding forest resilience in a changing climate, including the influence of fire and drought. A range of knowledge-sharing activities are prioritised, including regular open-access research publications, symposiums, public site tours and student education programs.

Alcoa is accelerating the pace and raising the bar for mine rehabilitation. The newly established Forest Research Centre, a company-wide collaborative initiative, is driving this transformation. Central to its mission is the integration of Aboriginal cultural knowledge, recognising the irreplaceable value of traditional ecological practices for sustainable and culturally sensitive outcomes. This paper presents initiatives showcasing how the integration of research, innovation, and cultural considerations leads to improved mine rehabilitation outcomes and redefined best practices.

INTRODUCTION

As the global shift towards a carbon-neutral future accelerates, Alcoa's operations play a crucial role in supporting the energy transition through responsible resource management and sustainable rehabilitation efforts. Mining activities, while essential for providing critical materials, pose socio-environmental challenges that demand research and innovative solutions.

Progressive rehabilitation is a key strategy to address these concerns, aiming not only to restore landscapes but also to enhance their resilience and ecological function. Alcoa's approach to land rehabilitation underscores the importance of integrating research, innovation, and increasingly, Aboriginal cultural knowledge into rehabilitation practices within specific environments such as the Northern Jarrah Forest, where the Huntly and Willowdale bauxite mines are located (Commander et al, 2024).

Alcoa's long-term commitment to this region for over six decades and 50 years of research reflect a sustained dedication to establishing and returning a self-sustaining jarrah forest ecosystem that aligns with agreed forest values. This approach integrates adaptive strategies to create a resilient and biodiverse restored forest, where biodiversity-focused practices – guided by scientific research and technological advancements – have driven key improvements in ecosystem health, advancing significant progress towards rehabilitation goals.

This paper explores the role of research, innovation, and the integration of Aboriginal cultural knowledge in advancing progressive rehabilitation, highlighting case studies, emerging technologies, and evolving methodologies that contribute to improved environmental outcomes and redefined best practices. By prioritising sustainable mining practices and rehabilitation efforts, Alcoa is setting new benchmarks for responsible resource management. Through collaborative efforts and continued investment in cutting-edge solutions, the path toward a carbon-neutral future becomes increasingly attainable – where economic development aligns with environmental stewardship and ecological preservation.

ALCOA'S MINING AND REHABILITATION APPROACH

Alcoa operates two bauxite mines in Western Australia – Huntly and Willowdale. Both mines are located within the Northern Jarrah Forest subregion, which is characterised by annual rainfall ranging from 600–1300 mm, predominantly occurring during the winter months.

The Northern Jarrah Forest ecosystem features key overstorey species such as jarrah (*Eucalyptus marginata*) and marri (*Corymbia calophylla*), alongside a diverse understory. It supports a remarkable biodiversity, including 29 mammal species, 150 bird species, and 45 reptile species, some of which hold conservation significance (Nichols and Muir, 1989). Many of the forest's plant species are adapted to fire, exhibiting traits such as resprouting and reseeding. Some plants even produce seeds that germinate in response to smoke (Bell, 2001; Norman *et al*, 2006; Roche, Koch and Dixon, 1997).

Alcoa employs a pod-mining method to extract relatively shallow bauxite ore deposits, typically located within the top 4–5 m of the soil profile. This method targets discrete areas of ore, allowing mining activities to cease once the ore in a specific pod is depleted. Importantly, Alcoa restricts its mining operations to areas that were previously logged, deliberately avoiding old-growth forests, national parks, and locations with high conservation value (Alcoa, 2023b).

Following the completion of ore extraction in a mined area, Alcoa undertakes a rigorous and progressive rehabilitation process (Figure 1) aimed at establishing and returning a self-sustaining jarrah forest ecosystem that aligns with agreed forest values. These current efforts are guided by continuous research and evolving rehabilitation completion criteria, developed in consultation with and approved by regulators. Alcoa is transitioning future mine approvals to a more contemporary process, with two current referrals currently released by the EPA for 12 weeks of public consultation.

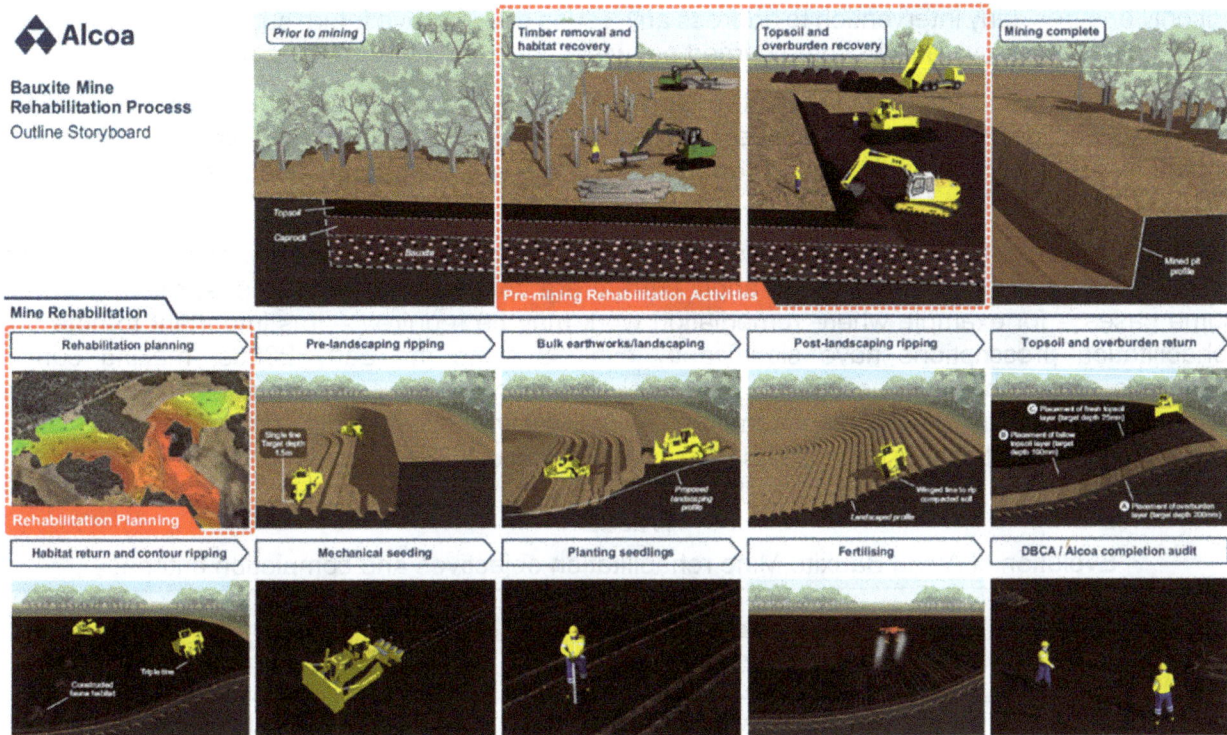

FIG 1 – Bauxite mine rehabilitation process.

EVOLUTION OF ALCOA BAUXITE MINE REHABILITATION OBJECTIVES AND COMPLETION CRITERIA

Alcoa has been successfully mining and undertaken rehabilitation in the Northern Jarrah Forest since the 1960s. Rehabilitation performance is measured against the Rehabilitation Completion Criteria developed in consultation with, and approved by the Bauxite Strategic Executive Committee (previously known as the Mining and Management Program Liaison Group).

As Alcoa's rehabilitation practices evolve over time – informed by improvements in research, technologies and shifting community expectations – the Rehabilitation Completion Criteria is periodically reviewed to reflect improved and modern practices. Due to this process, different Rehabilitation Completion Criteria and assessments of success are applicable to different periods of rehabilitation establishment.

Alcoa in collaboration with the Mining and Management Program Liaison Group drafted the first set of completion criteria. Rehabilitated areas established up to 1987 (referred to as Early Era rehabilitation) used non-local species including pine trees and Eucalyptus species from Eastern Australia. This reflected the agreed Post Mining Land Use requirement, which at that time was to support the establishment of a non-jarrah forest ecosystem. Initially, there was a sole objective of supporting a future timber industry and this contributed to species being selected for rehabilitation in this period based on their resilience to the soil-born pathogen *Phytophthora cinnamomii.* Planting practices were also based on facilitating high-yield timber harvesting in the future.

Since 1988 (referred to as Current Era rehabilitation) only local jarrah forest species have been used in rehabilitation. Key Rehabilitation Completion Criteria have been related to the establishment of a native species overstorey, which is the primary indicator of vegetation cover and primary productivity in a forest ecosystem, and native understorey species, which are the predominant indicator of floristic diversity in the jarrah forest. While support for a future timber industry remained one of the objectives, other objectives have been introduced over time.

Each criterion has an associated standard, including quantitative targets against which Alcoa monitors and reports rehabilitation performance.

Assessment of rehabilitation against the Rehabilitation Completion Criteria is applied throughout each stage of rehabilitation operations and during the early years of ecosystem development. This not only ensures early intervention to address any issues and improve future success, but also allows any required corrective actions to be carried out while operations are ongoing nearby.

Alcoa's rehabilitation approach under the current Rehabilitation Completion Criteria (endorsed in 2015 and applied from 2016) is to restore a self-sustaining jarrah forest ecosystem planned to enhance or maintain water, timber, recreation, conservation and/or nominated forest values (Alcoa, 2015).

Assessing Alcoa's rehabilitation success requires the application of era-specific completion criteria. While Alcoa looks at opportunities to apply current era completion criteria to past rehabilitation in some cases – for example where remediation work may be required – it is important to note that rehabilitation prescriptions have significantly evolved since the early 2000s, making current standards unsuitable for assessing older rehabilitation eras. This progression from basic revegetation towards establishing complex, self-sustaining ecosystems, driven by decades of research and operational experience, is highlighted in Table 1.

TABLE 1

Evolution of Alcoa Bauxite Mine rehabilitation objectives and completion criteria.

Rehabilitation establishment year(s)	Applicable completion criteria standard (and approval date)	Main objectives/characteristics of the era
1966–1987	Early Era (pre-1988) Rehabilitation. Formally accepted by the MMPLG on 5 June 2002.	Support the establishment of a non-Jarrah Forest ecosystem. Species selected for rehabilitation in this period were chosen based on their resilience to the soil-born pathogen *Phytophthora cinnamomi*.
1988–2004	Current Era (post-1988) Rehabilitation Formally accepted by MMPLG on 14 October 1998.	Shift towards using key native species like Jarrah and Marri. Introduction of specific stocking density targets influenced from forestry end-use goals. Increasing focus on native understorey.
2005–2015	Completion Criteria for 2005 onwards – Current Era Rehabilitation. The review was completed and approved by MMPLG on 15 March 2007.	Reduction in target tree stocking densities and decreased fertiliser application to better mimic natural Jarrah Forest conditions and nutrient cycling.
2016–2026	Completion Criteria for 2016 Onwards – Alcoa's Bauxite Mine Rehabilitation Program Completion Criteria and Overview of Area Certification Process, 2015 Revision approved by MMPLG on 6 February 2015.	Restore a self-sustaining Jarrah Forest ecosystem planned to enhance or maintain water, timber, recreation, conservation and/or nominated forest values.

Note: Mining and Management Program Liaison Group = MMPLG.

ENDURING RESEARCH COMMITMENT – THE FOREST RESEARCH CENTRE

For decades, Alcoa has maintained a strong commitment to investing in and developing research to advance the science and practice of mine rehabilitation. This enduring dedication has fostered numerous significant improvements and innovations in the rehabilitation of the jarrah forest. Building on this extensive history, in 2024 Alcoa announced the establishment of the Forest Research Centre (Alcoa, 2024), a company-wide collaborative initiative, which will serve as a central hub for scientific research, collaborative projects, and the development of cutting-edge techniques, playing a pivotal role in Alcoa's progressive approach to creating resilient and biodiverse rehabilitated ecosystems. Recognising the irreplaceable value of traditional ecological knowledge for achieving truly sustainable and culturally sensitive rehabilitation outcomes, the integration of Traditional Owners' unique knowledge is central to its mission. Research will address five core pillars: First Nations Two-

Way Science, Fauna Protection and Return, Enhancing Forest Flora Knowledge, Rehabilitation Execution, Water Stewardship.

SNAPSHOT OF ALCOA'S REHABILITATION ADVANCES AND INNOVATIONS

Optimising soil conditions for ecosystem recovery

Deep ripping

The process of extracting bauxite has provided opportunities for innovative rehabilitation techniques. While heavy machinery movement can lead to soil compaction, addressing this compaction effectively can enhance water infiltration, improve soil aeration, and support the growth and penetration of plant roots improving the establishment and growth of native vegetation.

Alcoa has significantly advanced rehabilitation techniques by pioneering the application of deep ripping, transforming soil rehabilitation practices. Early efforts in deep ripping utilised conventional chisel tines capable of reaching depths of 1.4 m, laying the groundwork for future advancements. Building on these early efforts, Alcoa invested in cutting-edge research that led to the development of the winged tine in the 1980s, a breakthrough that revolutionised soil fragmentation. Decades of research and development have shown that winged tines outperform conventional straight tines by delivering superior soil disruption while avoiding the depth limitations often caused by soil type or moisture content.

This landmark innovation significantly enhanced deep tillage by lifting and fracturing a greater volume of soil, dramatically improving aeration and reducing bulk density with unparalleled efficiency (Croton and Ainsworth, 2007). Alcoa's continuous commitment to research and development led to various optimisations of winged tine design, ensuring maximum soil fragmentation and durability over time (Figure 2). Through rigorous testing, Alcoa perfected its design, ultimately adopting an optimal wingspan to maximise results.

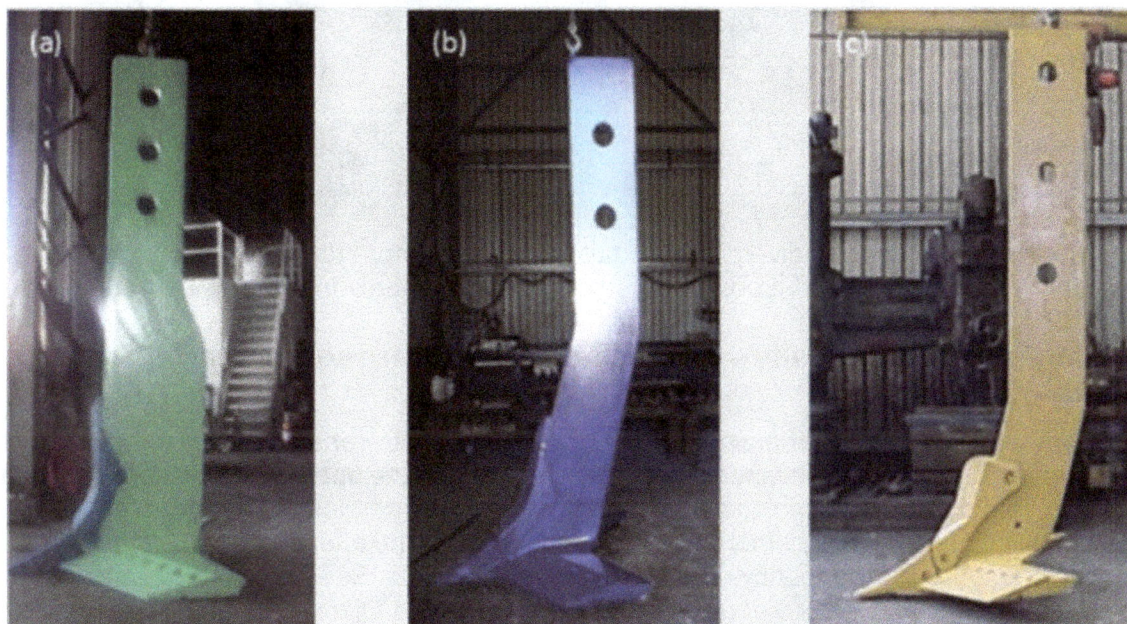

FIG 2 – Alcoa research and development optimisation of winged tine design for maximum soil fragmentation and durability: (a) 1980–1990; (b) 1990–2000; (c) post-2000.

Topsoil management

Recognising the critical importance of topsoil for successful forest rehabilitation, Alcoa employs a strategic and research-driven approach to its management. A key element of this approach is the pioneering double-stripping technique (Tacey and Glossop, 1980). This method involves separately removing the uppermost layer of soil, typically to a depth of approximately 50–75 mm (though depths can vary), during the dry summer months and storing it for less than three months. Research indicates this shallow layer contains the majority of viable native seeds within the jarrah forest topsoil, including seeds of geosporous (soil-stored) species, as well as beneficial micro-organisms (Liddicoat *et al,* 2022) and fungi (Ducki, 2020). A second layer of soil termed as overburden – the material beneath the seed-rich topsoil layer – is removed and separately stored. This consists of beneficial organic material and is returned to rehabilitation before the seed-rich topsoil layer.

This targeted removal of the seed-rich topsoil significantly enhances seedling emergence in rehabilitated areas compared to single stripping or prolonged stockpiling of the entire topsoil layer, ultimately capturing nutrients and beneficial organisms that contribute approximately 60 per cent of species diversity in the rehabilitation area (Figure 3). The remaining species diversity is achieved through approximately 30 per cent direct seeding and 10 per cent planting of recalcitrant species. It is important to note that every mine pit receives this vital topsoil. This practice helps to restore the soil profile structure and, importantly, contributes to root zone reconstruction by providing sufficient depth and a suitable physical environment for root development. The combination of this stable foundation and the nutrient-rich topsoil layer provides a more favourable environment for both plant establishment and the development of a resilient long-term ecosystem.

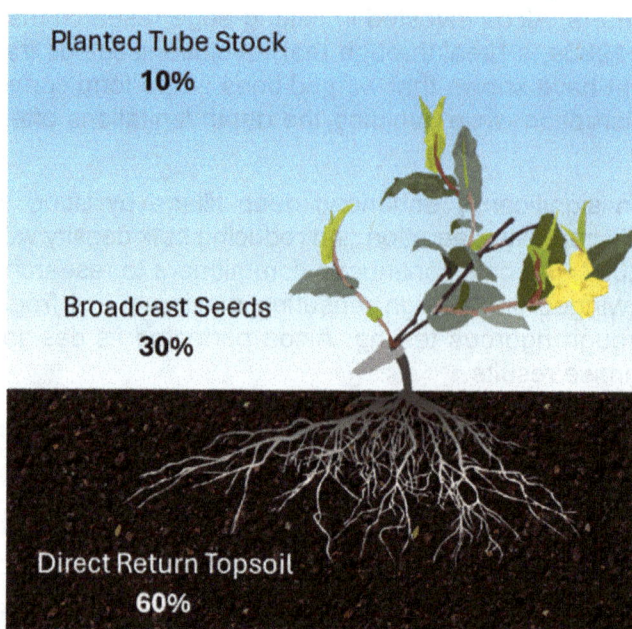

FIG 3 – Percentage of species diversity contributed by direct return topsoil, broadcast seeding, and planted tubestock (percentages are approximate).

Restoring plant biodiversity – tailored approach through research and innovation

Alcoa prioritises the re-establishment of plant biodiversity in its rehabilitated areas, employing a comprehensive strategy that combines extensive research with the application of tailored seed mixes and innovative propagation techniques. This commitment is underpinned by a deep understanding of seed behaviour, informed by numerous studies investigating the effects of heat, temperature, light, burial depth, smoke, and dormancy (Bell, Vlahos and Watson, 1987; Bell and Bellairs 1992; Bell 1994; Grant *et al,* 1996; Norman *et al,* 2006; Roche, Koch and Dixon, 1997). This robust body of knowledge directly guides the seed pre-treatments utilised by Alcoa to maximise germination and establishment success. Most plant species are introduced in the first two years of rehabilitation execution activities, however, if early targets are not met, remediation including additional planting occurs.

Alcoa determines the most effective methods for returning plant life to rehabilitated areas by carefully considering their functional traits. Seed storage characteristics are key. Species with seeds that naturally reside in the soil (geosporous) are returned via topsoil application, while those that store seeds in their canopy (serotinous) are reintroduced through direct seeding (Figure 4). The way seeds are dispersed also provides valuable information, with wind-dispersed species expected to naturally colonise the area over time. Certain species, such as clonal plants, reproduce through means other than seeds. These are typically propagated using cuttings or tissue culture and then planted by hand. Before undertaking large-scale planting, each species undergoes field trials to rigorously assess survival rates.

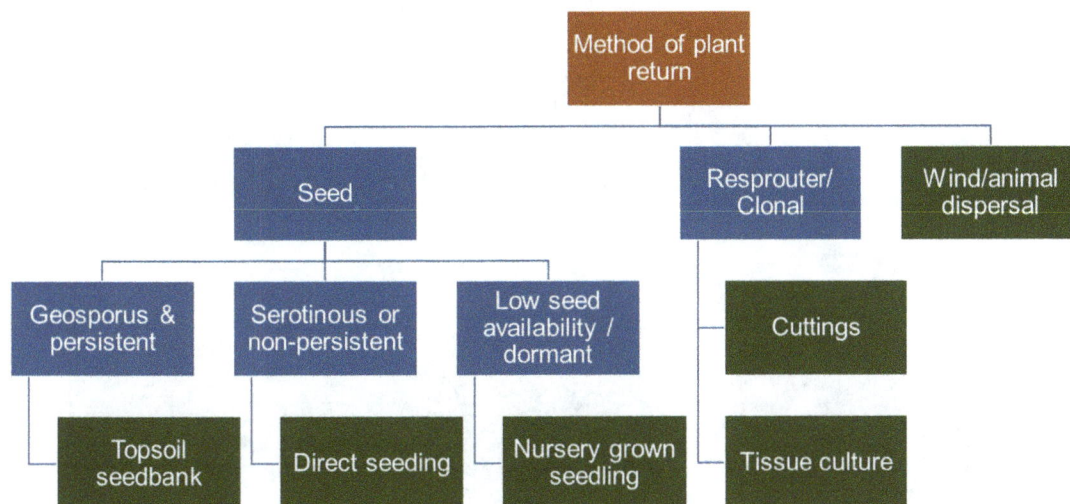

FIG 4 – Method of plant species return considering species reproduction traits (Commander *et al,* 2023).

A cornerstone of Alcoa's progressive rehabilitation strategy is the development and meticulous application of tailored seed mixes that are subject to continuous review and adaptation. To ensure the genetic integrity of the restored ecosystems, seeds are collected from designated seed collection zones (Figure 5). These provenance zones, which are specific geographic areas with defined environmental conditions for seed collection were developed based on DNA analyses and the geographic genetic differentiation of various plant species, as well as the mine cycle progresses. Alcoa's seed mixes typically incorporate a diverse array of 40 to 50 native species contributing approximately 30 per cent to the overall plant species of rehabilitated areas and complementing the 60 per cent derived from topsoil.

Recognising that some species are challenging to establish solely from seed, Alcoa has also invested in innovative propagation techniques. In the 1990s, a tissue culture facility was established to propagate 'recalcitrant' species which either fail to produce seeds or produce them in very low quantities, preventing propagation through seeds, enabling the reintroduction of otherwise poorly represented flora. This led to the development of the recalcitrant program in the 2000s, which by 2025 is reaching an annual production of up to 750 000 seedlings, planted at an average density of ~1300/ha. Current survival of these recalcitrant species in the rehabilitation is around 76 per cent for monocots and 64 per cent for dicots used regularly in the program after the second summer. Approximately 10 per cent of plant species found in rehabilitated areas are returned through tubestock plantation.

Current research continues to focus on improving the seed germination of dormant species, such as *Hibbertia* and *Ericaceae*, through the development of targeted pre-treatments. Collaborations with engineers are driving technological advancements in direct seeding machinery to improve seed placement, survival rates, and subsequent germination success. Finally, a long-term study on challenging plant species is providing valuable insights into optimal pot sizes and planting treatments to maximise the survival of planted seedlings and cuttings, with direct applications in refining rehabilitation practices.

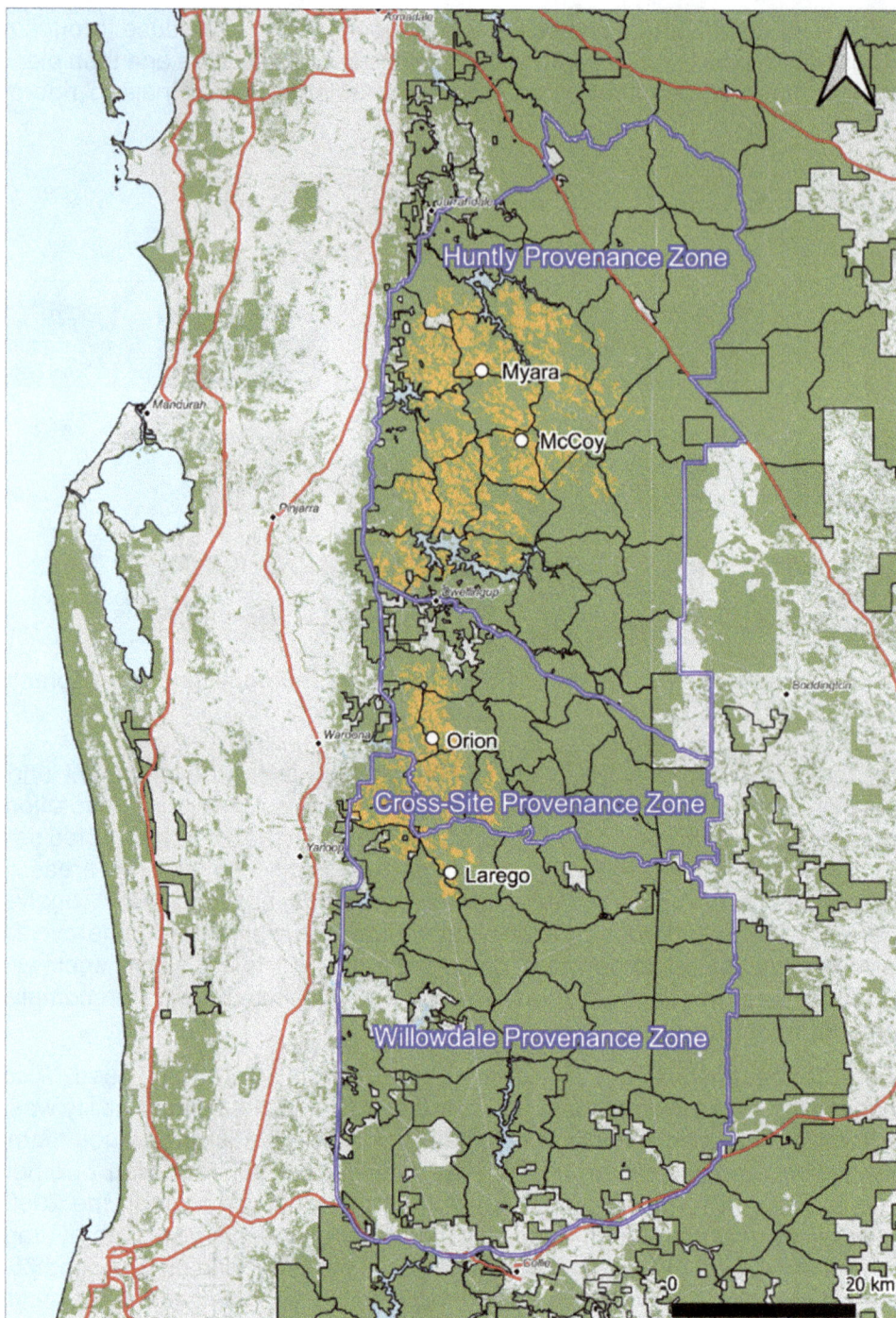

FIG 5 – Alcoa seed provenance zones.

Remote sensing technologies

Optimising contour ripping accuracy

Remote sensing tools are proving to be valuable for the early detection of erosion within rehabilitated mine sites. To further enhance landform management, Alcoa has developed an advanced technology to optimise the precision of contour ripping in mine rehabilitation. This internally engineered tool provides a powerful means of enhancing water management and erosion control in rehabilitated areas. By accurately assessing surfaces using High-resolution imagery along with Structure from Motion (SfM) modelling techniques, Alcoa can ensure surface ripping is completed on contour and generates insightful metrics on the impact of ripping to surface hydrology (Figure 6). This allows Alcoa to gain a significant advantage in proactively identifying areas for optimisation. This innovative approach enables early intervention to safeguard soil health, maximise surface water capture for infiltration, and ultimately contribute to the long-term stability and resilience of the rehabilitated ecosystem by identifying potential points of failure before soil loss occurs.

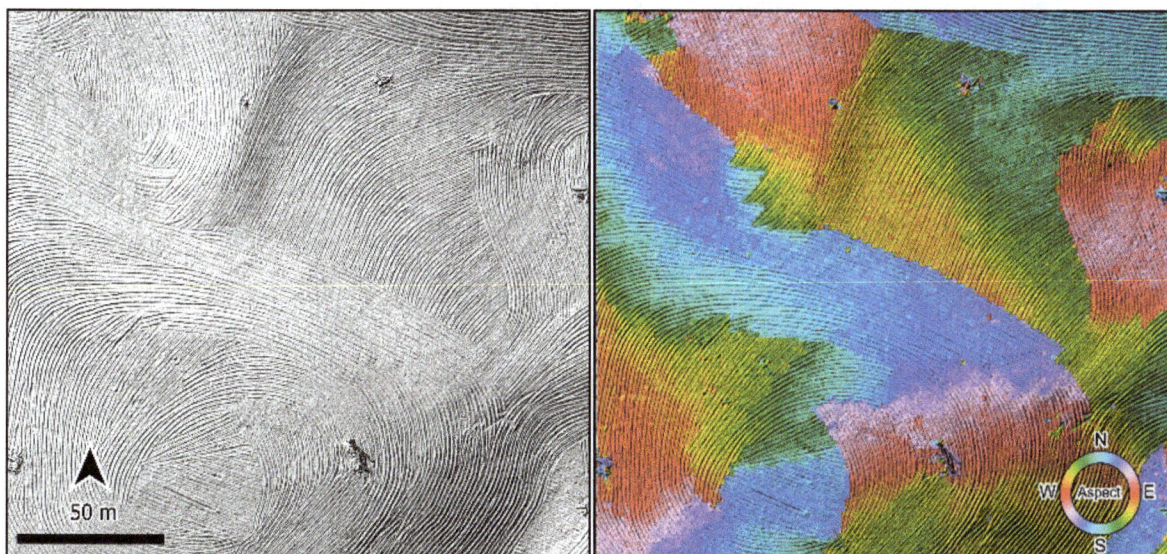

FIG 6 – Results of a surface ripping compliance check algorithm, demonstrating a freshly ripped pit surface (left), and the results of the algorithm (right).

Early erosion detection – surface modelling and landform stability analysis

Recognising that erosion is a natural process in dynamic landscapes, Alcoa employs advanced technologies for the early detection and management of potential erosion within rehabilitated mine sites. Utilising Unmanned Aerial Vehicle (UAV) technology and Digital Elevation Models (DEMs) (Figure 7), Alcoa can proactively identify areas where erosion gullies may be forming. These techniques also provide detailed and highly accurate estimates of landform metrics that aid in predicting erosion patterns, conducting thorough root cause analysis, and developing site-specific remediation strategies. This commitment to early detection enables Alcoa to implement timely interventions and remedial actions to minimise erosion of the rehabilitated areas. This proactive approach is essential for ensuring the long-term physical stability and ecological integrity of the restored forest ecosystems.

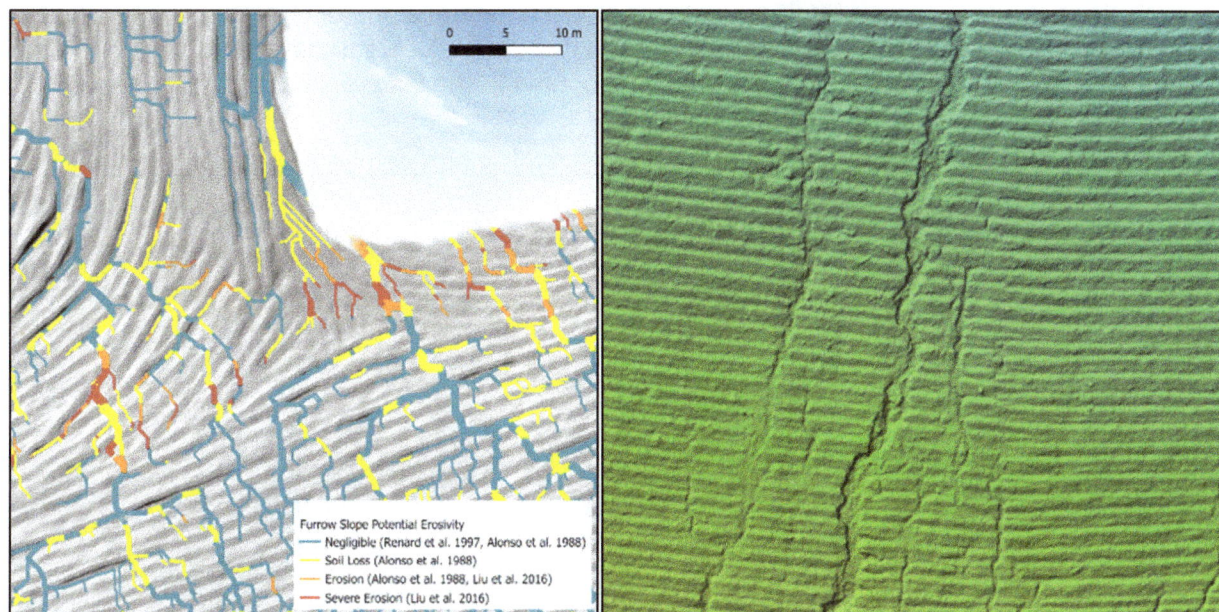

FIG 7 – Structure from Motion generated high resolution Digital Elevation Model of erosion in rehabilitation can form the basis of sophisticated hydrological modelling and erosion prediction (left), and then precisely quantify modes of failure for root cause analysis (right).

Assessing long-term forest integrity using satellite and LiDAR data

Alcoa utilises Landsat satellite based NIR imagery to analyse long-term trends in vegetation indices (Figure 8). This historical data, dating back several decades, allows the company to assess the resilience of rehabilitated areas to natural disturbances such as fire and drought over extended time frames, providing a valuable perspective on the long-term sustainability of rehabilitation efforts.

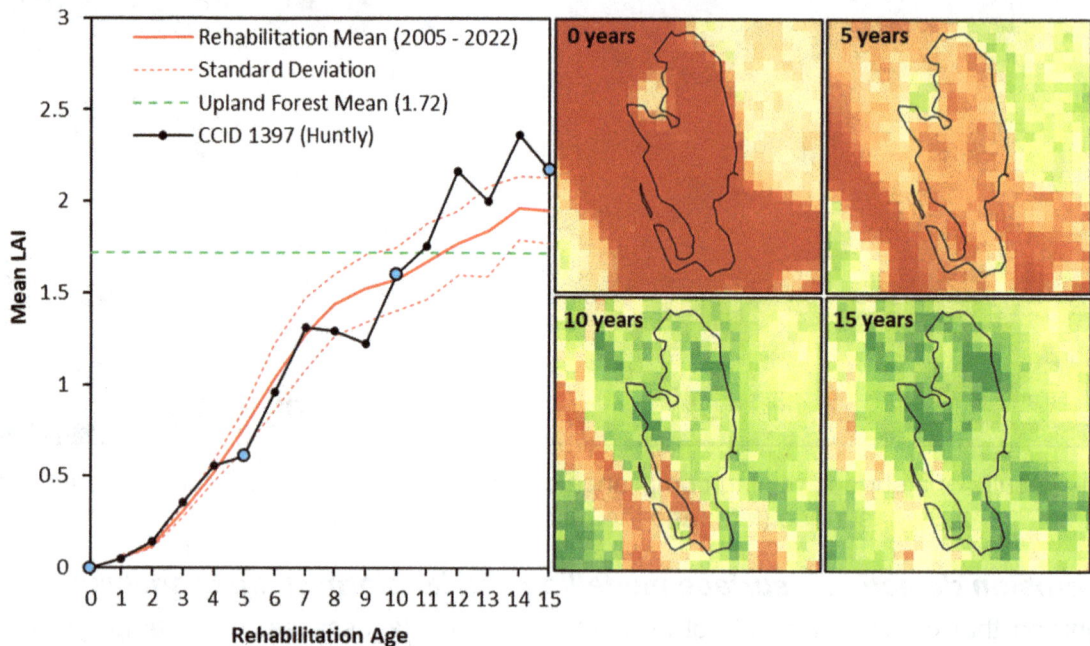

FIG 8 – Long-term LAI trajectory of a single rehabilitation pit over 15 years post rehabilitation, compared with the 15 year mean LAI trajectory of all rehabilitation pits rehabilitated between 2005 and 2022.

Light Detection and Ranging (LiDAR) and other remote sensing technologies also play a crucial role in Alcoa's assessments of rehabilitation integrity, by providing detailed topographic information and vegetation structure data (Figure 9). The presence of significant bare ground, often detectable through these methods, can indicate potential issues with vegetation establishment, soil erosion, or weed invasion, prompting further investigation and potential remedial actions. The assessment of vegetation structure is crucial to understanding rehabilitation development, as vertical distributions can provide sub-canopy metrics without the need for extensive foot traverses. This advanced monitoring enables Alcoa to proactively manage potential challenges and ensure the long-term success and resilience of its rehabilitated ecosystems.

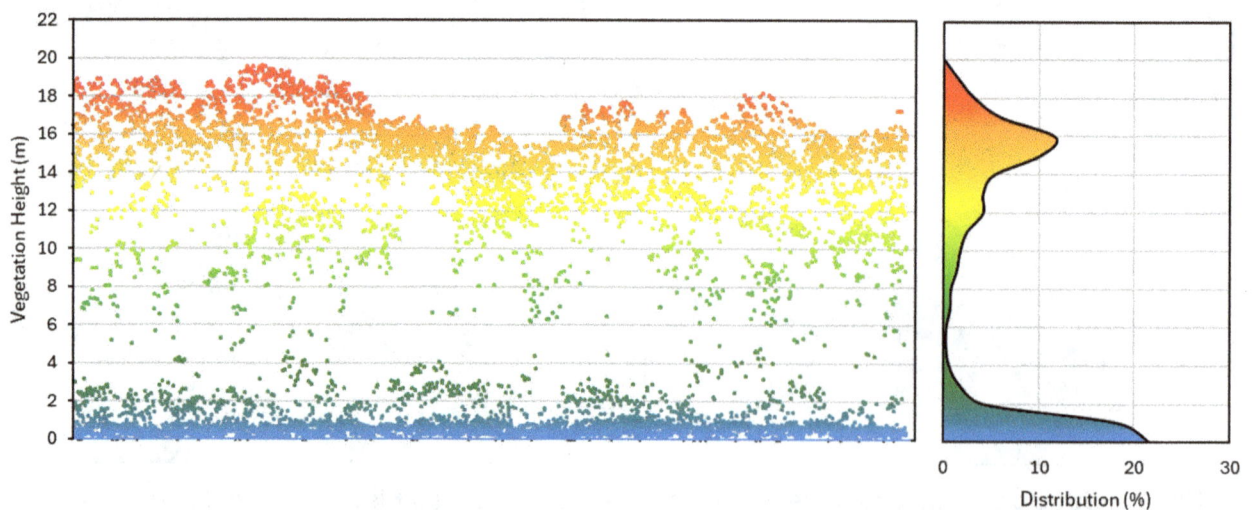

FIG 9 – Vertical distribution of vegetation derived from multiple LiDAR returns in a 30 m section of 36-year-old rehabilitation.

Integrating technology with traditional on-ground monitoring

Alcoa employs a comprehensive monitoring strategy that integrates broad-scale and efficient remote sensing data with detailed traditional on-ground monitoring methods to ensure the long-term success of the rehabilitation areas. These ground-based surveys typically involve the establishment of representative plots where ecologists assess various parameters, including plant species richness, density, and percentage cover, among others. The synergistic combination of remote sensing and on-ground data provides a more comprehensive and robust assessment of the overall integrity and ecological trajectory of Alcoa's rehabilitated areas. Reflecting this enduring commitment, Alcoa has established more than 4700 floristic plots throughout the rehabilitation areas since 1990, creating a valuable long-term data set for understanding ecosystem recovery.

Research insights on ecological recovery

The successful recolonisation of fauna, pollinators, and microbial communities serves as a key measure of ecological recovery in post-mining rehabilitation. These organisms provide essential ecosystem services such as pollination, seed dispersal, and nutrient cycling.

Fauna recolonisation

The return of birds, reptiles, mammals, and invertebrates to rehabilitated sites, reflects the gradual re-establishment of species based on habitat requirements. While all mammal species eventually recolonise Alcoa rehabilitated areas, their return varies by rehabilitation stages of growth. Rehabilitation sites have been found to provide essential habitat for threatened mammals such as the quokka (Craig *et al*, 2017) and western quoll (chuditch) (McGregor, Stokes and Craig, 2014). Bird populations also demonstrate successful recolonisation, with approximately 90 per cent of species recorded in rehabilitated areas. However, species reliant on hollow trees for nesting initially use these areas primarily for foraging, as observed in black cockatoos, which have been recorded foraging in rehabilitation sites as early as four years post-establishment (Doherty *et al*, 2016). Similarly, reptile recolonisation follows successional trends, with generalist species returning first before habitat specialists (Craig *et al*, 2018).

Pollinators and insect communities

Nearly a decade after rehabilitation, insect assemblages, including pollinators, demonstrate compositional similarity to those in adjacent unmined forests (Tudor, Cross and Tomlinson, 2023). The natural reassembly of these communities following the reestablishment of native vegetation strengthens pollination networks, facilitating biodiversity recovery and enhancing ecosystem stability.

Microbial recovery

Soil microbiota play a fundamental role in ecosystem recovery by supporting key ecological functions and plant communities. Liddicoat *et al* (2022), using eDNA, found that rehabilitation trajectory plots indicated recovery, displaying a general pattern of increasing similarity to reference forest sites as rehabilitation age progressed.

Management interventions to support fauna recolonisation

Key management strategies have been implemented to accelerate the return of fauna to rehabilitated areas. Alcoa constructs fauna habitats using logs, stumps, and rocks. Strategically positioned in the landscape following soil return (Nichols and Grant 2007), the.se structures actively draw animals back from the adjacent native forest and are critical for accelerating recolonisation, particularly for coarse woody debris-dependent fauna (Craig *et al*, 2014). Additionally, the Western Shield control program, a landscape-scale initiative managed by the Western Australian Department of Biodiversity, Conservation and Attractions (DBCA) and partially funded by Alcoa, was implemented following studies on fox baiting efficacy. This program now targets both red foxes and feral cats across the Northern Jarrah Forest.

Ongoing fauna research and adaptive management

Alcoa continues to focus research efforts on key fauna groups, including birds, short-range endemic species, mammals, and feral animals. Avian conservation research is evaluating the effectiveness of retained nest trees and artificial hollows for black cockatoos to inform habitat protection strategies. Research on short-range endemic Mygalomorph spiders has led to the development of ethical trapping methods that allow identification and relocation without burrow excavation. Camera trap studies on endemic mammals and feral species (cats, foxes, and pigs) provide data on population distributions, informing mine planning and targeted feral animal control efforts.

Designed to address key management questions, these research initiatives contribute directly to Alcoa's adaptive rehabilitation strategies, ensuring continued refinement of best practices to support ecological recovery and faunal recolonisation in post-mining landscapes.

Integrating cultural and ecological values: the case of grasstrees

Xanthorrhoea preissii, commonly known as the grasstree or balga, holds significant ecological and cultural importance within the jarrah forest ecosystem. These iconic Australian plants contribute to the forest's structural diversity, provide essential habitat and food for native fauna, and are recognised as indicators of ecological maturity and long-term stability due to their exceptionally slow growth (estimated at around 10 mm per annum) and remarkable longevity, with some individuals living for centuries.

Beyond their ecological role, grasstrees are deeply significant to the Noongar people, the Traditional Owners of the South-west region of WA. Traditionally used for food, medicine, and resin production, grasstrees are also integral to Noongar stories and cultural practices. Consequently, the successful re-establishment of grasstrees in rehabilitated areas is vital not only for ecological recovery but also for their cultural significance.

Despite their importance, establishing grasstrees in rehabilitation areas presents considerable ecological challenges. These include low seed production, difficulty in collecting viable seeds, extremely slow growth rates, grazing pressure from native herbivores including kangaroos, and specific preferences for sandier soil types.

Recognising these challenges, Alcoa actively incorporates *Xanthorrhoea* species into the standard rehabilitation seed mixes. While initial germination rates can be promising (around 42 per cent for *Xanthorrhoea preissii*), high seedling mortality often results in limited numbers of mature plants. Research has highlighted the detrimental effects of kangaroo grazing, with artificial protection significantly improving seedling survival and growth (Koch, Richardson and Lamont, 2004). Alcoa's newly established Forest Research Centre is further strengthening these efforts by embracing Traditional Owners' cultural values through a two-way science program. This initiative aims to foster collaboration with Noongar knowledge holders, integrating traditional ecological knowledge into research and management practices to improve the establishment and long-term survival of culturally significant species such as grasstrees.

Reflecting this commitment, Alcoa is undertaking a project focused on transplanting ancient grasstrees to rehabilitated areas. A small-scale trial in 2004, involving both direct and nursery-stabilised transplants, showed moderate survival rates after seven years (28 per cent and 59 per cent respectively). Building on this experience, Alcoa is currently designing a larger trial to improve transplantation techniques by incorporating Traditional Owner knowledge.

Leveraging remote sensing for comprehensive rehabilitation monitoring and management

Alcoa has increasingly integrated advanced technologies, particularly remote sensing tools, into its rehabilitation monitoring and management practices. The application of these technologies enhances the efficiency, accuracy, and spatial coverage of the monitoring efforts, providing valuable data for assessing the progress and integrity of rehabilitated areas.

Enhancing long-term ecosystem resilience in the face of climate change

To restore a self-sustaining jarrah forest ecosystem, rehabilitation must be free from key ecological threats, such as invasive plant species. Effectively managing these threats is a crucial step in building ecosystem resilience, enabling the rehabilitated forest to better withstand the challenges posed by a changing climate. Weed species in Alcoa's rehabilitation since 2016 represent under 1.6 per cent of vegetation cover and less than 5 per cent of species richness. Over 98 per cent of weeds in Alcoa's rehabilitation are annual herbs or grasses. As Alcoa's rehabilitation matures, these weeds are outcompeted by native species and ultimately decline.

The Northern Jarrah Forest is largely adapted to fire, with flora biomass recovering from fire generally within three to five years. Alcoa's rehabilitation is considered similarly resilient to fire, with numerous research studies published over the last three decades focusing on post-fire recovery of vegetation in rehabilitation (Grigg and Grant 2009; Grant 2003; Grant and Loneragan 1999). Although this collection of research presents a comprehensive picture of post-fire recovery of vegetation across rehabilitation completed prior to 2000, Alcoa's ongoing research is focusing on understanding the recovery trajectory of land rehabilitated using more recent methods of rehabilitation prescription.

Recognising the importance of both managing existing threats and preparing for future climate impacts, Alcoa is also actively investigating climate-adapted strategies for its rehabilitation efforts. Currently, through its partnership with the Cooperative Research Centre for Transformations in Mining Economies (CRC TiME), Alcoa is participating in a field trial installed at the Willowdale mine. This trial is part of the nationwide project Evidence for effective climate-adapted seed sourcing strategies for revegetation success and transition to mine closure in a changing climate. The project aims to provide evidence needed to guide effective mine closure that confers greater climate resilience and reduces the risk of post-mine revegetation failure, both now and under future changes. Through these integrated approaches, focusing on both current best practices in weed management and proactive research into climate-resilient seed sourcing, Alcoa is actively working to enhance the long-term ecological resilience of its rehabilitated jarrah forest ecosystems in the face of a changing climate.

COMMITMENT TO KNOWLEDGE EXCHANGE AND COLLABORATIVE RESEARCH

Alcoa recognises the critical importance of sharing the extensive knowledge and expertise gained through its long history of mine rehabilitation in the jarrah forest. To this end, Alcoa actively engages in a range of knowledge-sharing activities aimed at disseminating best practices and fostering collaboration within the mining industry, the scientific community, and the broader public. A key aspect of this commitment is the regular publication of research findings in open-access scientific journals and detailed technical reports, ensuring research methodologies and outcomes are widely accessible. Alcoa also actively participates in and hosts various symposiums and conferences, including the annual Environmental Symposium, providing valuable platforms for researchers to share their latest findings and engage with other experts. To promote transparency and foster community understanding, Alcoa conducts regular open-access public site tours. Recognising the importance of building capacity, Alcoa prioritises student education programs, offering graduate opportunities, and vacation placements.

Collaboration is a cornerstone of Alcoa's approach to mine rehabilitation. The company actively works in partnership with government agencies, research institutions, and other relevant stakeholders to develop and implement leading-edge rehabilitation programs, including participation in initiatives like CRC TiME. Alcoa's long-standing commitment to environmental excellence and successful rehabilitation has earned significant recognition, including being the first mining company acknowledged by the United Nations' Global 500 Roll of Honour in 1990 and receiving multiple conservation and sustainability awards in Western Australia (Alcoa, 2023a).

CONCLUSION

Alcoa's journey of mine rehabilitation in the jarrah forest exemplifies a long-term commitment to progressive practices, driven by continuous research and innovation. Alcoa has made significant improvements in optimising soil conditions through advanced deep ripping techniques and strategic topsoil management, ensuring a foundation for successful ecosystem recovery. The development and application of tailored seed mixes, incorporating a diverse array of native and ecologically significant plant species, coupled with innovative tissue culture and germination pre-treatment methods, have contributed to achieving rehabilitation completion criteria targets. Furthermore, the documented return of a substantial proportion of key fauna species, including mammals, birds, reptiles, and even invertebrates, provides compelling evidence of a positive recovery trajectory on Alcoa's rehabilitated mine sites.

Moving forward, future research should prioritise enhanced strategies for the successful establishment and long-term survival of culturally significant species like *Xanthorrhoea preissii*, which will warrant further investigation, incorporating Traditional Owners' unique knowledge. Moreover, continued research into optimising rehabilitation practices to maximise the resilience of restored areas to the anticipated impacts of future environmental changes will be essential. Alcoa's enduring commitment to research, coupled with its prioritisation of knowledge sharing through open publications, active participation in symposiums, facilitation of public tours, and investment in student education programs, are vital for advancing the science and practice of mine rehabilitation. Sustained collaboration with government agencies, research institutions, and Traditional Owners will be crucial for effectively addressing the remaining challenges and striving towards more holistic and truly sustainable rehabilitation outcomes in the Jarrah Forest and in other biodiversity-rich mining regions worldwide.

ACKNOWLEDGEMENTS

The authors respectfully acknowledge the Traditional Owners and Custodians of the lands where Alcoa operates, with respect paid to Elders—past, present, and emerging. We gratefully acknowledge the collective efforts of individuals within Alcoa, alongside contractors, operators, and collaborators from various research institutions whose have driven advancement and innovation in bauxite mine rehabilitation in the Northern Jarrah Forest.

REFERENCES

Alcoa, 2015. Alcoa's Bauxite Mine Rehabilitation Program, Completion Criteria and Overview of Area Certification Process, 2015 Revision, Alcoa World Alumina (Australia). Available from: <https://www.alcoa.com/australia/en/pdf/mining-operations-rehabilitation-program-completion-criteria-overview-area-certification-process.pdf>

Alcoa, 2023a. Australian awards, Alcoa World Alumina (Australia). Available from: <https://www.alcoa.com/australia/en/pdf/awards.pdf>

Alcoa, 2023b. Alcoa commits to no mining zone to protect environmental and social values around Dwellingup, Alcoa World Alumina (Australia). Available from: <https://www.alcoa.com/australia/en/news/releases?id=2023/06/alcoa-commits-to-no-mining-zone-to-protect-environmental-and-social-values-around-dwellingup&year=y2023>

Alcoa, 2024. Alcoa Announces $15 Million for New Forest Research Centre, Alcoa World Alumina (Australia). Available from: <https://www.alcoa.com/australia/en/news/releases?id=2024/05/alcoa-announces-15-million-for-new-forest-research-centre&year=y2024>

Bell, D T and Bellairs, S M, 1992. Effects of temperature on the germination of selected Australian native species used in the rehabilitation of bauxite mining disturbances in Western Australia, *Seed Science and Technology*, 20(1):47–55.

Bell, D T, 1994. Interaction of fire, temperature and light in the germination response of 16 species from the Eucalyptus marginata forest of south-western Western Australia, *Australian Journal of Botany*, 42(5):501–509.

Bell, D T, 2001. Ecological response syndromes in the flora of Southwestern Western Australia: fire resprouters versus reseeders, *Botanical Review*, 67(4):417–440.

Bell, D T, Vlahos, S and Watson, L E, 1987. Stimulation of seed germination of understorey species of the northern jarrah forest of Western Australia, *Australian Journal of Botany*, 35(5):593–599.

Commander, L E, Blackburn, C, Mullins, G, Lilly, A and Barker, J, 2023. Large-scale, multi-species plant translocation in the Jarrah Forest, in 'Learning from the past, adapting to the future': Third International Conservation Translocations Conference, 13–15 November 2023. Fremantle, Australia.

Commander, L, Barker, J, Blackburn, C, Grigg, A, Mullins, G and Pattinson, A, 2024. Research-led adaptive management in rehabilitation, in *Mine Closure 2024: Proceedings of the 17th International Conference on Mine Closure* (eds: A B Fourie, M Tibbett and G Boggs), pp 415–426 (Australian Centre for Geomechanics: Perth). https://doi.org/10.36487/ACG_repo/2415_30

Craig, M D, Grigg, A H, Hobbs, R J and Hardy, G E S J, 2014. Does coarse woody debris density and volume influence the terrestrial vertebrate community in restored bauxite mines?, *Forest Ecology and Management*, 318:142–150.

Craig, M D, Smith, M E, Stokes, V L, Hardy, G E S and Hobbs, R J, 2018. Temporal longevity of unidirectional and dynamic filters to faunal recolonization in post-mining forest restoration, *Austral Ecology*, 43(8):973–988.

Craig, M D, White, D A, Stokes, V L and Prince, J, 2017. Can postmining revegetation create habitat for a threatened mammal?, *Ecological Management and Restoration*, 18(2):149–155.

Croton, J T and Ainsworth, G L, 2007. Development of a winged tine to relieve mining-related soil compaction after bauxite mining in Western Australia, *Restoration Ecology*, 15(s4):S48–S53.

Doherty, T S, Wingfield, B N, Stokes, V L, Craig, M D, Lee, J G H, Finn, H C and Calver, M C, 2016. Successional changes in feeding activity by threatened cockatoos in revegetated mine sites, *Wildlife Research*, 43(2):93–104.

Ducki, L, 2020. Soil fungi, but not bacteria, track vegetation reassembly across a 30-year restoration chronosequence in the northern jarrah forest, Western Australia, Honours dissertation, Murdoch University.

Grant, C D and Loneragan, W A, 1999. The Effects of Burning on the Understorey Composition of 11–13 Year-Old Rehabilitated Bauxite Mines in Western Australia – Vegetation Characteristics, *Plant Ecology*, 145(2):291–305.

Grant, C D, 2003. Post-Burn Vegetation Development of Rehabilitated Bauxite Mines in Western Australia, *Forest Ecology and Management*, 186(1):147–157.

Grant, C D, Bell, D T, Koch, J M and Loneragan, W A, 1996. Implications of seedling emergence to site restoration following bauxite mining in Western Australia, *Restoration Ecology*, 4(2):146–154

Grigg, A H and Grant, C D, 2009. Overstorey Growth Response to Thinning, Burning and Fertiliser in 10–13-Year-Old Rehabilitated Jarrah (Eucalyptus Marginata) Forest after Bauxite Mining in South-Western Australia, *Australian Forestry*, 72(2):80–86.

Koch, J M, Richardson, J and Lamont, B B, 2004. Grazing by Kangaroos Limits the Establishment of the Grass Trees *Xanthorrhoea gracilis* and *X, preissii* in Restored Bauxite Mines in Eucalypt Forest of Southwestern Australia, *Restoration Ecology*, 12:297–305. https://doi.org/10.1111/j.1061-2971.2004.00335.x

Liddicoat, C, Krauss, S L, Bissett, A, Borrett, R J, Ducki, L C, Peddle, S D, Bullock, P, Dobrowolski, M P, Grigg, A, Tibbett, M and Breed, M F, 2022. Next generation restoration metrics: Using soil eDNA bacterial community data to measure trajectories towards rehabilitation targets, *Journal of Environmental Management*, 310:114748.

McGregor, R A, Stokes, V L and Craig, M D, 2014. Does forest restoration in fragmented landscapes provide habitat for a wide-ranging carnivore?, *Animal Conservation*, 17(5):467–475.

Nichols, O and Muir, B, 1989. Vertebrates of the jarrah forest, in *The Jarrah Forest: A Complex Mediterranean Ecosystem* (eds: B Dell, J Havel and N Malajczuk), (Kluwer Academic Publishers: Netherlands).

Nichols, O G and Grant, C D, 2007. Vertebrate fauna recolonization of restored bauxite mines—Key findings from almost 30 years of monitoring and research, *Restoration Ecology*, 15(S4):S116–S126.

Norman, M A, Plummer, J A, Koch, J M and Mullins, G R, 2006. Optimising smoke treatments for jarrah (Eucalyptus marginata) forest rehabilitation, *Australian Journal of Botany*, 54(6):571–581.

Roche, S, Koch, J M and Dixon, K W, 1997. Smoke enhanced seed germination for mine rehabilitation in the southwest of Western Australia, *Restoration Ecology*, 5(3):191–203.

Tacey, W H and Glossop, B L, 1980. Assessment of Topsoil Handling Techniques for Rehabilitation of Sites Mined for Bauxite within the Jarrah Forest of Western Australia, *Journal of Applied Ecology*, 17(1):195–202.

Tudor, E, Cross, A T and Tomlinson, S, 2023. Insect community reassembly in a spatiotemporally heterogenous restoration landscape, *Landscape Ecology*, 38:2763–2778. https://doi.org/10.1007/s10980-023-01747-2

Feedback on the integrity of multi-linear drainage geocomposites installed on final covers after 12 years of operation

P Saunier[1], C Ciuperca[2] and S Fourmont[3]

1. Business Development Manager North-America and Pacific, AFITEX-Texel, Québec G6E 1G8, Canada. Email: psaunier@afitextexel.com
2. Project Manager, Sperling Hansen Associates, North Vancouver V7J IJ3, Canada. Email: cciuperca@sperlinghansen.com
3. Business Development Manager North-America, AFITEX-Texel, Québec G6E 1G8, Canada. Email: sfourmont@afitextexel.com

ABSTRACT

Drainage geocomposites are now widely used in the mining industry. Their use on final covers and orphan sites remediation makes it possible to drain surface water and guarantee the sustainability of the cover over time, as long as we can monitor their behaviour. However, it is difficult to validate the actual durability of these materials other than through accelerated aging tests performed in the laboratory. The best way to assess the long-term integrity of drainage geocomposites is to exhume them after several years of operation and observe the evolution of their mechanical and hydraulic properties. This paper presents two case studies with the exhumation of two Multi-linear drainage geocomposites installed as a drainage layer in final covers after ten years of operation in France and 12 years of operation in Canada. The samples were inspected and then analysed in the laboratory and the residual properties were compared with those of the original product.

INTRODUCTION

Multi-linear drainage geocomposites have been widely used for more than 30 years for liquids drainage (water, leachate etc) and gas collection (LFG, Radon, VOCs etc).

While laboratory studies provide insights into long-term behaviour, field exhumations offer direct validation of durability. This paper presents two exhumation case studies assessing the integrity of multi-linear drainage geocomposites used in landfill final covers as surface water drainage layers. The first exhumation conducted at Lapouyade Landfill in France, examined a geocomposite installed for ten years. The second, at Vancouver Landfill, Canada, investigated a geocomposite in service for 12 years.

The two multi-linear drainage geocomposites are representative of the range of products used in landfill final covers. The product used at the Lapouyade landfill has embedded perforated mini-pipes with a diameter of 20 mm, while the product used at the Vancouver landfill has perforated mini-pipes with a diameter of 25 mm.

LAPOUYADE LANDFILL, FRANCE

Lapouyade Landfill in France has been receiving only non-hazardous waste since 1996. It is located in the north-east of the Gironde department, in the Nouvelle Aquitaine region, 50 km from Bordeaux. The site is operated by Veolia, a long-established landfill operator with many sites in France and around the world.

The Lapouyade landfill covers an area of 105 hectares and has a treatment capacity of 430 000 t of waste per annum. Waste comes from local authorities and industry of the region. Once each cell has been closed, Veolia will provide 30 years of post-operational monitoring to ensure that the site's environment is controlled and protected.

Final cover design in 2004

Each 5000 m^2 compartment is filled and then gradually closed, before being landscaped. This allows the various landfills to be gradually revegetated. All the leachate and biogas collection systems put in place during the operation of the landfill continue to be collected, monitored and/or recovered.

The landfill cell #3 presented in this paper has been closed for 20 years. The landfill cover consists of a semi-impermeable clay layer, followed by a multi-linear drainage geocomposite and a cover layer of 0.80 m of topsoil. The geocomposite replaces a traditional 0.30 to 0.50 m thick layer of granular drainage material.

The images below show the installation of the geocomposite multi-linear drainage system on the site in 2004 and the vegetated and monitored cover (Figure 1).

FIG 1 – Placement of topsoil over multi-linear geocomposite drainage and final vegetation cover.

This typical cross-section (Figure 2) has been used on several of the site's landfills since 2004.

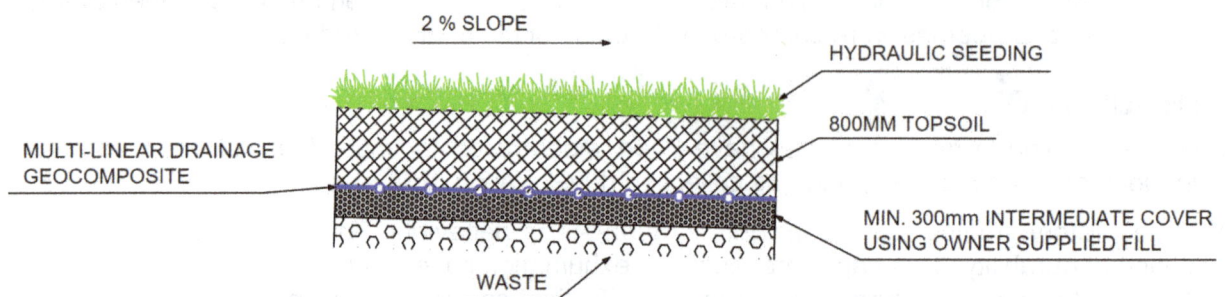

FIG 2 – Lapouyade landfill – cross-section of the final cover design.

Exhumation of the drainage geocomposite in 2014

In order to determine whether the multi-linear drainage geocomposite has retained its mechanical and hydraulic characteristics after ten years of operation, a 1.50 m × 2.00 m sample was taken *in situ* to carry out test controls.

The 800 mm topsoil was gently removed without damaging the drainage geocomposite (Figure 3).

FIG 3 – Removal of the topsoil on top of the multi-linear drainage geocomposite.

Prior to sampling, observation of the top filter layer of the product showed no area of clogging, despite the sandy-clay nature of the topsoil. The bottom section of the drainage geocomposite also appeared to be very clean, as did the underlying clay layer (Figure 4).

FIG 4 – Manual exposition of the two faces of the geocomposite.

The uniform surface condition showed that there was no seepage over this area and that the infiltrated precipitation has been completely drained away by the multi-linear drainage geocomposite. The perfectly clean area of the lateral overlap (> 150 mm) between two rolls confirmed that there was no contamination of the product from the sides and through the overlap (Figure 5).

FIG 5 – Observation of a lateral overlap of product.

The exhumation has been repaired using another piece of the same product, connecting the mini-pipes with couplers before backfilling again.

Comparative testing in the laboratory

In addition to a visual inspection and photos of the product before sampling, tests were carried out in the laboratory. A sample of the original geocomposite, stored in the laboratory for ten years, was compared with the sample taken on-site.

Visual observations on the virgin geocomposite and the geocomposite exposed on-site

The visual appearance of the two products was very similar, only the mini-pipes looks dirtier without being clogged or crushed (Figures 6 and 7). The durability of the drainage geocomposite was visually validated after a ten year period of use as a landfill cover.

FIG 6 – Visual inspection of the virgin geocomposite (geotextile layers and mini-pipes.

FIG 7 – Visual inspection of the exposed geocomposite (geotextile layers and mini-pipes).

Testing in the laboratory

The tests were carried out in accordance with the standards applicable to geotextiles and related products. The benchmarks are current AFNOR standards and the ASQUAL 'Geotextile and related product test method data' 2014 version.

The exhumed product was tested for mechanical and hydraulic properties. The results were compared with the original values tested on the product before installation. As shown in Table 1, after ten years the product shows no evidence of rapid degradation or any sensible change in its original properties.

TABLE 1

Comparison of the residual values with the initial values.

	Characteristic	Standard	Ratio residual/initial	Unit
Geotextile Layers	Mass per Unit area	NF EN 9864	197%	g/m^2
	Grab tensile Strength	NF EN 9863-1	98%	mm
	Grab Elongation	NF EN 10319	92%	kN/m
	Puncture Resistance	NF EN 10319	69%	%
Geocomposite	In-Plane flow capacity *i=0.1 / normal load = 20 kPa*	NF EN ISO 12958-1	84%	m^2/s
	In-Plane flow capacity *i=0.1 / normal load = 100 kPa*		92%	m^2/s
Filter	Opening Size	NF EN ISO 12956	100%	micron

Residual mass per Unit area is higher due to dirt presence in the fibres.

The properties tested on the products are the most important mechanical and hydraulic properties for the function of the geocomposite on this landfill cover. Thickness and tensile strength properties are related to the behaviour of the product under long-term compression, flow capacity is related to the drainage function and the opening size of the filter to prevent clogging. All these properties are maintained after ten years of use of the geocomposite as a drainage layer on the final landfill cover.

VANCOUVER LANDFILL, CANADA

Vancouver Landfill, located in Delta, serves as a regional waste disposal site. Operated by Metro Vancouver, it is the region's largest landfills. In 2012, Sperling Hansen Associates, the engineer of record, incorporated multi-linear drainage geocomposites into the final cover to replace gravel drainage layers, initially as a pilot project.

Landfill cover design (2012)

To reduce environmental impact, maximise landfill capacity, and minimise construction costs and time, engineers implemented multi-linear drainage geocomposites as a substitute for gravel in the surface water collection system of the final cover. Starting in 2012, the geocomposite was installed on a section of the north slope of Phase 2 and monitored for several years. This pilot project informed the subsequent large-scale application on the Western 40 cover, which took place a couple of years later.

The geocomposite effectively managed rainwater drainage (above the geomembrane) and landfill gas (LFG) collection (below the geomembrane) following the W40 construction implementation in 2017, 2018 and 2019.

The Draintube multi-linear drainage geocomposite is represented in blue in Figure 8.

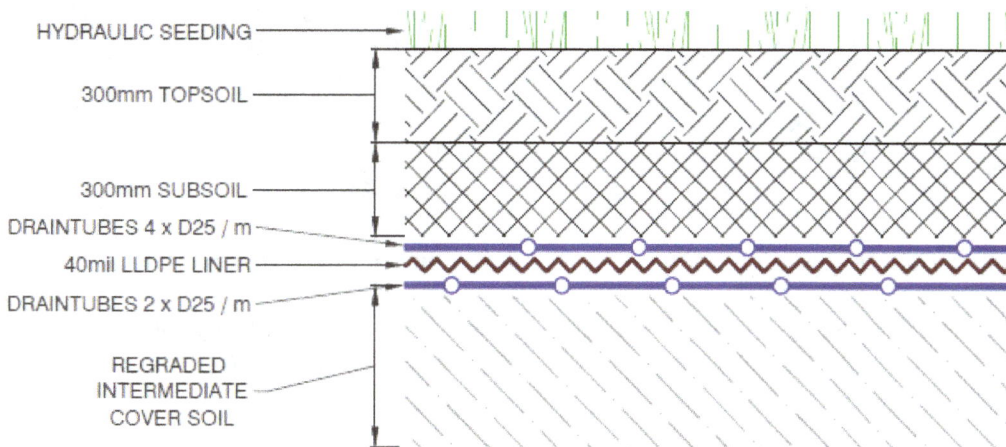

FIG 8 – Vancouver landfill – typical cover system including multi-linear drainage geocomposites.

Phase 5 final cover (2024)

Recently, as part of the conceptual design of Phase 5, Sperling Hansen Associates explored the option of extending the use of multi-linear drainage geocomposites beyond the crest areas, where they had been successfully implemented and operated in the W40 phase, to the landfill slopes. This provided an ideal opportunity to assess the material's condition after 12 years of continuous use. As a result, the exhumation of the geocomposite installed on the slope of Phase 2 in 2012 was planned and scheduled for May 2024.

Exhumation process (2024)

A multidisciplinary team, including geosynthetics manufacturers (Afitex-Texel), the site engineer of record (Sperling Hansen Associates), geosynthetics installer (Western Tank and Linings) and independent laboratories (Sageos CCT Group), oversaw the exhumation. The process involved:

- Careful soil removal to avoid damaging the underlying geomembrane (Figure 9).

- Manual exposure and sampling of the geocomposite (Figure 10).

- Endoscopic inspection of mini-pipes, revealing no clogging or deformation (Figure 11).

FIG 9 – Location of the exhumation and removal of the soil on top of the geocomposite.

FIG 10 – A large sample is collected to be sent for laboratory testing.

Preliminary inspection showed that while some roots were present in the geotextile layer, none had penetrated the mini-pipes. The liner remained clean after 12 years, with no clogging in the mini-pipes.

FIG 11 – Exposition of the geocomposite and the liner. Endoscopic inspection of the mini-pipes.

An endoscopic camera was inserted into each mini-pipe to assess internal conditions, confirming no deformation, root intrusion, or debris accumulation (Figure 12). The geocomposite maintained stable flow conditions, demonstrating the effectiveness of its geotextile filter.

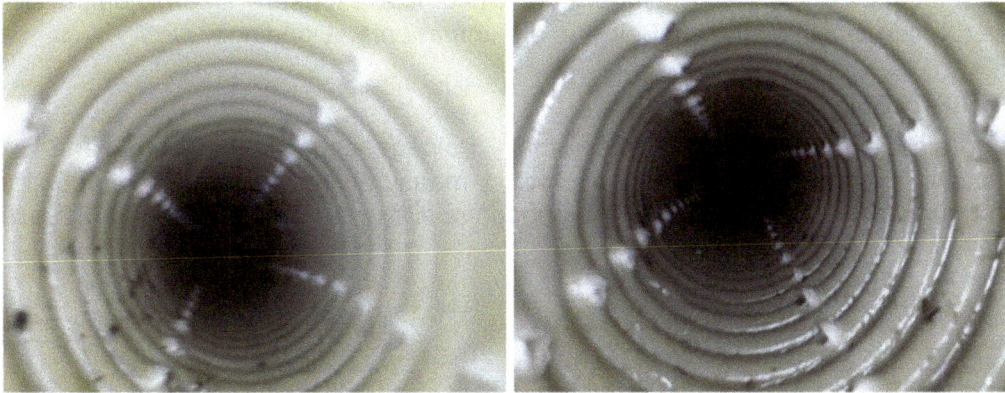

FIG 12 – Inspection of the mini-pipes up-grade (left) and down-grade (right).

Following exhumation, repairs were conducted by installing a new section of multi-linear drainage geocomposite, reconnecting mini-pipes using couplers before backfilling the area.

Laboratory testing

A comparison between the exhumed and original geocomposite revealed:

- Mechanical properties (tensile strength, puncture resistance) remained stable (≥95 per cent).

- Mini-pipe stiffness increased (142 per cent), indicating no loss of structural integrity.

- Flow capacity retained 105 per cent of original values.

- Filtration properties remained intact, preventing soil intrusion.

These findings demonstrate the geocomposite's resilience under real-world landfill conditions.

TABLE 2

Comparison of the residual values with the initial values.

	Characteristic	Standard	Ratio residual/initial	Unit
Geotextile Layers	Mass per unit area	ASTM D5261	100%	g/m²
	Grab tensile strength	ASTM D4632	95%	N
	Grab elongation	ASTM D4632	100%	%
	Puncture resistance	ASTM D6241	102%	N
Mini-pipes	Outside diameter	ASTM D2412	0%	mm
	Pipe stiffness at 5%	ASTM D2412	142%	
Geocomposite	Transmissivity *i = 0.1 / normal load = 400 kPa / Seating time = 100 hr*	ASTM D4716	105%	m²/s
Filter	FOS	CAN 148.1 No10	114%	micron
	Perittivity	ASTM D4491	109%	s⁻¹

Initial values are the ones published on the technical data sheet. They might be higher than the measured initial value.

EXTERNAL STRESSES PERFORMANCE

Multi-linear drainage geocomposites have been extensively used for over 30 years for liquid drainage (water, leachate) and gas collection (LFG, Radon, VOCs). Their performance under external stresses such as soil load, extreme temperatures, chemical exposure, and biological clogging has been investigated in multiple studies. Below is a summary of key findings.

Load resistance

Multi-linear drainage geocomposites retain their drainage capacity over time and under load. Figure 13 shows the transmissivity of the product measured for loads up to 2400 kPa and the variation over time of the transmissivity of the geocomposite under 2400 kPa for 1000 hrs, for several gradients.

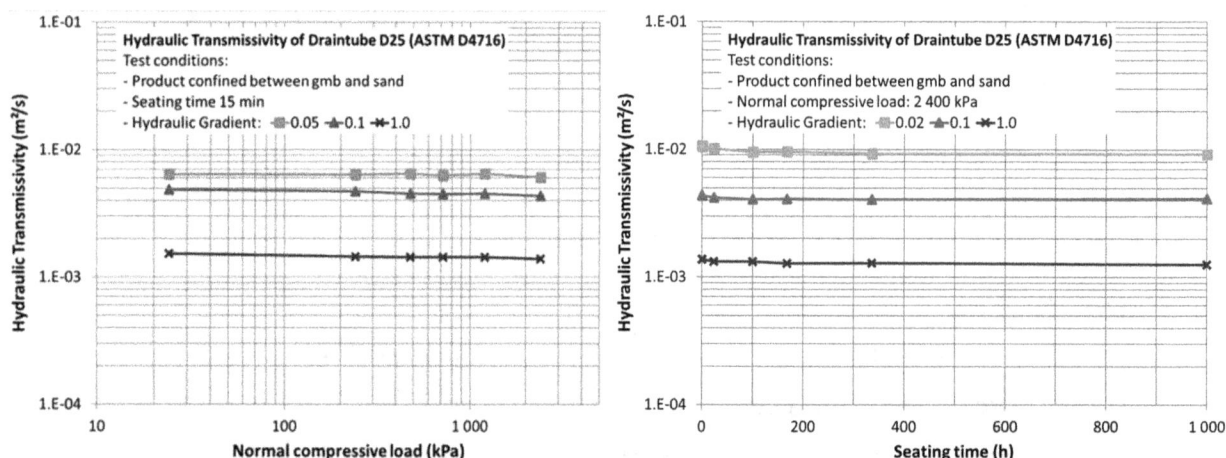

FIG 13 – Hydraulic transmissivity of the geocomposite function of load intensity and time.

It can be observed that the product is not load sensitive when confined between a geomembrane and a soil layer. They demonstrate resistance to creep compression, with a reduction factor for creep (RF_{CR}) of 1.0 when confined in soil under loads up to 2400 kPa. Laboratory tests show no significant reduction in transmissivity over extended periods (Saunier, Ragen and Blond, 2010).

Cold temperature performance

Multi-linear drainage geocomposites are made with polypropylene corrugated perforated mini-pipes as drainage conduits. Unlike HDPE, polypropylene is not sensitive to environmental stress cracking.

The apparent modulus of elasticity of the mini-pipes have been tested as well as the unrolling test, under several temperatures. Table 3 shows the main results of tests performed with temperatures up to -70°C. The apparent modulus of elasticity is almost three times higher at -30°C compared to 23°C. That change has no consequence on the behaviour of the mini-pipe when it is unrolled as demonstrated by the ASTM D5636 (2017) that simulates the unrolling of the mini-pipe around a 150 mm diameter mandrel. The test was performed at -70°C and no crack or any other failure was observed on the mini-pipe (Fourmont and Saunier, 2015).

TABLE 3

Apparent modulus of elasticity of the mini-pipe.

Temperature	+23°C	-30°C	-70°C
Apparent modulus of elasticity (Mpa) ASTM D790 (2017)	8.99	24.54	-
Unrolling test – Mandrel diameter 150 mm ASTM D5636 (2017)	-	-	OK no cracks

Resistance to biological clogging

Since 2013, several field tests have been conducted using fresh leachate circulation systems to assess the long-term behaviour of multi-linear drainage geocomposites in landfill leachate environments. One such study, carried out in Pennsylvania, USA, by the Geosynthetic Research Institute (GRI) over three years, compared multi-linear drainage geocomposites with biplanar geonet geocomposites.

The study concluded that...

> the needle-punched nonwoven geotextile performed the best when placed over the tubular drainage composite. It is well designed with respect to the concrete sand's gradation to avoid piping and is open enough to resist long-term clogging. This is demonstrated by its ability to remain free-flowing with leachate as a permeant for over three years of testing. (Fourmont and Koerner, 2017)

Figure 14 illustrates the system flow rate over time for four different geotextile configurations tested during the field study: NPNW (Needle-Punched Non-Woven geotextile), WM (Woven Monofilament geotextile), and HBNW (Heat Bonded Non-Woven geotextile).

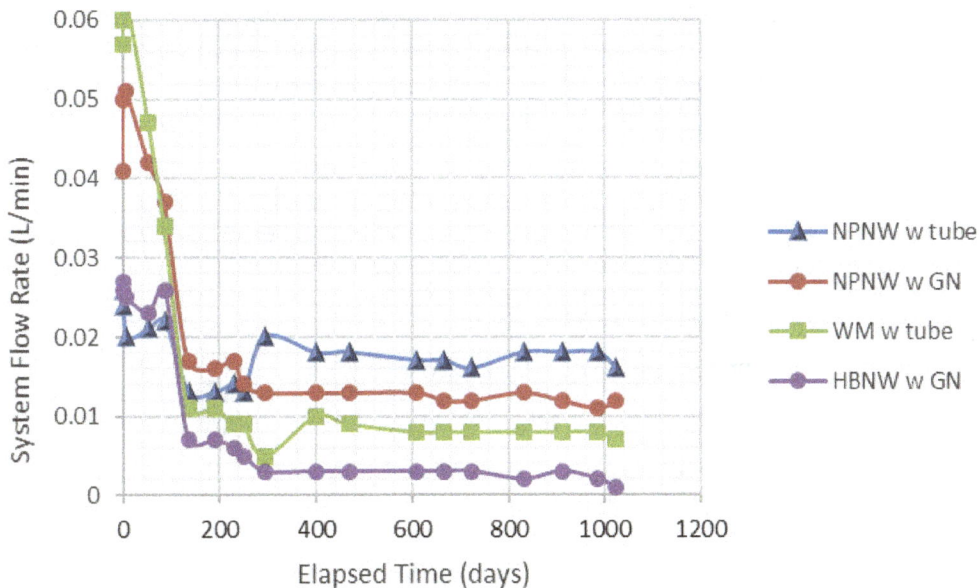

FIG 14 – Combined long-term flow curves for the four different geotextiles with two different drainage components.

DESIGN CONSIDERATIONS

During the system design process, engineers address the long-term behaviour of geocomposites by applying Reduction Factors (RFs) to critical product properties. For drainage geocomposites, these reduction factors are typically applied to drainage capacity (in-plane flow rate or transmissivity). Comparisons are conducted around those factors to determine equivalencies between different technology types (Diamiano and Steinhauser, 2020).

Field measurements and laboratory studies enhance confidence in the product's performance and confirm the Reduction Factors currently in use for this type of product. Table 4 outlines the recommended reduction factors for multi-linear drainage geocomposites.

TABLE 4

Recommended RFs for multi-linear drainage geocomposites.

Applications	Requirements			
	RF_{GI}	RF_{CR}	RF_{CC}	RF_{BC}
Landfill leachate collection	1.0	1.0	1.5 to 2.0	1.0* to 1.3
Retaining walls	1.0	1.0	1.1 to 1.5	1.0 to 1.2
Sport fields	1.0	1.0	1.0 to 1.2	1.0* to 1.3
Landfill covers	1.0	1.0	1.0 to 1.2	1.0* to 3.5

* In cases when using DRAINTUBE ACB, which contains a non-leachable, silver-based biocide treatment.

Environmental impacts

Recent studies indicate that the embodied carbon of a traditional gravel drainage layer is approximately three times higher than that of a drainage geocomposite used as a replacement (Durkheim and Fourmont, 2015). Furthermore, when evaluating the full Life Cycle Assessment (LCA) from cradle to grave, the substitution of gravel with a drainage geocomposite can lead to a reduction in CO_2 emissions of over 90 per cent, depending on the location and transport distance from the gravel source. Additionally, the use of a multi-linear drainage geocomposite can further reduce CO_2 emissions by up to 40 per cent compared to conventional geocomposite drainage products.

CONCLUSIONS

The exhumation and analysis of multi-linear drainage geocomposites after 10 and 12 years confirm their long-term durability in landfill final covers. Key findings include:

- No clogging or significant degradation.
- High retention of mechanical and hydraulic properties.
- Resistance to root intrusion, bacterial clogging, and extreme temperatures.

These results, combined with extensive laboratory studies, reinforce the suitability of multi-linear drainage geocomposites as a reliable solution for landfill final covers. Future research should explore longer-term performance beyond 12 years.

REFERENCES

ASTM International, 2017. ASTM D5636M-94 – Standard Test Method for Low Temperature Unrolling of Felt or Sheet Roofing and Waterproofing Materials, ASTM International, West Conshohocken. https://doi.org/10.1520/D5636_D5636M-94R17

ASTM International, 2017. ASTM D790 – Standard Test Methods for Flexural Properties of Unreinforced and Reinforced Plastics and Electrical Insulating Materials, ASTM International, West Conshohocken. https://doi.org/10.1520/D0790-17

Diamiano, L and Steinhauser, E, 2020. A guide for specifying drainage geocomposites, *Geosynthetics Magazine, 8–10, April.*

Durkheim, Y and Fourmont, S, 2015. Drainage geocomposites: a considerable potential for the reduction of greenhouse gas emission, Afitex Group. Available from: <https://texel.afitex.com/public/Medias/publications/en/>

Fourmont, F and Koerner, G, 2017. Determining the Long-Term Transmissivity of Selected Drainage Geocomposites to Landfill Leachate, in Proceedings of the Geotechnical Frontiers 2017, Orlando, Florida, USA. Available from: <https://doi.org/10.1061/9780784480434.029>

Fourmont, S and Saunier, P, 2015. Behavior of Drain Tubes planar drainage geocomposite under extreme cold temperatures, in GéoQuebec Conference, Québec, Canada.

Saunier, P, Ragen, W and Blond, E, 2010. Assessment of the resistance of drain tube planar drainage geocomposites to high compressive loads, in Proceedings of the 9th International Conference on Geosynthetics, vol 3, Guarujá, Brazil.

Case study – event-based spatial accounting of the effectiveness of sediment control activities and erosion risk modelling using remote-sensed RUSLE

B Silverwood[1], A Yates[2], A Costin[3], B Wehr[4] and G Dale[5,6,7]

1. Senior GIS and Systems Engineer, Verterra Ecological Engineering, Brisbane Qld 4001. Email: ben.silverwood@verterra.com.au
2. Land Resource Specialist, Verterra Ecological Engineering, Brisbane Qld 4001. Email: andrew.yates@verterra.com.au
3. Environmental Engineer, Verterra Ecological Engineering, Brisbane Qld 4001. Email: adam.costin@verterra.com.au
4. Principal Scientist, Verterra Ecological Engineering, Brisbane Qld 4001. Email: bernhard.wehr@verterra.com.au
5. Chief Technical Officer, Verterra Ecological Engineering, Brisbane Qld 4001. Email: glenn.dale@verterra.com.au
6. Adjunct Associate Professor, Faculty of Engineering, School of Civil and Environmental Engineering, Queensland University of Technology.
7. Adjunct Associate Professor, Centre for Sustainable Agricultural Systems, University of Southern Queensland.

EXTENDED ABSTRACT

Management and quantification of the interaction between erosive rainfall and landforms, altered or otherwise, has long been recognised as a complicated and dynamic issue. Much has been made of resolving the ambiguity between guidelines informing erosion and sediment control (ESC) plans and the relevant competing objectives in the resources and other industries (Chrystal et al, 2021). Local, spatially explicit methods to account for the outcomes from implementation of ESC plans has been hindered by available tools focusing on longer term averages and generalised inputs to relevant models.

We present an application of a spatially and temporally explicit method to quantify erosion relative to expected background levels. Originally developed and approved as a peer-reviewed, accredited method under the Reef Credit Standard to account for sediment export change in response to changed grazing management practice, the method has been adapted for application to the mine rehabilitation industry to monitor and manage disturbance areas.

The method can provide managers with an auditable erosion budgeting tool that can quantify the impact of interventions relative to baseline erosion rates, helping to drive more effective rehabilitation and management decisions that increase value for every dollar invested in environmental works. Features include output based on measured vegetation cover factors, use of readily accessible monthly satellite data sets, and capacity to benchmark against erosion in surrounding land uses (reference sites). The method can be used to monitor and manage land uses such as grazing on rehabilitated sites or maintain visibility on lessees' activities and impacts on company-owned land in buffers surrounding operational areas.

Based on the long-established and widely applied Revised Universal Soil Loss Equation (RUSLE) (Renard et al, 1997), the addition of monthly accounting from local rainfall data moves the modelling to an event basis that can capture local extremes rather than longer term trends. The novel application of Dynamic Reference Cover Modelling (DRCM) (Bastin et al, 2012) to the ground cover factor allows the model to compensate for temporal and seasonal changes in vegetation cover, enabling management interventions that increase or decrease erosion to be accurately quantified.

The critical difficulty in accounting for the impact of management interventions on sediment export from large systems with diffuse erosion sources has been the capacity to identify and select reference areas to remove the influence of intra- and inter-annual climatic variability. Incorporation of DRCM provides an automatic and locally relevant means to provide a seasonally interpreted reference for expected ground in the absence of an intervention at any specific site without the need for separate and expensive local control plots.

Utilising mine-specific landform input layers (LS Factor) from LiDAR or design DEMs, along with site-specific soil erodibility layers (K Factor) from soil laboratory data, processed monthly satellite imagery of groundcover (C Factor) and local rainfall information (R Factor), two different spatially explicit erosion settings can be created. The primary spatial layer, consisting of the modelled sediment export for the given calendar month and containing a combination of seasonality, site conditions and influence of sediment control interventions presents the current situation on the ground (Figure 1a). The alternate spatially modelled layer then uses DRCM and the surrounding reference sites to calculate baseline ground cover (and hence RUSLE modelled sediment export) without intervention for each calendar month (Figure 1b).

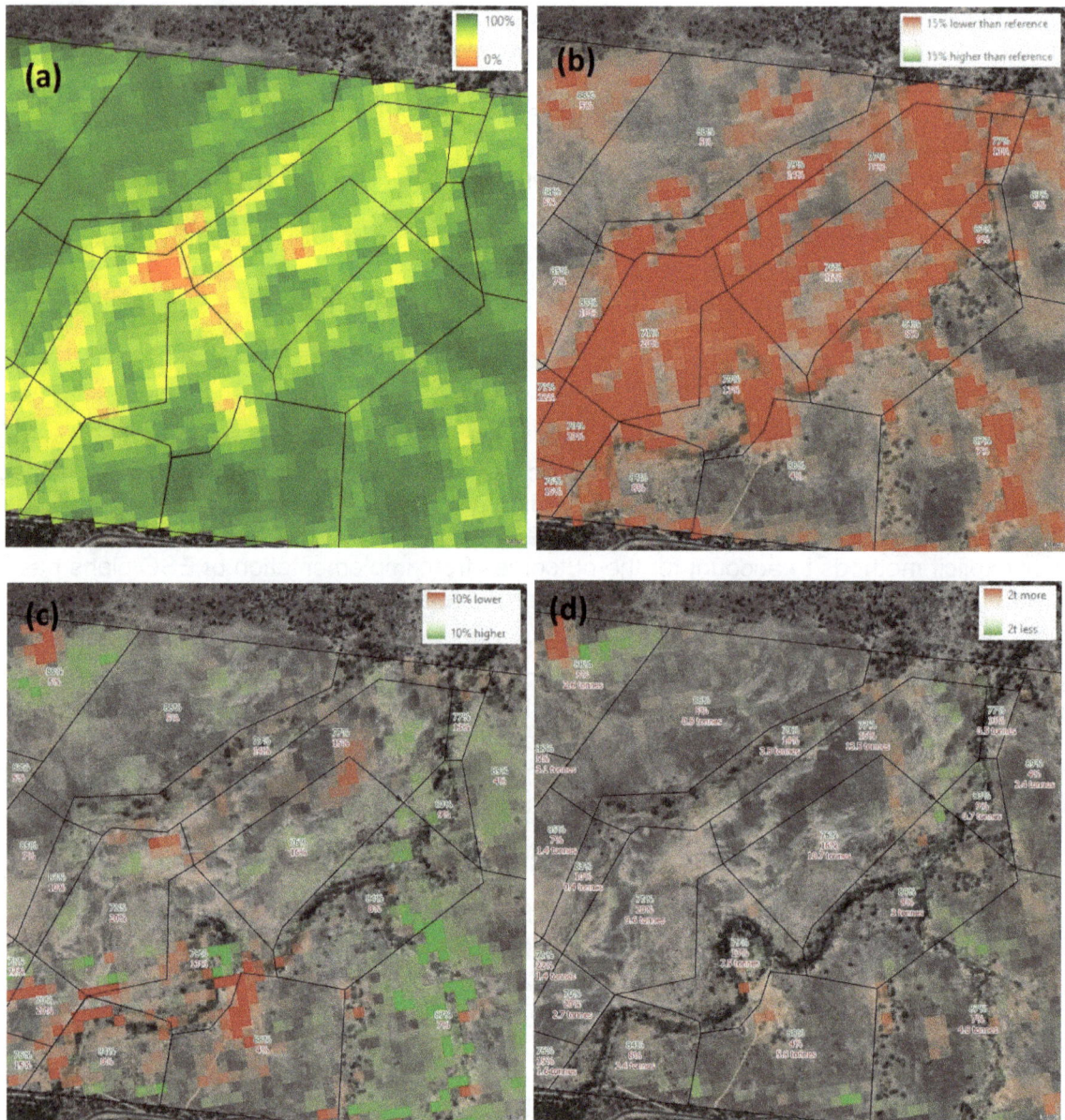

FIG 1 – (a) Average monthly vegetation cover as determined by drone NDVI determined for the period January 2017 to December 2024; (b) monthly difference in vegetation cover relative to reference area; (c) average improvement in vegetation cover (%) since land management change; (d) Change in total tonnes of sediment exported, estimated from spatial RUSLE since land management change.

The locally relevant nature of the reference sites with their matching land types has a key benefit in recognising and accounting for the often-difficult soils commonly encountered in mining regions of the Bowen Basin, giving managers confidence that any impacts or improvements are assessed against a baseline relevant to their operations.

The combination of the calendar month RUSLE erosion raster layers, actual and predicted from prior practice, allows the model to quantify the success of interventions, providing mine managers with the capacity to view and account for erosion from site and erosion control effectiveness (Figure 1c and Figure 1d). Derivatives of these two key layers can also provide insights into:

- Actual values for sediment savings for erosion events and practices in tonnes for cost-effectiveness calculation and reporting.

- Areas of current high erosion rates for investigation and mitigation.

- Areas that show the greatest improvement in erosion rates or declines in condition from what was expected prior to disturbance or intervention, independent of seasonality.

- Visibility and accounting for the practices of lessees on mine-controlled land.

- Extraction of trends, positive or negative, that might warrant investigation for broader application across the landscape or early intervention to arrest any decline before it becomes larger and more expensive issue.

Further development of ecosystem service markets within the broader economic landscape may also present opportunities for resource companies to evidence their social licence to operate, along with an accounting system for offsetting impacts from activities.

REFERENCES

Bastin, G, Scarth, P, Chewings, V, Sparrow, A, Denham, R, Schmidt, M, O'Reagain, P, Shepherd, R and Abbott, B, 2012. Separating grazing and rainfall effects at regional scale using remote sensing imagery: a dynamic reference-cover method, *Remote Sensing of Environment,* 121:443–457.

Chrystal, R, Côte, C, Asmussen, P and Shaygan, M, 2021. Erosion and sediment control framework for Queensland Mines, Sustainable Minerals Institute, The University of Queensland.

Renard, K G, Foster, G R, Weesies, G A, McCool, D K and Yoder, D C, 1997. Predicting soil erosion by water: A guide to conservation planning with the Revised Universal Soil Loss Equation (RUSLE), United States Government Printing Office.

Progressive rehabilitation certification at Oaky Creek Coal Mine – a case study

P W S Tett[1], B J Radloff[2] and J May[3]

1. Technical Director, SLR Consulting Australia Pty Ltd, Mackay Qld 4740.
   Email: ptett@slrconsulting.com
2. Technical Director, SLR Consulting Australia Pty Ltd, Mackay Qld 4740.
   Email: bradloff@slrconsulting.com
3. Environment and Community Manager, Oaky Creek Coal Pty Ltd, Tierri Qld 4709.
   Email: joel.may@glencore.com.au

EXTENDED ABSTRACT

This paper presents the rehabilitation and monitoring methods leading to the progressive rehabilitation certification (PRC) as per the *Queensland Environmental Protection Act 1994* (Department of Environment, Science, Tourism and Innovation (DETSI, 1994) and DETSI (2024). To date, approximately 2933 ha of open cut and underground mining areas at the Oaky Creek Coal Mine (OCC) in central Queensland have been certified by DETSI (Figure 1).

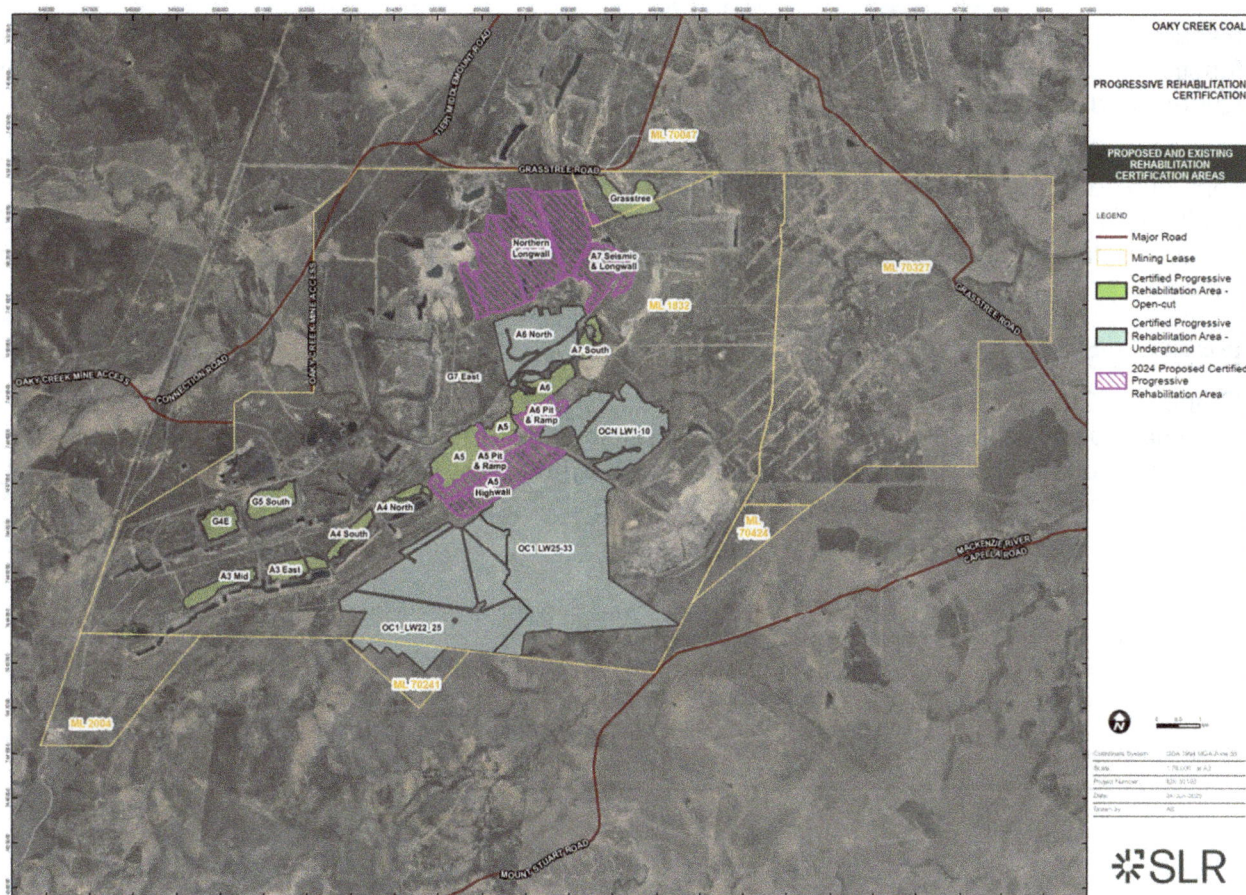

FIG 1 – Oaky Creek Coal Pty Ltd – progressive rehabilitation certification areas.

METHODS

Between 1982 and 2006 traditional open cut dragline methods were employed to win coal from within 24 open cut pits. In 1989, underground longwall mining was introduced to supplement open cut operations and win deeper coal within the mining lease. Underground coal reserves have been mined by longwall methods since 1989, with the forecast life-of-mine currently 2033.

Progressive rehabilitation of open cut areas to the approved post mining land use (PMLU) of native vegetation and has been occurring since the late 1980s and the rehabilitation of underground mining areas to the approved PMLU of beef cattle grazing has been occurring since the early 1990s. To date, approximately 5550 ha of the 8700 ha disturbance footprint has been rehabilitated.

Open cut rehabilitation methods included dozer push regrade to the final internally draining landform design. More erosion prone materials (sodic tertiary overburdens), coal tailings and rejects have been buried and capped with Permian overburden. After soil testing and rehabilitation trials conducted early in the mine life, open cut areas were largely rehabilitated without topsoil to reduce competition between seeded native trees and shrubs and introduced pasture grasses, including *Cenchrus ciliaris* (Buffel grass). Following soil testing for pH, electrical conductivity, cation exchange capacity, exchangeable sodium percentage and macronutrients. Seed bed preparation was undertaken by gypsum addition (as necessary), contour ripping (to 300 mm with 1.5 m max tyne spacings), fertilisation (eg Di-ammonium Phosphate) and direct seeding with native tree and local shrub seed.

Rehabilitation methods above the underground mining area consisted of the identification and ripping of substantial subsidence cracks (nominally >50 mm) in width. Within associated exploration, seismic and roads and track disturbance areas where topsoil stripping and adjacent stockpiling occurred, rehabilitation methods included replacing topsoil and seeding (if required) with native and endemic pasture grasses.

Rehabilitation monitoring programs and success indicators were developed using landscape functional analysis (LFA) transect and quadrat methods in consultation with the University of Queensland (Eskine and Fletcher, 2014).

In recent years Artificial Intelligence (AI) supported remote sensing capable of identifying individual flora and fauna species, diversity, abundance and cover has been used to verify and support traditional monitoring methods. This included drone flights to obtain high resolution visual and spectral data and on ground monitoring by ecologists to correlate and validate the drone derived data.

RESULTS

Of the 2,933 hectares certified to date, 577 ha (20 per cent) has been open cut mining areas and 2356 ha (80 per cent) subsidence areas, exploration and seismic areas (Table 1). A further 880 ha is currently under application and being considered by the Queensland Department of Environment, Tourism, Science and Innovation.

TABLE 1

Oaky Creek Coal Pty Ltd – progressive rehabilitation certification.

Year	Open cut – native vegetation PMLU area (ha)	Underground and subsidence – grazing PMLU area (ha)	Total (ha)
2019	132.8	0	**132.8**
2021	200	232.7	**432.7**
2022	48	861	**909**
2023	196.55	1262.23	**1458.78**
Total certified	**577.35**	**2355.93**	**2933.28**
2024 application area	121.98	757.67	879.65
Total applications	**699.33**	**3113.6**	**3812.93**

DISCUSSION

Key challenges within the certification process included obtaining regulator acceptance of innovative rehabilitation and monitoring approaches. This was particularly the case with regulator acceptance of the use of AI based remote sensing technology, the sustainability of the native vegetation PMLU without topsoil and ephemeral ponded subsidence troughs to supplement grazing with ponded pasture and cattle drinking water.

With the correlation of LFA and remote sensing data over representative areas, drone flights and AI processing were used to accurately and cost-effectively acquire rehabilitation monitoring data across the entire PRC application area. It was found that the remote sensing data was more accurate than the LFA data in measuring rehabilitation success. This result was due to remote sensing analysing the entire rehabilitation block thus removing spatial sample variation bias inherent in the LFA methodology.

The hurdles of remnant native vegetation sustainability and ephemeral ponding within Grazing PMLU subsidence troughs were overcome by presenting monitoring data gathered over the 30-year rehabilitation history at site. Subsidence ponding was analysed over time using a time series of aerial photographic evidence to assess the temporal distribution and extent of ponding. Water quality results from the subsidence ponds were compared against the Australian and New Zealand Guidelines for Fresh and Marine Water Quality (ANZG, 2018) for livestock drinking water.

The sustainability of the native vegetation PMLU was validated by presenting LFA derived data for abundance, diversity, structure and recruitment over the monitoring history. It was demonstrated that native vegetation establishment without topsoil was successful depending on slope, overburden characteristics within the root zone and rehabilitation techniques. Competition between introduced grasses (particularly Buffel grass) and emergent native trees and shrubs has been repeatedly demonstrated to suppress native tree and shrub species establishment. Therefore, topsoil and/or high rates of fertilisation (particularly Nitrogen) should be used with caution when establishing native vegetation PMLUs.

KEY FINDINGS

Key findings from historic rehabilitation data analysis and introduction of AI methods used in the PRC process included:

- Starting rehabilitation and rehabilitation monitoring as early as possible. Commencing rehabilitation early and gathering reliable monitoring data throughout this period is critical to early certification and relinquishment. Reliable trend analysis for historical rehabilitation areas is valuable in demonstrating success trajectory for more recent rehabilitation areas.

- The use of remote sensing AI technologies for rehabilitation monitoring were proven to be cost-effective in providing a complete spatial data set suitable for supporting the rehabilitation maintenance and certification process and supplementing LFA derived data sets.

- For native vegetation PMLU establishment and sustainability on inherently stable overburden, topsoil and/or high rates of fertiliser application (particularly Nitrogen) should be used with caution to minimise grass (particularly Buffel grass) competition.

- With minimised grass ground cover and the shading effects of established native tree and shrub species, the native vegetation PMLU was suited to areas not requiring higher ground cover for erosion protection. This contrasts with the general misconception that all rehabilitation areas require topsoil or an analogous manufactured growth medium for success.

- From a landform stability and sustainability perspective, steeper areas with less inherent stability that require a higher degree of erosion protection (longer slopes, dispersive materials, drainage lines etc) should, in general, be topsoiled (or growth media added), ameliorated with a relatively high fertiliser rate and grassed. This contrasts with many Queensland Environmental Authority conditions for open cut rehabilitation areas where steeper slopes require rehabilitation to native vegetation and flatter areas are grassed for a Grazing PMLU.

- Transparent stakeholder engagement including early regulator and landholder consultation was key to achieving certification and overcoming the above challenges. On-site presentations supported by site derived data and site inspections with the regulator, were used to demonstrate that proposed PRC areas meet the required completion/milestone criteria and that both native vegetation and grazing PMLUs were sustainable.

REFERENCES

Australian and New Zealand Governments (ANZG), 2018. Australian and New Zealand Guidelines for Fresh and Marine Water Quality, Australian and New Zealand Governments and Australian state and territory governments, Canberra ACT, Australia. Available from: <https://www.waterquality.gov.au/anz-guidelines> [Accessed: 24th June 2025].

Department of Environment, Tourism, Science and Innovation (DETSI), 1994. Environmental Protection Act 1994 [online]. Available from: <https://www.legislation.qld.gov.au/view/html/inforce/current/act-1994-062#> [Accessed: 10th July 2025].

Department of Environment, Tourism, Science and Innovation (DETSI), 2024. Guideline – Progressive Certification of Resource Activities, ESR/2022/5900, Version 1.02 [online]. Available from: <https://www.des.qld.gov.au/policies?a=272936:policy_registry/rs-gl-progressive-certification.pdf> [Accessed: 24th June 2025].

Eskine, P E and Fletcher, A, 2014. Rehabilitation Monitoring Manual for Glencore Coal Qld.

Mine closure – a liability or an opportunity: a South African perspective

V R K Vadapalli[1], H Coetzee[2], M Solomon[3], S Gcasamba[4], B Mahlase[5], J Shongwe[6], L Nkabane[7], R Mahlaule[8], N Chere[9], J Mathekga[10], S Kekana[11] and L Mudau[12]

1. Chief Scientist, Council for Geoscience, Pretoria Gauteng 0184, South Africa. Email: vvadapalli@geoscience.org.za
2. Specialist Scientist, Council for Geoscience, Pretoria Gauteng 0184, South Africa. Email: henkc@geoscience.org.za
3. Professor, University of Cape Town, Cape Town Western Cape 7700, South Africa. Email: michael.solomon@uct.ac.za
4. Scientist, Council for Geoscience, Pretoria Gauteng 0184, South Africa. Email: sgcasamba@geoscience.org.za
5. Scientist, Council for Geoscience, Pretoria Gauteng 0184, South Africa. Email: bmahlase@geoscience.org.za
6. Chief Scientist, Council for Geoscience, Pretoria Gauteng 0184, South Africa. Email: jshongwe@geoscience.org.za
7. Scientist, Council for Geoscience, Pretoria Gauteng 0184, South Africa. Email: lnkabane@geoscience.org.za
8. Scientist, Council for Geoscience, Pretoria Gauteng 0184, South Africa. Email: rmahlaule@geoscience.org.za
9. Scientist, Council for Geoscience, Pretoria Gauteng 0184, South Africa. Email: nchere@geoscience.org.za
10. Deputy Director Mine Closure, Department of Mineral and Petroleum Resources, Pretoria Gauteng 0001, South Africa. Email: jacqueline.mathekga@dmre.gov.za
11. Deputy Director Mine Closure, Department of Mineral and Petroleum Resources, Pretoria Gauteng 0001, South Africa. Email: sebitsa.kekana@dmre.gov.za
12. Director Mine Closure, Department of Mineral and Petroleum Resources, Pretoria Gauteng 0001, South Africa. Email: lindelani.mudau@dmre.gov.za

ABSTRACT

Mine closure in South Africa is governed by mining, environmental and water management legislation, which provide a sound framework for responsible mine closure and post closure management. In recent years, South Africa's Department of Mineral and Petroleum Resources and Council for Geoscience have undertaken the development of a National Mine Closure Strategy which aims to promote sustainable mine closure by improving mine closure practices. The strategy encourages a process of life cycle mine closure planning, ensure fit-for-purpose environmental remediation and concurrent economic diversification aligned with post-mining economies through economic succession and improving the social impact of mine-closure related activities. Furthermore, it is also recommended that mine closure areas of interconnection and/or cumulative environmental and socio-economic impacts should be identified, and area-based strategies to be developed in conjunction with other interventions identified in the main document. This strategy therefore aims to incorporate socio-economic factors into mine closure planning and implementation and align the physical rehabilitation of mine sites with the long-term developmental priorities and local government initiatives. Successful implementation of the principles of the strategy is expected to ensure sustainable mine closure and change the perception around mine closure from liability to an opportunity. This paper presents the process towards the development of the National Mine Closure Strategy and its recommendations. The strategy development process has involved extensive multi-stakeholder engagement and is currently in an advanced stage of development towards approval (not a final version).

INTRODUCTION

Mine closure is a complex process that involves careful environmental and post-mining economic land use planning and is often regarded as a liability by regulators and mining companies, mainly due to the lack of incentives for mine closure and the perception that no opportunities exist for post-operational revenue generation as well as the need for mine operators to take responsibility for prolonged latent impacts. As in many jurisdictions, rigorous environmental regulation is a recent phenomenon, leading to numerous mining legacies (Figure 1). In most instances, physical rehabilitation including concurrent rehabilitation is prioritised, focusing on restoring the land to a state close to its original state. Moreover, Social and Labour Plan (SLP) projects are often not aligned with the local government initiatives resulting in inefficiencies and duplication of projects. Land rehabilitation often does not align with these programs. As a result, there is little coordination between different initiatives causing a depleted impact on mining communities, and opportunities for economic diversification are neglected. Furthermore, challenges around: (i) interconnected mines with cumulative and integrated impacts; and (ii) the requirement to provide for latent post-closure impacts, including long-term water management have led to a complex situation where sustainable mine closure is difficult to achieve. Therefore, it is critical to deal with long-term impacts and balancing post mining economic diversification with environmental remediation. Recently, the Department of Mineral and Petroleum Resources (DMPR) with the help of the Council for Geoscience (CGS) is developing a National Mine Closure Strategy (NMCS) to address the challenges outlined above. This paper explores and presents the process towards the development of the NMCS as well as the recommendations under development to better facilitate mine closure in South Africa. NMCS is being developed with the aim of better aligning mine closure practices with the imperative to develop viable post-mining economies and alternative land use, within the current and evolving legislative and regulatory framework.

FIG 1 – Consequences of poor historical mine closure practices including unsafe openings and illegal mining activities.

Process of strategy development

The current draft NMCS has been developed through an iterative process of strategy formulation and multi-stakeholder engagement.

A first draft of the strategy was presented in 2019, focusing on addressing the biophysical impacts of mining and approaches to mitigation of these impacts. Following this, an intensive review was undertaken to address the socio-economic aspects of mine closure and post closure management, introducing the concept of concurrent economic diversification in mining regions. This was to enable the development of a viable post-mining economy before mine closure and to better align mine closure processes with the United Nations Sustainable Development Goals (SDGs) (United Nations, 2024). Additional aspects of mine closure were also addressed including interconnected mines where their safety, health, social or environmental impacts are integrated which results in a cumulative impact, as regulated under Section 43(9) of the Mineral and Petroleum Resources Development Act (MPRDA) (Act 28 of 2002, as amended) (Republic of South Africa, 2002). The resulting draft document was published in the Government Gazette (Department of Mineral Resources and Energy (DMRE), 2021) for consultation.

The stakeholder engagement process on the strategy was initiated in 2021 during the COVID-19 pandemic, necessitating the extensive use of virtual consultative meetings. Separate consultative webinars were held with different stakeholder groups allowing extensive and incisive comments to be compiled which, together with written comments received, allowed the drafting of an improved version of the strategy. A second series of webinars, as well as some in-person consultations were then held to allow a high degree of consensus to be attained between the drafting team and diverse stakeholders. This strategy is in the process of finalisation, prior to approval and adoption.

CURRENT DRAFT OF THE STRATEGY

Pillars of the strategy

The strategy is based on four pillars:

1. Rigorous closure planning, throughout the mining life cycle.

2. Environmental remediation and rehabilitation.

3. Improved social impact of closure-related activities.

4. The development of a vibrant and diverse post-mining economy.

The integration of these pillars aims to promote the incorporation of the SDGs into the mine closure process (Figure 2).

FIG 2 – National Mine Closure Strategy vectors (modified after Solomon *et al*, 2020).

Legal framework

Mine closure is regulated by the DMPR under the MPRDA, with procedures and standards detailed in the National Environmental Management Act (NEMA) (Act 107 of 1998) (Republic of South Africa,

1998a). These two acts have been amended since 2014, harmonising mining and environmental legislation under the 'One Environmental System', with the DMPR assuming the responsibility of implementing both acts, as they apply to the mining industry. Water management in mining operations is regulated in terms of water use licenses (WULs), issued by the Department of Water and Sanitation (DWS) in terms of the National Water Act (NWA) (36 of 1998) (Republic of South Africa, 1998b). Stakeholder engagement and consultation, subject to prescribed standards, comment periods etc, underpins the process of mine closure planning and implementation.

Alignment of mine closure with social and labour plans

The development and implementation of mines' SLPs, as required under the MPRDA and defined in the relevant MPRDA regulations regulates the way that mines support local economic development. The National Mine Closure Strategy identifies SLPs as a vehicle to align mine closure planning and implementation with regulated social expenditure and with the developmental needs for economic diversification towards a post-mining economy. This also highlights the need for stakeholder engagement throughout the process. The strategy recommends aligning SLP expenditure with local government initiatives, as expressed in municipal Integrated Development Plans.

Interventions identified to improve mine closure practice

Specific interventions are proposed in the National Mine Closure Strategy to improve closure practices.

Closure planning throughout the mining life cycle, commencing before mining and continuing throughout the life of a mine is a critical requirement for mine closure. Legislation requires agreement between stakeholders on the end land-use and provides for the amendment of closure plans to align with changing conditions and circumstances throughout the life-of-mine.

In some areas, mines have integrated and/or cumulative environmental and socio-economic impacts. Three scenarios have been identified, where area-based closure strategies, using the procedures defined in Sect 43 (9–12) of the MPRDA (Republic of South Africa, 2002), will facilitate mine closure:

1. Areas of hydraulic interconnection, where inter-mine water flows and shared water management measures mean that the closure of any one mine can affect neighbouring mines and need to implement strategies to allow mines to close without disadvantaging other mines.

2. In areas where multiple mines exist within a single municipality and mining is a major contributor to local economic activity and employment, mines can enable the development of a coherent post-mining economy by aligning their identified post-mining land-use(s).

3. In special cases where multiple mines have a cumulative impact on a shared receiving environment, site-specific strategies may assist in mitigating this impact.

The scope of mine closure needs to be expanded from the traditional concept, which focuses strongly on the physical remediation of disturbed environments and mine residues to include the mitigation of the socio-economic impacts of closure. Economic diversification concurrent with mining is a critical enabler of the development of a post-mining economy. Meaningful engagement with stakeholders throughout the mining life cycle in key decision-seeking processes is critical to ensure optimal outcomes in mine rehabilitation which align with socio-economic considerations. A few stakeholders identified the need for partial mine closure, as an enabler for economic diversification during the life-of-mine. This has been received positively but would require, regulatory clarity, rigorous planning and strict control.

While the NMCS has prioritised alignment with existing legislation and regulations, a few areas of focus have been identified. These include:

- In cases where mining rights are transferred to new owners, the new operators need to demonstrate that they have the technical and financial capacity to operate and close the mine.

- Regular reporting and auditing, as required in regulations is needed to ensure that mines remain in compliance with legislation and their own closure plans. Specific area for attention is the adequacy of mines' financial provision for closure and the need to address premature mine closure.

- As water management has been identified as a central component of mine operation and closure, the alignment of mining authorisations and closure plans with the requirements of Water Use Licenses granted under the NWA needs to be ensured. The existing best practice guideline for water management in the mining industry (Munnik and Pulles, 2009) hierarchy aims at better closure outcomes, prioritising pollution prevention and water reclamation, with long-term treatment seen as an option of last resort (Figure 3).

- Inadequate mine closure has contributed to a national problem of illegal mining on abandoned mine sites. While this is a multi-faceted problem being addressed by several state departments and entities, the role of legitimate artisanal miners needs to be considered.

- Where appropriate, voluntary compliance with external guidelines and standards by mining companies, eg ISO standards, may assist in ensuring sustainable mine closure.

- Implementation of the strategy needs to acknowledge that different mines are at different stages within their life cycles. To accommodate all mines, a practical and pragmatic approach to the transition will be needed to improve mine closure practices.

- The implementation of the harmonised mining and environmental legislation represents a transition to a system with a strong process orientation. This provides an opportunity for the improvement of standard operating procedures to ensure sustainable mine closure.

- Regular reviews of the strategy will be required to apply the lessons learned during implementation and to adapt to changing circumstances.

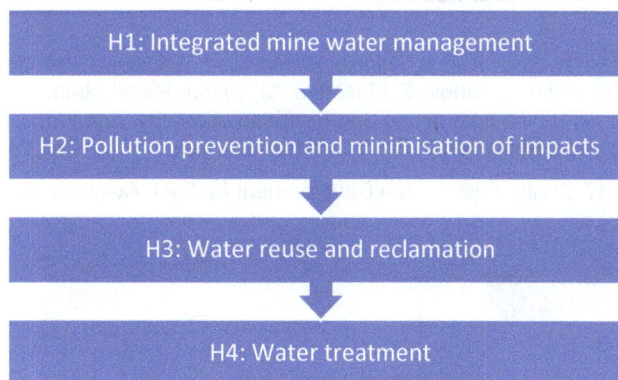

H1: Integrated mine water management

H2: Pollution prevention and minimisation of impacts

H3: Water reuse and reclamation

H4: Water treatment

FIG 3 – Hierarchy of water management for the South African mining industry (Coetzee *et al*, 2025; after Munnik and Pulles, 2009).

CONCLUSIONS

The draft NMCS, is at an advanced stage of development towards approval (not a final version), which has followed iterations of engagements with stakeholders of diverse backgrounds compiling expert knowledge. While time-consuming, this has ensured that most of the parties affected by the closure of mines have had multiple opportunities to engage with and contribute to the strategy.

The draft strategy attempts to align the need for better mine closure practice and sustained mine closure in South Africa within the existing legislation, in particular the MPRDA, NEMA and the NWA and the relevant regulations promulgated under these acts.

The essence of the strategy is to improve the planning and implementation of mine closure through number of interventions identified, such as implementation of interconnected and/or cumulative environmental and socio-economic area-based closure strategies, closure planning throughout the mining life cycle, alignment of SLP projects with local government initiatives and fostering post-mining economies through concurrent economic diversification, aiming towards a vibrant post-mining economy. This aims to achieve both the physical rehabilitation of mining-affected environments and avoid the socio-economic impacts often associated with mine closure.

This paper reflects the current state of a strategy which is under development as part of an ongoing project, undertaken by the DMPR and the CGS, which has not yet been formally adopted by the Department and do not represent a finalised strategy.

ACKNOWLEDGEMENTS

The DMPR is acknowledged for their support for the process of strategy development. The stakeholders who participated in multiple rounds of consultation and made important written and oral comments and contributions have enabled the development of a strategy which takes consideration of a broad range of interests and opinions.

REFERENCES

Coetzee, H, Vadapalli, V, Gcasamba, S, Mahlase, B, Nkabane, L, Mahlaule, R, Chere, N, Shongwe, J and Solomon, M, 2025. Integrated Research into Mine Closure Programme: National Strategy for Mine Closure and the Management of Derelict and Ownerless Mines, CGS report 2024–0073.

Department of Mineral Resources and Energy (DMRE), 2021. Mineral and Petroleum Resources Development Act: National Mine Closure Strategy [online]. Available from: <http://www.gov.za/sites/default/files/gcis_document /202105/44607gen446.pdf> [Accessed: 4 June 2025].

Munnik, R and Pulles, W, 2009. The implementation of the recently developed Best Practice Guidelines for Water Resource Protection in the South African Mining Industry, in *Proceedings International Mine Water Association Conference 2009*, pp 22–28 (International Mine Water Association).

Republic of South Africa, 1998a. National Environmental Management Act (as amended), No. 107 of 1998, November 1998.

Republic of South Africa, 1998b, National Water Act (as amended), No. 36 of 1998, August 1998.

Republic of South Africa, 2002. Mineral and Petroleum Resources Development Act (as amended), No. 28 of 2002, October 2002.

Solomon, M, Vadapalli, V, Coetzee, H, Sogayise, S, Hanise, B, Mahlase, B, Dube, G, Modise, M, Mtyelwa, O, Masindi, M, Lekoadu, S, Gcasamba, S, Kgari, T, Ntholi, T, Masenya, M, Ugwu, P and Mathekga, J, 2020. Draft National Mine Closure Strategy on planning for the postmining South African economy: Economic succession planning as a basis for mine closure planning and certification in South Africa, CGS report 2019–026.

United Nations (UN), 2024. The 17 Goals: Sustainable Development [online]. Available from: <https://sdgs.un.org/goals> [Accessed: 5 March 2025].

Assessment of coal spoil piles with respect to soil cover requirements at closure

W Zhang[1], N Thompson[2], M Stimpfl[3] and M Edraki[4]

1. Research fellow, The University of Queensland, St Lucia Qld 4072.
   Email: wenqiang.zhang@uq.edu.au
2. Specialist Environmental Geochemist, BHP, Brisbane Qld 4001.
   Email: natasha.thompson@bhp.com
3. Superintendent Geo-Environmental, BHP, Perth WA 6000. Email: marilena.stimpfl@bhp.com
4. Professor, The University of Queensland, St Lucia Qld 4072. Email: m.edraki@uq.edu.au

ABSTRACT

The rehabilitation plans for spoil piles often include expensive soil cover systems. However, we suggest that unlike hard rock mining, the natural degradation of most sedimentary spoil materials over time will significantly reduce oxygen diffusion and the infiltration of rainfall, preventing the generation and transport of acidic, metalliferous and saline drainage. To investigate the hydrological and geochemical behaviour of spoil piles over time, two large, instrumented columns (1.2 m in height and 0.4 m in diameter) were established in a glasshouse at The University of Queensland. Two types of spoil, a soil-like and a rock-like spoil, were sourced from a coalmine site in Queensland, with various rock fragment sizes. Comprehensive mineralogical and geochemical characterisation, including quantitative X-ray diffraction analysis, acid-base accounting, and major and trace element analysis, have been conducted on nine representative samples. Almost all tested spoil materials were classified as non-acid forming (NAF). The spoil samples were packed into the two columns, with mudstone-to-sandstone ratios of 1:3 and 3:1, respectively. Changes in volumetric moisture content, matric potential, temperature, and oxygen concentration were monitored over depth, along with pore water and pore gas samplers. Additionally, a time-lapse camera was installed to continuously record weathering patterns, and a weather station was set-up to calculate evaporation rates which helps estimate the overall water balance within the columns. Results indicate that the spoil with a higher percentage of mudstone exhibits more progressive desiccation and oxygen diffusion patterns at depths above 300 mm but demonstrates higher water-holding capacity at depths below 700 mm, compared to the spoil with a high percentage of sandstone. The experimental data over wetting-drying cycles help understand the long-term weathering behaviour of spoil piles. These insights will inform progressive rehabilitation and closure plans, aiming to reduce, optimise or eliminate the need for an engineered soil cover system at closure.

INTRODUCTION

Coal spoil materials are potential sources of acid and metalliferous drainage (AMD) including saline drainage. The rehabilitation plans for spoil piles often include expensive soil cover systems. Soil cover is mostly used at mine closure to protect the surrounding environment by ensuring the long-term chemical and physical stability of mine waste storage facilities (eg tailings storage facility and waste rock dump) and provide a suitable substrate for revegetation. Global Acid Rock Drainage (GARD) Guide (International Network for Acid Prevention (INAP), 2009) has summarised common cover system designs based on site-specific climate, ranging from simple (eg monolithic soil cover) to more complex (involving several layers of materials with a capillary break to prevent moisture movement between cover soil and mine waste). However, there is limited data that are available to assess the long-term performance of these designs in the various climatic zones of Australia, particularly in the subtropical to tropical regions of Queensland, characterised by extremely variable rainfall with alternating wet and dry periods. Establishing a robust and cost-effective cover system requires a thorough understanding of environmental conditions and the properties of the materials available for building a cover. A successful outcome relies on extensive material characterisation, design work supported by numerical modelling, and field trials. Cover designs need to be trialled and monitored for sufficient durations at each site's context. However, large-scale field trials are logistically challenging and often do not align with mining schedules during the operation's life.

Considering the rehabilitation cost and availability of soil sources, we suggest that, unlike hard rock mining, the natural degradation of most sedimentary spoil materials over time will significantly reduce oxygen diffusion and the infiltration of rainfall, preventing the generation and transport of acidic, metalliferous and saline drainage (Park, Edraki and Baumgartl, 2017; Edraki *et al*, 2017). The weathering pattern and resulting mineralogy of coal spoils may reduce the need for complex and expensive soil cover systems at closure, provided reactive waste is buried within the landform at some depth during operation. This study aims to understand the long-term weathering behaviour of coal spoil piles and the environmental benefits of spoil cover in view of maximising those benefits as spoil piles are built during the operational phase of mining.

METHODOLOGY

To investigate the hydrological and geochemical behaviour of spoil piles over time, two large, instrumented columns (1.2 m in height and 0.4 m in diameter) were established in a glasshouse at The University of Queensland. Two types of spoil, mudstone (soil-like) and sandstone (rock-like) spoil, were sourced from a coalmine site in Queensland, with various rock fragment sizes. Comprehensive mineralogical and geochemical characterisation, including quantitative X-ray diffraction (QXRD) analysis, acid-base accounting (ABA), and major and trace element analysis, have been conducted on nine representative samples. The QXRD results indicated that the tested samples are rich in quartz and contain high amounts of clay minerals such as mica, mixed-layered illite-smectite and kaolinite. Such mineralogy suggests that the spoil samples have a potential of relatively high water-holding capacity when weathered, which is an important property of cover materials to limit water/gas ingress to mine waste. Almost all the tested spoil materials were classified as non-acid forming (NAF) based on the ABA results. The spoil samples were crushed down below 42 mm and packed into the two columns, with mudstone-to-sandstone ratios of 1:3 and 3:1, respectively. Given the relatively large scale of the leaching columns and the inherent heterogeneity of the spoil materials, replicates were not considered in this study. Changes in volumetric moisture content, matric potential, temperature, and oxygen concentration were monitored over depth, along with pore water and pore gas samplers. Additionally, a time-lapse camera was installed to continuously record weathering patterns, and a weather station was set-up to calculate evaporation rates which helps estimate the overall water balance within the columns. Figure 1 shows the final set-up of the two instrumented columns. The irrigation regime followed the rainfall record from the mine site to simulate wet and dry seasons. Leachate was collected every month, as well as porewater at different depths, to analyse electrical conductivity, pH and major and trace element concentrations.

FIG 1 – Instrumented column leaching apparatus for coal spoil piles.

RESULTS

Figures 2–3 show the monitored results from the instruments in two columns, including volumetric water content, temperature, bulk electrical conductivity (EC), matric potential and oxygen concentrations over depth. So far, two major irrigation scenarios have been applied to the columns: 1) flushing two columns with denoised water in mid-November 2024; and 2) simulating the wet season based on the site climate in mid-February 2025. Table 1 summarises the concentrations of major cations and anions in the flushing leachates from the two columns. Leachate from column 1 exhibits higher concentrations of all listed components compared to column 2, consistent with the general trends of bulk EC in Figures 2–3. Figure 4 shows volumetric water content profiles over depth in the two columns. The monitored results indicate that the spoil with a higher percentage of mudstone exhibits more progressive desiccation at depths above 300 mm but demonstrates higher water-holding capacity at depths below 700 mm, compared to the spoil with a high percentage of sandstone. The drier conditions at shallow depths in column 1 result in more frequent fluctuations in oxygen concentration (between 11 per cent and 17 per cent), whereas column 2 shows relatively stable oxygen levels of ~16 per cent at shallow depths. Variations in water content, matric potential and oxygen concentration over depth suggest that an effective thickness of spoil material for limiting water infiltration should be more than 700 mm.

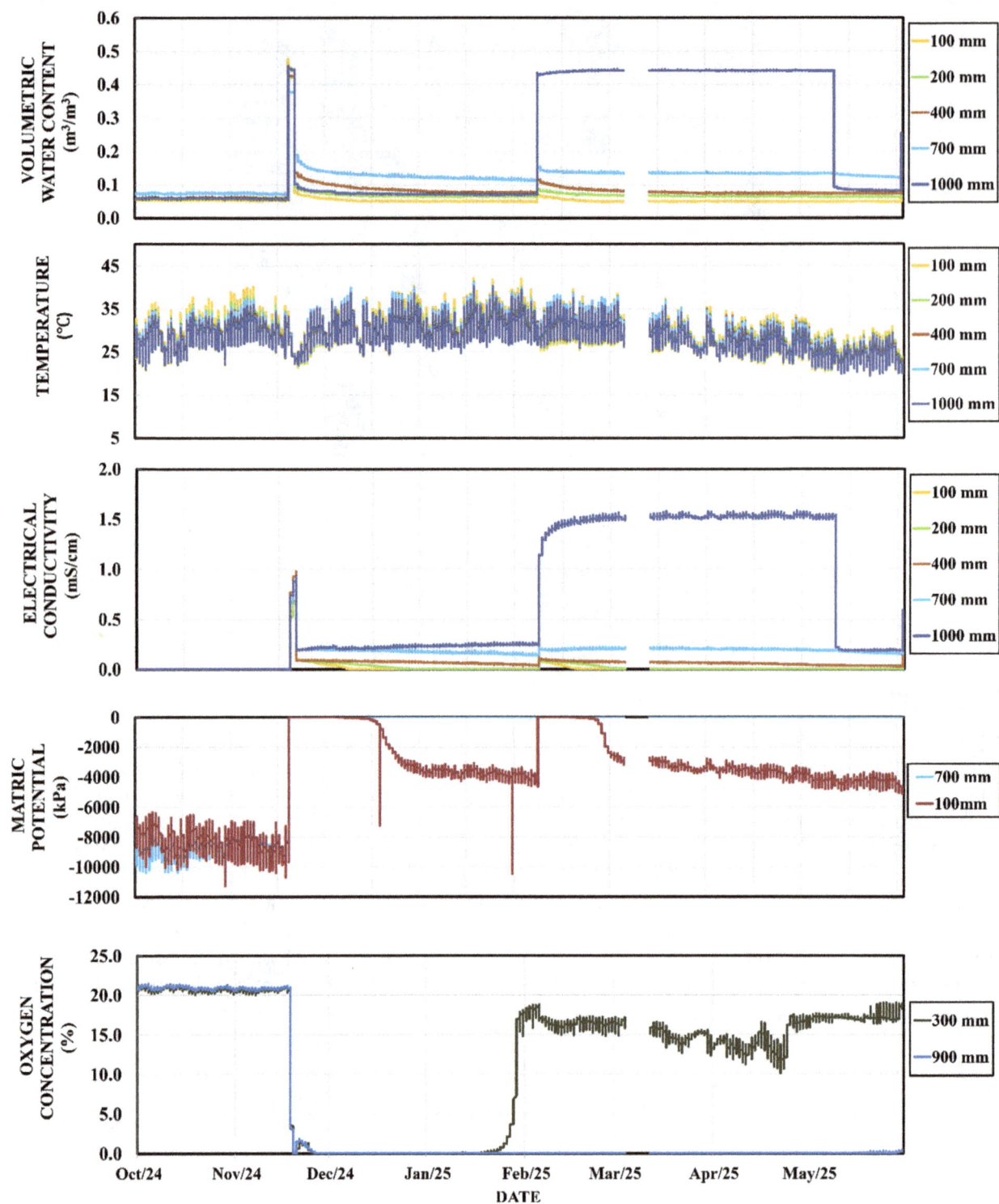

FIG 2 – Monitoring results of column 1 (mudstone to sandstone 1:3).

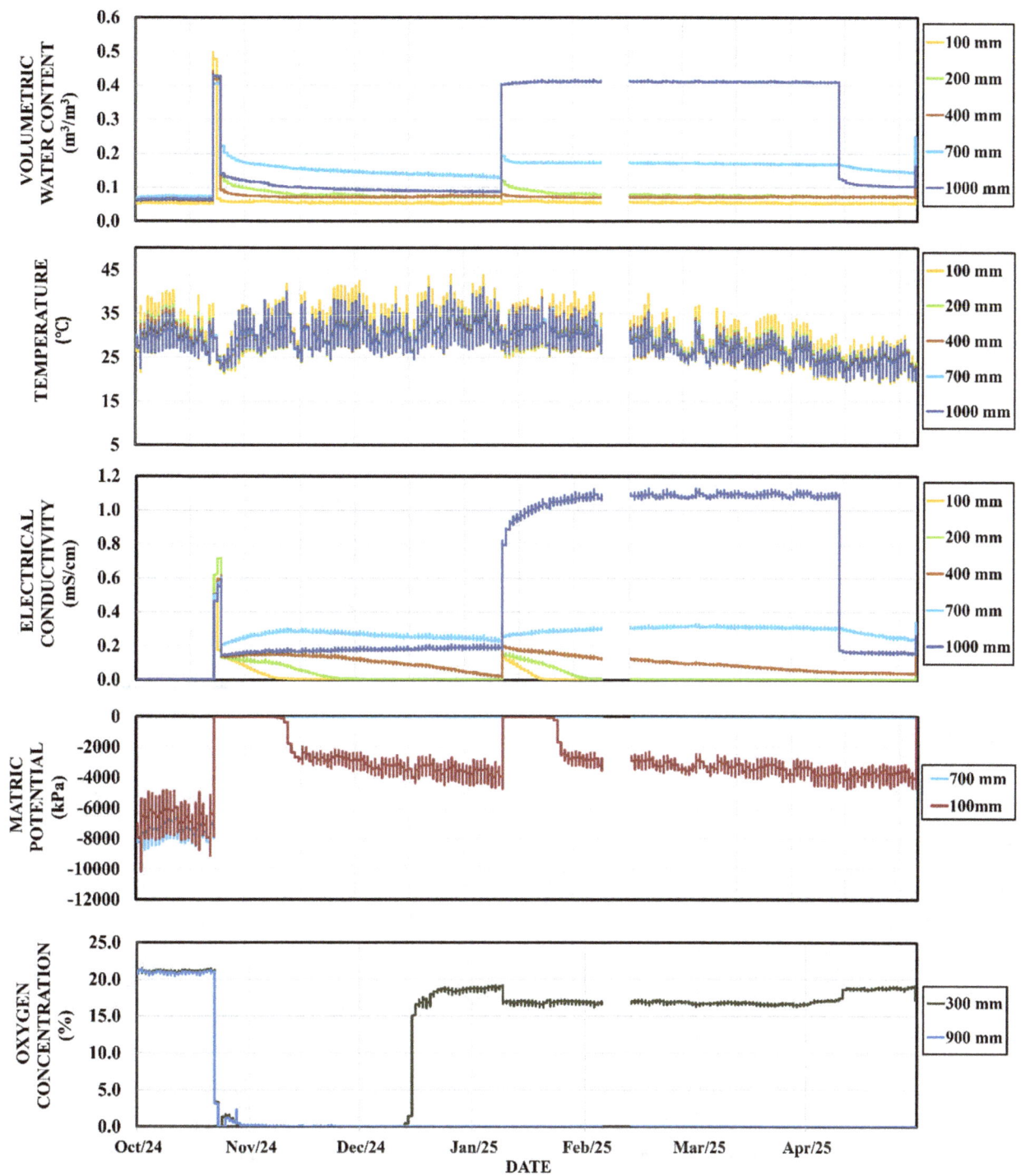

FIG 3 – Monitoring results of column 2 (mudstone to sandstone 3:1).

TABLE 1

Flushing leachate chemistry of two columns.

Component	Concentrations (mg/L)	
	Column 1	Column 2
Cl	138	95.3
SO4	385	240
Ca	12.1	4.8
Mg	9.1	2.5
K	3	1.1
Si	2.59	2.38
Na	351	247
Fe	0.02	0.09
Zn	0.02	0.02
Cu	0.01	0.01
Se	0.1	0.1
Sr	0.67	0.32

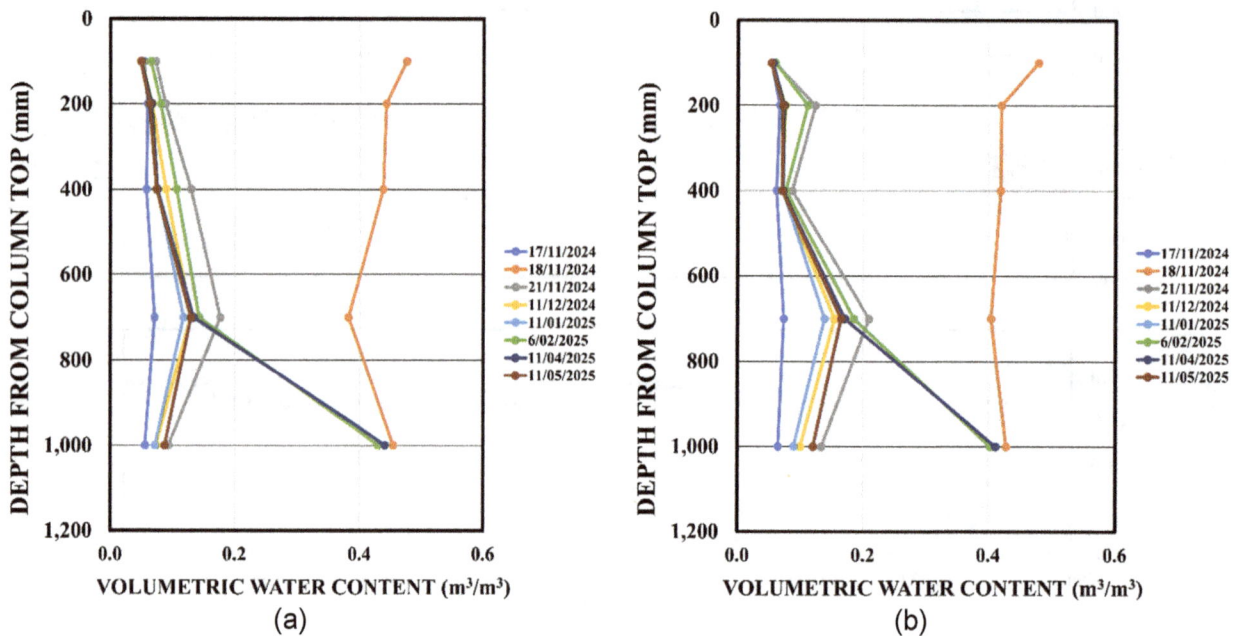

FIG 4 – Volumetric water content profiles over depths of: (a) column 1; and (b) column 2.

The instrumented large columns demonstrate the capacity to capture the evolution of spoil parameters, especially the unsaturated zone. The experimental data helps understand the long-term weathering behaviour of sedimentary coal spoil under wetting-drying cycles, contributing to assessing its environmental benefits. These findings will inform progressive rehabilitation and closure plans, aiming to reduce, optimise or eliminate the need for an engineered soil cover system at closure. As the studied spoil materials were initially fresh, it is recommended that the experiment continues to monitor ongoing weathering patterns, as well as hydrological and geochemical responses under extreme weather conditions, eg prolonged droughts or intense rainfall events.

REFERENCES

Edraki, M, Baumgartl, T, Mulligan, D, Fegan, W and Munawar, A, 2017. Geochemical characteristics of rehabilitated tailings and associated seepages at Kidston gold mine, Queensland, Australia, *International Journal of Mining, Reclamation and Environment*, 33(2):133–147. https://doi.org/10.1080/17480930.2017.1362542

International Network for Acid Prevention (INAP), 2009. *Global Acid Rock Drainage Guide* (GARD Guide). Available from: <http://www.gardguide.com/>

Park, J H, Edraki, M and Baumgartl, T, 2017. A practical testing approach to predict the geochemical hazards of in-pit coal mine tailings and rejects, *Catena*, 148:3–10.

Integrated strategic mine planning

Breaking barriers to enabling post-mining transitions – six key challenges and strategies

G Boggs[1], J D'Urso[2], T Measham[3] and B Kelly[4]

1. Chief Executive Officer, Cooperative Research Centre for Transformations in Mining Economies, Perth WA 6000. Email: guy.boggs@crctime.com.au
2. External Relations and Impact Director, Cooperative Research Centre for Transformations in Mining Economies, Perth WA 6000.
3. Research Director, Cooperative Research Centre for Transformations in Mining Economies, Brisbane Qld 4072. Email: tom.measham@crctime.com.au
4. Board Chair, Cooperative Research Centre for Transformations in Mining Economies, Melbourne Vic 3000. Email: bruce.kelley013@gmail.com

INTRODUCTION

We sit at a time of significant potential to reimagine and transform post-mine transitions. Significant opportunities are emerging with an estimated 240 mines expected to close across Australia by 2040, peaking around 2030. A key mental shift is required to realise opportunities is to think beyond closure and returning to the pre-mining state to the focus on post-mining outcomes (Measham *et al*, 2024).

In 2024, the Cooperative Research Centre for Transformation in Mining Economies (CRC TiME) undertook a review of research completed to date. This included more than A$30 million of collective investment across 46 projects, informed by more than 100 industry, government, regional, First Nations and research leaders throughout delivery. To build on what we learnt and reflect and shape this fast-changing context, we sought to understand both the key constraints preventing transformative change and outcomes and where we could contribute most value over the next five-year period.

Six key areas emerged from this process (CRC TiME, 2024).

KEY AREAS

Supporting corporate leadership in valuing post-mine transitions

Can we develop a comprehensive business case for post-mining transition?

There still remain a number of areas that challenge the business case for early and sustained investment in interventions that enable post-mine transitions. This includes the need for decision-makers to consider a wider set of inputs in the application of net present value (NPV) with regard to different post-mine transition scenarios. Studies have also highlighted discrepancies between forecasted mine closure costs and the actual expenses incurred, while Moody's recently reported US$78B of reclamation obligations for the 24 largest rated mining companies. This has future implications for accessing capital and potential escalation in credit risk.

A comprehensive business case for transition based approaches to closure also requires advancement in our ability to accurately forecast closure costs, and better assess and integrate intangible values of different post-mine transitions. Creating and demonstrating the value of opportunities for earlier transition of land, infrastructure and wastes to next use/s is required to address confidence in an area of limited examples of transition. This is critical to address rising costs and risks associated with closure of operational mines and growing legacy asset portfolios.

Operationalising mine closure transition opportunities within mine planning processes

How do we provide access to leading knowledge, tools and resources to better integrate mine closure and post-mine transitions in asset-level and operational decision-making?

The integration of progressive activities that inform, enable and position mined land and assets for transition generally remains an issue. A priority is the identification and integration of post-mine transition KPIs into short-term planning processes. We identified Natural Capital Accounting (NCA) systems as a mechanism for transparently and consistently assessing and valuing changes over the life cycle. Integrating NCA systems with forecast models provides mine planners and mine closure specialists with a more effective tool to evaluate alternative designs as well as track progress.

Key biophysical risk points require targeted and cross industry research and innovation programs to develop solutions. Water management and treatment is a significant cost, and limitation for post-mine use. We highlight priority research on Acid and Metalliferous Drainage (AMD) source control and treatment as well as mine pit/void design and hydrological understanding to create opportunities for novel post-mine economic or nature positive use of the lakes created.

Creating self-sustaining ecosystems is the second identified risk point for mine sites, with seed and soil availability key issues. Work is required to develop seed supply chains and optimise seed use to meet the scale of rehabilitation required at a national scale. Equally, creating guidance and establishing networked trials is important to support the sectors' ability to develop functional novel soils in soil constrained environments.

Establishing the business case for asset transfer

What are the top key contributors to levels of residual risk across all aspects of mining, ESG, cultural and regulatory domains? How do we identify acceptable levels of risk for post-mine stakeholders?

Transition, distinct from closure, generally requires the transfer of assets between mining and post mining actors. Focused effort is required both on articulating the different aspects of residual risk and fit for purpose mechanisms that provide confidence between parties in the transfer or sharing of future risk post asset transfer (Maybee *et al*, 2024). While work has focused on reducing risk, a growing body of work is developing a deeper understanding of the opportunities associated with post-mine land, waste and asset use.

Asset transfer requires focused and trusted mechanisms to evaluate the value proposition for different uses and build the business case for post-mine stakeholders to acquire the asset. More transparent and wider access to information, through atlas products such as the Australian Mine Waste Atlas, can help the private, public and community sectors to identify opportunities suited to different site characteristics. Finally, uncertainty associated with asset transfer presents an opportunity to explore securitisation mechanisms, with this opportunity growing as more mines are taken through a full transition process.

Contribute to policy and regulation that minimises negative impacts and maximises benefits

How can we assist governments to include greater optionality in policy and regulation, while managing risk, to move beyond return to prior state as the presumed best outcome?

The tension that exists between regulation of post-mine risk with the exploration and execution of value based post-mine uses is a key issue for policy makers and governments, as discussed in the recent New South Wales (NSW) inquiry into Post-mine Land Use (NSW Legislative Council, 2025). Beyond residual risk noted above, key issues that have been identified include tenure models that limit optionality and constrain post-mine investment and ways of building consensus around potential post-mine transition options. Exploring models to address these issues is needed.

At the same time, an opportunity exists across Australia for multiple jurisdictions to identify and develop shared principles and a roadmap for mine closure and post-mine transitions. This recognises the shared challenge and opportunity being explored in each part of Australia as more mines reach maturity, First Nations and other rightsholders expression through transition is realised and recognition grows of circular economy based productivity from mined land and assets.

Moving from mine closure to transition is a fundamental shift and principles are needed to provide a framework for new policies, ensuring consistency, clarity, and alignment to statutory goals, a key gap identified in previous reviews (Hamblin, Gardner and Haigh, 2022). Reliance on 'soft law' through guidance documents remains central, but in a growing environment of documentation a framework and toolkit that draws on evidence-based guidance can support transition-based policy implementation.

Enable regional scale outcomes

How do companies with multiple stakeholders and rights holders deliver transitions that align with regional scale goals and planning?

Whether mines are isolated or co-located within a region, their effects are felt through regional economies, communities and landscapes. Regional Cumulative Effects Assessment and Management (RCEAM) is crucial for mine closure because it helps to understand and manage the cumulative environmental, social, cultural, and economic impacts that can arise from mine closure. This systematic approach ensures informed decision-making, supports the development of post-mining land use, and enables better planning for socio-economic transitions in mining regions.

A second priority is ensuring there is greater understanding of the future economic pathways for regions to help prepare communities for the loss of a key economic driver and ensure a sustainable transition to alternative industries. This knowledge allows for proactive planning, diversification, and the development of strategies to mitigate the social and economic impacts of mine closure.

Finally, collaborative planning approaches are central to enabling an enduring and supported post-mine transitions. Developing a shared vision and understanding across relevant parties including communities, industry, government agencies and First Nations groups, and recognising the importance of intergenerational engagement, helps reduce conflict and build buy in to ensure transitions are supported (Finucane, 2024).

These approaches need to be paired with growing the capability of rightsholders and stakeholders to engage in planning processes. Cross regional knowledge exchange mechanisms provide opportunities for different stakeholders to learn from each other, build networks and gain lessons for different regions undergoing transitions.

Support supply chain, education, training and workforce development for closure and post-mine transitions

How do we build a connected post-mine transition and closure supply chain and ensure the future mining and mining equipment, technology and services workforce has capacity to deliver on mine closure solutions and post-mine transitions?

In recent years, reports have highlighted the scale and forecasted growth in mine closure and transitions, the diversity of skills and activities required and the need to develop a dedicated workforce strategy and implementation plan. In 2024, a review commissioned by CRC TiME and undertaken by the Mining and Automotive Skills Alliance (AUSMASA) and Business Skills Victoria (BSV) noted how crucial mine closure education and training program development is to the successful transition of mine sites post closure, while highlighting that a nationally consistent mine closure curriculum or program that addresses the complexities of mine closure does not exist.

Education and training opportunities are required that are scalable, including micro credential, VET, higher education and postgraduate opportunities. This recognises the upskilling and reskilling required in core areas to support closure integration, such as through finance, mine planning or on ground roles such as earthworks execution, as well as emerging areas such as natural capital accounting and environmental stewardship.

Addressing skill gaps is also critical, highlighting the diversity of technical skills in socio-economic and biophysical aspects of closure throughout and beyond the life-of-mine. First Nations communities, often deeply affected by mining activities, are at the centre of vocational programs that combine traditional knowledge with contemporary skills to contribute to and realise opportunities presented through post-mine transitions.

CONCLUSION

These six themes, representing the key constraints that hinder effective post-mine transitions, reflect a set of considerable challenges. It is important to recognise that all six themes are interconnected, and we need to approach them collectively. Supporting Corporate leadership in valuing post-mine transitions by presenting a well-informed rationale and business case for investing in activities that enable post-mine transitions, will benefit from establishing the business case for asset transfer, enabling regional transition planning and a policy environment that minimises negative impacts and maximises benefits.

ACKNOWLEDGEMENTS

The process undertaken to review and develop priorities involved many people. We acknowledge the CRC TiME Board, Staff, Project Teams and Partners for their contribution to the knowledge base developed and consultation process undertaken in 2024 to develop and refine priorities captured in this extended abstract.

REFERENCES

CRC TiME, 2024. CRC TiME Strategic Plan 2025–2027, CRC TiME Limited, Perth, Australia.

Finucane, S J, 2024. Ready, set, close! Assessing social values and community readiness for mine closure and post-closure transitions, in *Mine Closure 2024: Proceedings of the 17th International Conference on Mine Closure* (eds: A B Fourie, M Tibbett and G Boggs), pp 213–230 (Australian Centre for Geomechanics, Perth). https://doi.org/10.36487/ACG_repo/2415_14

Hamblin, L, Gardner, A and Haigh, Y, 2022. Mapping the Regulatory Framework of Mine Closure, CRC TiME Limited, Perth, Australia.

Maybee, B, Boggs, G, Stevens, R and Scrase, A, 2024. Who is responsible for the residual risk and how can it be shared or transferred to optimize post-mine outcomes?, *Research Directions: Mine Closure and Transitions*, 1:e6. https://doi.org/10.1017/mcl.2024.2

Measham, T, Walker, J, McKenzie, F H, Kirby, J, Williams, C, D'Urso, J, Littleboy, A, Samper, A, Rey, R, Maybee, B, Brereton, D and Boggs, G, 2024. Beyond closure: A literature review and research agenda for post-mining transitions, *Resources Policy*, 90:104859. https://doi.org/10.1016/j.resourpol.2024.104859

New South Wales Legislative Council, 2025. Standing Committee on State Development Report no. 53 Beneficial and productive post-mining land use, New South Wales Parliament.

Adapting the mining sector to climate change – a web-tool for climate-resilient resource development

N Bulovic[1], N McIntyre[2], R Trancoso[3,4] and R McGloin[5]

1. Research Fellow, The University of Queensland, St Lucia Qld 4072.
   Email: n.bulovic@uq.edu.au
2. Professor of Water Resources, The University of Queensland, St Lucia Qld 4072.
   Email: n.mcintyre@uq.edu.au
3. Science Leader Climate Projections and Services, Queensland Treasury, Queensland Government, Brisbane Qld 4102.
4. Adjunct Associate Professor of Climate Change, The University of Queensland, St Lucia Qld 4072. Email: r.trancoso@uq.edu.au
5. Senior Hydroclimate Scientist, Queensland Treasury, Queensland Government, Brisbane Qld 4102. Email: ryan.mcgloin@des.qld.gov.au

INTRODUCTION

Climate change is a growing reality, manifesting through unprecedented rainfall extremes, intensifying heatwaves, droughts and other physical impacts. Even if global carbon emissions are reduced and energy systems rapidly decarbonised, some physical impacts and risks from climate change are already locked in and are likely to persist for centuries (MacDougall *et al*, 2020). This will challenge the mining sector to adapt to significant climate-related physical risks across all stages of the mine life cycle – from designing resilient infrastructure (eg mine waste storage facilities capable of withstanding extreme rainfall events) to planning for closure (eg selecting vegetation suited to evolving climate conditions). To ensure long-term sustainability and resilience, it is crucial to integrate climate risk considerations into mine planning and operations.

This presentation introduces a new online web-tool developed to support climate-resilient resource development in Queensland. As part of the Queensland Future Climate (QFC) science program's *Adapting to Future Climate* case studies, the educational tool serves two primary purposes:

1. Educating a range of stakeholders on the role of the mining industry in global decarbonisation efforts, and improving understanding of which physical climate risks are relevant to the mining industry and how they may change in the future; and

2. Providing a foundation for discussions on industry preparedness for future climate conditions.

The web-tool translates the findings of two key research papers (Bulovic, McIntyre and Trancoso 2024; Bulovic *et al*, 2024) into an easy-to-read format, featuring interactive conceptual diagrams, time-lapse videos and an interactive dashboard with maps and charts teasing out future impacts across Queensland's mining regions. Below we provide a brief overview of the web-tool structure, content, and next steps in advancing this work.

'DEVELOPING CLIMATE RESILIENT MINES' CASE STUDY

Background and overview

The Queensland Future Climate's *Adapting to Future Climate* series aims to translate complex climate change science, data, and projections into accessible and engaging case studies that support regional understanding and adaptation (<https://www.longpaddock.qld.gov.au/qld-future-climate/adapting/>). Building on previous case studies focused on heatwaves and water security, a new addition – 'Developing Climate Resilient Mines' – explores the intersection of climate change and the mining sector, and is structured around five key components:

1. Mining and decarbonisation in Australia.

2. The climate sensitive mine life cycle.

3. Assessing climate change risks to mining regions.

4. Understanding future climate projections across mine sites.

5. Climate resilient mine site planning and design.

The aim of this case study approach is not to provide detailed or specific advice on how the industry is to adapt to individual climate risks or how to directly incorporate the climate projections in mine site planning. The aim is rather to educate the user about the range of potential future changes in climate variables and how this data can be used to screen for future risks/vulnerabilities.

Mining and decarbonisation in Australia

The first component of the web-tool explores the Australian mining sector's role in the low-carbon transition. It provides informative resources from Federal and State government agencies (eg Geoscience Australia) that highlight the scale of mining operations and the growing importance of critical minerals for the energy transition. Interactive maps of mine sites across Queensland and Australia are used to visualise the current landscape of mining activity and illustrate how it may evolve in response to future demand for low-emission technologies.

The climate sensitive mine life cycle

The next component introduces general stakeholders to the four different stages of the mine life cycle (exploration, development, production, and closure), linking typical mine site features/activities to climate hazards which are described through pop-up information boxes (Figure 1).

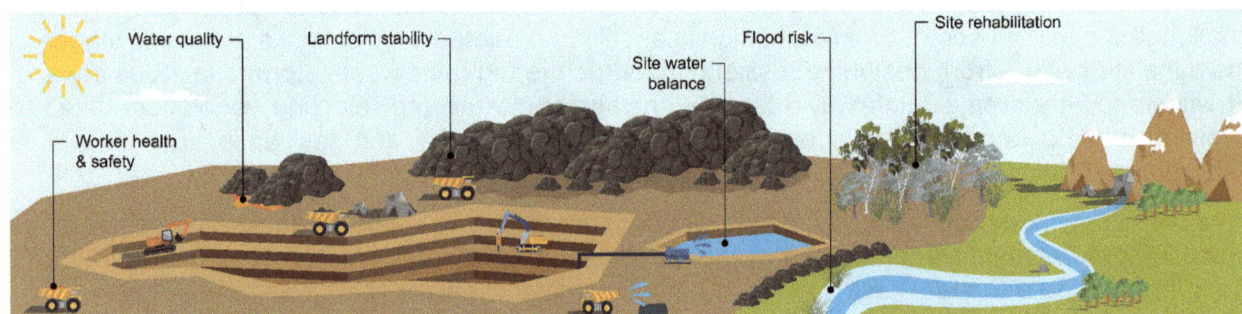

FIG 1 – Key interactive web-tool figure used to illustrate a typical Australian mine site, highlighting key climate-sensitive mine site element. By clicking on a text label, more detailed information on that element is provided, including: background information, example picture, how climate hazards are important to that element, and how the impacts are managed and mitigated by the sector.

Assessing climate change risks to mining regions

The third component of the web-tool focuses on the future climate projection data sets, explains the role of emissions scenarios and outlines an approach for quantifying climate change risks. High-resolution Queensland Future Climate projection data sets (Syktus et al, 2020) form the basis of the risk analysis, which are dynamically downscaled climate simulations from the coupled model intercomparison project phase 5 (CMIP5) ensemble. Building on the methodology developed by Bulovic, McIntyre and Trancoso (2024), demonstrates how regional climate risks targeted at specific mine site elements (Figure 1) were assessed through a step-by-step process visualised using flow charts:

- Identify key biological and physical factors influencing the target mine site element.

- Select and quantify the relevant climate index for estimating physical climate risk.

This structured approach allows users to trace the logic from mine site element to measurable change in climate risk through a proxy climate index.

Understanding future climate projections across mine sites

Climate change-driven shifts to five climate indices are presented here, focusing on:

1. Site water balance, estimated by the aridity index.

2. Flood risk, estimated by RX1day.

3. Landform stability/erosion, estimated by R-factor.

4. Site revegetation, estimated by extreme drought frequency.

5. Water quality, estimated by drying period duration.

A storytelling approach is used to present the main results for each climate index across Queensland to the user. An interactive drop-down tool (Figure 2) enables the user to further explore likely changes to specific climate indices across specific mining regions. For example, in Figure 2 the following selection was made: climate index 'R-factor', emission scenario 'RCP8.5', time slice 'End century (2070–2099)', and mining region 'Charters Towers'. The drop-down tool generated a representative map of expected changes to R-factor across the Charters Towers region, and also showed a summary figure of the range in projected changes based on the climate model ensemble.

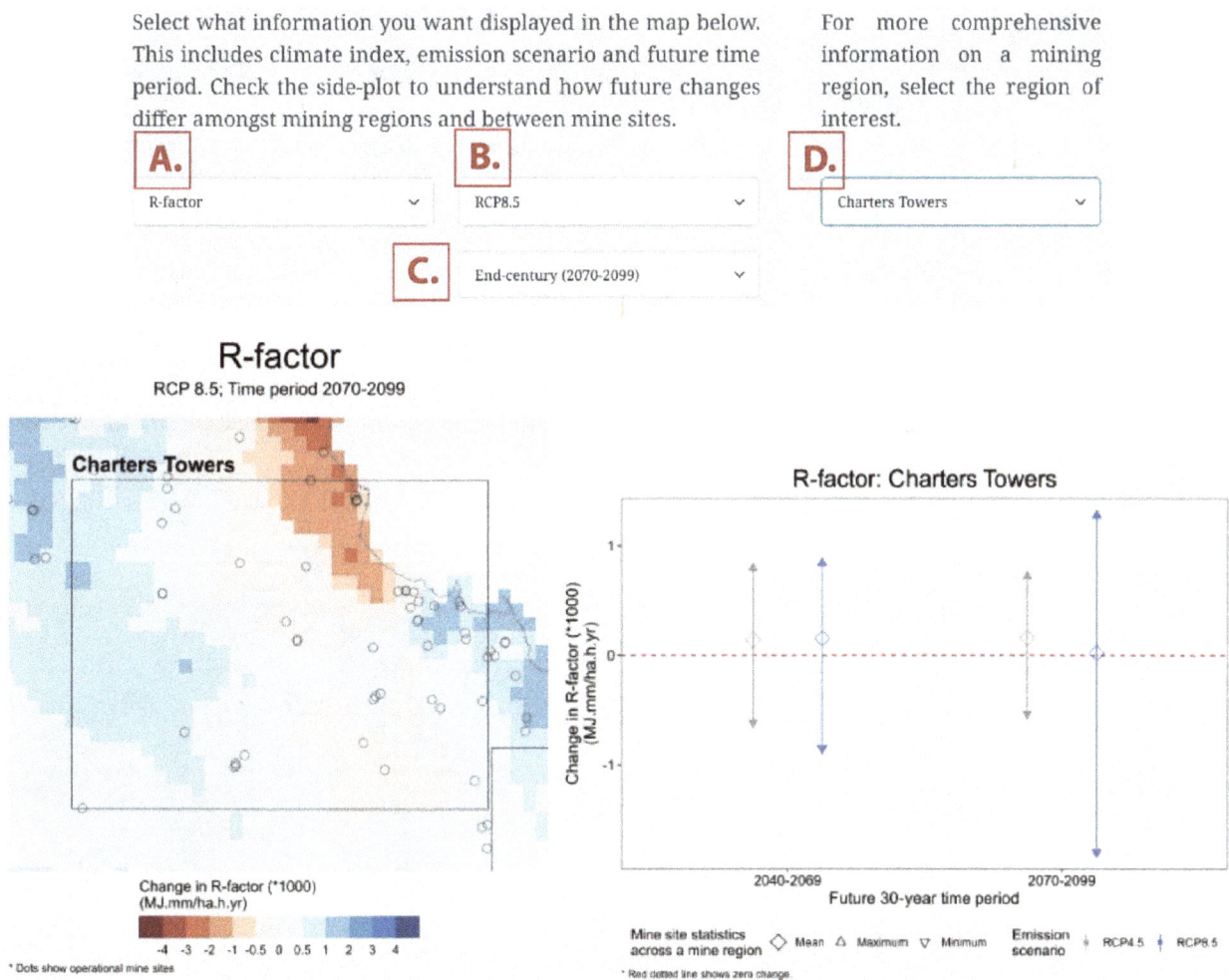

FIG 2 – Interactive drop-down tool that enables exploration of selected: (a) future climate index projections, across selected; (b) emission scenarios; (c) time slices; and (d) mining regions. Once the four variables are selected (A to D, shown in red), the tool generates the respective map, and a summary plot of projected changes across all emission scenarios and time slices for the selected climate index and mining region.

Climate resilient mine site planning and design

The focus of the final component is to provide information and links to relevant industry guidelines, guidance, standards, and data sets that can be leveraged to help climate adaptation in the sector.

CLOSING REMARKS

Finally, we discuss ongoing efforts to expand the tool's geographical applicability and to update the climate projections. In particular, the second version of the online web-tool will:

- Expand from a Queensland-centric focus to also incorporate data for important mining regions across Australia, such as the Pilbara, Central NSW, and so on.

- Use the latest state-of-the-art climate projections and emission scenarios in line with the latest International Panel on Climate Change (IPCC) Assessment Report 6. Practically this means that the CMIP5-based climate indices (based on data from the IPCC Assessment Report 5) will be updated and recalculated using the latest CMIP6-based climate data.

ACKNOWLEDGEMENTS

The authors acknowledge project funding from the former Queensland Government Department of Energy and Climate (now Queensland Treasury), and the University of Queensland's Resourcing Decarbonisation program (Project name: Tools to support climate resilient resource development). Kaan Arslan (University of Queensland) is thanked for his work in developing the website.

REFERENCES

Bulovic, N, McIntyre, N and Trancoso, R, 2024. Climate change risks to mine closure, *Journal of Cleaner Production*, 465:142697. https://doi.org/10.1016/j.jclepro.2024.142697

Bulovic, N, McIntyre, N, Trancoso, R, Bolz, P and Shaygan, M, 2024. Downscaled climate model erosivity projections and drivers of change across distinct climate regions, *CATENA*, 244:108250. https://doi.org/10.1016/j.catena.2024.108250

MacDougall, A H, Frölicher, T L, Jones, C D, Rogelj, J, Matthews, H D, Zickfeld, K, Arora, V K, Barrett, N J, Brovkin, V, Burger, F A, Eby, M, Eliseev, A V, Hajima, T, Holden, P B, Jeltsch-Thömmes, A, Koven, C, Mengis, N, Menviel, L, Michou, M, Mokhov, I I, Oka, A, Schwinger, J, Séférian, R, Shaffer, G, Sokolov, A, Tachiiri, K, Tjiputra, J, Wiltshire, A and Ziehn, T, 2020. Is there warming in the pipeline? A multi-model analysis of the Zero Emissions Commitment from CO$_2$, *Biogeosciences*, 17(11):2987–3016. https://doi.org/10.5194/bg-17-2987-2020

Syktus, J, Trancoso, R, Ahrens, D, Toombs, N and Wong, K, 2020. Queensland Future Climate Dashboard: Downscaled CMIP5 climate projections for Queensland. Available from: <https://www.longpaddock.qld.gov.au/qld-future-climate/>

Speaking a common language – mine closure and the MLRA Vocabulary

N Gardiner[1] and A Scrase[2]

1. Technical Specialist, MLRA, Morwell Vic 3840. Email: nathan.gardiner@mineland.vic.gov.au
2. Technical Director, MLRA, Morwell Vic 3840. Email: antonia.scrase@mineland.vic.gov.au

ABSTRACT

The Australian Standards (2021) AS ISO 20305:2021 – Mine Closure and Reclamation Vocabulary, addressed a lack of a consistent set of terms for mine closure. The ISO terms, however, do not account for variances between different mining regions around the world and the complete document must be purchased, although an online version can be freely accessed.

The lack of consistent closure terminology in Victoria, and between Victoria and other states and countries, has created confusion for government, industry and communities. The Mine Land Rehabilitation Authority (MLRA) developed and published its first version of the MLRA Vocabulary in 2023 (MLRA, 2023), to enable better communication and provide a common language for mine closure in Victoria. The MLRA's aim was to provide a free, publicly accessible vocabulary, by collating commonly used rehabilitation and closure terms (including terms specific to Victorian legislation for closure) and refining them for general use.

The Vocabulary currently comprises over 100 terms and provides a basis for conversations about mine closure in Victoria. It does not replace or alter existing definitions in Victorian legislation, but rather incorporates and simplifies terminology, providing a broader resource for industry, government and community to engage in mine closure conversations with greater consistency and clarity.

The Vocabulary was accessed almost 500 times across Australia and in six other countries worldwide in its first full year, 2024. The next revision of the Vocabulary is underway and will comprise additional terminology, identified through ongoing stakeholder interactions, and is expected to be released in 2025. This paper describes the methodology the MLRA used to develop its list of terms and definitions, including the consultation process undertaken with key Victorian stakeholders, both for the first revision and the planned update.

INTRODUCTION

Glossaries, lexicons or vocabularies all describe similar tools that help aid understanding and comprehension. A list of terms and their definitions, or vocabulary (hereafter), provides a reader with information on concepts and terms they may not be familiar with.

Vocabularies provide common definitions where the subject matter is complex and/or is relevant to parties from diverse backgrounds. Even though parties may not agree on a particular definition, vocabularies nevertheless help avoid confusion by establishing a shared understanding.

The Mine Land Rehabilitation Authority (MLRA) helps to facilitate good rehabilitation and closure outcomes for three 'declared', or high risk, coalmines in the Latrobe Valley, Victoria (Figure 1). Mine closure, and the range of terms it entails, is generally an unfamiliar process for most stakeholders the MLRA consults with, which notably includes communities living in townships within 10 km of the mines.

Stakeholders in Victoria are currently exposed to mine closure terminology via a range of processes, including statutory engagement on Declared Mine Rehabilitation Plans, Mineral Resources (Sustainable Development) Act (Parliament of Victoria, 1990) and Environmental Effects Statements, Ministerial guidelines for assessment of environmental effects (see Department of Transport and Planning, 2023).

Many industry bodies provide vocabularies for different regions and stakeholders, but none of those reviewed were fit for purpose in the Victorian context. The MLRA identified a need to establish a resource for the Latrobe Valley and the wider Victorian community that included frequently used terms specific to declared mine rehabilitation and closure. The MLRA scope was for the vocabulary be a concise list of terms relevant to declared mines in the Latrobe Valley, drafted in clear language that was comprehensible to high-school level readers.

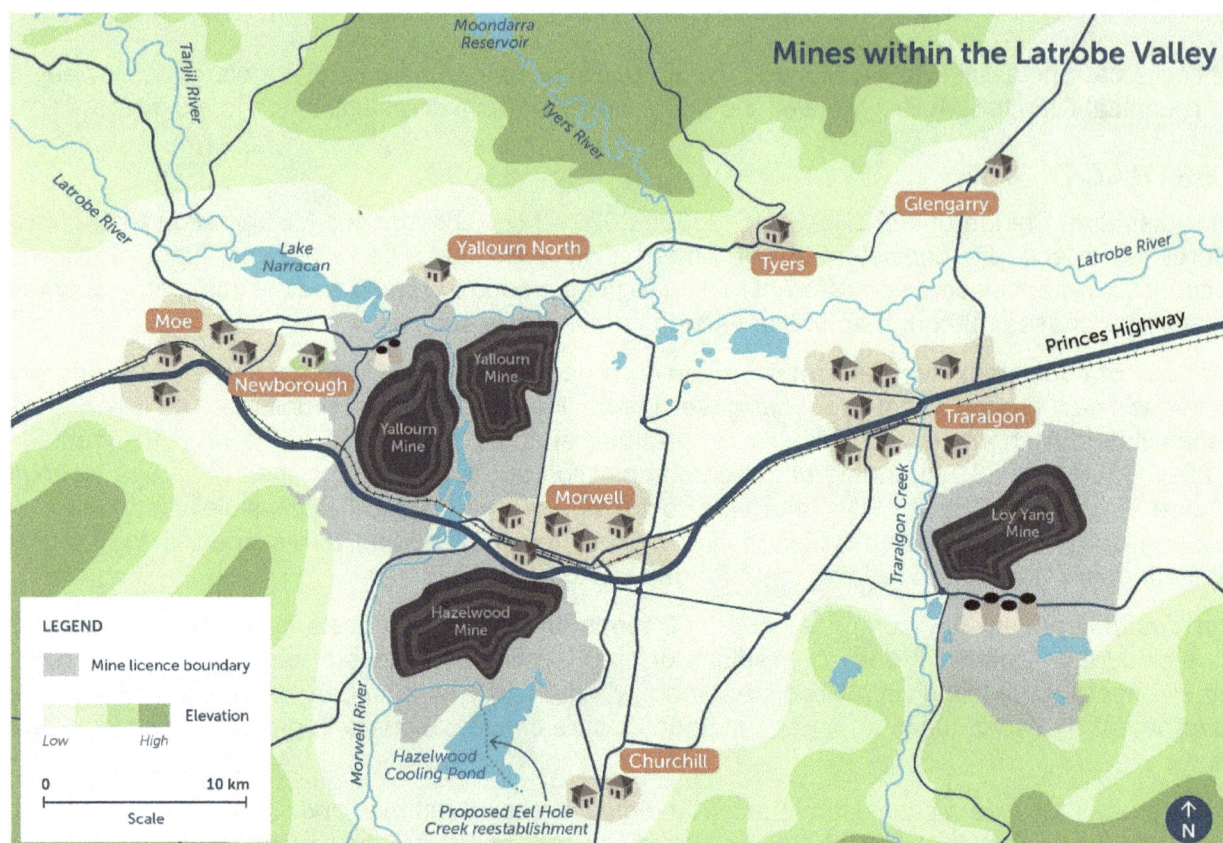

FIG 1 – The Latrobe Valley coalmines exist in close proximity to communities (pale orange). Morwell is located approximately 150 km east south-east of Melbourne via the Princes Highway, which passes close to all three sites.

The MLRA drafted its vocabulary through targeted stakeholder participation to provide a public resource, to help gain alignment between government, industry stakeholders, and affected communities. The first version of the MLRA Vocabulary was published late in 2023 (MLRA, 2023). The MLRA is currently preparing a second version to include additional relevant terms, which are being identified as planning and closure for Victoria's declared mines progresses and evolves.

This paper describes the context and rationale the MLRA uses to develop its list of terms and definitions, the processes employed to reach consensus with key stakeholders and provide a resource to the public, and its proposed update schedule.

METHODOLOGY

An initial list of proposed terms requiring definition were compiled during various communications the MLRA was undertaking, or was aware of, with community, industry and government. Terms that consistently arose during these ongoing communications – verbally, written, or otherwise – were added to the list.

The MLRA's aim was to draw on established resources and avoid (where possible) creating new definitions. However, as the aim of the MLRA vocabulary was primarily to assist Latrobe Valley communities, it was necessary to alter some of the established definitions to incorporate nuances relating to the region where the three declared mines are located.

The MLRA draft vocabulary comprised of an initial list of terms, which were then matched to relevant definitions from a wide range of reputable sources. These sources included the Victorian government, such as legislation covering mineral resources and water, other documents developed by state and regional water administrators, peak industry bodies, academic and governmental bodies, published vocabularies and formally defined terms. For example, the initial list included the definition of 'sustainability' derived from the United Nations' Brundtland Commission (Brundtland, 1987). In other cases terms from English language or legal dictionaries unrelated to mining or declared mines, for example, were altered to reflect local nuances or usage. In all such cases the MLRA sought to retain a term with minimum modification, to promote a broader understanding as part of a public resource.

The terms were then subdivided into 'general' and 'technical' categories for easier reference, denoted '*Gen*' and '*Tech*', respectively. General terms include those relating to policy, administration or principles relevant to closure and rehabilitation, while technical terms derive from science or engineering disciplines.

The MLRA then convened a group of participants to test the list of proposed terms and respective definitions. This group included key local stakeholders, and subject matter experts from the declared mines and relevant government bodies. The MLRA conducted a series of initial briefings to outline the aims of the project, intended workflows and timelines. Participants subsequently provided feedback on the proposed terms within individual worksheets over the course of several weeks. Participants principally provided feedback on terms of specific interest, or terms with which they had specific expertise, and provided refinements of the draft terms or proposed alternative definitions. In some cases participants also suggested additional resources for the MLRA to consult.

Two rounds of review were conducted on the draft vocabulary, with targeted discussions with participants to resolve cases where alternative definitions were suggested. The MLRA engaged with the respective parties to resolve these collaboratively, providing the rationale for including the term and the specific language for the proposed definition. In most such cases, the adopted term and definition were arrived at through consensus between participants. In rare cases the MLRA exercised final editorial discretion for the draft terms.

The MLRA collated feedback on the revised definitions into the draft MLRA vocabulary and fed back the outputs of this consultation to the participants.

FINDINGS

The MLRA's aim was to provide a list of key terms grounded in mining and mine closure, science, and relevant legislation and policy. These terms were also required to relate specifically to the declared mines in the Latrobe Valley, to serve as a reference for community, industry and government stakeholders. The initial list comprised 114 proposed terms – 70 general and 44 technical.

Of the initial 114 terms, nine were defined by the MLRA. This process utilised internal expertise within the MLRA, or a reformulation of an established term in more accessible or explicit language, to assist both public understanding and the declared mine licensees' implementation of the legislation. An example of this is the compound definition of '*safe, stable and sustainable*' – a key term within the Victorian legislation relating to declared mine rehabilitation. In this instance, the MLRA extended the definition of *stable* to include chemical and ecological stability. The MLRA also drafted a definition for *sustainable* that incorporates aspects of the United Nations definition of sustainable development (Brundtland, 1987) and a University of California (Los Angeles) definition of sustainability (UCLA Sustainability, 2023).

Participants were given the opportunity to provide a final round of feedback on the draft MLRA Vocabulary, which was accepted by the group in early September 2023. The final MLRA Vocabulary 2023 (revision 1) is comprised of 109 terms and their definitions, of which 64 are general and 45 technical. The MLRA Vocabulary was initially circulated to participants in September 2023, and published online a week later (MLRA, 2023). It is available as a public resource and is searchable online <https://www.mineland.vic.gov.au/learn/vocabulary/>, or can be downloaded as a PDF file <https://www.mineland.vic.gov.au/wp-content/uploads/2024/06/MLRA-Vocabulary-V1June24.pdf>; (Figure 2).

Definitions

Note: *Unique reference numbers (Ref.) are provided for each definition. These are either of the form;*

1) Gen001 = A term used in a general context, typically a term in common usage or from legislation, or;

2) Tech001 = A technical term used explicitly in a scientific or engineering context

10 ⌄ entries per page   Search:

Ref.	Term	Definition (June 2024)	Link 1
Gen001	**Advice (MLRA provided)**	As the statutory authority for declared mine rehabilitation, the MLRA must provide formal advice to the Minister under the Mineral Resources (Sustainable Development) Act. Advice is either provided to inform the Minister on rehabilitation, requested by the Minister to facilitate decision making, or as findings of an investigation	
Gen002	**Beneficial uses**	A use to the environment, or a segment of the environment, that leads to public benefit, welfare, safety, health or aesthetic enjoyment and which requires protection from the effects of waste discharges, emissions or deposits. A beneficial use may be an existing or potential use. A resource may have more than one beneficial use	
Gen003	**Care and maintenance/ temporary mine closure**	Phase following a temporary cessation of operations, when infrastructure, plant and equipment remain intact and are maintained in anticipation of production recommencing. Such a site may be	

FIG 2 – Example terms from the published MLRA Vocabulary (MLRA, 2023).

Principal references of the MLRA Vocabulary

The following sources provided terms and definitions adopted for the MLRA Vocabulary (note – all references are Victorian, unless explicitly stated):

- Environment Protection Act (Parliament of Victoria, 2017).

- Mineral Resources (Sustainable Development) Act, 1990 (Parliament of Victoria, 1990).

- Gippsland Groundwater Atlas (Southern Rural Water, 2012).

- Water dictionary (Victorian Water Register, 2025).

- Glossary, Central and Gippsland Region Sustainable Water Strategy (Department of Environment, Land, Water and Planning, 2022).

- Mine closure and reclamation – Vocabulary (ISO 20305:2021; Australian Standards, 2021).

- Public engagement framework 2021–2025: Definitions (Victorian State Government, 2023).

The MLRA has received written and verbal feedback on the value of the MLRA Vocabulary in providing a shared understanding of complex concepts and processes. This feedback has come from a wide variety of parties including declared mine licensees, consultants working in mine closure, members of the community, advocacy groups, government, and international mine closure specialists. The MLRA has witnessed firsthand Victorian stakeholders referencing the MLRA Vocabulary in difficult conversations, assisting engaged parties by providing a shared understanding of terms used during these discussions.

The MLRA Vocabulary provides a broad central resource for industry, government and community to engage in mine closure conversations with greater consistency and clarity. The Vocabulary was accessed almost 500 times across Australia, and worldwide in six other countries, in its first full year (2024). To the MLRA's knowledge users includes students learning about mine closure and community groups in the Latrobe Valley engaging in consultation with government and industry.

The MLRA is currently planning an update of the MLRA Vocabulary, to propose additional terms identified during the two years since publication of the initial version. As these terms have been identified, they have been collected and will be included in the updated vocabulary. The update process will follow the same consultative methodology described above, with the exception that only new terms will be reviewed by the participant group. The MLRA aims to repeat this update process at least every two years, to support public understanding as declared mine closure evolves and matures.

The second revision of the vocabulary will include additional terms as rehabilitation conversation progresses and further needs are identified. The MLRA is currently preparing to engage key stakeholders on version two.

CONCLUSIONS

The MLRA Vocabulary is an example of where value has been added by enabling a clearer understanding of complex terminology and processes for a variety of stakeholders. An initial draft vocabulary was developed and refined in consultation with subject matter experts, drawing on established definitions where possible. The MLRA Vocabulary was published online in September 2023 and contains over 100 terms relevant to declared mine closure in the Latrobe Valley, Victoria.

While the Vocabulary in no way replaces legislated terms and responsibilities, it provides a useful reference for licensees, Traditional Owners, government and communities to engage in discussions around mine closure in the Latrobe Valley with a shared understanding of key terms and concepts.

The MLRA Vocabulary has been met with appreciation from Victorian stakeholders, and the MLRA has been informed of its use on multiple occasions and has witnessed it being used in conversations and legislated documents. In addition, a range of parties have been proactive in providing the MLRA with additional terms for inclusion in the next update. The MLRA has also received positive feedback on the Vocabulary from outside of Victoria, and is aware of usage both nationally and internationally, further indicating the value and utility of the work beyond its original intended audience.

ACKNOWLEDGEMENTS

The MLRA would like to acknowledge the participation of the following key people in the development and publication of the MLRA Vocabulary, Version 1:

- Aron Crane (AGL)
- Dr Daniel Mainville (Alinta)
- Cassandra Tolsma, Nikki Jenkins, Paul Metlikovec, Rhonda Hastie (Energy Australia)
- Adam Moran, David McGavin (Engie)
- Ian McLeod (Earth Resources Regulation, Vic)
- Prof Thomas Baumgartl (Federation University)
- Paul Young (Gippsland Water)
- Chris Buckingham (Latrobe Valley Authority, Vic)
- Chris McCauley (Latrobe Valley Regional Rehabilitation Strategy, Vic)
- Terry Flynn (Southern Rural Water)
- Dr Andrea Ballinger (DEECA – Water and Catchments, Vic).

REFERENCES

Australian Standards, 2021. AS ISO 20305:2021: Mine Closure and Reclamation – Vocabulary. Available from: <https://www.iso.org/obp/ui/#iso:std:iso:20305:ed-1:v1:en> [Accessed: 20 June 2025].

Brundtland, G H, 1987. Our Common Future: Report of the World Commission on Environment and Development, Geneva. Available from: <http://www.un-documents.net/our-common-future.pdf> [Accessed: 10 April 2025].

Department of Transport and Planning, 2023. Ministerial guidelines for assessment of environmental effects under the Environment Effects Act 1978, 8th edn.

Department of Environment, Land, Water and Planning, 2022. Central and Gippsland Region Sustainable Water Strategy Final Strategy.

Mine Land Rehabilitation Authority (MLRA), 2023. Vocabulary, Mine Land Rehabilitation Authority. Available from: <https://www.mineland.vic.gov.au/learn/vocabulary/#Definitions> [Accessed: 14 February 2024].

Parliament of Victoria, 1990. Mineral Resources (Sustainable Development) Act, v.126. Available from: <https://www.legislation.vic.gov.au/in-force/acts/mineral-resources-sustainable-development-act-1990/126> [Accessed: 13 April 2022].

Parliament of Victoria, 2017. Environment Protection Act, v.005. Available from: <https://www.legislation.vic.gov.au/in-force/acts/environment-protection-act-2017/019>

Southern Rural Water, 2012. Gippsland Groundwater Atlas, 60 p (Southern Rural Water: Victoria).

UCLA Sustainability, 2023. What is Sustainability? Available from: <https://sustain.ucla.edu/what-is-sustainability/> [Accessed: 10 April 2023].

Victorian State Government, 2023. Public engagement framework 2021–2025: Definitions. Available from: <https://www.vic.gov.au/public-engagement-framework-2021-2025/definitions> [Accessed: 20 June 2025].

Victorian Water Register, 2025. Water Dictionary. Available from: <https://waterregister.vic.gov.au/water-dictionary?start=80> [Accessed: 20 June 2025].

Yolŋu master plan – cultivating economic empowerment and food security post-mine closure

M J Kauthen[1] and K James[2]

1. Senior Manager, The Palladium Group, Melbourne Vic 3000.
   Email: meg.kauthen@thepalladimgoupr.com
2. Director, The Palladium Group, Sydney New South Wales 2000.
   Email: karen.james@thepalladiumgroup.com

ABSTRACT

The closure of Rio Tinto's Gove mine by the end of the decade will bring economic and social impacts to the local community. It also presents an opportunity to reshape the region's future. To navigate challenges and embrace emerging opportunities, integrated strategic mine closure planning is essential. To address potential impacts of a socio-economic transition and support integrated strategic mine closure planning, in 2023 Rio Tinto invested in an Agribusiness Options Study (Study) to identify businesses that could enhance food security and economic empowerment. This Study was guided by a local Steering Committee (SteerCo) led by Palladium. It included Traditional Owner corporations – Gumatj Corporation Ltd, Rirratjingu Aboriginal Corporation, and North-East Arnhem Aboriginal Corporation, along with support from Rio Tinto, Developing East Arnhem Ltd, the Northern Land Council, and the Northern Territory Government.

From the Study, SteerCo identified the Yolŋu Master Plan (YMP) as a sustainable agribusiness focused on producing fresh produce and bushfoods. The YMP aims to grow local fresh food, support local employment, learning, and training, and help the Homelands develop their own Market Gardens. Originally designed to supply local markets, the YMP has the potential to expand to external markets (eg Darwin) and develop agritourism opportunities.

Due to the modest scale of agriculture in the region, the YMP was divided into three Horizons to align capital spending with community readiness. In 2024 SteerCo initiated Horizon 1a to grow fruits and vegetables and explore bushfood options. This six-month trial provided key insights and empowered Traditional Owner Boards with the information required to guide them for the long-term implementation of the YMP.

This paper will examine how having strong relationships with Traditional Owners and incorporating multiple perspectives ensured that the YMP aligned with community priorities for a post-mine economy. Through effective governance, a balance was struck across cultural, social, and economic objectives. As a result, the YMP serves as an approach to developing post-mine economies that prioritise sustainable Indigenous-led development.

INTRODUCTION

The closure of the Gove Mine in remote Nhulunbuy, Northern Territory, Australia, represents both a significant challenge and opportunity, and it is a shared responsibility among all stakeholders. Rio Tinto, Traditional Owner Corporations, local businesses, and the Northern Territory Government need to work together to develop sustainable, non-mining livelihoods for Yolŋu Indigenous communities beyond the life-of-mine.

With strategic intervention and shared decision-making, socio-economic transitions can be navigated more smoothly, helping to prevent challenges such as food insecurity and ensuring a more stable and resilient outcome. Recognising this, Rio Tinto proactively invests in solutions, following good governance practices and a structured approach to drive economic diversification and strengthen food security. Rio Tinto has collaborated with local partners to support the establishment of Developing East Arnhem Ltd., which focuses on promoting local development, including tourism. Additionally, Rio Tinto is a member of the Gove Peninsula Futures Reference Group, which aims to collectively facilitate the socio-economic transition of the Gove Peninsula as it shifts away from mining activities.

In 2023 Rio Tinto engaged The Palladium Group (Palladium) to conduct an Agribusiness Options Study (Study), exploring agriculture and aquaculture interventions to reduce food insecurity and provide alternative non-mining livelihoods. A Steering Committee (SteerCo) was established based on current governance structures used in the region, and to ensure the Study was community-led and driven. It included representatives from:

- Gumatj Corporation Ltd (GCL)
- Rirratjingu Aboriginal Corporation (RAC)
- North-East Arnhem Land Aboriginal Corporation (NEAL)
- Developing East Arnhem Ltd (DEAL)
- Northern Territory Government (NTG)
- Northern Land Council (NLC)
- Rio Tinto.

The SteerCo followed a structured process and provided a platform to align corporate, government, and community perspectives while embedding Yolŋu priorities in decision-making. Throughout the year of developing the Study, the SteerCo met biweekly to ensure that the research aligned with local needs. Through three dedicated face-to-face workshops and one online workshop, the SteerCo contributed invaluable on-the-ground insights. Additionally, at key decision-making points, the Traditional Owner Boards from GCL, RAC, and NEAL provided guidance on preferred choices and culturally unsuitable options. As a result, Traditional Owner governance protocols were honoured.

With SteerCo guidance, Palladium conducted an external environmental analysis to hypothesise the likely beyond-life-of-mine environment for an agribusiness or aquabusiness. The team reviewed ten investment options across agriculture and aquaculture value chains, narrowing them down to three for a detailed value chain analysis and financial assessment. The Study resulted in the creation of the Yolŋu Master Plan (YMP), a three-horizon strategy aimed at developing Yolŋu Indigenous-led agribusiness that supports growing fresh produce, cultivating bushfoods and supporting learning and training in agriculture, which is owned and led by the Yolŋu people. Effectively, the creation of an agribusiness hub or precinct.

This dedicated approach was instrumental in aligning potential options with the agreed-upon purpose and meeting the Traditional Owners' desires. Consequently, it empowered SteerCo to develop a viable, feasible, and desirable YMP (Figure 1).

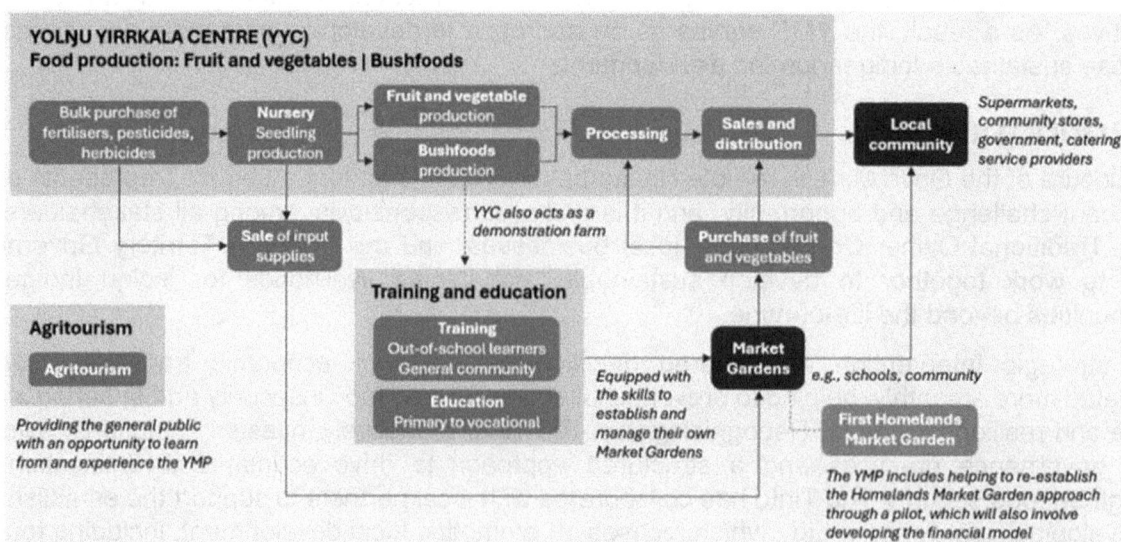

FIG 1 – The Yolŋu Master Plan (YMP).

This paper explores the YMP's co-design and early implementation. It highlights how incorporating multiple perspectives, especially those of Traditional Owners, following a strong governance process, and including cross-sector partnerships can foster culturally appropriate, inclusive and resilient economic development in the context of mine closure. Momentum and trust have been achieved by bringing together industry, government, and Yolŋu stakeholders, ensuring that long-term economic benefits are both scalable and sustainable.

REGIONAL ECONOMIC CONTEXT

North-East Arnhem Land is home to the Yolŋu peoples—Traditional Owners with deep cultural, linguistic, and ancestral ties to Country (Christie and Greatorex, 2004). The region is characterised by remoteness, a mixed economy, and a strong desire among Yolŋu leaders to reduce dependency on external systems while maintaining cultural integrity (Gove Peninsula Futures Reference Group, 2021; Pearson and Helms, 2013). This involves initiatives to strengthen local governance, education, and economic opportunities that align with Yolŋu values and traditions.

The Gove Mine has provided employment and infrastructure for over 50 years. With closure expected by the decade's end, the established community and economy face a period of transition. Rio Tinto's closure planning extends beyond environmental rehabilitation to support a socio-economic transition for nearby Yolŋu communities.

Given the upcoming impact of closure, SteerCo defined economic empowerment (Figure 2) as achieving it through community leadership, working with key stakeholders.

'Together, they can provide:

- *Purpose that the business is to serve the Yolŋu people;*
- *Education, training and mentorship;*
- *Information to the community so they can make informed decisions;*
- *A range of business models to align with community desires;*
- *Meaningful action to achieve the Traditional Owner vision;*
- *Tiers of sustainability to build community capacities over time; and*
- *Income through employment of revenue from the business.*

As a result, the community are enabled to transition off welfare and achieve economic development, which provides communities choice, shared wealth, the support to live the way of life the community wants and independence.' (SteerCo, 2023)

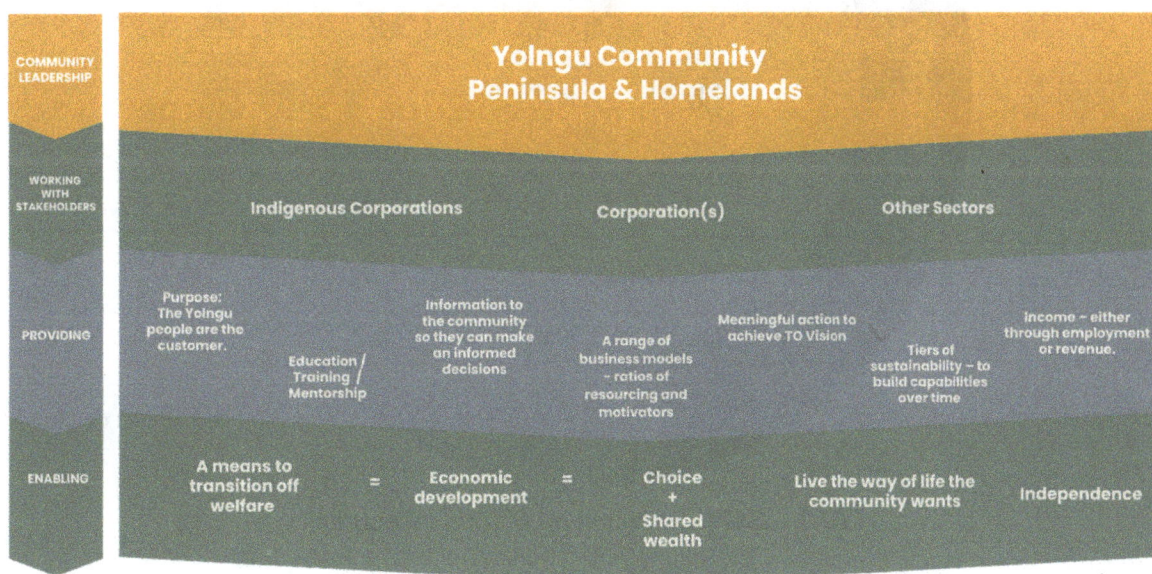

FIG 2 – SteerCo definition of economic empowerment.

REGIONAL FOOD SECURITY CONTEXT

The cost of food in remote Indigenous communities is very high relative to cities and towns near stronger supply chains (Australian Government, 2020). The NT Market Basket Survey (Northern Territory Government, 2019) revealed that, on average, a Healthy Food Basket (HFB) in remote community stores is 56 per cent more expensive than major supermarkets in a district centre and 6 per cent higher than corner stores in the district centres. In addition, on average, a Current Diet Basket (CDB) in remote community stores is 40 per cent more expensive than major supermarkets and 8 per cent higher than corner stores in a district centre. The high cost of food and groceries in remote communities is broadly caused by two key factors: lower commercial purchasing power of remote suppliers and higher operational costs, including freight and repair costs (Australian Government, 2020).

The closure of the Gove mine highlights the importance of food security through the economic transition. In response, the SteerCo is exploring ways to support access to affordable and nutritious foods. SteerCo recognised these potential challenges and defined food security (Figure 3) as:

> 'There are two distinct markets that need to be served: local consumption (including Peninsula and Homelands) and Commercial markets.
>
> These markets (both local consumption and commercial) require nutritious, accessible, desirable, reliable and sustainable access to food. Food needs to be supplied in balance with the nature market (eg seasonality of yam production) and support counter seasonal production to ensure food security is achieved throughout the year.
>
> To maintain production balance, it is also essential to complement the current market in terms of supply and demand of produce.' (SteerCo, 2023)

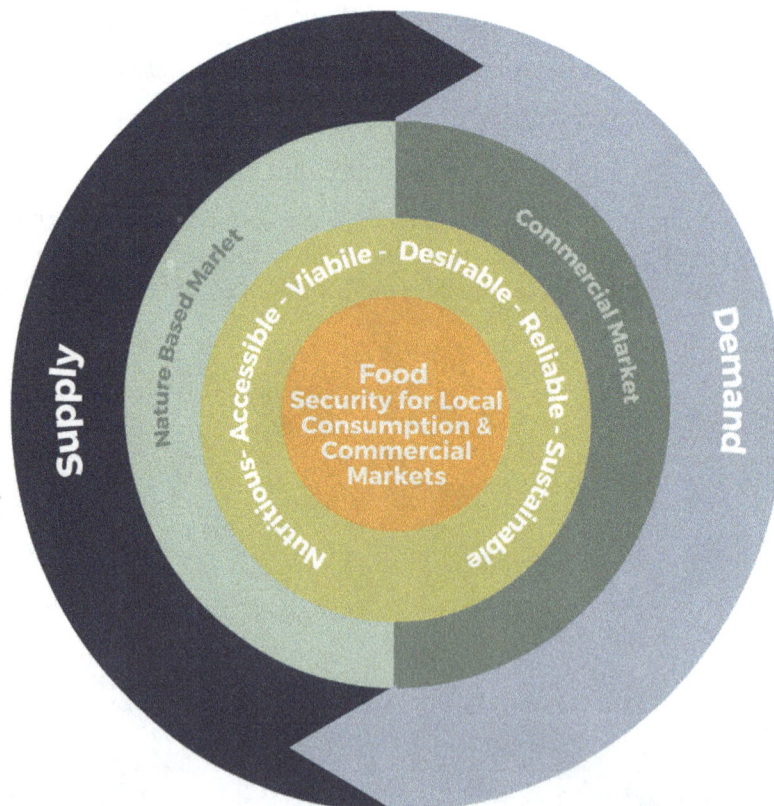

FIG 3 – SteerCo definition of food insecurity.

THE YOLŊU MASTER PLAN – PURPOSE, STAGED IMPLEMENTATION AND GOVERNANCE

Purpose

Given the impact of closure on both economic empowerment and food security, the SteerCo agreed on the following purpose to guide the YMP:

> 'Our purpose is to achieve economic empowerment for Rirratjingu, Gumatj, and surrounding Yolŋu communities and support food security in North-East Arnhem Land. This agribusiness must have a community mandate, a Yolŋu focus, and a Yolŋu worldview.' (SteerCo, 2023)

Horizon-based implementation

Recognising the region's limited agricultural scale, the YMP systematically expands fresh produce, integrates agriculture training into local education systems, supports market gardens in remote communities called the Homelands, and develops the bushfoods industry by supporting rehabilitated mine land to grow two bushfoods – Green Plum (*Buchanania obovata*) and Red Bush Apple (*Syzygium suborbiculare*).

To manage risk and respect community readiness, the YMP was divided into three Horizons.

Horizon 1a (January– October 2024): During Horizon 1a, the YMP established a strong foundation for long-term agribusiness development. Palladium contracted two agronomists (each executing three-month contracts) to develop the 0.1-hectare site at the Yirrkala Farm (Farm), which had been left fallow since 2011 (Figure 4). Here, two crop cycles were tested, assessing the viability of different produce and evaluating potential profitability. A Monitoring, Evaluation, and Learning (MEL) framework was collaboratively developed with the Traditional Owner corporations and approved by their respective boards to guide the analysis of land productivity, local labour interest, and economic potential. Community engagement focused on workforce development through partnerships with the NEAL Community Development Program (CDP) participants and the Dajtala Work Camp. At the same time, local schools and education stakeholders were consulted on learning and training opportunities.

FIG 4 – The Yirrkala Farm, which had not been used to grow food since 2011.

Significant progress was also made in market development, with potential buyers—including Sodexo, Simon George and Sons, and local stores—being engaged early. In addition, Palladium supported the Traditional Owner corporations to secure investment from the National Indigenous

Australians Agency ($100 000), an Australian bank ($25 000), and technical support from the Northern Territory Department of Agriculture and Fisheries.

In December 2024, the agronomics of Green Plum and Red Bush Apple were reviewed. In early 2025, fruit sourced from the region was analysed based on its nutritional and bioactive properties to understand the value proposition of the fruit. Based on the nutritional profiles of fruits, potential bushfood markets were reviewed, and interviews with key Indigenous cooperatives were conducted. Green plum is recognised as one of the highest sources of folate globally. Additionally, Red Bush Apple seeds boast a remarkably high total phenolic content and antioxidant capacity, even surpassing that of the Kakadu Plum. This indicates a strong bioactive potential for both fruits, highlighting their significant market potential.

As part of Horizon 1a, the Financial Model from the Study was updated with the Monitoring, Evaluation and Learning (MEL) data to provide a pragmatic, tested forecast of potential profit over the Horizons. In addition, the Traditional Owner corporations worked towards establishing an entity structure and funding mix for the YMP. These efforts culminated in drafting the Horizon 1b workplan and positioning the YMP for locally-led scaled implementation. Importantly, the majority of Horizon 1a was funded by DEAL, with financial support from GCL, RAC, NEAL and NTG. In addition, Rio Tinto provided in-kind support to cover flights and accommodation requirements.

Horizon 1b (2025–2026): This Horizon will see Traditional Owner corporations lead in delivering the next phase of the YMP, with planned activities focused on building infrastructure, including key works such as site clean-up, installing irrigation, regenerating soil, and planting crops and bushfoods. Traditional Owner corporations will finalise the entity structure, manage workforce development, integrate education opportunities aligned with Yolŋu pedagogy, support Homeland Market Gardens, and establish central processing functions at the Farm. These planned actions reflect a shift towards local ownership and delivery, positioning the YMP as a catalyst for a Yolŋu-led regional food economy.

Horizon 2 (2027–2028): During this Horizon, the YMP will improve agricultural practices and continue to support Homeland Market Gardens. Planned activities include crop rotation, local fertiliser use, pest and biosecurity management, and bushfood trials scaling on rehabilitated mine land. Learning, training and central support will continue, with potential expansion through local partners.

Horizon 3 (2029 onwards): By 2029, the YMP will focus on full commercial operations, with bushfoods and produce generating income. The Farm will act as a central hub, potentially expanding to larger farming areas and joint ventures with other Traditional Owner groups, establishing fully Indigenous-owned regional agribusiness opportunities. These efforts aim to build a sustainable, scalable Yolŋu-led food economy.

By applying a structured scale-up to the agribusiness and its services in learning, training, and job opportunities, the YMP allows the local community to grow as it scales over time.

INCORPORATING MULTIPLE PERSPECTIVES

One of the key strengths of the YMP lies in its commitment to inclusive governance and decision-making, grounded in the diverse perspectives of Traditional Owners, industry, government, and community stakeholders. This began with the Study, where over 100 interviews were conducted across Australia to ensure a deep understanding of commercial opportunities aligned with community aspirations, agronomic potential, and economic opportunities. The process prioritised Traditional Owner leadership while also drawing on technical, market, and policy expertise to shape a strategy that was both culturally grounded and economically viable (Hunt, 2013; Altman, 2001).

This multi-perspective approach was carried into implementing Horizon 1a of the YMP. The governance model centred around regular biweekly SteerCo meetings. Stakeholder engagement expanded further during this phase, with more than 47 funders, impact investors, banks, and government agencies consulted to ensure the YMP was financially sound and positioned for long-term impact. Rather than a top-down model, the process was designed to elevate Yolŋu voices, balance economic, cultural, and social priorities, and build shared accountability (Smith, 2008; Foley, 2003).

Tools such as a Monitoring, Evaluation, and Learning (MEL) framework, ongoing feedback loops, and on-country engagement by the agronomists ensured cultural protocols were respected, and community leadership through the SteerCo was central. This approach not only strengthened relationships and transparency but also ensured the YMP remained aligned with Yolŋu values and the broader goal of a sustainable, Indigenous-led post-mine economy.

LEARNINGS FROM HORIZON 1A

Horizon 1a provided critical operational insights that helped shape the future of the YMP. Agronomically, the Farm successfully identified crops suited to local growing conditions, including watermelon, rocket, baby green leaf and sweet potatoes. Early agronomic analysis of bushfoods also showed strong potential for future cultivation on rehabilitated mine land and commercialisation. These findings offered early proof that the region's soil and climate could support food security in the region through a combination of fresh produce and bushfoods at a meaningful scale. In Horizon 1a the demand analysis was out of scope, and this was conducted locally in Horizon 1b.

This Horizon also revealed valuable information about local capacity. Through engagement with CDP participants, the project identified interest in horticulture as a job opportunity and a clear need for targeted training. This highlighted the importance of embedding learning, training and employment pathways within the broader agribusiness strategy.

From a governance perspective, Horizon 1a strengthened confidence in SteerCo's joint decision-making approach. Regular meetings and active participation from Traditional Owner corporations helped build trust, reinforcing the shared accountability underpinning the YMP. Market engagement was also a key outcome, with conversations initiated with regional buyers laying the groundwork for future commercial partnerships.

Horizon 1a also demonstrated the value of taking a methodical, data-informed approach to implementation. The project could track progress, assess viability, and generate evidence to guide decision-making by applying the MEL framework throughout the trial. This transparent, step-by-step process ensured that Traditional Owner boards had the information they needed to make informed choices with confidence. The structured approach helped de-risk the early phases of the YMP and created a sense of clarity and momentum, reinforcing the Traditional Owner corporations' trust in the process and their readiness to lead future stages of the program.

A critical principle embedded in this Horizon was the concept of 'build to leave.' Palladium intentionally designed the program so Traditional Owner corporations could assume full control and management at any point across the Horizons. This approach recognises long-term success lies in building local capacity and leadership, not external dependency. The aim is to work oneself out of a job—creating a foundation where Yolŋu can lead, own, and grow the agribusiness in ways that align with their values, priorities, and aspirations.

BROADER IMPLICATIONS FOR MINE CLOSURE PLANNING

The YMP presents a forward-looking model for a post-mine transition that centres on Indigenous agency, long-term economic development, and inclusive governance. First and foremost, it prioritises Indigenous-led development, recognising Traditional Owner corporations and their boards as central decision-makers in the transition from mining to a diversified regional economy. This reflects global best practices outlined by the International Council on Mining and Metals (ICMM), which advocates for meaningful community participation and shared responsibility throughout the mine closure life cycle (ICMM, 2019).

Using a phased implementation strategy, the YMP aligns investment with community capacity. This staged approach allows for de-risked growth while respecting cultural readiness and ensuring communities lead at a pace that suits their aspirations. Such planning is consistent with the Australian Mine Closure Guidelines, which emphasise progressive, flexible closure planning that responds to evolving social, cultural, and economic needs (Department of Industry, Science and Resources (DISR), 2023).

In complex mine closure contexts such as the Gove Mine, it is essential to draw on multiple local and national perspectives to inform decision-making. No single organisation, discipline, or individual holds all the answers when navigating the social, cultural, environmental, and economic dimensions of closure. Traditional Owners bring deep place-based knowledge, cultural protocols, and long-term stewardship values; local service providers offer practical insights into infrastructure and delivery, while local government, technical experts, and investors contribute strategic, policy, and financial perspectives. The strength of the YMP lies in its ability to synthesise these diverse voices into a coherent, community-led roadmap. As recommended in both the ICMM Integrated Mine Closure Guide and the Australian Mine Closure Guidelines, successful closure planning requires inclusive, adaptive governance that enables all stakeholders to contribute meaningfully, ensuring outcomes that are not only technically viable but socially and culturally legitimate.

In doing so, the YMP also addresses ESG and closure obligations in a technically sound and socially transformative way. Rather than focusing solely on environmental rehabilitation or short-term job creation, the YMP invests in building Indigenous capabilities, infrastructure, and enterprise systems that can support a post-mine economy for generations. It demonstrates that mine closure strategies must not only 'leave no one behind,' but actively invest in the futures that communities envision for themselves—rooted in culture, self-determination, and economic opportunity.

CONCLUSION AND RECOMMENDATIONS

The Yolŋu Master Plan (YMP) demonstrates that mine closure, when approached with purpose, respect, and collaboration, can catalyse long-term regional transformation. More than an agribusiness strategy, the YMP initiative is rooted in Traditional Owner leadership, informed by diverse perspectives, and designed to build sustainable, culturally grounded economic futures.

The success of the YMP reflects the power of strong partnerships and inclusive governance. This is exemplified by the collaboration with the Yirrkala Farm Partners (Gumatj Corporation, Rirratjingu Aboriginal Corporation, and North-East Arnhem Land Aboriginal Corporation), who advised:

> 'Palladium led a 2023 agribusiness options study and a 2024 pilot study for the Yirrkala Farm for a Steering Committee comprising Traditional Owners corporations, Rio Tinto, Northern Territory Government representatives and Developing East Arnhem Limited (DEAL). The pilot study effectively assessed the performance of a wide range of crops on the former Yirrkala Banana Farm and **provided critical evidence about the potential to successfully grow crops in this location**. Gumatj Corporation, Rirratjingu Aboriginal Corporation and North-East Arnhem Land Aboriginal Corporation have developed a business case and are implementing the Yirrkala Farm to start production in the second half of 2025. **Palladium, through these studies, provided critical inputs for scoping the operational version of the farm in the business case.**' (Gumatj Corporation, Rirratjingu Aboriginal Corporation, and North-East Arnhem Land Aboriginal Corporation, 2025)

Through its methodical, data-informed, and community-led approach, Palladium helped establish the foundation for Traditional Owner corporations to access new revenue streams, create local employment opportunities, and enhance food security in alignment with Yolŋu cultural priorities and territory and national policy frameworks. The YMP also exemplifies how flexible, scalable models can be adapted to other mine closure contexts across Australia.

The recommendations are:

- Connect and work with local governance and partnership structures.

- Begin closure transition planning early with multiple perspectives to build trust and reduce risk.

- Create and leverage governance structures that embed Traditional Owner leadership, ensuring cultural legitimacy and long-term ownership.

- Use flexible, phased approaches that align with community readiness, allowing for staged investment and learning.

- Recognise closure as a generational opportunity, not just a regulatory obligation, and commit to leaving a positive legacy shaped by those who call the region home.

- Ensure partners develop Studies and Implementation plans with the intention for handover to locally led organisations – translated into action with collaboration and documentation.

As mining companies, governments, and communities navigate the complexities of mine closure, the Yolŋu experience offers valuable lessons. Closure should not be seen as the end of value creation but rather as the beginning of a new, community-defined economy. This requires investment in people, place, and potential.

REFERENCES

Altman, J C, 2001. *Indigenous communities and business: Three perspectives 1998–2000,* Canberra: Centre for Aboriginal Economic Policy Research (CAEPR), Australian National University.

Australian Government, 2020. Australian Government response to the House of Representatives Standing Committee on Indigenous Affairs report: Inquiry into food pricing and food security in remote Indigenous communities [online], National Indigenous Australians Agency (NIAA), Resource Centre, Health and Wellbeing. Available from: <https://www.niaa.gov.au/resource-centre/australian-government-response-house-representatives-standing-committee-0>

Christie, M and Greatorex, J, 2004. Yolngu Life in the Northern Territory of Australia: The Significance of Community and Social Capital, *Asia Pacific Journal of Public Administration,* 26(1):55–69. https://doi.org/10.1080/23276665.2004.10779285

Department of Industry, Science and Resources (DISR), 2023. Australian Mine Closure Guidelines, Canberra: Australian Government. Available from: <https://www.industry.gov.au/publications/leading-practice-handbooks-sustainable-mining/mine-closure>

Foley, D, 2003. An examination of Indigenous Australian entrepreneurs, *Journal of Developmental Entrepreneurship,* 8(2):133–151.

Gove Peninsula Future Reference Group, 2021. A New Journey Together – Traditional Owners' Vision for the Future of Nhulunbuy and the Gove Peninsula, Available from: <https://govefutures.nt.gov.au/__data/assets/pdf_file/0017/1145033/traditional-owner-vision-new-journey-together.pdf>

Gumatj Corporation, Rirratjingu Aboriginal Corporation and North-East Arnhem Land Aboriginal Corporation, 2025. Statement on partnership and governance, Yirrkala Farm Partners.

Hunt, J, 2013. Engaging with Indigenous Australia—Exploring the conditions for effective relationships with Aboriginal and Torres Strait Islander communities, Issues paper 5, Closing the Gap Clearinghouse, Canberra: AIHW and AIFS.

International Council on Mining and Metals (ICMM), 2019. *Integrated Mine Closure: Good Practice Guide,* 2nd edn, London: International Council on Mining and Metals. Available from: <https://www.icmm.com/integrated-mine-closure

Northern Territory Government, 2019. 2019 NT Market Basket Survey. Available from: <https://data.nt.gov.au/dataset/nt-market-basket-survey-2019>

Pearson, C A L and Helms, K, 2013. Indigenous Social Entrepreneurship: The Gumatj Clan Enterprise in East Arnhem Land, *The Journal of Entrepreneurship,* 22(1):43–70. https://doi.org/10.1177/0971355712469185

Smith, D E, 2008. From collaboration to co-optation: Prospects for Indigenous partnerships in natural resource management, in *Engaging Indigenous economies: Debating diverse approaches* (eds: D Ritter, M Rowse and J Altman), Canberra: ANU Press.

SteerCo, 2023. Workshop summary: SteerCo session, The Palladium Group, Nhulunbuy, Northern Territory.

Mapping the legislative pathway to mine closure in the Latrobe Valley, Victoria

T Mok[1] and A Scrase[2]

1. Technical Advisor, Mine Land Rehabilitation Authority, Morwell Vic 3840.
   Email: tanya.mok@mineland.vic.gov.au
2. Technical Director, Mine Land Rehabilitation Authority, Morwell Vic 3840.
   Email: antonia.scrase@mineland.vic.gov.au

ABSTRACT

High risk mines in Victoria are assigned 'declared' status. These are mines which have geotechnical, hydrogeological, water quality or hydrological factors that may be deemed to pose significant risk of harm to the community, environment and infrastructure. There are currently three declared mines in Victoria, all located in the Latrobe Valley. These mines are, and soon will be, transitioning to rehabilitation and closure. The Mine Land Rehabilitation Authority's (MLRA) role is to facilitate good rehabilitation outcomes for the Latrobe Valley and, as part of this, assist in understanding the complex and changing Victorian regulatory regime that governs the closure of these mines.

Victorian mine licensees are required to operate and close in accordance with numerous Acts, Regulations and policies. The key Acts the mine licensees must navigate include the *Mineral Resources (Sustainable Development) Act 1990* (MRSDA; Government of Victoria, 1990), *Environment Effects Act 1978*, and others.

The MRSDA has undergone rapid changes in recent years to provide more focus on the rehabilitation and closure of the declared mines. It is, however, untested and the rapid changes in the MRSDA, driven by the region's shift to mine closure, has resulted in potential gaps within this Act. This is further complicated by the overlay of multiple regulatory processes, raising questions about which assessment process goes first: the approval of each site's mine closure plan or an assessment of the site's environmental impacts. These go hand-in-hand but are assessed and governed under different Acts.

As the Latrobe Valley begins to shift away from mining and towards closure, both government and operators face the challenges of navigating through the complexities of mine closure and the associated regulatory framework. This paper will explore the problems identified above and discuss some of the initiatives the MLRA has taken to map out the regulatory pathways to mine closure in the Latrobe Valley.

INTRODUCTION

The three open cut brown coalmines in the Latrobe Valley have been extracting coal and generating power for Victoria over the last 100 years. The first of the three mines, the Yallourn Mine and the first of Yallourn's power stations, was commissioned by the state-owned State Electricity Commission Victoria (SECV) in the 1920s (EnergyAustralia, 2020). This was followed by the opening of the Hazelwood Mine (then known as the Morwell Open Cut) and the Loy Yang Mine in the 1950s and 1970s respectively (Hazelwood Mine Fire Inquiry, 2016). Like Yallourn Mine, both Hazelwood Mine and Loy Yang Mine, and their associated power stations, were initially owned and operated by the SECV. All three mine sites have since been privatised (circa 1990s).

After decades of operations, Victoria seeks to transition away from coal and towards more renewable energy sources. The closure of the Latrobe Valley brown coalmines is impending. Extraction of coal at the Hazelwood Mine ceased in March 2017 and closure dates for the Yallourn and Loy Yang mines have been announced, with Yallourn intending to cease mining operations in mid-2028 and Loy Yang in 2035.

Extraction of coal at these sites has created large open voids in the landscape. As of 2023, the combined mined (pit) area of Hazelwood, Yallourn and Loy Yang mines exceeds 50 hectares, with the depths of the pits reaching as deep as 200 m below the original ground level (Mine Land Rehabilitation Authority (MLRA), 2023). These three mines are also located close to large towns, infrastructure and waterways, and are considered high risk mines (or 'declared' mines in Victoria) due to their potential to pose significant risk of harm to the community, environment and infrastructure. As of 2025, they are the only three mines which have been 'declared' in Victoria. Figure 1 depicts the location of the declared mines and their proximity to infrastructure, waterways and the neighbouring towns of Moe, Newborough, Morwell, Yallourn North and Traralgon.

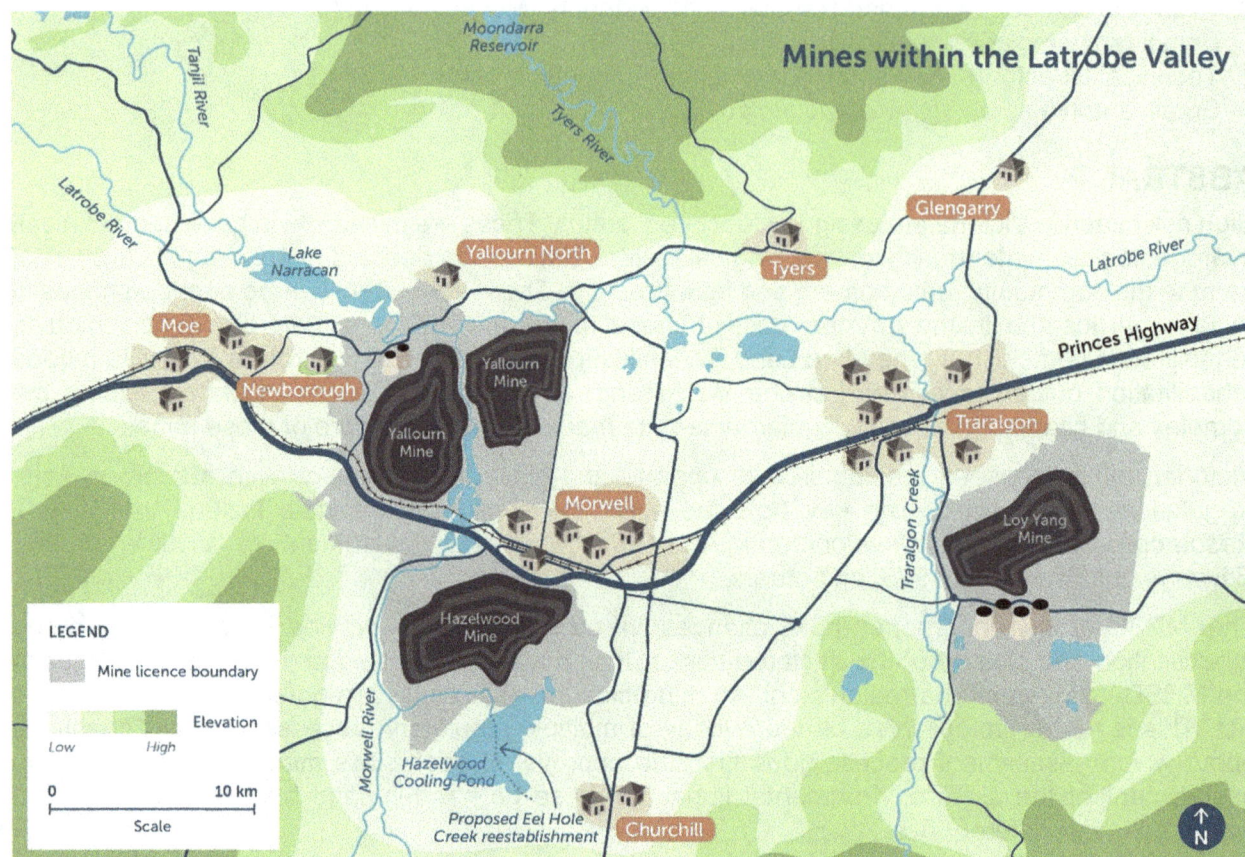

FIG 1 – Locality plan of declared mines.

At the cessation of operations, these large voids and their surrounding areas will need to be rehabilitated by the licensees to safe, stable and sustainable landforms. However, the journey to rehabilitation and closure is not a simple one.

The key piece of legislation governing the mining, rehabilitation and closure of the Latrobe Valley mines (the *Mineral Resources (Sustainable Development) Act 1990 (Vic)* (MRSDA; Government of Victoria, 1990)) has undergone rapid changes in recent years. It is, however, untested and the rapid changes in the MRSDA and supporting Regulations, driven by the region's shift to mine closure, has potentially resulted in gaps or ambiguities within the Act and Regulations. This is further complicated by the overlay of multiple regulatory processes and site-specific challenges.

The Mine Land Rehabilitation Authority's (MLRA) role is to facilitate good rehabilitation outcomes for the Latrobe Valley and as part of this, assist in understanding the complex and changing Victorian regulatory regime that governs the closure of these mines. This paper explores the current regulatory framework which governs the closure of the Latrobe Valley mines and discusses some of the initiatives led by the MLRA, and worked in collaboration with other government departments, to assist with identifying and navigating through potential challenges.

HISTORY OF MINE REHABILITATION IN THE LATROBE VALLEY

Despite the long history of mining in the Latrobe Valley, there was little consideration of mine rehabilitation by both government and mine operators up until this century (MacKay *et al*, 2019). Whilst some consideration of mine rehabilitation began in the 1970s to early 1980s by the SECV, the rehabilitation planning at this stage was noted to be insufficient and lacked a clear/structured approach to achieving environmental and rehabilitation objectives (ibid).

The conceptual and high-level nature of rehabilitation planning continued up until recent years, with rehabilitation plans included in approved Work Plans up until 2015/2016 still noted to be conceptual and lacking in detail (Hazelwood Mine Fire Inquiry, 2016). It was not until the Hazelwood Mine fire that the focus began to shift towards mine rehabilitation. NB: A Work Plan is a document in Victoria which outlines how mining operations will be conducted within a mining license area. The Work Plan also contains a rehabilitation plan, both of which are required to be approved by the Victorian mining regulator.

The Hazelwood Mine fire was a major incident in the Latrobe Valley. The coalmine fire, sparked by embers from a nearby bushfire, burned for 45 days in early 2014 and resulted in significant impacts on the adjacent town of Morwell (see Figure 1) and the greater Latrobe Valley community (Victoria State Government, 2016; Environment Protection Authority (EPA) Victoria, 2015). In light of the incident, the Victorian Government conducted a public inquiry known as the Hazelwood Mine Fire Inquiry (HMFI) in 2014 and again in 2015/2016. The inquiry identified the immaturity of rehabilitation and mine closure planning within the region, further noting the regulatory framework was '*ill-suited to contemporary needs*' and identified the need for more work to be done to improve the regulatory framework, to ensure that rehabilitation is done successfully (Hazelwood Mine Fire Inquiry, 2016).

Since then, the regulatory framework for mine rehabilitation has undergone significant change, with major changes to the MRSDA and associated regulations in 2019 and 2022 respectively, strengthening the approach to mine rehabilitation. It also triggered the development of a regional strategy, the Latrobe Valley Regional Rehabilitation Strategy (LVRRS) in 2020, which '*outlines policy and provides guidance to progress mine rehabilitation planning*' (Department of Jobs, Precincts and Regions (DJPR), 2020).

VICTORIA'S CHANGING MINING ACT

The MRSDA (Government of Victoria, 1990) in Victoria is the central piece of legislation which governs mining and mine rehabilitation in Victoria. In response to the findings and recommendations of the HMFI, more stringent requirements were introduced into the MRSDA in 2019, providing more focus on rehabilitation, improving rehabilitation planning readiness and closing the gap towards good practice mine closure. The Regulations accompanying the 2019 amendments to the MRSDA were also updated and took effect in 2022. New provisions in the Act and Regulations related to the rehabilitation of the declared mines included:

- The establishment of the MLRA, an independent body created to provide oversight, facilitate and promote good rehabilitation outcomes for the declared mines.

- The introduction of the Declared Mine Rehabilitation Plan (DMRP). A dedicated rehabilitation plan which all declared mine licensees must prepare and submit to the State for approval by 1 October 2025.

  o The DMRP is intended to be an iterative plan. It acknowledges the ongoing process of mine closure and is designed to drive continuous improvement, allowing for the DMRP to be updated as technical studies, trials and data is collected over time to address uncertainties and knowledge gaps.

- Within the DMRP, declared mine licensees are also required to prepare a Post-Closure Plan which outlines the required monitoring and maintenance works after the mine license is relinquished. Declared mine licensees are also required to cost out their Post-Closure Plan and upon license relinquishment will be required to pay this sum into a fund (the Declared Mine Fund) to cover the ongoing monitoring and maintenance costs of the land.

  o The Post-Closure Plan and Declared Mine Fund has been set-up in recognition that there will likely be some degree of monitoring and maintenance required to manage residual risks and liabilities after the surrender of the mining license (DJPR, 2022). The landowner (which may be the mine licensee) and/or MLRA will be responsible for the management of land post-closure (post relinquishment of the mining license). The MLRA will have a role in administering the fund.

- The establishment of a register to allow for the registration of declared mine land after the surrender of the mining license. The register will include the registration of land, any conditions or prescribed matters that apply to the land and the post-closure plan.

As noted by Scrase and Brereton (2024):

> … the 2019 Act amendments and 2022 regulations have fundamentally changed the approach of the Victorian regulation for mine rehabilitation from the prescriptive rules-based, tick box approach to one that is aiming for proponent-led, outcomes-focused and risk-based results… This approach aims to reduce the risk to government and enables flexibility and innovation by the licensees; if implemented appropriately by government and industry, it should lead to better outcomes for the Victorian community.

Whilst the recent changes in the legislation has led to a more robust approach to rehabilitation, the legislative framework is new and untested, and as such there is the potential for unforeseen gaps or areas in the legislation which may be ambiguous and require further clarity. Government and licensees must also learn to regulate and operate within the new framework. This will take time and may impact on the approval and implementation of the DMRP's. The legislative framework is also anticipated to continue to change in the coming years, with the introduction of a duty-based framework in July 2027 (Resource Victoria, 2024) and both government and industry will need to be adaptive to these changes.

The recent changes to the MRSDA and associated regulations have been made in the twilight years of Yallourn and Loy Yang mines and after the closure of Hazelwood Mine, potentially causing difficulties in applying some of the regulatory requirements and increasing the cost of rehabilitation for the licensees.

All three mines have largely missed opportunities to integrate good mine closure planning practices throughout their operations, which can lead to an increase in rehabilitation costs, loss of rehabilitation options and other opportunities (International Council on Mining and Metals (ICMM), 2025).

Due to the very long history of mining, licensees and government may face the challenges of dealing with legacy issues not being able to achieve current regulatory requirements. This has been seen at some legacy tailings-storage facilities, with significant work required to restore and rehabilitate the land in-line with today's standards or in some cases, community, government and operators are faced with the fact that restoring the land is infeasible (Weinig and Crouse, 2024) and as such accepted as a larger than desired liability requiring long-term monitoring and maintenance.

As of April 2025, all three declared mine licensees are currently in the process of preparing their DMRPs for submission in October 2025 (as legislatively required under the MRSDA). The development, assessment and approval of each of the declared mines first DMRPs is crucial in moving towards a structured, risk managed and tangible approach to mine rehabilitation. It provides certainty to the community, government and licensee on the envisaged final landform and potential future uses of the land, stepping away from the conceptual and high-level nature of the rehabilitation plans which currently exist for the sites.

THE BROADER REGULATORY ENVIRONMENT

Whilst the MRSDA is central to rehabilitation, it is not the only legislative requirement that mine licensees must satisfy to successfully rehabilitate their mines. Licensees must also:

- Understand their obligations under the *Environment Effects Act 1978* and the *Environmental Protection and Conservation Act 1999 (Cth)* (EPBC Act).

- Seek approvals under the *Water Act (Vic) 1989* in order to use and access water for the purpose of rehabilitation.

- Comply with the duties and requirements under the *Environment Protection Act (Vic) 2017*, including duties to manage contaminated land, etc.

- Obtain and operate in accordance with the relevant planning permits under the *Planning and Environment Act (Vic) 1987*.

- Obtain and operate in accordance with the relevant requirements under the *Aboriginal Heritage Act 2006*.

It should be noted that the above is not an exhaustive list and is only intended to highlight some of the many legislative requirements that declared mine licensees must operate and rehabilitate within.

The remainder of this paper further explores the declared mine licensees' requirements under the *Water Act* and *Environment Effects Act*, noting that assessments and approvals under these two Acts have generally been identified to be on the critical path to successful rehabilitation and relinquishment.

Water Act 1989

The three Latrobe Valley Mine licensees are pursuing water-based rehabilitation concepts (pit lakes) (AGL, 2024; ENGIE, 2023; EnergyAustralia, 2021) and as such will likely require water access under the *Water Act*.

Loy Yang and Yallourn Mine currently have access to water via a bulk water entitlement (a water access instrument under the *Water Act*) strictly for the purpose of power generation. Should the mine licensees seek to use water for mine rehabilitation, they can apply to the Victorian Minister for Water for new entitlements (The State of Victoria Department of Energy, Environment and Climate Action (DEECA), 2025). The LVRRS Amendment (DEECA, 2023) outlines indicative conditions that will govern access to water under any new entitlement, to help mitigate the expected future impacts of a drier climate future and climate variability.

As of April 2025, AGL (owner and operator of the Loy Yang Mine and Loy Yang A Power Station) has submitted a bulk water application for the purpose of rehabilitation, with the application currently pending assessment by the Minister for Water (DEECA, 2025).

Environment Effects Act 1978

Where projects are deemed to have potential impacts or significant effects on the environment, proponents may be required to prepare an environment effects statement (EES) under the *Environment Effects Act 1978*.

To determine if an EES is required, the project is referred to the Victorian Minister for Planning by the proponent or a statutory body. The project can also be 'called in' by the Minister for Planning themselves. The Minister for Planning then decides whether an EES (or alternative assessment process under the *Environment Effects Act*) is required. Whilst a decision is being made, the Minister for Planning has the authority to direct other decision-makers/statutory bodies to not make further decisions on the project until the Minister has given advice on the referral, and if relevant, an assessment of the EES is complete.

Therefore, in the case that the works associated with the rehabilitation of the declared mine are determined to have the potential to result in significant impacts or effects on the environment, licensees may be directed to prepare an EES and decision-making on the DMRP and other processes will likely be put on hold. Notably, Hazelwood referred their rehabilitation project to the Minister for Planning in November 2021 and in February 2022, the decision was made that an EES would be required. As of April 2025, ENGIE continue to proceed with preparing their EES. Until the EES is completed, and the Minister for Planning completes their assessment on the project, approval of Hazelwood's DMRP will be on hold.

It is worth noting that assessments undertaken under the *Environment Effects Act* do not constitute the 'approval' of a project. Instead, it provides non-binding recommendations which can be used to inform decision-makers on the approval of other statutory processes relating to the project.

The LVRRS Amendment states that the '*requirement for referrals under the EE Act [Environment Effects Act] and EPBC Act at Hazelwood has created a presumption that referrals are likely to be required for the remaining mines to inform submission of DMRPs by their due date of October 2025*'

(DEECA, 2023). With the estimated time frame to complete an EES being in the order of 3 to 5 years (a presumption based on Hazelwood's current EES progress), it raises the question on how this assessment process will sequence alongside the DMRP and its implications to the overarching time frame and pathway to closure.

MAPPING OUT REGULATORY PATHWAYS

From the above, it is evident that the pathway to mine closure and relinquishment is not a simple one. Government and licensees must navigate through a complex, changing regulatory framework and must manage and coordinate the assessment of various regulatory processes. Notably, time is quickly becoming a limiting factor, with the closure dates of the coal fire power stations impending, reducing opportunities to integrate mine rehabilitation during the operations.

To assist in understanding the complex and changing Victorian regulatory environment, the MLRA has been undertaking a body of work which maps out the legislative requirements/processes for mine closure and to license relinquishment. This was undertaken in two parts:

The mapping firstly focused on understanding the flow and key decision points of the MRSDA and accompanying regulations, to determine potential pinch points that may cause issues to achieving good rehabilitation outcomes.

Site-specific time frames and regulatory processes under the MRSDA, *Water Act* and *Environment Effects Act* were then overlayed, and the interaction/sequencing of these processes were considered, with the intent to determine potential pinch points in the overlapping processes and develop suitable solutions/mitigations.

Mapping out the MRSDA

In the first part of this project, the MLRA firstly focused on the mapping of the MRSDA and accompanying regulations, considering the relevant legislative processes from rehabilitation (including progressive rehabilitation through the operation phase) through to the point of mine license relinquishment. The intent of this process was to gain a wholistic understanding of the key assessment and approval processes which were required to enable successful rehabilitation of the declared mines. It also aimed to identify any potential gaps or sections within the Act and Regulations which require further clarity and/or could inadvertently impact the outcomes of rehabilitation.

To aid in understanding and identifying gaps in the legislation, detailed visual process maps were created. These maps identified the key process flow, decision points and responsibilities (including points in the process which required input from external stakeholders such as other government departments). Workshops were then held with the relevant government agencies which included representatives from the Victorian mining regulator and the associated policy division, to seek input on the correctness of the visual maps, and to confirm and discuss the identified gaps.

From the visual maps the following were then identified:

- potential gaps or parts of legislation which require further clarity and/or may present challenges during implementation

- parts of legislation which are ambiguous, which could allow flexibility and adaptability from licensees and discretionary powers by the decision-maker.

Through the workshops, it was identified that a number of gaps had already been mitigated, however some were outstanding and required further consideration.

Sequencing considerations

In addition to considering legislative processes in a standalone manner, the MLRA identified the need to consider the processes collectively, and as such considered and mapped out the interaction and potential sequencing scenarios between the MRSDA, *Water Act* and *Environment Effects Act.*

A number of sequencing scenarios were developed and tested, starting firstly with a 'default' sequencing scenario pathway which mirrored the pathway that Hazelwood is currently undergoing ie referral and completion of EES ahead of assessment and approval of DMRP. Potential issues or hinderances for this scenario were then identified, with potential solutions and/or alternative sequencing scenarios developed to address these.

Notably, the above process was undertaken in a collaborative manner. Key stakeholders from the Resources and Water departments of the government were again brought together in a series of workshops to identify potential issues and hinderances with the sequencing scenarios and brainstorm potential solutions. The workshops discussed both process-specific issues (eg those identified in the first stage of the project) and broader overarching and regional issues. Solutions were tailored to the severity and type of issue, with solutions ranging from more administrative fixes (such as the development of practice notes to supplement legislation or the drafting and publishing of explanatory notes for the purpose of keeping the community informed) through to legislative amendments were considered. The collaborative process allowed for diverse perspectives to be shared and innovative solutions to be developed.

It is noted that the project is still ongoing with further work required to develop solutions and a clearer pathway towards mine closure. This requires continuing and ongoing collaboration with key stakeholders. The MLRA seeks to engage with broader government stakeholder groups to share learnings and facilitate good rehabilitation outcomes for the Latrobe Valley mines.

CONCLUSION

In recent years, the regulatory framework for mine closure in the Latrobe Valley has undergone rapid change, with the Victorian government recognising the need improve rehabilitation preparedness ahead of the closure of the three declared mines. The regulatory framework has now shifted towards an outcomes-focused and risk-based approach, with the aim provide a robust framework which will lead to better rehabilitation outcomes for the Victorian community.

To better understand the Victorian regulatory environment, the MLRA initiated a project which aimed to map out the various regulatory processes related to mine rehabilitation, identify and close any potential gaps in legislation and map out potential regulatory pathways to mine closure.

Undertaking the mapping project has allowed for an increase in the general understanding of the Victorian regulatory environment. The project facilitated open discussions among government stakeholders to identify potential and upcoming issues/challenges which could impact on the rehabilitation of the declared mines, and has created a mutual understanding of pinch points, potential gaps and mitigations. Based on these issues, potential solutions were developed noting that further development and refinement of these solutions, with input and collaboration from other stakeholders, is required. This project has provided a stepping stone towards continual improvement of the regulatory processes in Victoria for the declared mines. Ultimately, the project aims to guide future decision-making and pave a clearer pathway forward for the rehabilitation of the Latrobe Valley mines.

REFERENCES

AGL, 2024. Water for mine rehabilitation, Factsheet #2. Available from: <https://www.agl.com.au/content/dam/digital/agl/documents/about-agl/how-we-source-energy/loy-yang-power-station/241003-agl-factsheet-waterforminerehabilitation.pdf> [Accessed: 27 May 2025].

Department of Jobs, Precincts and Regions (DJPR), 2020. Latrobe Valley Regional Rehabilitation Strategy. Available from: <https://earthresources.vic.gov.au/__data/assets/pdf_file/0011/558884/Latrobe-Valley-Regional-Rehabilitation-Strategy.pdf> [Accessed: 04 April 2025].

Department of Jobs, Precincts and Regions (DJPR), 2022. Regulatory Impact Statement – Proposed Mineral Resources (Sustainable Development) (Mineral Industries) Amendment Regulations 2022. Available from: <https://resources.vic.gov.au/__data/assets/pdf_file/0004/895693/Regulatory-Impact-Statement-Proposed-Mineral-Resources-Sustainable-Development-Mineral-Industries-Amendment-Regulations-2022.pdf> [Accessed: 07 April 2025].

EnergyAustralia, 2020. History – A century of powering the present, and planning for the future. Available from: <https://www.energyaustralia.com.au/about-us/energy-generation/yallourn-power-station/history#:~:text=The%20story%20of%20Yallourn%20Power,the%20station%20we%20know%20today> [Accessed: 04 April 2025].

EnergyAustralia, 2021. Delivering a community asset fact sheet. Available from: <https://www.energyaustralia.com.au/sites/default/files/2021-03/EA_DeliveringACommunityAsset_vF.pdf> [Accessed: 27 May 2025].

ENGIE, 2023. Hazelwood Rehabilitation Project. Available from: <https://media.caapp.com.au/pdf/mvl761/b08f8d15-848d-458b-9ce4-95ad9c2046dc/Hazelwood%20project%20factsheet%20July%202023.pdf> [Accessed: 27 May 2025].

Environment Protection Authority (EPA) Victoria, 2015. Summarising the air monitoring and conditions during the Hazelwood mine fire, 9 February to 31 March 2014, EPA Publication: 1598.

Government of Victoria, 1990. Mineral Resources (Sustainable Development) Act 1990, Australia.

Hazelwood Mine Fire Inquiry, 2016. Hazelwood Mine Fire Inquiry Report 2015–2016, volume IV – Mine Rehabilitation, Melbourne.

International Council on Mining and Metals (ICMM), 2025. *Integrated Mine Closure: Good Practice Guide*, 3rd edn (London).

Mackay, R, Hastie, R, Lilley, H and Mathew, M, 2019. Mine rehabilitation in the Latrobe Valley, the start of a long journey: the Commissioner's role, in *Mine Closure 2019: Proceedings of the 13th International Conference on Mine Closure* (eds: A B Fourie and M Tibbett), pp 803–816 (Australian Centre for Geomechanics: Perth). https://doi.org/10.36487/ACG_rep/1915_65_Mackay

Mine Land Rehabilitation Authority (MLRA), 2023. Latrobe Valley. Available from: <https://www.mineland.vic.gov.au/learn/latrobe-valley/> [Accessed: 04 April 2025].

Resource Victoria, 2024. Victoria's approach to regulating mines and quarries is changing. Available from: <https://resources.vic.gov.au/legislation-and-regulations/regulation-review-and-reform/victorias-approach-to-regulating-mines-and-quarries-is-changing> [Accessed: 07 April 2025).

Scrase, A and Brereton, J, 2024. The perfect storm: mine closure in the Latrobe Valley, Victoria, in *Mine Closure 2024: Proceedings of the 17th International Conference on Mine Closure* (eds: A B Fourie, M Tibbett and G Boggs), pp 1297–1310 (Australian Centre for Geomechanics, Perth). https://doi.org/10.36487/ACG_repo/2415_94

State of Victoria Department of Energy, Environment and Climate Action, The (DEECA), 2023. Latrobe Valley Regional Rehabilitation Strategy – Amendment. Available from: <https://resources.vic.gov.au/__data/assets/pdf_file/0004/984082/Latrobe-Valley-Regional-Rehabilitation-Strategy-Amendment.pdf> [Accessed: 08 April 2025].

State of Victoria Department of Energy, Environment and Climate Action, The (DEECA), 2025. Potential water access for Latrobe mine rehabilitation. Available from: <https://engage.vic.gov.au/potential-water-access-for-latrobe-mine-rehabilitation> [Accessed: 08 April 2025].

Victoria State Government, 2016. Hazelwood Mine Fire Inquiry: Victorian Government Implementation Plan, Melbourne. Available from: <https://www.healthassembly.org.au/wp-content/uploads/2024/08/Hazelwood-Mine-Fire-Inquiry-Implementation-Plan.pdf>

Weinig, W and Crouse, P, 2024. Developing sustainable mine-closure plans for legacy tailings-storage facility management. Available from: <https://www.stantec.com/en/ideas/topic/energy-resources/developing-sustainable-mine-closure-plans-legacy-tailings-storage-facility-management> [Accessed: 09 April 2024].

Applying a control framework approach to improve mine planning for sustainable operations

P Standish[1]

1. Consulting Director, Risk Mentor Pty Ltd, Canberra, ACT, 2614.
   Email: peter.standish@riskmentor.com.au

ABSTRACT

Successful mining operations that achieve smooth transitions across the life cycle are built on strong early-stage planning. A robust approach to managing environmental impacts, engaging stakeholders, and meeting regulatory requirements is essential. This paper introduces the Control Framework (CFw)—a validated, good-practice model that meets and exceeds the regulated requirements for conducting risk assessments.

Conventional processes often focus on generating a long list of potential issues, giving these risk scores, and referencing studies or controls that will address them and measuring impacts after they occur. In contrast, the Control Framework uses a leading practice model to exhaustively identify threats, design proactive mitigation measures, and prioritising progressive rehabilitation to a compliant and value adding final landform. It bridges the gap between detailed Environmental Impact Statement (EIS) findings and real-world operations, making complex information accessible and actionable at all phases of the mining life cycle.

The approach works for all types of mining operations in all jurisdictions and the demonstration of conformance with requirements is demonstrated for mines in New South Wales. The Control Framework also makes risk analyses more robust and transparent. This approach helps achieve smoother approval pathways, clearer transitions from planning to construction, and effective operational controls. The process of having subject matter experts critically review the framework during initial Environmental or Rehabilitation Risk Analyses ensures a rigorous and practical outcome is described.

The CFw approach provides a structured approach to mine design, operational planning, and performance monitoring that integrates strategic decision-making with practical outcomes. By focusing on proactive management of impacts, progressive rehabilitation, and broader consultation, it delivers real value for mine owners and operators striving for efficient, compliant, and sustainable operations.

INTRODUCTION

This paper covers two key areas – understanding the risks to achieving an integrated mine closure and implementation of business activities (controls) which will help to achieve acceptable outcomes.

When considering the approach to understanding the risks there is a contrast between current approaches and the application of a control framework model.

Current approaches work with teams of people to breakdown the subject area into tasks, aspects, impacts, hazards, causes and controls – typically in a spreadsheet format. Time is then invested in ranking the risks using subjective tools and stating the controls intended to address them – and then re-ranking the issue with the controls in place.

The CFw approach inverts the traditional process—prioritising controls based on credible failure modes and consequence severity, rather than perceived inherent risk. This leads to the team review being much more productive and generating value adding improvements to business activities (controls).

The paper also discusses how the controls identified can be best considered as business inputs – activities which prevent losing required operating states and yield desired outcomes. Adopting this approach to a mining operation can yield multiple benefits:

- Consistent conduct of activities – sound planning of the mine and mining to the plan.

- Simplifying management systems and improving conformance.

- Generating early signals that activities are not occurring in line with site requirements.

- Providing workers at all organisational levels with the information they need to perform their tasks.

LIFE-OF-MINE (AND CLOSURE) RISK ASSESSMENT

Many jurisdictions require that a risk assessment be conducted as part of preparing closure plans and considering mining project life cycle phases. Most of these requirements use a qualitative matrix style approach to determining the risks present.

Matrix based risk assessments

An unstated goal of the risk assessments prepared is to confirm for the proponent, typically the mining company, as well as the regulator and interested third parties that issues have been considered, and the planned work presents a tolerable level of risk.

Many of these analyses (eg Western Australia Government, 2025) guide towards ranking an issue in its inherent state and again after considering the relevant risk treatments or controls. Industry guidance, as provided by ICMM (2019, Tool 8) also considers risk analysis as a key aspect of preparing a closure plan.

There are multiple problems with this approach. From a facilitation perspective the rankings are so fuzzy that the process of conducting a risk assessment frequently becomes more a case of mediating a dispute than a sober assessment of a real measure. Hubbard (2009) provides a detailed, and frequently scathing, assessment of these processes. Some key failings are that:

- inputs are subjective and so are error prone

- scoring magnifies the input errors

- the approach fails to consider basic mathematics.

The basic mathematics in question is that there are essentially four ways that anything can be measured:

1. Nominal – this is a basic grouping such as Male or Female, Preferred to Not-preferred, Alive or Dead, etc.

2. Ordinal – this is a scale of qualitative merit or impact for example a 5-star hotel, or a 4-tomato film. All risk matrices in common use adopt this approach with a 1 to 3, 4, 5, 6 or 7 scale invoked – and some prose to describe the basis of the ranking provided. Frequently, the prose will include a reference to a frequency or a percentage impact.

3. Scalar – measurable quantities that start from a notional base position – for example degrees Centigrade or degrees Fahrenheit.

4. Real – measurable quantities with a zero start point – for example degrees Kelvin, where 0°K represents the absence of atomic movement. Other real quantities such as distance, speed or acceleration which can be measured in agreed units from a nominated baseline.

The segregation is important as it determines what mathematical functions can be performed. For the list above:

1. Nominal – no mathematical functions are valid

2. Ordinal – no mathematical functions are valid

3. Scalar – addition and subtraction can be performed, and

4. Real – all mathematical functions can be performed in line with preservation of unitary coherence (which just means that the result can be expressed in the relevant units or clearly describe a feature in percentage terms eg m/sec for speed and m/m=% for elongation or strain).

Risk, by definition, is the product of probability and consequence. Actuaries apply this approach extensively to determine the appropriate level of insurance premium that applies to an unwanted outcome (that can be presented in $ terms).

The accepted approach to risk management that prevails in our industry and certainly amongst our regulators is based on the flawed combination of ordinal numbers to produce a risk score or level upon which decisions can be made.

It is for this (and other) reason(s) that well-meaning professionals and work team members end up in fruitless debates over what level to score a risk at. The measure is meaningless. As a case *ad absurdum* consider my recent night out where I went to a 3-star restaurant and then went to the movies and saw a 2-star film. I did NOT have a 6-star night. I had a reasonable meal and suffered through a worse than average movie.

To put this in perspective consider Figure 1. This is the suggested risk matrix for considering threats to achieving mine closure outcomes. It is based on textual prose around what constitutes likelihood with Improbable being less than 3 per cent, Unlikely being 3–10 per cent, up to Almost Certain at greater than 90 per cent. Similarly for Consequence where impacts are Insignificant at less than 1 per cent, Minor at 1–3 per cent, up to Major at >30 per cent. The matrix in Figure 1 taken from the original reference shows each of the risk zones of equal size and offers a numerical score of the risk from 1 to 25.

Likelihood	Consequence scale				
	Insignificant	Minor	Moderate	High	Major
Almost certain >90%	11 (Medium)	16 (Significant)	20 (Significant)	23 (High)	25 (High)
Likely 30%–90%	7 (Medium)	12 (Medium)	17 (Significant)	21 (High)	24 (High)
Possible 10%–30%	4 (Low)	8 (Medium)	13 (Significant)	18 (Significant)	22 (High)
Unlikely 3%–10%	2 (Low)	5 (Low)	9 (Medium)	14 (Significant)	19 (Significant)
Improbable <3%	1 (Low)	3 (Low)	6 (Medium)	10 (Medium)	15 (Significant)

FIG 1 – The ICMM Tool 8 Scoring Matrix (courtesy ICMM (2019, p 107)).

Figure 2 on the other hand presents the same risk zones shown at a linear scale. In many of the risk matrices in use the scale is effectively log-log so that small values close to zero appear the same as larger values further from the zero point. This must contribute to the confusion – and it is alarming that so many issues rated by teams inhabit the vanishingly small section at the bottom left of these matrices. Rating items Low or Medium risk can lead decision-makers to assume that they require no follow-up or ongoing monitoring of controls. This is frequently a flawed assumption.

Figures 1 and 2 illustrate the contrast between conventional risk matrices and a linear, to-scale representation.

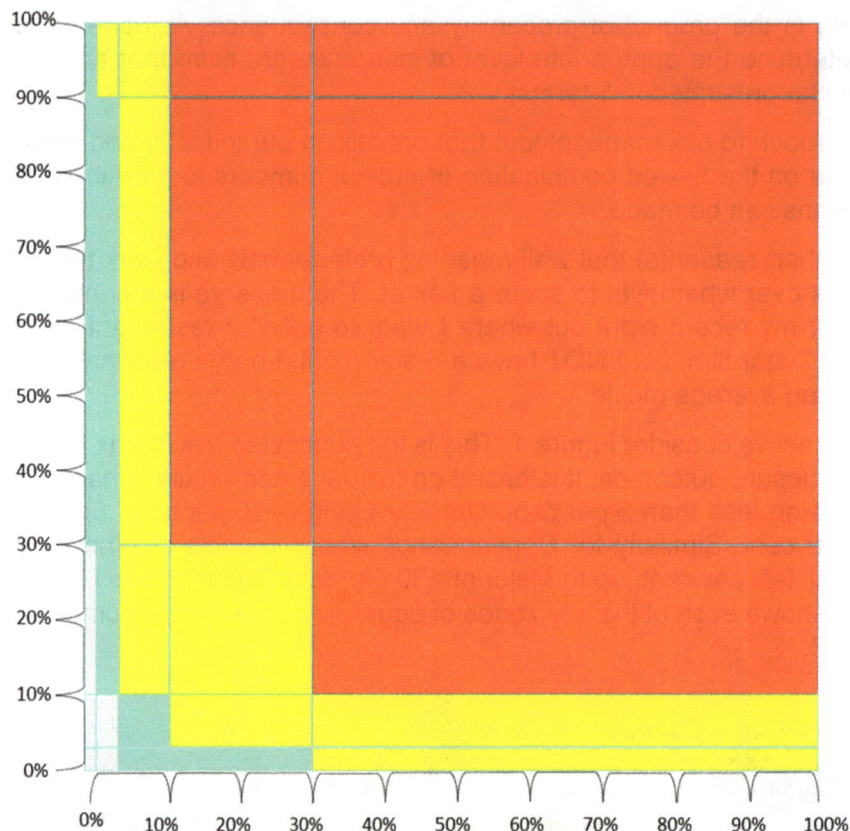

FIG 2 – Linear representation of the ICMM Tool 8 Scoring Matrix (100% × 100%).

It is also of interest to note that using these tools means that risk assessment scores vary significantly according to the make-up of the team which was in place at the time of the assessment to consider the subject. The scores can also be strongly influenced by the facilitator. In extreme cases this lends itself to a more correct term being 'facipulator' – a portmanteau of facilitator and manipulator – with groups guided to arrive at a pre-determined outcome (eg everything is low risk).

Control framework assessments

A control framework (CFw) assessment is based on the principle that certain operating states must be maintained to meet the company's objectives. This approach can apply across many subject areas – and the following hierarchy of states exhaustively describe what is needed to achieve integrated mine closure goals for a mining operation.

- R01 Rehabilitation works create stable landforms:
  - R01.01 Dumps are stable, non-polluting and matched to the surrounding landforms.
  - R01.02 Excavations are filled, recontoured, sealed, or stabilised.
  - R01.03 Changes in level post-mining provide a long-term safe and stable landform.
- R02 Water movement at the site does not cause harm:
  - R02.01 Chemically active materials do not produce contamination of waters.
  - R02.02 Water ways and storages provide on-site benefits and prevent off-site sediment or contaminant movements.
  - R02.03 Excavations do not detract from water table levels off-site.
- R03 Rehabilitated lands are suited to beneficial final land uses:
  - R03.01 Established vegetation matches long-term prevailing climate.
  - R03.02 Landforms are stable and non-polluting.

- R03.03 Final site topography is to design and matched to intended land uses.
- R04 Final landforms and closure processes meet community expectations:
  - R04.01 Landforms match consolidated requirements from stakeholders.
  - R04.02 Closure processes transition local communities to an optimum post-closure state.
  - R04.03 Recipients of the site benefit from the transfer.

These operating states outline key elements necessary for successful business operations without specifying controls. The Required Operating States (ROS) form the backbone of the CFw model and support a structured breakdown of control expectations, as illustrated in Figure 3.

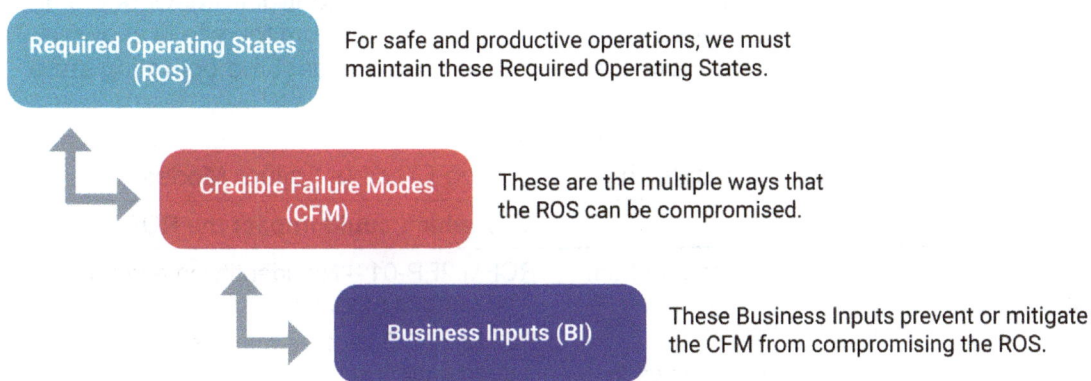

FIG 3 – Graphical representation of a control framework.

The reason the Required Operating States (ROS) are useful is that they provide a way of clearly articulating a robust set of business activities – which need to be in place to address the Credible Failure Modes which can impact this operating state.

Figure 3 shows the relationship between operating states, failure modes, and business input activities.

A key aspect to the Control Framework for a subject area is the exhaustive way in which it has been prepared. The approach draws on:

- Defined using the MECE principle (Mutually Exclusive, Collectively Exhaustive) after Minto (2009; see also McKinsey Quarterly (2025)), the operating states are structured to avoid overlap and ensure completeness. They also consider the requirements of critical control thinking after ICMM (2015).

- Developing Credible Failure Modes based on the collected knowledge from incident data and the principles contained in IEC (2018).

- Clearly described Business Input activities (controls) which are presented in sufficient detail to allow a business to confirm that the measures will be valid and maintained. To this end, activities have a nominated accountable role as the owner, with one or more responsible roles tasked with execution and they have these four elements:

  - Expectation statement – which gives the why of the activity.

  - Specification – providing detail about what is required to meet the expectation.

  - Implement details – which describes how the specification is operationalised eg the training, skills, engineering controls (maintenance etc), construction activities and procedures to follow.

  - Monitoring – the measures by which the quality and quantity of the 'work-as-done' is communicated within the company.

- Management of the whole through the generation of a 'single source of truth' (as popularised by The Open Group (2018)) where elements of the business inputs provide:

- o Components of standards, equipment and design specifications, management plans, procedures, training modules, and assessment tools.

- o Governance measures – with links to requirements of internal standards and external statute and standards.

- o Linkage to other business systems.

- o Support for successful moves to wider digitisation of business process guidance and monitoring.

Samples of a control framework for integrated closure management

Risk Mentor (2025) provides a detailed explanation and list of the control framework elements. The top level required operating states related to water flows to, on and from the site is shown in Table 1. This table also presents the Credible Failure Modes which could impact the operating state.

TABLE 1

Required operating state R02 and related Credible Failure Modes.

ROS Name	Detail	CFM which could impact the ROS
ROS R02 – Water movement at the site does not cause harm	Water flows into, on and from the site are managed. Management achieves: • optimum surface flows to the downstream catchment • containment of any contaminated waters • minimum depression and ultimate recovery of groundwater available to off-site users • removal of sediments from water discharging from the site	RCFM.2EP-01 – Not identifying and removing hazardous materials RCFM.3VD-03 – Tree species not suited to riparian conditions RCFM.3VD-11 – Designs not free draining RCFM.3VD-14 – Not designing for capping of reactive materials RCFM.3VD-15 – Failing to store capping materials during mining operations RCFM.3VD-18 – Failing to isolate final mining voids RCFM.3VD-19 – Water removed from catchment RCFM.3VD-20 – Water from voids discharging to the environment RCFM.3VP-01 – Generation of wastes during decommissioning RCFM.3VP-03 – Sediment release from final landform RCFM.3VP-07 – Insufficient cover allowed for in final landform RCFM.3VP-13 – Contamination of water due to site operational or rehabilitation activities RCFM.3VP-14 – Groundwater seeping from voids RCFM.3VP-15 – Poor water quality draining from site RCFM.3VP-19 – Excavations draw ground and surface water out of catchment RCFM.3VP-20 – Cumulative impact on water from multiple operations RCFM.3VT-01 – Failure to identify contaminated lands RCFM.3VT-03 – Exposing reactive materials during operations

The 18 credible failure modes listed in Table 1 are addressed by one or more Business Input activities. Table 2, presents a subset of these failure modes and the names of some of the Business Inputs which address them.

TABLE 2

Sample of the Failure Modes and the Business Input activities which address them.

Credible Failure Mode	Detail	Related Business Inputs
RCFM.2EP-01 – Not identifying and removing hazardous materials	Rehabilitation and decommissioning process fails to identify and remove hazardous materials such as hydro-carbons, hazardous wastes, rubbish and other wastes etc. This includes establishing species mix or vegetation densities which pose an unacceptably high bushfire risk.	RBI-LRA.41.50 – Surveyed plans of all mine features RBI-SDG.92.80 – Waste Management and Planning for Decommissioning and Rehabilitation Works RBI-SPA.85 – Contractor Management, Plans and Processes
RCFM.3VD-03 – Tree species not suited to riparian conditions	Ecology – flora and fauna related. Tree species established along diversions or other watercourses are not suited to riparian environment. Leads to rehabilitation being inadequate, requiring further works.	RBI-LDG.46.30 – Revegetation planning and associated activities RBI-LPG.55.30 – Cleared vegetation is processed to improve growth media RBI-LRG.41.55 – Database (GIS) of Heritage Items and other features RBI-SDG.92.85 – Mine life considerations included in mine and rehabilitation planning RBI-SRG.91.80 – Monitoring and Adaptive Management protocols RBI-SRG.97.80 – Rehabilitation completion milestones are identified and tracked
RCFM.3VD-11 – Designs not free draining	Final Landform Landform (excluding final void domains) not free draining. This includes a lack of, or inappropriately sited contour banks and other flow paths.	RBI-LDG.46.15 – Erosion and Sediment controls included in Operational and Final Landform Designs RBI-LDG.46.17 – Water management plans prevent release (bunds or mine voids) RBI-LPG.46.10 – Water Management design and implementation RBI-LPG.46.15 – Site drainage lines minimise impact to surrounding environment RBI-LRG.55.20 – Longitudinal studies of rehabilitation effectiveness RBI-SDG.92.20 – Mine Planning and Approval to Mine
RCFM.3VD-14 – Not designing for capping of reactive materials	Rehabilitation Management Inadequate capping or removal carbonaceous or reactive material and subsequently generating ongoing spontaneous combustion or acid drainage events.	RBI-LDG.46.20 – Mine plans include provision of materials for sensitive structures and encapsulation RBI-LDG.46.25 – Mine design considers reactive material in dumps RBI-LPG.46.36 – Stripping and stockpiling procedures RBI-LPG.46.52 – Capping materials are placed on TSF to achieve a long-term stable landform RBI-MDG.75.40 – Spontaneous Combustion Management (Characterisation and Management Processes) RBI-MPG.73.20 – Topsoil Management Plan

Table 3 presents a number of the Business Inputs noted in Table 2. The key point to note with these Business Inputs is that they not only address the failure modes – they also respond to and meet the requirements of relevant license conditions, standards, and clauses of statute. Table 3 demonstrates how selected Business Inputs map to regulatory requirements, showcasing the CFw's alignment with statutory obligations. The rightmost column provides a sample mapping of selected requirements from New South Wales (Australia) standards. References marked **RRA** relate to the NSW Rehabilitation Risk Assessment Guideline (NSW RR, 2021b), while those marked **RRC** refer to the NSW Resources Regulator Guideline on Rehabilitation Controls (NSW RR, 2021a).

TABLE 3

Subset of the Business Inputs which sustain the required operating state and the clauses they address.

BI name	Expectation	Specify	Related clauses
RBI-03D.30 – Subsidence Modelling	Detailed modelling of mining identifies conservative locations and amounts of predicted vertical subsidence, horizontal movement, and strains.	Subsidence predictions are developed to meet site requirements with consideration of: • leading practice modelling for similar strata/rock masses • geological structures and geotechnical (stress levels etc) conditions in the mining area • observations of earlier mining subsidence from current or nearby workings • rigorous surveys and audits of the overlying strata and surface features	RRA08.05 – Subsidence impacts on flora (tree roots etc)
RBI-LDG.46.15 – Erosion and Sediment controls included in Operational and Final Landform Designs	Site final landform designs implemented to include variable slopes (to minimise erosion) and generate slow flow locations on the site (to settle out entrained solids) in water flows leaving the area.	Site requirements for erosion and sediment include: • Mine Technical Services, Environment and Rehabilitation Project leads having accountability for establishing and implementing water management, and erosion and sediment management plans. • Design plans developed and issued that optimise the control of erosion/sediment during active operations and from the final landform. • Operational … [Note: more information available – please follow the link in the Risk Mentor (2025) reference]	RRA04.01 – Poor QA/QC leading to unstable landforms RRA07.07 – Erosion of rehabilitated lands RRC-04.07 – Implement water management to control sediment and optimise outcomes for flora and fauna RRC-04.08 – Any diversions in final landform should be consistent with those watercourses in the locality

To keep Table 3 concise and focused, the full implementation and monitoring details for each listed Business Input have not been included. Interested readers can refer to Risk Mentor (2025) for comprehensive descriptions and operational guidance.

The benefit of applying a Control Framework approach include:

• Building on a structured knowledge base to identify potential problems and workable solutions (business input activities = controls). (This compares to identifying controls based on the knowledge base of the Risk Assessment workshop team).

- Control Frameworks are acquisitive, digital and readily modified.

Creating the bridge to move from matrices to frameworks

Much time, effort, and scholarly writing has been invested in qualitative risk assessments of different flavours. They all suffer the same failings of:

- inconsistent and non-repeatable results (risk scores)
- subjective errors based on the biases of the individuals involved
- limiting the time that a group spends on considering the integrity and resilience of the business activities (controls) that need to be in place to prevent unwanted outcomes.

The Control Framework approach provides a structured method for assessing a site's current control effectiveness and systematically identifying improvement opportunities that align with, or exceed, established good practice benchmarks. For long-lived activities, such as an Integrated Mine Closure project commencing during the active mining phase, multiple reviews of the Control Framework during the project can help to identify where progress is being made and where activities are drifting from good practice.

Executing this rigorous process within the broader Integrated Mine Closure project can lead to some significant outcome benefits. In particular:

- Identifying areas of potential shortfalls in the application and monitoring of Business Input activities.
- Cross fertilising information, approaches and ideas across multiple departments and business functions at the site.
- Demonstrable conformance with regulations, licence conditions and third-party requirements.
- Enhancing resilience in rehabilitation and closure activities to support a smooth and orderly transition of the site to long-term custodianship.

The difference between Control Framework and traditional risk analyses is significant but does not need to generate challenges in demonstrating compliance with the intent of regulator documented requirements. Risk Mentor (2025) contains a sample of a bow tie that aligns with the material presented in NSW RR (2022) where the preventative and mitigating controls are the Required Operating States shared in the section on Control Framework Assessments earlier in this paper.

Operationalising control frameworks in an integrated mine closure project

A sample of the work breakdown structure (WBS) within which the Control Framework process operates is shown in Figure 4. The complete WBS from which this is sourced is available at Risk Mentor (2025).

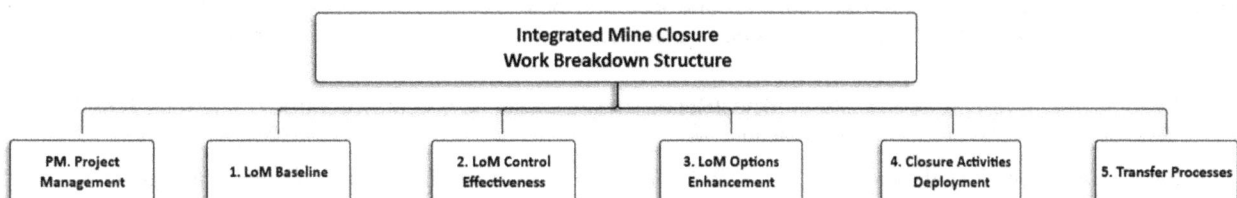

FIG 4 – High-level WBS for an integrated mine closure project.

The adoption of sound project management techniques when delivering the Business Input activities (controls) in the logical project steps that address all of the key site and stakeholder requirements makes delivery of an exemplar mine closure project much more likely. Phase 1, LoM Baseline is where the CFw is tailored to match the site's documented requirements. In Phase 2, LoM Control Effectiveness this documented information is critically reviewed and identified gaps generate opportunities for improvement which are addressed in subsequent project phases.

CONCLUSIONS

The Control Framework (CFw) approach enhances the way risk assessments are conducted across the mining life cycle—particularly for rehabilitation and closure. By focusing on Required Operating States and systematically identifying Credible Failure Modes and Business Inputs, the CFw promotes transparency, repeatability, and better decision-making.

Moving from traditional risk matrices to a CFw approach aligns mining practices more closely with stakeholder expectations, regulatory intent, and industry-leading sustainability standards. Integrating CFw early in the planning process enables operators to proactively manage risks, achieve regulatory compliance, and build resilience into rehabilitation and closure outcomes.

As mining operations face increasing scrutiny and complexity, structured, scalable, operationally integrated frameworks such as CFw that can evolve and iterate over time are essential tools for delivering on both business and societal expectations.

ACKNOWLEDGEMENTS

The author acknowledges the contributions of clients, site personnel, and regulatory colleagues whose engagement and feedback helped shape the Control Framework approach. Particular thanks go to the ICMM, their working groups and peer collaborators whose support continues to enhance the development of structured, practical tools for life-of-mine risk management.

REFERENCES

Hubbard, D, 2009. *The Failure of Risk Management: Why It's Broken and How to Fix It*, pp 68–71; Exhibit 4.2, p 74 (John Wiley and Sons).

International Council of Mines and Metals (ICMM), 2015. *Health and Safety Critical Control Management: Good Practice Guide* (ICMM: London).

International Council of Mines and Metals (ICMM), 2019. *Integrated Mine Closure, Good Practice Guide*, 2nd edition (ICMM: London).

International Electrotechnical Commission (IEC), 2018. IEC 60812:2018 – Failure Modes and Effects analysis (FMEA and FMECA) Standard, edition 3.0.

McKinsey Quarterly, 2025, 16 May. Barbara Minto: 'MECE: I invented it, so I get to say how to pronounce it' [online], Alumni News, McKinsey Quarterly. Available from: <https://www.mckinsey.com/alumni/news-and-events/global-news/alumni-news/barbara-minto-mece-i-invented-it-so-i-get-to-say-how-to-pronounce-it> [Accessed: April 2025].

Minto, B, 2009. *The Pyramid Principle: Logic in Writing and Thinking* (Financial Times/Prentice Hall).

NSW Resources Regulator (RR), 2021a. Guideline: Rehabilitation Controls, NSW Resources Regulator, published 2 July 2021. Available from: <https://www.resources.nsw.gov.au/sites/default/files/documents/guideline-rehabilitation-controls.pdf> [Accessed: March 2025].

NSW Resources Regulator (RR), 2021b. Guideline: Rehabilitation risk assessment, NSW Resources Regulator, published 2 July 2021. Available from: <https://www.resources.nsw.gov.au/sites/default/files/documents/guideline-rehabilitation-risk-assessment.pdf> [Accessed: March 2025].

NSW Resources Regulator (RR), 2022. Material and landform bow tie, NSW Resources Regulator. Available from: <https://www.resources.nsw.gov.au/sites/default/files/documents/rehab-bowtie-1-mining-phase.pdf> [Accessed: April 2025].

Risk Mentor, 2025. Overview of the Life of Mine (Rehabilitation and Mine Closure) Control Framework, Risk Mentor. Available from: <https://riskmentor.com.au/LoM-control-framework/> [Accessed: April 2025].

The Open Group, 2018. TOGAF, version 9.2, The Open Group Architecture Framework (TOGAF®) Standard.

Western Australia Government, 2025. DEMIRS Risk Assessment Template [online], Department of Energy, Mines, Industry Regulation and Safety. Available from: <https://www.wa.gov.au/government/publications/demirs-risk-assessment-template> [Accessed: March 2025].

Integrating renewables and resources – governance challenges and opportunities

C Tomlin[1], M Press[2] and J Maunder[3]

1. Partner, ERM, Brisbane Qld 4000. Email: charissa.tomlin@erm.com
2. Managing Consultant, ERM, Brisbane Qld 4000. Email: mifanwy.press@erm.com
3. Principal Consultant, ERM, Brisbane Qld 4000. Email: josh.maunder@erm.com

ABSTRACT

Co-locating renewable energy assets on mine sites presents an emerging opportunity for the resource sector to decarbonise and strive towards net zero targets. In addition to the notable environmental and social benefits, economic opportunities may also be extracted to create additional revenue streams for mining operations. In Queensland, approval requirements for resource projects present an obstacle to this opportunity being realised. The current governance arrangements that regulate mining and non-mining development on mining tenements provide challenges to planning and executing co-located decarbonisation projects. This paper explores the complexities of regulatory frameworks where the renewable and resource sectors meet and highlights alternative approval pathways that can be used to unlock these opportunities. In particular, we draw on how this approach is being practically applied to a wind farm project, which is co-located within an operational metal mine in North-west Queensland.

INTRODUCTION

Renewable energy assets, whilst not a new concept, are a reasonably new land use in Queensland. Legislation has been playing catch-up for the past several years, with successive State Governments making ongoing changes to the way these projects are assessed and approved. Whilst clarity in the assessment process has emerged through State Codes and clearer land use definitions within legislation, there exists an impending tension between traditional development approval processes and those relevant to resource projects. Significant changes in the way wind and solar projects are assessed and approved occurred in early 2025 adding a new complexity to the process.

Resource projects in Queensland are typically approved through two key pieces of legislation, the *Mineral Resources Act 1989* (MR Act) and the *Environmental Protection Act 1994* (EP Act). The MR Act grants the relevant resource tenure (eg Mining Lease), whilst the EP Act grants the overarching Environmental Authority (EA). Under the MR Act, resource projects benefit from an exemption which negates traditional development approval processes assessed under the *Planning Act 2016* (PA) from applying. This exemption has resulted in a lack of consideration for how non-resource aspects, including renewable energy assets, can be integrated into resource approvals. Up until the last 5–10 years, the thought of a renewable energy asset being co-located with a resource project was likely never anticipated and as such, has presented a number of constraints never before considered. As the resource sector continues to implement decarbonisation initiatives, this consideration requires an urgent re-think. Rethinking the approach to co-location of renewable energy assets with resource projects is an important consideration for decarbonisation and has direct implications for energy security and closure opportunities.

CURRENT CHALLENGES

The Queensland Environmental Regulator, who is responsible for assessing and approving EA's for resource projects and governing the corresponding legislation, is not resourced to assess renewable energy projects. Ultimately, whilst it is a broad and powerful Act, the EP Act simply does not provide a clear and articulated mechanism for the Regulator to accurately and clearly assess these sorts of applications. For example, applying an approval framework that is designed to regulate emissions and environmental harm, to infrastructure that is better regulated through the land use approvals process, poses challenges for not only the Regulator, but also overestimates the operational risk that is presented by renewable energy infrastructure. There are corresponding compliance-based risks which also create complexity and uncertainty. There are additional considerations relating to closure, progressive rehabilitation and financial liabilities.

It is appreciated that the intent of providing a mechanism which facilitates the location of renewable energy infrastructure on Mining Leases (MLs) and held on EAs, was to streamline process for mines to progress their decarbonisation ambitions. However, due to the limitations of this process, the full benefit is not currently being realised. The ultimate result of the MR Act and EP Act process is that projects that have embarked down this pathway have been met with highly prescriptive Conditions and complex rehabilitation liabilities tying the renewable energy asset to the mines closure, in some cases, making them unviable or economically unfeasible.

Part of the reason for this is, from a land use and legality perspective, is that the MR Act does not provide the lawful mechanism for the export of power through a resource tenure. That is to say, the tenure does not allow energy export and ultimately requires a proponent to find an alternative avenue. This limits not only the economic benefit for the proponent, but also the ability of the project to contribute to State and Commonwealth decarbonisation commitments during mine operation and beyond the mine's operational life. The solution to this is to obtain a development approval for the renewables project via the PA, which involves resolving land tenure through the *Land Act 1994* (Land Act), whereas in most cases, the State of Queensland is the landowner for land where resources projects are located. Processes under the Land Act must also consider Native Title, which presents another layer of complexity. Given the vast majority of resource projects being located on State Land that are subject to leaseholder arrangements, this creates additional complications and one that is difficult, costly and timely to navigate.

Alternatively, there are benefits that this process provides, which include the ability to export electricity to the grid, which can generate an additional revenue source for the mine and decouple assets. This allows for potential future re-energisation of the renewable asset life for continued revenue generation and/or sale of the asset as a complete package, as to not impact mine closure timing or the estimated rehabilitation cost. Broadly speaking, the co-location of renewable energy assets with resource projects has the potential to strengthen and enhance the economic and social benefits flowing into remote and regional communities.

Whilst it is acknowledged that recent legislative changes to the PA have been implemented bringing increased alignment between the EP Act and PA process with both now subject to public exhibition/notification, the benefits provided by the PA process have not been realised in the EA process. This is not to say that outside of the governance arrangements, there are not challenges that are inherent to the co-location of renewables and mining activities.

The nature of renewable energy assets require any application to be supported by a design that has a defined project footprint, with limited ability to build in flexibility. This poses a challenge as the need for flexibility exists to ensure projects can be refined through the detailed design stage in such a way that the co-location of the long-term viability of the resource project itself (ie sterilisation of any resource) is not compromised. The complexities from a structural engineering front alone are significantly more complex than that of a traditional project, given the ongoing potential for vibration impacts from blasting as an example, meaning:

- The location of any renewable energy project needs to consider the mine footprint, including locations of potential future extension.

- Engineering details for any proposed renewable project must consider the mining activity to ensure structural stability of the new infrastructure.

Furthermore, whilst mining operations hold a significant depth of information relating to the environmental, socio-economic and cultural values within their ML, disconnects exist between what surveys and monitoring have been undertaken for the purpose of an EA, and what is required to be submitted with an application for renewable energy (in this instance, wind farms). For example, bird and bat utilisation surveys, landscape and visual impact assessments and electromagnetic interference assessments are a common information gap for these types of projects and would not typically be required for a mining approval. However, such information can cause wind farm project approval delays, if not considered or collected early.

In response to the current governance and information challenges, there are a number of alternative actions that can be implemented and undertaken for proponents navigating a similar pathway and progressing towards mine decarbonation efforts. These have been further explored in the renewables roadmap outlined below.

RENEWABLES ROADMAP

The integration of renewable energy projects within MLs presents a significant opportunity to reduce carbon emissions, lower energy costs, provide an additional revenue source and enhance the sustainability of mining operations. However, careful planning and execution are crucial to ensure successful implementation and minimise potential conflicts with ongoing mining activities and to select the approval pathway of greatest benefit for operations. This roadmap presented outlines key considerations and steps in the planning and decision-making process to facilitate approvals success and utilises the pathway taken by ERM in the planning process (Figure 1).

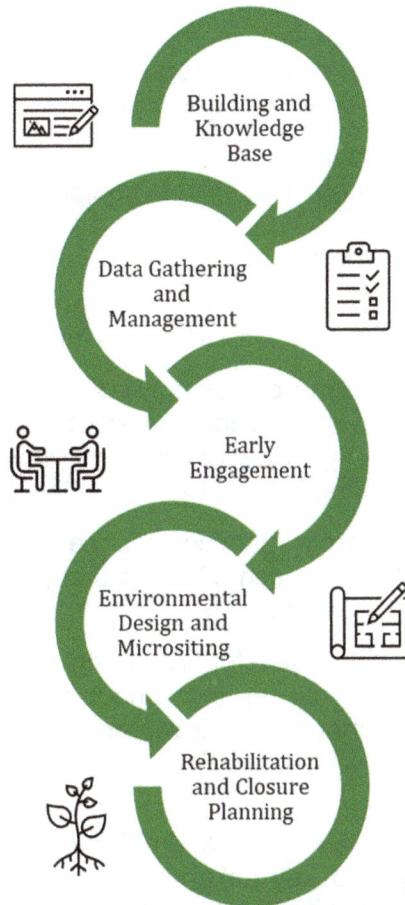

FIG 1 – The Renewables Roadmap.

Building a knowledge base

It is broadly accepted that a mining operation collects of suite of environmental, socio-economic and cultural information from the onset to support approvals processes and mine development. To account and resolve the potential information gaps between information the mine holds and application requirements, understanding of the existing and needed knowledge base is recommended. The following actions can be implemented to progress this stage:

- **Early-Stage Research**: Before any project development, establish a comprehensive understanding of the following:

    o *Renewable Energy Potential:* Assess the specific renewable resources available on or near the mining lease, including solar, wind, geothermal, and biomass. Conduct resource assessments to determine the viability and potential energy output of each resource.

- *Energy Demand Profile*: Analyse the current and future energy needs of the mining operation. This includes electricity consumption, heating and cooling requirements, and fuel usage for mobile equipment.

- *Regulatory Framework:* Identify all relevant regulations, permits, and approvals required for renewable energy projects within the specific jurisdiction. This includes environmental regulations, land use policies, and grid connection requirements.

- *Technological Options:* Evaluate the various renewable energy technologies available, considering their suitability for the specific site conditions, energy demand profile, and economic feasibility.

- *Land availability:* Evaluate the location, extent and availability of land that may house renewable energy without conflicting with mining operations.

- **Approval pathway selection:** Based on the site need, energy demand profile, energy generation potential and grid connection, opportunity determination can be made on the most appropriate approvals approach which will generate the best outcome for operations.

- **Gaps Analysis:** Based on approvals approach, determine what data is available from mining operations and any information gaps that exist to comply with renewable specific guidelines.

Data gathering and management

When the generalised approvals pathway is determined and preliminary 'study' area has been determined, information should be gathered to fill identified information deficits. This stage involves:

- **Infield data gathering:** Undertaking contemporised baseline assessments of the proposed project area (including the area around the project) to fill identified data gaps that will be required to be filled for supporting future applications. Opportunities exist to align this data gathering to site existing monitoring activities to streamline costs of mobilisation.

- **Impact modelling:** If available, preliminary compliance modelling can be undertaken to guide project design including the identification of areas that pose potential compliance issues or could be considered sensitive receptors.

- **Data Management:** Implement a system for organising and managing the collected data. This will facilitate informed decision-making throughout the project life cycle and create a comprehensive data repository that can be drawn on.

Early engagement

Upon further certainty of the project description and selected approvals pathway, identification of land tenure requirements and mapping of key stakeholders should be undertaken. Early consultation on both matters will ensure stakeholder buy-in, as well as early identification of conflict factors so that time and resources are not wasted on progressing a design and project description that is not agreeable:

- **Land Tenure:** Clearly identify all landholders within and adjacent to the mining lease area. This includes:

  - The mining company holding the lease.

  - Traditional Owners or Indigenous communities with native title rights. Due to the ongoing mining operations, it is likely a relationship exists with the Traditional Owners or Indigenous communities which can be utilised to commence conversations regarding any proposed project through upfront and informal mechanisms, prior to formalised negotiations occurring.

  - Government agencies responsible for land management (the government agency is likely to be approvals pathway specific).

  - Adjoining landowners and landholders, such as farmers or pastoralists.

- **Overlapping Rights:** Determine if there are any overlapping land rights or interests, such as resource tenures, existing easements, grazing rights, or cultural heritage sites.

- **Key stakeholder identification:** Identify key stakeholders for the project, including:
  - Leaseholders and/or landholders.
  - Traditional Owners and Indigenous communities.
  - Government agencies.

Wider stakeholder engagement

Engagement with the wider stakeholder community is critical to success of the project due to the nature of public exhibition and notification. Whilst it is generally recommended to consult early and honestly, timing considerations should be provided to the level and maturity of information available to be provided to communities. Engaging stakeholders too early may raise the risk of producing more questions answers, and it is not recommended that key stakeholder groups discover project information as part of wider consultation. The existing relationships and platforms that mining companies have with the local stakeholders can facilitate ease of these conversations.

At the time of publishing this paper, stakeholder engagement requirements for wind and solar projects in Queensland are subject to proposed legislative changes mandating the completion of a Social Impact Assessment and Community Benefit Agreement. Should these reforms be passed as currently proposed, the implementation of stakeholder engagement is likely to become one of the most critical aspects of any project, if not the most critical. The proposed legislative changes provide a more refined mechanism for Ministerial input into wind and solar projects, effectively providing an avenue for a refusal to be given on a community perception basis, irrespective of the broader impact assessment outcomes.

To ensure engagement proceeds effectively, the below considerations should be made:

- **Principles of Engagement:** Adopt a proactive and inclusive approach to stakeholder engagement, based on the principles of:
  - Early and ongoing communication.
  - Transparency and information sharing.
  - Respect for diverse values and perspectives.
  - Collaborative problem-solving.
  - Free, Prior, and Informed Consent (FPIC) with Indigenous communities.
- **Stakeholder Identification:** Identify all relevant stakeholders, including:
  - Local communities.
  - Key locations within line-of-sight of the proposed project.
  - Grid operator and communications link owners.
  - Environmental organisations.
  - Industry associations.
  - Local businesses and facility operators (eg airports).
- **Engagement Activities:** Implement a range of engagement activities, such as:
  - Initial meetings and consultations to introduce the project and gather feedback.
  - Regular progress updates and information sessions.
  - Workshops and focus groups to discuss specific issues and concerns.
  - Review and discussion regarding potential new or amended Indigenous Land Use Agreements (ILUAs).
  - Formal consultation processes as required by regulations.
  - Grievance mechanisms to address any complaints or disputes.

Environmental design and micrositing

Project layout refinement and design flexibility through the detailed design stage of the project, should be considered in the context of all available data and information available from consultation activities. The iterative design process will also assist with the demonstration of avoid-minimise/mitigate hierarchy aspects fundamental to any impact assessment, regardless of the approvals pathway that is chosen for the project. To ensure appropriate implementation, the following steps will be key:

- **Environmental Impact Assessment (EIA):** Conduct a comprehensive EIA to identify and assess the potential environmental impacts of the renewable energy project. This should include:

  o Impacts on biodiversity, including flora and fauna.

  o Impacts on water resources.

  o Impacts on soil and land use (specifically impacts to agricultural land).

  o Impacts on cultural heritage sites.

  o Visual and noise impacts.

- **Mitigation Measures:** Develop and implement appropriate mitigation measures to minimise or avoid negative environmental impacts. This may include, but is not limited to:

  o Careful site selection to avoid sensitive areas.

  o Implementation of erosion and sediment control measures.

  o Noise reduction strategies.

  o Habitat restoration and rehabilitation.

  o Cultural heritage management plans.

- **Micrositing:** Optimising the location and layout of the renewable energy infrastructure to minimise environmental impacts and maximise energy production. This involves:

  o Considering the topography of the site.

  o Avoiding areas of high ecological value.

  o Minimising visual impacts on the surrounding landscape.

  o Optimising already disturbed areas, access roads and transmission lines.

By following this roadmap and prioritising early planning, stakeholder engagement, and environmental design, mining companies can successfully integrate renewable energy projects into their operations and optimise the governance complexity, to best suit their operations. This in turn, can create a more sustainable and economically viable project and future of the operations.

Micrositing is also an important design consideration when co-locating renewable energy assets with resource projects to ensure the sterilisation of available resources does not occur. As the Life-of-mine Plan evolves, as is the case for every resource project, flexibility in the mining operation is necessary to allow projects to pivot in response to operational requirements. Sensitivity in the location of power generating assets is therefore a key consideration of any project and must be undertaken at the early stages of a project.

Consideration of rehabilitation and closure obligations

Rehabilitation and closure of resource projects remains one of the largest and most complex undertakings mining operations must complete. Over the past few years, changes to the rehabilitation framework in Queensland through the implementation of Progressive Rehabilitation and Closure Plans (PRCPs), means closure and rehabilitation objectives are front of mind during the operational life of a resource project. Consideration should be given if a mine already has an approved PRCP and whether a renewables project would likely require a major amendment to the existing PRCP, as well as Environmental Authority (EA). This is because it introduces new disturbance areas, infrastructure, revisions of rehabilitation areas and potential alterations to the post-mining land use for portions of the site. Furthermore, the Queensland Wind Farm Code (effective 3 February 2025) requires detailed decommissioning plans to be prepared, describing how decommissioning activities will ensure no adverse impacts on individuals, communities, or the natural environment. This directly impacts the PRCP, as the decommissioning plan will form a critical part of the overall closure strategy.

Co-location of renewable energy assets presents both opportunities and challenges that need to be overcome to ensure there are no net-negative impacts on closure and rehabilitation requirements. It will require comprehensive updates to the PRCP to encompass the new disturbance and rehabilitation commitments, expands the scope of rehabilitation obligations to include specialised wind farm decommissioning, and could increase the Estimated Rehabilitation Cost (ERC), potentially leading to a larger financial provision under Queensland's regulatory framework. It is imperative that the long-term implications on closure and rehabilitation are quantified and delineated to ensure a clear separation of liability between the resource project and the renewable energy asset.

CONCLUSION

The resource sector provides significant opportunities and advancements for decarbonisation initiatives. Most medium to large-scale resource companies have actively pursued and continue to pursue a decarbonisation agenda, supported by substantial information for their application processes. However, the very governance processes intended to drive renewable transition paradoxically make securing project approvals administratively complex and challenging. As companies increasingly recognise the value of the untapped renewable resources they can access, a clearly defined approach to navigating this process will be highly beneficial.

Whilst opportunities exist to refine legislation to resolve this issue, such a solution may not be sufficiently timely given the committed decarbonisation timelines and the lead times associated with approvals and construction. To navigate the existing regulatory environment and regimes, the presented roadmap offers a framework of considerations that can be adopted early to provide some process certainty.

REFERENCES

Department of Environment, Tourism, Science and Innovation, 1994. *Environmental Protection Act 1994.*

Department of Natural Resources and Mines, Manufacturing and Regional and Rural Development, 1994. *Land Act 1994.*

Department of Natural Resources and Mines, Manufacturing and Regional and Rural Development, 1989. *Mineral Resources Act 1989.*

Department of State Development, Infrastructure and Planning, 2016. *Planning Act 2016.*

Above and below the residual risk iceberg – insidious risks of life-of-mine, mine waste and tailings

C J Unger[1]

1. Research Fellow, Centre for Social Responsibility in Mining, Sustainable Minerals Institute, The University of Queensland, St Lucia 4067; and UQ Business School. Email: c.unger@business.uq.edu.au

EXTENDED ABSTRACT

Insidious risks are those risks that develop and worsen slowly and non-obviously during the operating life of a mine but become visible, obvious and urgent as closure draws near. At this time, the work required and associated costs of rectifying and containing insidious risks becomes clearer and quantifiable. While containment of insidious risks greatly reduces the magnitude of culminating residual risks at closure, they can persist after mine closure. Consider the following examples of insidious risks and their consequences for residual risk after mining ceases; risk of an invasive pest or weed species moving into restored mine land ecosystem destabilising it; a post-mining land use that has long been the agreed regulatory requirement but is not aligned with community and land owner expectations; a tailings facility that is stable while actively managed but cannot be left unmanaged, even when capped and closed, requiring in perpetuity management of cover erosion and seepage; adverse water quality (eg acid and metalliferous drainage and/or saline) is worse than predicted and in greater volumes than anticipated, adding to closure costs, management requirements, and delaying sign-off (ICMM, 2025).

Commonly adopted risk assessment frameworks address prevention of familiar risks that can suddenly materialise to harm in an accident or crisis (Reason, 2013). Even if they include environmental and social risks they are presented as sudden events (eg spills or conflicts). However, these tools and methods do not facilitate detection or management of slow developing and less obvious mining environmental and social risks within the flow of everyday work. Having developed incrementally over many years even decades insidious risks can suddenly become problematic at closure creating large residual risks. We label as 'insidious' these risks that have been studied in other contexts as latent, creeping or incubating risks (Boin, Ekengren and Rhinard, 2020; Ramanujam and Goodman, 2003; Turner, 1978). We view residual risk as the outcome of insidious risk management (IRM) processes over the whole life of a mine and use the iceberg analogy to distinguish between the residual risk we can see and quantify at the end of mining from the prior intangible IRM organising processes that create it.

After mine rehabilitation and closure are completed, residual risks can threaten safety, stability, pollution containment, sustainability and acceptability of post-mining land uses. Regulators are increasingly looking for methods of evaluating residual risk as part of the relinquishment process of mining tenure, to ensure these unmanaged risks are resourced. However, this measurable outcome or liability after mining, may come at a time when it is too late to intervene if the residual risk is much higher than anticipated. Hence, this research explains the process of managing insidious risks during the active life of a mine when residual risk is being created. Our research provides mining companies with early insights on the trajectory of IRM processes, toward qualitatively high or low residual risk. Armed with this understanding companies can intervene and redirect IRM processes toward low residual risk. This means recognising hidden dimensions of insidious risks and liabilities well before cessation of operations, and in an ongoing way, to meet relinquishment obligations with low residual risk.

The iceberg analogy highlights the invisible and gradual development of insidious risk below the surface that is only partially visible to scientific methods commonly applied to life-of-mine rehabilitation, closure and waste management. By applying this organisational, social science, lens to these critical aspects of mining sustainable development, this research identifies the eight sets of activities that comprise IRM and what they accomplish in terms of making insidious risks more, or less, visible, their consequence more, or less, clearly understood and methods of containment more, or less, effective. This study takes account of what practitioners do interacting with each other within and beyond the organisation, with regulators and stakeholders, and with their environment.

INSIDIOUS RISK RELATED LITERATURE

The gradual process of insidious risk materialising to harm is neglected in the literature. Only a few scholars have moved beyond the sudden unwanted crisis when risk is made obvious, to study the gradual and less obvious process of risk. Each argues that they reveal a different type of risk that is either latent (Ramanujam and Goodman, 2003), incubating (Turner, 1978), creeping and expanding (Beamish, 2002; Boin, Ekengren and Rhinard, 2020), slow-developing catastrophic (Nursimulu, 2015) or novel (Hardy and Maguire, 2020). Rather than separate risk phenomena, this research proposes they are instead related features of insidious risk and that a study of insidious risk might also explain the difficulties in explaining other slow developing risks. Risk assessment tools designed for sudden risks materialising to harm are ineffective for slow-developing and less obvious risks (eg safety tools for health risks; Cliff *et al*, 2017; Hopkinson and Lunt, 2014). In this study the term *insidious risk* is used for slow-developing and less obvious risks that involve practitioners using equipment and tools and interacting with the environment. In the case of mining, insidious risks of land disturbance, mine affected water, voids and waste management for instance, must be managed over the full life of a mine and sometimes beyond, to effectively transition to sustainable and beneficial post-mining uses and meet societal and regulatory expectations. These risks and their management are not confined to the organisation but involve interactions beyond the boundaries. Insidious risks are not constant but evolve over time with potential to be crisis-creating.

RESEARCH METHOD

This study examined a variety of ways insidious risk is managed to reveal the complex and less visible social process that culminates in particular outcomes. In order to understand insidious risk we undertook two studies exploring three cases. In the first, we analysed documented accounts of six examples where insidious risks were managed in contrasting ways: i) those culminating in socio-environmental crises with inquiry reports we could analyse; as well as ii) those successfully avoiding a socio-environmental catastrophe or 'leading practice' as a second case. Analysis and interpretation then identified the activities common to both cases, to show how the different outcomes resulted. Then in a second study, the author observed environmental management of land disturbance and mine affected water at an open cut coalmine, interviewed participants and analysed documents that spanned several decades. This gave a contemporary and longitudinal understanding of evolving IRM organising processes.

FINDINGS

From the two studies and three cases, we found not just one way of managing insidious risks but, three ways situated on a spectrum: blinkered, law-abiding and attentive. The three examples of blinkered IRM culminated in a tailings dam failure, a coalmine fire and misclassified metalliferous wastes, while three examples of attentive IRM created sustainable new landforms for biodiversity and community recreational use, and ensured long-term containment of tailings in below-ground storage. When insidious risks of land disturbance, mine affected water, mine waste and tailings, community relations and closure expectations are managed attentively, these risks are detected early, thoroughly understood, monitored and contained to create low residual risk and avoid crises. But the opposite is true of blinkered IRM that allows insidious risks to worsen and develop imperceptibly, with often surprising negative consequences of a sudden crisis and high residual risk.

Then in the field-case at a coalmine in Queensland I studied IRM organising processes of rehabilitation of land disturbance and managing mine-affected water. With fieldwork undertaken on two occasions 15 months apart during an intensifying phase of law-abiding IRM, it was possible to see how insidious risk became more visible, how their unknown consequences were examined, how insidious risks extended beyond the temporal, spatial and social boundaries of organisations, and finally how otherwise hidden value, for the organisation and stakeholders, was uncovered. This intermediate IRM process fluctuated in intensity over several decades: sometimes drifting toward blinkered and later intensifying to become attentive in response to regulatory reforms and legal deadlines on mining tenure.

In addition to detecting trajectories of IRM process we identified the sets of activities that comprised the organising process. For instance, we noted that, while *planning* comprises a set of activities regulators and industry rely upon for mine rehabilitation and closure in seeking low residual risk, we found that planning co-exists with seven other sets of activities in IRM processes. Further, all eight sets of activities together, are crucial for managing socio-environmental insidious risks. Generalising from case studies were are able to show how these sets of activities are already being carried out by mining operations, but in different ways. Importantly, we show how these eight sets of activities are carried out within the spectrum from blinkered to attentive, and can detect trajectories toward high or low residual risk and explain what is required to achieve low residual risk during ongoing life-of-mine operational activities.

DISCUSSION AND CONCLUSIONS

This study identified a spectrum of ways of managing insidious socio-environmental risks of mining that explained how different ways of executing key activities gave divergent outcomes. The ways of managing spanned: i) leading practice that creates beneficial and sustainable post mining uses – 'attentive'; to ii) sub-standard practice that creates growing liability and socio-environmental harms leaving persistent negative mining legacies – 'blinkered'; and in between iii) a law-abiding IRM that merely complied with regulatory requirements, and nothing more. The IRM process during mining and large, long-term, rehabilitation and closure processes, incrementally delivers post-rehabilitation outcomes that create from small to large residual risks. While the rigorous and responsible conduct of attentive IRM is most likely to deliver sustainable development post-mining with low residual risk, blinkered IRM is short-term and narrowly focused so its sets of activities are on a trajectory to an unsustainable crisis such as large unfunded liabilities at the end of mine life, persistent impacts to be managed in perpetuity, and/or an abandoned mine and related mining-dependent communities. The circumspect, regimented approach of the law-abiding way of managing insidious risks highlights the importance of regulatory standards and enforcement to protect against socio-environmental harms from mining while also showing that low standards will allow insidious risks to accumulate and worsen, while high standards afford greater protection from harm for those companies seeking only to comply. Nevertheless, it is noted that law-abiding IRM resists regulatory reforms that require higher closure standards, arguing that governments are 'shifting the goal posts' whereas attentive IRM goes beyond compliance requirements. Consequently, the study shows the value of high internal organisational standards rather than reliance upon government-imposed regulations. This research advances recognition that risk management does not fall into only two categories of success or failure (Corvellec, 2009) and instead is explained by a full spectrum of ways risks can be managed that yield different outcomes.

Instead of risk assessment matrices and tools that omit IRM, we recommend auditing and review of how key activities are carried out within IRM to anticipate outcomes in advance, rather than wait until closure. The myopic and cursory way typical of blinkered IRM is a crisis-in-the-making. In contrast, attentive IRM processes will perform activities in a conscientious and meticulous way avoiding crises and yielding low residual risk. Law-abiding IRM processes may find themselves needing to rapidly accelerate efforts to meet closure requirements having delayed rehabilitation or completed easy areas leaving difficult pits until near the end. We urge further inter-disciplinary and longitudinal perspectives of what mining practitioners are doing and whether they seek to collectively detect and contain the inconspicuous insidious risks by looking beyond organisational boundaries and participants to stakeholder insights on risk in IRM processes and staying ahead of regulatory requirements.

This understanding of how insidious risks develop over decades under collective management and through routine operations prompts new questions for risk managers. No longer is a toxic culture, careless worker, faulty equipment, tipping point or 'freakish' event the problem (Taleb, 2007). Rather organisations should examine how enviro-social insidious risks emerge, who, within and beyond the organisation, decides what is a risk or not, and identifies risks, which risks and over what time frames are prioritised by management and what may be contributing to overlooking certain insidious risks. Risk assessment frameworks are inadvertently prioritising sudden and familiar risks while overlooking insidious risk detection, assessment and management. Going beyond compliance and being constantly alert to small anomalies and investigating them, allowing sufficient time, space and resources to managing insidious risks and respecting the values of others related to these risks is likely to be rewarded with social approval to mine in new locations as expanded mining is heralded to meet the demands for critical raw materials in the green energy transition.

ACKNOWLEDGEMENTS

PhD supervisors, Professor Jorgen Sandberg, UQ Business School and Dr Jo-Anne Everingham, CSRM.

Note: this extended abstract is based on a paper under review for a journal and as such cannot be written as a full paper for this conference.

FUNDING

QRC and Queensland Government Coal Mine Minesite Rehabilitation Trust Fund postgraduate scholarship.

REFERENCES

Beamish, T D, 2002. *Silent Spill: the Organization of an Industrial Crisis* (MIT Press).

Boin, R A, Ekengren, M and Rhinard, M, 2020. Hiding in plain sight: conceptualizing the creeping crisis, *Risk, Hazards and Crisis in Public Policy*, 11(2):116–138.

Cliff, D, Harris, J, Bofinger, C and Lynas, D, 2017. The Application of Risk Management Methods to the Control of Respirable Dust in Underground Mines, presented at the Minesafe International 2017 Conference, Perth (AusIMM).

Corvellec, H, 2009. The practice of risk management: Silence is not absence, *Risk Management*, 11(3–4):285.

Hardy, C and Maguire, S, 2020. Organizations, Risk Translations and the Ecology of Risks: The Discursive Construction of a Novel Risk, *Academy of Management Journal*, 63(3):685–716.

Hopkinson, J and Lunt, J, 2014. In the same breath, *Safety and Health Practitioner*, 32(4):31.

International Council on Mining and Metals (ICMM), 2025. *Integrated Mine Closure, Good Practice Guide*, 3rd edition.

Nursimulu, A, 2015. Governance of slow-developing catastrophic risks: Fostering complex adaptive system and resilience thinking, 75th International Risk Governance Council.

Ramanujam, R and Goodman, P S, 2003. Latent errors and adverse organizational consequences: a conceptualization, *Journal of Organizational Behavior*, 24(7):815–836.

Reason, J, 2013. *A Life in Error: From Little Slips to Big Disasters*, (Ashgate Publishing Ltd).

Taleb, N N, 2007. *The Black Swan: the Impact of the Highly Improbable*, 1st edn (New York: Random House).

Turner, B A, 1978. *Man-made disasters* (London: Wykeham Publications Ltd).

Uncertainty analysis of drainage flow modelling in goafing zone for underground mine development

X Wang[1]

1. Lead Groundwater Modeller, KCB, Brisbane Qld 4000. Email: xwang@klohn.com

ABSTRACT

Modelling pore pressure changes and drainage flow-through the fractured overburden above longwall panels is essential for subsidence assessment and dewatering planning in underground mine development. However, the fracturing and hydraulic characteristics during the mining process remain uncertain. This paper examines uncertainties in specifying hydraulic properties in the goafing zone above longwall panels within groundwater modelling during longwall operations. Parameter changes across the height above the panels and throughout mining stress periods were implemented within plausible ranges and constrained by parameter bounds from the literature. PEST-IES, an effective uncertainty analysis approach, was used to estimate parameter probability distributions across both space (height) and time. Uncertainty analysis was conducted by quantifying the percentiles of modelled pore pressure and dewatering rates to parameter variations in the goafing zone. Additionally, a detailed uncertainty and risk assessment was performed using an ensemble of model outcomes to support site operations and water management planning.

INTRODUCTION

Dewatering in underground coalmines has been widely studied and is now recognised as being closely linked to surface water loading in shallow aquifers and the fracturing of overburden above mine workings in many Australian coalfields (Adhikary, Poulsen and Khanal, 2017; Frenelus, Peng and Zhang, 2021). Significant ground deformation and water inflows may occur if the interface between active aquifers and mining zones is not clearly defined or if mine water is poorly managed. High uncertainty in understanding of this interface could lead to severe consequences, such as environmental and constructive issues that imply high costs in the quarry, and social alarm. Therefore, accurate simulation and prediction of mining-induced water inflows and aquifer interference are critical for ensuring mining safety and economic viability.

Recent studies in groundwater modelling have allowed simulation of rock fracturing and stress redistribution around longwall panels with increasing confidence (Ditton and Merrick, 2014; Merrick, 2017). However, simulation of groundwater behaviour in such fracturing processes is still challenging work, since the variation of hydraulic properties being identified is highly dynamic and localised (Klenowski, 2000; Adhikary, Poulsen and Khanal, 2017). The characterisation of hydraulic connectivity and storativity in fracturing zones is key, but remains highly uncertain in the simulation (Tammetta, 2015; Merrick, 2017; Turnadge, Mallants and Peeters, 2018). Recent studies in parameter estimation (PEST) have shown that the quantification of this uncertainty could improve the representation of groundwater behaviour within fracturing zone and the simulation of flow drainage in longwall panels (Doherty and Hunt, 2010; Moore et al, 2015; Middlemis and Peeters, 2023).

The objective of this paper is to present a numerical trial evaluating how groundwater inflow-related factors may mitigate the impacts of fracturing associated with longwall extraction. Simulation results are compared against field observations and published data.

MODEL SET-UP AND PURPOSE OF THE ANALYSIS

A groundwater model was developed for assessing the impact on the groundwater system from the development of a mine that including open cut pits and underground mining (longwall panels) in the Bowen Basin (KCB, 2024). The model was built in a MODFLOW-USG model with 22 layers and 1 210 880 active cells in a regular gird. The model domain and hydrogeological conceptualisation for this trial were based on the regional groundwater model.

The model was calibrated with regional groundwater monitoring data in both steady state and transient cases. The statistics from the transient calibration are illustrated in Table 1. Based on

Australia Groundwater Modelling Guideline (Barnett *et al,* 2012), the metrics in the table are within the acceptable bounds of criteria. Hence, the calibrated model was deemed to be adapted for assessing regional groundwater impacts of mining operations, and groundwater recovery following cessation of mining activities within the hydrogeological study area. However, the maximum residual of 25.2 m between modelled and observed groundwater levels in vibrating wire piezometers (VWPs) above longwall panels showed that the uncertainty in groundwater simulation in this area has remained.

TABLE 1

Summary of regional model calibration performance.

Statistical metric	Transient calibration result (units)
Number of primary head calibration targets	592
Mean Sum of Residuals (MSR)	0.72 (m)
Root Mean Square Error (RMS)	6.9 (m)
Scaled RMS (SRMS)	3.9 (%)
Correlation coefficient (R^2)	97.6 (%)
Maximum Residual	25.2 (m)
Water balance error (over all)	<0.01 (%)

Key assumptions made for simulating groundwater flow responses due to underground mining in the regional model include:

- Mine dewatering is simulated with active drainage boundary during the panel excavation.

- Fracturing unit above the longwall panel is defined with Time Variable Material (TVM) package based on the hydraulic properties of the zoned hard rocks.

- Hydraulic properties (hydraulic conductivity and storage) vary with depth to the longwall panel and are represented by TVM factors.

The fracturing process above the longwall has been identified to vary dynamically (Adhikary, Poulsen and Khanal, 2017). A significant uncertainty in concept is when, and how, the fracturing zone is developed for various longwall panels. To address this uncertainty, this study investigated which stress period for changing hydraulic properties provides the closest match to the observed groundwater levels and dewatering rates, and what the preferred set-up is for assigning factors of hydraulic conductivities related to fracturing process. To achieve this, an effective iterative ensemble smoother (IES) methodology implemented by PESTPP-IES code (White, Doherty and Hughes, 2014) was used to estimate parameter (multiple-factor) distributions. Three PEST trials are implemented in this study:

- Base scenario – fracturing zone above the longwall panel is defined with TVM factors activated once mining in the panel has been completed (used for model calibration in project study).

- Scenario 1 – fracturing zone is defined with TVM factors activated to align with the commencement of mining in each panel.

- Scenario 2 – fracturing zone is defined with TVM factors activated one year after mining is completed in each panel.

To specifically investigate the variation of hydraulic properties of the fracturing process, some basic hydrogeological features were carried over from the regional model as outlined below:

- Basic model settings (material properties, recharge rates, boundary conditions) are extended from calibrated models.

- The stress periods are consistent with the operational model.

For this uncertainty analysis, the groundwater levels obtained in two multi-level VWPs in the overburden above an area of longwall panel and the dewatering rates from longwall panels measured between 2022 and 2024 were used as modelling targets for historical matching in PEST model.

GOAFING ZONE AND THE PARAMETER PRIOR

Significant changes in lithology could cause localised fracturing of the overburden strata once the longwall mining occurs. To assist in understanding this response, a number of approaches may be adopted for defining the extent and strength of the fracturing interfaces. In this paper a general approach was used that entails creating a sequence of zones of strata where deformation occurs in the fracturing units, reflected to the longwall operations, which are termed as a goafing zone. Associated with this, hydraulic conductivity is varied across several zones, from a lower zone of connective-cracking (immediately in and above the mined-out panel) to a less impacted upper zone of disconnected-cracking (consistent with Merrick, 2017). By using similar approach as Modflow-USG TVM package, the variation of hydraulic conductivity for each layer of goafing zone is applied with multiple factors corresponding to hydraulic conductivity of the original rock units.

To assist in the determination of goafing zone in the simulation, the algorithm developed by Ditton and Merrick (2014) was used to estimate connective fracture height of the zone. Based on the overburden geology and the shape of longwall panels, the height of goafing zone could be between 150 m and 180 m above the longwall panels at this site. Related to geological layering in the model, eight layers above the excavated panels were defined to represent the goaf zone. The details and the hydraulic conductivities of goafing zones from the calibrated model are listed in Table 2. In this table and the following tables, Kh and Kv represent the horizontal and vertical hydraulic conductivities, respectively.

TABLE 2

Parameters adopted for goafing zone (KCB, 2024).

Layer	Goafing zone (geology unit)	Average thickness (m)	Height (m)	Base rock Kh (m/d)	Base rock Kv (m/d)	Multiple Kh-factor	Multiple Kv-factor	Log (Kh-factor)	Log (Kv-factor)
11	GC Fm	33.8	165.8	7.1E-03	7.1E-04	5	30	0.70	1.48
12	Aquila Seam	3.0	132.0	1.7E-02	8.6E-03	10	76	1.00	1.88
13	GC Fm	35.1	129.0	5.2E-03	5.2E-04	12	100	1.08	2.00
14	Tieri Seam	2.7	93.9	1.7E-02	8.6E-03	20	240	1.30	2.38
15	GC Fm	43.7	91.2	4.1E-03	4.1E-04	22	480	1.35	2.68
16	Corvus Seam	4.6	47.5	1.7E-02	8.6E-03	34	340	1.53	2.53
17	GC Fm	39.9	42.9	2.3E-03	2.3E-04	68	1120	1.83	3.05
18	GC Seam	3.0	3.0	1.7E-02	8.6E-03	294	1163	2.47	3.07

To effectively represent the uncertainty of parameters with PESTPP-IES, the critical part of this framework is the definition of the Prior parameter distribution (White *et al*, 2020). We used the calibrated log-factors as the first moment (median) of lognormal distribution of the Prior and the half of ranges in log-factors as a declined sequence of standard deviation (SD) from Layer 18 to Layer 11 for the Prior. Parameter ranges were defined based on the results from CSIRO research (Adhikary, Poulsen and Khanal, 2017), which are showed in Figure 1. The parameter Priors for each layer are illustrated in Table 3.

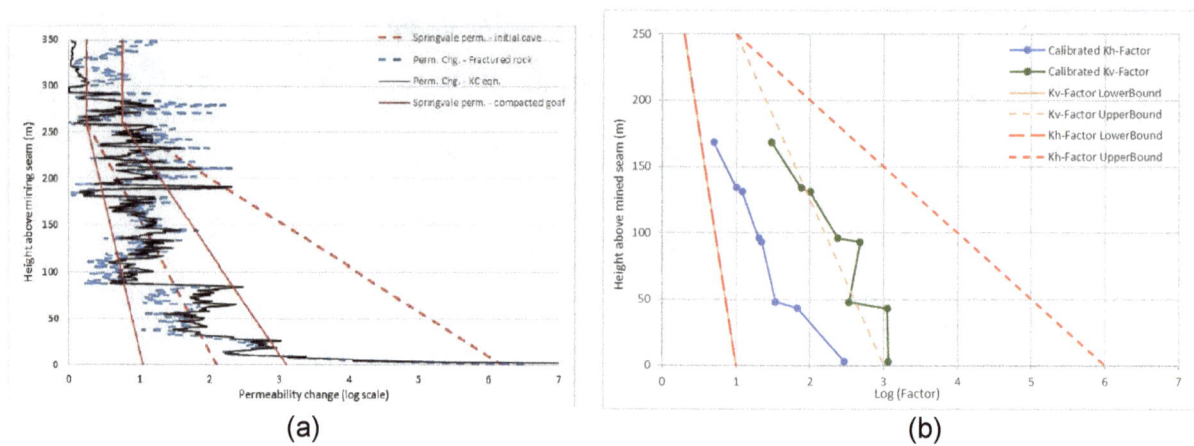

(a)   (b)

FIG 1 – Ranges and values of permeabilities of goaf zones: (a) Permeability change relationship above longwall panels (Adhikary, Poulsen and Khanal, 2017); (b) Ranges and calibrated values of multiple factors of permeability in goafing zones.

TABLE 3

Prior distribution of multiple factors (log) adopted for goafing zone.

Layer	Goafing zone (geo-unit)	Kh multiple factor (log)		Kv multiple factor (log)	
		Median	Standard deviation (SD)	Median	Standard deviation (SD)
11	GC Fm	0.70	0.35	1.48	0.35
12	Aquila Seam	1.00	0.44	1.88	0.66
13	GC Fm	1.08	0.54	2.00	0.96
14	Tieri Seam	1.30	0.63	2.38	1.27
15	GC Fm	1.35	0.72	2.68	1.58
16	Corvus Seam	1.53	0.81	2.53	1.89
17	GC Fm	1.83	0.91	3.05	2.19
18	GC Seam	2.47	1.00	3.07	2.50

IMPLEMENTATION AND RESULTS

The PEST control file, as well as template files and instruction files, were constructed with PEST utilities (Doherty, 2007). Additionally, a parameter list processor (PLPROC) was used to manipulate multiple factors of the goafing zone at each layer, and to write the outcomes of these manipulations into a model input file – layer property flow package (.lpf) for each single PEST run.

The prior distributions of multiple factors with median and standard deviation (SD) for each layer of goafing zone listed in Table 3 were selected as same initial inputs, and 100 realisations were implemented with PESTPP-IES for each scenario.

Posteriors of multiple factors for goafing zones obtained from the 100 realisations are shown in Table 4. The medians of posteriors all fall within the plausible parameter ranges for all three scenarios (refer to Figure 1). The standard deviation (SD) of posteriors indicates that the uncertainty of parameters could be less if the SD value is getting smaller. Compared with the SDs of Priors shown in Table 3, more than 90 per cent of SDs of posteriors are smaller than the SDs of Priors assigned to goafing zones. That means the uncertainty from hydraulic conductivities used for goafing zones in the model could be reduced through local historical-matching with PESTPP-IES approach. The lowest standard deviations (SD) shown in the table demonstrate the fracturing assumptions made in Base Scenario have less uncertainty with parameter estimations. Scenario 1 (goafing occurs as mining in each panel commences) provides the highest potential of uncertainty with less reduction of SD of multiple factors from Priors if it is used for prediction of the measured dewatering rates and the observed heads in this longwall mine.

TABLE 4

Posterior distribution of multiple factors (log) obtained from three scenarios.

| Goaf zone | Base scenario posterior | | | | Scenario 1 posterior | | | | Scenario 2 posterior | | | |
| | Kh-factor | | Kv-factor | | Kh-factor | | Kv-factor | | Kh-factor | | Kv-factor | |
	median	SD	median	SD	median	SD	median	SD	median	SD	median	SD
GC Fm	0.78	0.24	1.45	0.18	0.48	0.41	1.36	0.46	0.85	0.26	1.48	0.32
Aquila Seam	0.90	0.33	1.66	0.51	0.90	0.49	1.58	0.75	0.95	0.33	1.80	0.61
GC Fm	1.04	0.21	1.95	0.54	1.00	0.47	1.65	0.85	1.30	0.41	1.79	0.56
Tieri Seam	1.26	0.51	2.37	0.92	1.26	0.51	2.27	0.96	1.38	0.35	2.72	0.97
GC Fm	1.40	0.43	2.67	1.03	1.30	0.49	2.39	1.57	2.03	0.46	3.01	1.11
Corvus Seam	1.56	0.56	2.51	1.14	1.45	0.79	2.41	1.19	1.65	0.68	2.92	1.24
GC Fm	1.86	0.63	3.06	1.42	1.79	0.86	2.93	1.87	2.41	0.57	3.35	1.07
GC Seam	2.48	0.58	3.08	1.01	1.94	0.88	2.70	1.25	3.23	0.36	4.06	0.85

The hydrographs of the history-matching for the 5th percentile, median, and 95th percentile of posterior ensembles bracket at multi-level VWPs above the longwall panel are presented in Figure 2. In the figure, VWP_A is located in the shallow aquifer (40 m below surface), while VWP_D is about 15 m above the coal seam. The figure showed groundwater levels at shallow aquifer VWP_A were not greatly affected by the fracturing assumptions made in the three cases. However, the simulation of groundwater levels at VWP_D in fracturing zone could result in more than 50 m difference in 95th percentile depending on which assumption regarding the timing of fracturing is made. Additionally, the simulated water level gap between the 95th and 5th percentiles in Scenario 1 is 30 m larger than the gap in Base Scenario and 10 m larger than the gap in Scenario 2. This further demonstrated that the uncertainty in prediction of groundwater level with Scenario 1 would be higher than other two scenarios.

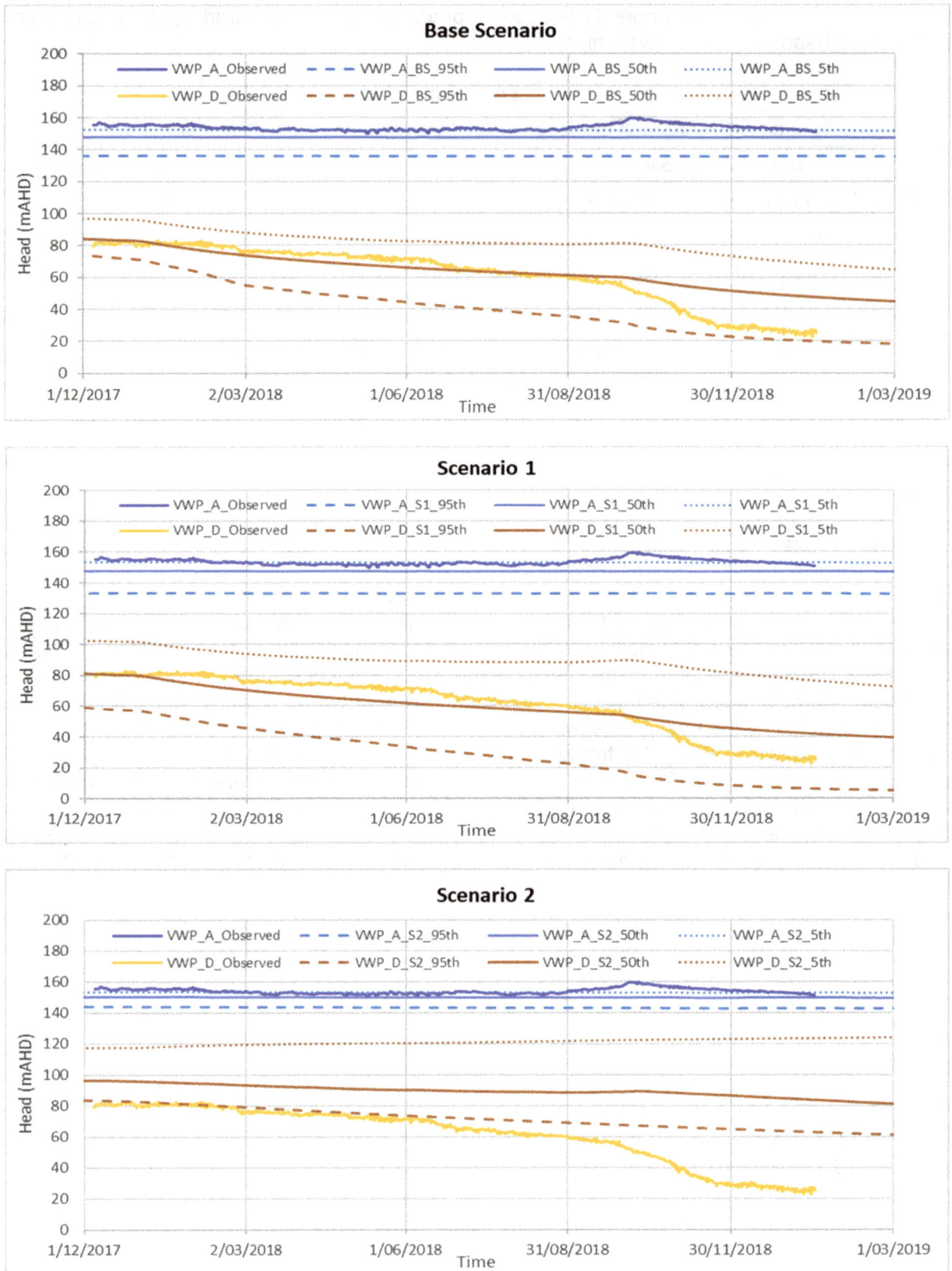

FIG 2 – Variation of simulated groundwater levels at shallow and deep VWPs in three scenarios.

The observed and simulated groundwater abstraction rates from underground mine are illustrated in Table 5. The table showed the range of the simulated rates between the 95th and 5th percentiles in Base Scenario was about 5 L/s smaller than the range in Scenario 1 and 1~5 L/s smaller than the range in Scenario 2 over the three years. This demonstrated that the uncertainty in the simulation of groundwater behaviour with the Base Scenario would be less than the other two scenarios. The medians of the simulated flow rates tightly clustered the observed rates in the Base Scenario further demonstrated that the fracture networks in the overburden could be generated at the time right after mining terminated in the longwall panel for this mine site.

TABLE 5
Observed and simulated flow rates from underground mine.

| | Flow rate from UGM (L/s) | | | | | | | | |
| Scenario | Y2022 – percentile | | | Y2023 – percentile | | | Y2024 – percentile | | |
	95th	50th	5th	95th	50th	5th	95th	50th	5th
Observed		24.32			24.31			10.65	
Calibrated		27.9			25.4			9.3	
Base scenario	33.0	24.7	21.6	31.1	23.4	20.1	16.9	10.9	6.5
Scenario 1	43.1	29.3	25.6	39.9	27.0	23.5	21.1	12.0	7.2
Scenario 2	31.7	23.2	16.1	30.5	22.6	13.9	15.7	8.8	4.8

CONCLUSION AND DISCUSSION

The models were evaluated against field monitoring data and this trial has enhanced the understanding of fracturing processes associated with underground coal mining operations. Records of inflow events and multi-level VWP data above the longwall panel enabled a comparative review of different modelling assumptions on the timing of simulation of goafing impacts, particularly regarding the staged activation of fracturing within groundwater models. Selecting the most appropriate scenario for each underground mine could reduce the uncertainty in simulating the inflow and groundwater responses above the mining. This is important for using groundwater modelling results as an indication for mine dewatering plans and groundwater management. This also allows for better constraints when conducting subsidence assessment with reduction of uncertainty in characterising changing hydraulic conductivities within the goafing zone.

This trial has shown the importance of considering the timing in which goafing affects the parameters in groundwater modelling of underground coalmines. Assessing which assumption provides the closest match to measured data as part of uncertainty analysis offers notable benefits, particularly in improving transparency and providing increases in predictions such as aligning monthly mining schedules with predicted inflows to support decision-making and mine water management.

However, it is essential to note that the model assumes each goafing unit is homogeneous. As a result, the uncertainty analysis does not fully capture the natural variability in hydraulic properties within these zones.

Future work could expand on this approach by incorporating pilot points within each goafing zone and simulating dynamic changes in TVM factors over time. This would support a more advanced conceptual understanding of the fracturing process and improve the representation of heterogeneity in hydraulic behaviour.

REFERENCES

Adhikary, D P, Poulsen, B A and Khanal, M, 2017. Assessment of Longwall Mining Induced Connective Fracturing of Overburden Strata, ACARP Report C24020, CSIRO, Australia.

Barnett, B, Townley, L R, Post, V, Evans, R E, Hunt, R J, Peeters, L, Richardson, S, Werner, A D, Knapton, A and Boronkay, A, 2012. Australian groundwater modelling guidelines, Waterlines report, National Water Commission, Australian Government.

Ditton, S and Merrick, N, 2014. A new subsurface fracture height prediction model for longwall mines in the NSW coalfields, *Geological Society of Australia, 2014 Australian Earth Sciences Convention (AESC)*, Newcastle, NSW, pp 136.

Doherty, J E and Hunt, R J, 2010. Approaches to highly parameterized inversion – A guide to using PEST for groundwater-model calibration, US Geological Survey Scientific Investigations Report 2010-5169, 59 p.

Doherty, J, 2007. PEST surface water modelling utilities, Watermark Numerical Computing.

Frenelus, W, Peng, H and Zhang, J, 2021. Evaluation methods for groundwater inflows into rock tunnels: a state-of-the-art review, *Int J Hydro,* 5(4):152–168. https://doi.org/10.15406/ijh.2021.05.00277

KCB Australia Pty Ltd (KCB), 2024. German Creek Mine Complex PCRP, Hydrogeological Assessment.

Klenowski, G, 2000. The influence of subsidence cracking on longwall extraction beneath water courses, aquifers, open cut voids and spoil piles, ACARP Project C5106. August 2000.

Merrick, N P, 2017. Recent Advances in Groundwater Modelling for Coal Mines, in *Proceedings of the 10th Triennial Conference on Mine Subsidence 'Adaptive Innovation for Managing Challenges'*, pp 29–35 (Mine Subsidence Technology Society).

Middlemis, H and Peeters, L J M, 2023. Uncertainty analysis—Guidance for groundwater modelling within a risk management framework, A report prepared for the Independent Expert Scientific Committee on Coal Seam Gas and Large Coal Mining Development through the Department of the Environment and Energy, Commonwealth of Australia.

Moore, C R, Doherty, J, Howell and Erriah, L, 2015. Some challenges posed by coal bed methane regional assessment modeling, *Groundwater,* 53(5):737–747.

Tammetta, P, 2015. Estimation of the Change in Hydraulic Conductivity Above Mined Longwall Panels, *Groundwater,* 53(1):122–129. https://doi.org/10.1111/gwat.12153

Turnadge, C, Mallants, D and Peeters, L, 2018. Sensitivity and uncertainty analysis of a regional-scale groundwater flow system stressed by coal seam gas extraction, CSIRO, Australia.

White, J T, Doherty, J E and Hughes, J D, 2014. Quantifying the predictive consequences of model error with linear subspace analysis, *Water Resources Research,* 50:1152–1173. https://doi.org/10.1002/2013WR014767

White, J T, Foster, L K, Fienen, M N, Knowling, M J, Hemmings, B and Winterle, J R, 2020. Toward Reproducible Environmental Modeling for Decision Support: A Worked Example, *Frontiers in Earth Science,* 8:50. https://doi.org/10.3389/feart.2020.00050

Life of mine waste and tailings management

Why we should look beyond social license to operate – toward a care-based framework for corporate social responsibility in mining

A A Andrade[1]

1. MPhil candidate, The University of Queensland, Brisbane Qld 4169.
   Email: a.dasilvaandrade@uq.edu.au

ABSTRACT

The mining industry has long relied on the concept of a social licence to operate (SLO) to guide its relationship with local communities. Introduced in the late 1990s, SLO was intended to go beyond legal compliance by aligning corporate actions with public expectations. However, its effectiveness has come under increasing scrutiny. High-profile cases of corporate social irresponsibility (CSI)—such as the tailings dam disasters in Brumadinho and Mariana, Brazil, and the destruction of Juukan Gorge in Australia—highlight that companies often maintain legitimacy while neglecting deeper responsibilities to people and places.

This paper argues that the mining sector must move from seeking legitimacy to practising responsibility. Corporate social responsibility (CSR) has the potential to support trust-building, environmental stewardship, and ethical governance. Yet when driven by metrics or branding, CSR risks becoming another form of procedural engagement. To reframe CSR as participatory ethical governance, this paper introduces the ethics of care, positioning CSR as a practice of co-stewardship rooted in long-term community well-being.

The discussion combines theory with personal reflection. Drawing on the author's lived experience as an affected person in Brumadinho, the paper illustrates how care-based practices gradually emerged through sustained community mobilisation. The change, though incremental and limited, demonstrated a shift toward care-centred engagement.

A care-based CSR model reframes communities as partners in shared futures instead of mere stakeholders to be consulted. It moves beyond the limits of SLO, addressing the ethical gaps in conventional CSR. By adopting care as a guiding principle, the mining sector can act with greater responsibility, relational accountability, and respect in an increasingly fragile world.

INTRODUCTION

The mining sector has traditionally operated on a transactional model: extract resources, create jobs, and expect local acceptance. In the early 1990s, most mining companies regarded environmental degradation and social disruption as external to their core business responsibilities. Many 'lacked a sense of responsibility towards the communities in which they operated' (Dashwood, 2012). As the decade progressed, the industry came under increasing public scrutiny due to ongoing high-profile tailings dam failures, toxic spills, and growing tensions with local communities. These events changed how environmental and social risks were perceived and eroded the sector's social legitimacy (Hitch and Barakos, 2021; Thomson and Boutilier, 2011). By the end of the 1990s, mounting public scrutiny and pressure for greater accountability were prompting a shift in how mining companies responded to their social and environmental impacts. It was becoming clear that the old ways of doing business, operating with little regard for social and environmental impacts, could no longer be justified. If the industry was to regain trust and remain viable, it had to change (Prno and Slocombe, 2012). In this changing landscape, CSR began to gain traction as a framework through which companies could demonstrate broader commitments to ethical governance and sustainability. Dashwood (2012) highlights the leadership role taken by some major mining firms, particularly those headquartered in advanced industrialised economies, in shaping and promoting CSR norms across the global sector.

It was in this context that the SLO emerged. Coined in 1997 by Canadian mining executive Jim Cooney, the term reflected a growing recognition that securing legal permits was no longer sufficient (Lopes and Demajorovic, 2020). Companies also needed to gain and maintain the informal approval of local communities, without which mining projects could face delays, disruption, or even cancellation (Prno and Slocombe, 2012). The concept captured a new reality: public acceptance and local support had become essential to project viability. SLO was soon taken up by the International Council on Mining and Metals (ICMM), founded in 2001, where it became a central principle guiding industry commitments to social and environmental responsibility (Hitch and Barakos, 2021). Since its introduction, the concept of SLO has been widely adopted across the global mining sector. It has become a core element of industry discourse, CSR policies, and sustainability frameworks, regularly cited by companies, regulators, and international bodies as a marker of ethical engagement and social accountability. SLO was originally framed as a relational, trust-based process involving the ongoing approval of local communities, where societal expectations of corporations are met (Thomson and Boutilier, 2011; Prno, and Slocombe, 2012). Over time, however, its application has shifted. Rather than fostering shared decision-making or addressing power imbalances, SLO is often used as a procedural tool to protect operational continuity and manage reputational risk (Owen and Kemp, 2013). This concept, increasingly out of step with contemporary expectations, has come under scrutiny amid further environmental and social mining disasters.

More recent high-profile mining-related incidents, such as the tailings dam collapses in Mariana and Brumadinho, the destruction of Juukan Gorge, and conflicts in Bougainville and Madagascar, have revealed that companies can retain a social licence while perpetuating serious harm. In Mariana (2015) and Brumadinho (2019), catastrophic tailings dam failures resulted in vast environmental devastation, the loss of hundreds of lives, and deep and long-lasting social impact. In 2020, Rio Tinto legally destroyed the Juukan Gorge rock shelters, sacred Aboriginal heritage sites, ignoring the cultural and spiritual significance communicated by Traditional Owners. In Bougainville, decades of conflict linked to the Panguna Mine have left lasting social scars and environmental degradation. In Madagascar, QIT Madagascar Minerals' operations have sparked unrest over land access, pollution, and community marginalisation. These are not just operational failures but ethical ones and are cases of corporate social irresponsibility (CSI). As Kemp and Owen (2022) note, cases of CSI expose how mining companies can maintain social legitimacy through procedural compliance or reputational management while sidestepping deeper ethical obligations to affected communities. CSI reveals the limitations of current governance models, particularly when corporate responsibility is decoupled from relational accountability and justice. The poignant examples of CSI mentioned herein point to the need for a framework that moves from seeking legitimacy to practising social and environmental responsibility.

METHODS

This paper combines theoretical analysis and personal reflection grounded in the author's lived experience as a person affected by the disaster in Brumadinho in 2019. The central aim is to contribute to a more relational and justice-oriented understanding of CSR in the mining sector, one that moves beyond procedural legitimacy toward ethical responsibility and community co-stewardship.

The analytical framework is based on Tronto's (1993, 2013) ethics of care, a model developed in feminist moral and political theory that centres attentiveness, responsibility, competence, responsiveness, and solidarity as the foundations of ethical action. This framework is used to assess the moral implications of corporate behaviour in mining contexts, particularly where harm has occurred despite claims of legitimacy. The disaster in Brumadinho provides a particularly relevant anchor for this analysis for several reasons. Firstly, it was a globally visible example of mining-related harm, prompting widespread media coverage, legal proceedings, and public outrage. Secondly, the disaster involved Vale S.A., a member company of the ICMM, an organisation that publicly commits to upholding best practices in sustainability and stakeholder engagement. The incident thus raises pressing questions about how corporate actors reconcile, or fail to reconcile, their shareholder and stakeholder obligations, especially when operating under voluntary governance models such as CSR and SLO.

This reflective component draws on reflexive and narrative methodologies that position lived experience and researcher positionality as critical sources of insight in environmental justice, sustainability, and CSR research (Nastar, 2023; Sultana, 2015). These approaches are particularly relevant in the contexts of harm, trauma, and institutional failure, where experiential knowledge can reveal ethical blind spots and signal alternative paths toward accountability. Feminist and activist traditions have long argued that reflexivity enables deeper ethical analysis by uncovering the relational and structural dynamics that formal models often obscure (Sultana, 2015; Girschik, Svystunova and Lysova, 2022). Together, the theoretical and experiential lenses provide a means of evaluating the gap between corporate discourse and ethical practice.

By drawing from the author's experience of the disaster in Brumadinho, the paper explores how fragments of care, though delayed and often fragile, can begin to emerge, and how they might inform a more responsible, care-oriented model of CSR in the mining sector.

THE LIMITATIONS OF THE SOCIAL LICENCE TO OPERATE

The social license to operate concept was developed to enable resource companies to operate with community acceptance. It was initially envisioned as a trust-based and relational concept—a form of informal, ongoing community approval that extended beyond regulatory compliance. SLO was positioned as a progressive alternative to purely legalistic approaches to project legitimacy. Rather than treating communities as obstacles, the idea was that companies would earn and maintain social support through meaningful, participatory engagement. However, in practice, the operationalisation of SLO has often diverged from this ideal. Owen and Kemp (2013) argue that while companies have adopted the language of responsiveness and legitimacy, SLO is frequently used to manage reputational risk and reduce operational disruption, rather than foster ethical accountability or shared decision-making. Business risk-led governance tends to foster a conservative corporate culture that avoids uncertainty and downplays qualitative engagement. This narrow focus restricts opportunities for deeper learning, ethical reflection, and the development of alternative, more relational practices (Kemp, Owen and van de Graaff, 2012).

Three cases of CSI in the last decade where companies were seen to hold a social license, but failed to act responsibly, reveal this disconnect starkly. They illustrate the ethical limitations of SLO when it becomes a procedural or reputational mechanism rather than a foundation for mutual care and co-stewardship. In each instance, the ICMM member companies behind the incidents maintained a social license—formally or symbolically—while enacting or perpetuating serious harm. When SLO is implemented as a corporate risk strategy, its capacity to foster ethical practice and relationship is undermined. The result is a legitimacy gap: companies appear responsible on paper but act with procedural indifference in practice.

The collapse of the Fundão tailings dam at the Samarco mine in Mariana, Brazil, in 2015, released a torrent of mining waste into the Doce River, killing 19 people and causing severe environmental and social harm across two states. Operated jointly by BHP and Vale, the project had previously been viewed as a flagship for responsible mining, with Samarco actively promoting its socio-environmental commitments (Lopes and Demajorovic, 2020). Yet the post-disaster response was plagued by jurisdictional confusion, legal delays, and a lack of clarity about corporate responsibility (Primo et al, 2021). The Renova Foundation, established to implement resettlement and recovery initiatives, has faced widespread criticism for its ethical shortcomings (Fonseca et al, 2024). The complexity of the institutional arrangement—featuring multiple stakeholders, overlapping regulatory bodies, and opaque recovery processes—allowed accountability to be diffused and delayed. Maher (2022, as cited in Fonseca et al, 2024) 'suggests that Samarco deployed the formally independent multi-stakeholder initiative structure of the Renova Foundation to shirk the company's responsibilities associated with their duties towards the care and respect of their victims' human rights.'

Four years after the Fundão failure, Vale was again at the centre of catastrophe when the Córrego do Feijão tailings dam collapsed in Brumadinho, killing 272 people. Despite its public commitment to sustainability and CSR, the company failed to act on critical geotechnical warnings. In the aftermath of the disaster, Vale assumed responsibility for administering the remediation programs itself. Its initial response was largely defensive—focused more on controlling reputational damage than supporting affected communities (de Nardin Budó, França and Natali, 2023). As a result, affected people and civil society groups had to intervene to advocate for more inclusive processes. This case will be further discussed in the following sections.

In 2020, Rio Tinto detonated 46 000-year-old Aboriginal heritage rock shelters in Juukan Gorge, Western Australia—sites of profound cultural and spiritual significance to the Puutu Kunti Kurrama and Pinikura peoples. The destruction was legal: the company had secured government approval under outdated heritage laws. Yet it occurred despite ongoing objections from Traditional Owners and internal reports warning of the site's importance. The event triggered national and international outrage, leading to parliamentary inquiry and the resignation of several Rio Tinto executives (Parliament of Australia, 2020). The company's claim to social legitimacy was invalidated by its failure to respect Indigenous authority and cultural continuity. Juukan Gorge demonstrates the limits of procedural compliance and corporate governance mechanisms that fail to recognise Indigenous law, cultural landscapes, and the ethical responsibilities of consent-based engagement (Kemp, Kochan and Burton, 2022). Even when legal, such acts violate the moral premises that SLO asserts to uphold.

CSR – POTENTIAL AND PITFALLS

In response to the erosion of public trust in extractive industries during the 1990s, CSR emerged as a strategy to reposition companies as ethically aware, socially engaged, and environmentally accountable (Dashwood, 2012; Hamann, 2003). For mining companies, CSR offered a broader, more structured framework than SLO, incorporating sustainability principles, stakeholder engagement, and ethical governance into their business models (Prno and Slocombe, 2012). It also aligned with global agendas such as the predecessor to the UN SDGs, reinforcing the idea that corporations have a role to play in addressing structural social and ecological challenges (Frynas, 2008). Unlike legal compliance, which imposes minimum standards, CSR is largely voluntary. It invites companies to self-regulate, demonstrate transparency, and build legitimacy through proactive commitments (Carroll and Shabana, 2010). This flexibility allows companies to tailor initiatives to local contexts and stakeholder expectations. However, it also creates space for CSR to become decoupled from ethical practice. When driven by reputational risk or investor relations, CSR can shift from a tool for meaningful accountability to a strategy of brand protection (Carroll and Shabana, 2010; Nijhof and Jeurissen 2010).

In many cases, CSR has been shaped by strategic business concerns instead of ethical or relational imperatives. As Gond, Cabantous and Krikorian (2018) note, companies engage in a process of 'strategifying' CSR—transforming it into a tool for competitive advantage through selective framing, internal alignment with performance metrics, and integration into core strategic planning. This process reinforces the dominance of managerial logic within CSR, where ethical concerns are subordinated to instrumental outcomes. As a result, even well-intentioned CSR initiatives risk being evaluated not by the quality of their relationships or long-term social impacts, but by how well they serve corporate goals. Across sectors, the 'business case' for CSR has increasingly dominated its framing—privileging corporate-defined goals, competitive advantage, and investor appeal over transformative social or environmental justice (Carroll and Shabana, 2010). Nijhof and Jeurissen (2010) argue that this instrumentalisation of CSR creates a 'glass ceiling': companies will pursue social responsibility only to the point that it supports existing economic logics. They state that CSR nowadays 'is much more about new market opportunities and a business-wise approach to ecological and social problems' than responsibility.

In the mining sector, CSR often manifests through sustainability reports, impact assessments, and community investment programs. These can serve important functions, raising visibility, articulating intent, and setting internal targets, but they can also become performative. As Kemp, Owen and van de Graaff (2012) note, CSR in mining is often embedded in an 'audit culture' where the emphasis is on documentation, metrics, and demonstrable compliance, rather than relational accountability or structural change. The risk is that CSR becomes a managerial tool, where metrics substitute for moral responsibility and relational harms are left unaddressed. There is a growing demand for mining companies to adopt meaningful and accountable CSR practices (Kemp and Owen, 2020).

Furthermore, CSR's language of 'stakeholder engagement' often retains a top-down logic. Communities may be consulted, surveyed, or invited to workshops but rarely are they positioned as co-decision-makers in shaping development trajectories. This reinforces a transactional dynamic, in which companies retain control over timelines, definitions of success, and the terms of inclusion. The Samarco case exemplifies this dynamic. Multiple studies have shown that the governance structure of Renova excluded community members from shaping its mandate and planning processes, limiting their influence over key decisions in violation of best practice in disaster recovery and reparative justice (Euclydes, Pereira and Fonseca, 2022; Lavalle *et al*, 2022; Losekann and Milanez, 2023).

Fonseca *et al* (2024) describe how recovery programs were rolled out in top-down fashion, reinforcing existing hierarchies rather than redistributing power or addressing moral obligations. Affected people described feeling tired, disrespected, and sceptical due to engagement processes that were experienced as tokenistic and disempowering. Pereira (2021) notes that Samarco's CSR discourse persisted even as affected communities were left without clear recourse or timely support.

Girschik, Svystunova and Lysova (2022) contend that the transformative potential of CSR—particularly at the micro or practice level—is constrained precisely because CSR is defined, implemented, and evaluated from within the corporate system. This allows companies to maintain control over how responsibility is understood, measured, and enacted. As a result, CSR may entrench rather than challenge managerial dominance, serving to reinforce rather than disrupt business-as-usual. However, despite these limitations, CSR remains an important mechanism for intervention. Its institutional recognition and integration into global standards provide a foothold for reimagining how mining companies relate to people and place. For CSR to become a vehicle for ethical responsibility rather than reputational risk management, it must be reoriented. This paper argues that the ethics of care offers such a reorientation; one that shifts the focus from compliance to relationship, and from legitimacy to justice.

THE ETHICS OF CARE – FROM THEORY TO RESPONSIBILITY

Care ethics emerged from feminist moral philosophy, particularly the work of Carol Gilligan (1982) and Nel Noddings (1984), who challenged the dominance of abstract, rule-based moral reasoning by introducing the importance of relational context, empathy, and emotional understanding. They argued that moral development is grounded in the capacity to care for others and attend to human relationships. While early care theorists focused on interpersonal ethics, Joan Tronto (1993, 2013) expanded care ethics into the political and institutional domains, making it a powerful tool for evaluating systems of governance and organisational behaviour. In Moral Boundaries (Tronto, 1993) and Caring Democracy (Tronto, 2013), Tronto identifies five core elements of ethical care:

1. Attentiveness – recognising the needs of others.

2. Responsibility – taking ownership of the obligation to care.

3. Competence – ensuring care is delivered effectively.

4. Responsiveness – adjusting care in light of how it is received.

5. Solidarity – acting out of collective responsibility and justice.

These dimensions shift the focus of CSR from procedural legitimacy to relational ethics. Rather than asking whether companies have followed the right steps or produced the right reports, a care-based lens asks how corporate actions are received by those affected, whether they reflect a genuine understanding of local needs, and whether they contribute to justice and well-being over time. Rather than viewing communities as external stakeholders to be managed, care-based CSR recognises them as co-creators of meaning, priority, and future direction—essential partners in building just and resilient mining governance.

GROUNDING CARE IN LIVED EXPERIENCE – REFLECTIONS FROM BRUMADINHO

As someone personally affected by the tailings dam collapse in Brumadinho in 2019, the author witnessed firsthand the consequences of the absence of care and the potential of its presence. In the immediate aftermath, the institutional response was marked by procedural defensiveness and an overriding concern with legal liability. Support for affected communities was fragmented and formalistic, reflecting the absence of attentiveness, responsiveness, and moral responsibility. These early responses reflected what Tronto (1993) would describe as a failure of care at every level.

Over time, however, some practices began to shift. Community meetings increasingly prioritised listening over control. Grief was allowed to be expressed, rather than contained or managed. Most significantly, affected people were eventually included in co-defining priorities for remedy and reparations. These actions were far from perfect, but they felt relationally meaningful. They created the conditions for tentative trust, emotional repair, and a shared sense of direction. It is important to clarify that this reflection represents the author's point of view. Many affected people may hold different or opposing perspectives, particularly where justice and care remain unfulfilled. Still, from the author's point of view, these moments suggest the ethical potential of care-based engagement. They affirm that care is not only theoretically coherent—it can be ethically transformative in practice.

One dimension of this transformation was the deepening of dialogue as a form of accountability. Drawing on Roberts (1996), Cooper and Owen (2007) distinguish between discussion—where decisions are made—and dialogue, which seeks to explore complexity without rushing toward agreement. In this spirit, engagement processes began to reflect not only procedural fairness but a willingness to grapple with grief, dissent, and the emotional weight of what had occurred. As Kemp, Owen and van de Graaff (2012) argue, treating dialogue as both a process and a practice of accountability helps restore balance, preventing the instrumental pursuit of power and profit from proceeding without regard for social and environmental consequences.

This was not abstract theory—it was lived reality. When engagement began to reflect Tronto's (2013) care dimensions—especially attentiveness to the needs of affected people and responsiveness in seeking viable solutions—it became possible to speak, be heard, and experience some measure of repair. Yet the limits of care also became visible. One key request repeatedly voiced by affected families—the redesign of Vale's employee uniforms—was denied. For most families, the green uniform had become a symbol of trauma, as many victims were workers wearing that attire when they died. Its continued use remains a site of unresolved pain, revealing the boundaries of what corporate actors were willing to concede.

On the other hand, one hard-won achievement was the construction of the Memorial for the Victims in Brumadinho. This was not granted easily. It followed years of negotiation, protest, and insistence on a governance model independent of Vale. Ultimately, the memorial was realised with a structure and mandate defined by the victims' families. Yet it is important to acknowledge that much of the pain and conflict surrounding its establishment might have been avoided if the principles of care—attentiveness, responsiveness, and solidarity—had been meaningfully enacted earlier. The delays, refusals, and struggles that marked this process speak to what is lost when care is deferred: opportunities for healing, trust, and dignity.

As Fonseca *et al* (2024) note, meaningful stakeholder engagement (MSE) must address procedural, relational, and substantive dimensions: from transparency and consistency, to the perception of being genuinely heard, to whether the outcomes actually meet the needs of those most affected. In Brumadinho, some of these elements gradually came into being as engagement became more meaningful. This happened because affected people and civil society groups have insisted that the dimensions of care—and justice from the affected people's perspectives—must be part of how recovery is imagined and enacted.

FROM LEGITIMACY TO ETHICAL RELATIONSHIP – CONTRASTING MODELS

Using Tronto's (1993, 2013) five care dimensions, it is possible to contrast conventional SLO-based models with a care-oriented CSR framework as shown in Table 1.

TABLE 1

Relational comparison of SLO and care-based CSR models.

Dimension	SLO engagement	Care-based CSR
Attentiveness	Reactive, risk-based	Continuous, relational listening
Responsibility	Procedural, image-driven	Moral obligation grounded in presence
Competence	Often performative or public relations-driven	Resourced, context-specific
Responsiveness	Slow, defensive	Embedded feedback loops
Solidarity	Minimal or symbolic	Co-decision and justice as institutionalised standard

Care-based CSR recasts corporations as co-stewards of shared futures. This transformation demands deep structural change—one that reframes time horizons, power dynamics, and accountability mechanisms.

RECOMMENDATIONS – OPERATIONALISING CORPORATE CARE

Reorienting corporate social responsibility around care is not simply a matter of adding new vocabulary or metrics. It requires a deep structural and cultural shift in how mining companies understand their role in society, their relationships with affected communities, and their ethical obligations beyond compliance. If the sector is to move from rhetorical commitment to genuine transformation, several interrelated actions must be taken.

First, companies must co-design engagement processes with the communities they affect. As Pereira, Buhmann and Fonseca (2024, p 427) emphasise, 'the participation of affected stakeholders in the design of engagement processes is absolutely key to the legitimisation and meaningfulness of participatory processes.' This includes recognising the diverse temporal, spatial, and emotional realities in which people engage—especially in the aftermath of harm. Participatory processes that are imposed, rigid, or procedurally detached from the lived experience of affected groups are unlikely to generate trust or justice. In the case of the Samarco disaster, the Renova Foundation's failure to include affected people in the foundational design of its mandate has been a recurring source of frustration, resentment, and perceived injustice. A care-based model, by contrast, recognises that legitimacy flows not from procedural completion, but from relational and ethical grounding.

To support this shift, CSR frameworks must embed the five dimensions of care proposed by Tronto (2013)—attentiveness, responsibility, competence, responsiveness, and solidarity—into their evaluation metrics and organisational benchmarks. These dimensions offer not only a moral compass but a set of practical criteria for assessing whether CSR initiatives are building real relationships, responding to needs, and promoting justice. For example, is a company attentive to the stated needs and silences of affected communities? Does it take responsibility beyond legal liability? Is its care competent—adequately resourced, culturally informed, and sustained over time?

Further, companies must go beyond stakeholder engagement to create co-governance structures in which community members hold real authority. Shared decision-making platforms that include Indigenous leaders, women, youth, and displaced people must become a standard component of project planning, impact management, and remediation. These structures must also recognise the asymmetric power dynamics at play, providing capacity-building and cultural brokerage mechanisms where needed to ensure equitable participation.

This transformation also requires investment in relational ethics and cultural competency training for CSR teams and corporate leadership. A care-based approach cannot be implemented through checklists or consultant packages alone—it demands that those designing and leading corporate initiatives understand the historical, cultural, and emotional landscapes they are entering. This includes grappling with legacies of colonialism, systemic inequality, and environmental trauma, especially in jurisdictions where operations overlap with Indigenous lands or historically marginalised groups.

Lastly, given the increasingly transnational nature of mining operations, the role of ethical oversight across borders is also crucial. Many mining companies operate in jurisdictions with weak or uneven regulatory protections, yet claim alignment with international frameworks such as the UNGPs or the OECD Guidelines. Independent oversight bodies, including human rights observers, transnational civil society alliances, and ethics review panels, can play a key role in holding companies accountable to their stated commitments and in elevating community voices that might otherwise be sidelined.

CONCLUSION

The high-profile cases of CSI discussed in this paper—from Mariana and Brumadinho to Juukan Gorge—underscore the profound disconnect between corporate discourse and ethical practice in the mining sector. They expose the limits of frameworks like SLO, which, though originally conceived as a means of cultivating trust and mutual respect, has too often devolved into a performative tool of risk management. SLO, as currently practiced, often prioritises reputational legitimacy over relational accountability. Similarly, CSR, while offering a broader platform for ethical and environmental commitments, remains vulnerable to instrumentalisation when divorced from lived experience and genuine engagement.

This paper argues that if CSR is to remain relevant—and become transformative—it must be reframed around the ethics of care. Tronto's (1993, 2013) five dimensions of care provide a powerful framework to reimagine CSR not as a managerial checklist but as a relational practice grounded in moral obligation, contextual responsiveness, and shared stewardship. Care-based CSR does not ask whether companies have consulted stakeholders or published glossy sustainability reports. It asks whether companies have listened attentively, responded with competence, taken responsibility for past and ongoing harm, and acted in solidarity with those most affected.

The disaster in Brumadinho, as reflected upon through the author's own experience, reveals both the cost of care's absence and the possibilities of its emergence. While many engagement efforts initially failed to meet the needs and expectations of affected people, fragments of care-based practice gradually took shape. These included shifts toward meaningful dialogue, shared decision-making, and the eventual establishment of the memorial for the victims. Though partial and contested, these moments of relational repair suggest that more ethical forms of corporate engagement are possible—if companies are willing to reorient their priorities.

To move from rhetoric to practice, mining companies must fundamentally transform how they view their role in society. This requires a departure from extractive logics that frame communities as risks to be managed or markets to be cultivated, and instead embrace a governance model rooted in ethical relationship. It also demands an internal cultural shift: away from the dominance of audit culture and towards an ethics of presence, accountability, and relational care.

The stakes of this transformation are high. As mining continues to expand into ecologically sensitive and socially complex territories, the sector's legitimacy—and more importantly, its moral credibility—will depend not on its ability to perform responsibility, but to practise it. The path forward requires humility, collaboration, and the courage to centre the perspectives of those most impacted by mining operations.

A care-based CSR model is not a panacea. It will not prevent all harm, nor resolve the structural inequities that underpin extractive economies. But it offers a principled and practical starting point for doing better. Rather than discarding SLO entirely, care-based CSR should be understood as a renewal of its original ethical intent. It restores the focus on trust, consent, and mutual respect—but embeds them in deeper commitments to justice, co-governance, and long-term care. In doing so, care-based CSR provides a more credible and morally grounded pathway for the mining sector to regain public trust and contribute meaningfully to a just and sustainable future.

ACKNOWLEDGEMENTS

The author would like to thank Professor David Williams and Professor Deanna Kemp for their invaluable guidance and critical insights during the development of this paper. Their feedback helped strengthen the analytical depth and practical relevance of the work.

The author would also like to acknowledge the use of OpenAI's ChatGPT (GPT-4) to support the revision and editing of this paper. The tool was used under the author's direction to clarify wording, check alignment with submission guidelines, and refine structure, with all content critically reviewed and approved by the author.

REFERENCES

Carroll, A B and Shabana, K M, 2010. The business case for corporate social responsibility: A review of concepts, research and practice, *International Journal of Management Reviews*, 12(1):85–105.

Cooper, S M and Owen, D L, 2007. Corporate social reporting and stakeholder accountability: The missing link, *Accounting, Organizations and Society*, 32(7–8):649–667. https://doi.org/10.1016/j.aos.2007.02.001

Dashwood, H S, 2012. *The Rise of Global Corporate Social Responsibility: Mining and the Spread of Global Norms* (Cambridge University Press).

de Nardin Budó, M, França, K and Natali, L, 2023. Beyond retributive justice: Listening to environmental victims' demands in Brazil, in *Green crime in the global South* (ed: D R Goyes), (Cham: Palgrave Macmillan). https://doi.org/10.1007/978-3-031-27754-2_9

Euclydes, F M, Pereira, J J and Fonseca, F C P da, 2022. The collapse of the Fundão dam: An analysis of the marginalization of affected communities in the post-disaster governance process, *Revista de Contabilidade e Organizações*, 16:e186049. https://doi.org/10.11606/issn.1982-6486.rco.2022.186049

Fonseca, A, Marconi, C A, Buhmann, K and Miranda, R S N, 2024. Assessing meaningful stakeholder engagement through ethics standards: Lessons from the Samarco dam break and its operational-level remediation program, in *The Routledge Handbook on Meaningful Stakeholder Engagement* (eds: K Buhmann, A Fonseca, N Andrews and G Amatulli), 1st edn (London: Routledge). https://doi.org/10.4324/9781003388227-35

Frynas, J G, 2008. Corporate Social Responsibility and International Development: Critical Assessment, *Corporate Governance: An International Review*, 16(4):274–281. https://doi.org/10.1111/j.1467-8683.2008.00691.x

Gilligan, C, 1982. *In a Different Voice: Psychological Theory and Women's Development* (Cambridge, MA: Harvard University Press).

Girschik, V, Svystunova, L and Lysova, E I, 2022. Transforming corporate social responsibilities: Toward an intellectual activist research agenda for micro-CSR research, *Human Relations*, 75(1):3–32. https://doi.org/10.1177/0018726720970275

Gond, J-P, Cabantous, L and Krikorian, F, 2018. How do things become strategic? Strategizing corporate social responsibility, *Strategic Organization*, 16(3):241–272. https://doi.org/10.1177/1476127017702819

Hamann, R, 2003. Mining companies' role in sustainable development: The 'why' and 'how' of corporate social responsibility from a business perspective, *Development Southern Africa*, 20(2):237–254.

Hitch, M and Barakos, G, 2021. Virtuous natural resource development: The evolution and adaptation of social licence in the mining sector, *The Extractive Industries and Society*, 8(3):100838. https://doi.org/10.1016/j.exis.2021.100902

Kemp, D and Owen, J R, 2022. Corporate social irresponsibility, hostile organisations and global resource extraction, *Corporate Social Responsibility and Environmental Management*, 29(5):1816–1824. https://doi.org/10.1002/csr.2329

Kemp, D and Owen, J, 2020. Corporate affairs and the conquest of social performance in mining, *The Extractive Industries and Society*. https://doi.org/10.1016/j.exis.2020.06.012

Kemp, D, Kochan, K and Burton, J, 2022. Critical reflections on the Juukan Gorge parliamentary inquiry and prospects for industry change, *The Extractive Industries and Society*, 10:101063. https://doi.org/10.1016/j.exis.2022.101063

Kemp, D, Owen, J R and van de Graaff, S, 2012. Corporate social responsibility, mining and 'audit culture', *Journal of Cleaner Production*, 24:1–10. https://doi.org/10.1016/j.jclepro.2011.11.002

Lavalle, A G, Leirner, A, Albuquerque, M, do, C and Rodrigues, F P, 2022. The voice of the communities: the construction of problems and proposals in the territories over time [in Portuguese: A voz das comunidades: construção de problemas e propostas nos territórios ao longo do tempo], in *Disaster and misgovernance in the Rio Doce: affected peoples, institutions, and collective action [Desastre e desgovernança no Rio Doce: atingidos, instituições e ação coletiva]* (ed: E Carlos), 436 p (Rio de Janeiro: Garamond).

Lopes, J C and Demajorovic, J, 2020. Corporate social responsibility: A critical view from the case study of Samarco's socio-environmental tragedy, *Revista de Administração Pública*, 54(2):331–348. https://doi.org/10.1590/0034-761220180197

Losekann, C and Milanez, B, 2023. Mining disaster in the Doce River: Dilemma between governance and participation, *Current Sociology,* https://doi.org/10.1177/00113921211059224

Maher, R, 2022. Correction to: Deliberating or Stalling for Justice? Dynamics of Corporate Remediation and Victim Resistance Through the Lens of Parentalism: The Fundão dam Collapse and the Renova Foundation in Brazil, *J Bus Ethics,* 178:37. https://doi.org/10.1007/s10551-021-04880-7

Nastar, M, 2023. A Critical Realist Approach to Reflexivity in Sustainability Research, *Sustainability*, 15(3):2685.

Nijhof, A and Jeurissen, R, 2010. The glass ceiling of corporate social responsibility: Consequences of a business case approach, *International Journal of Sociology and Social Policy*, 30(11/12):618–631. https://doi.org/10.1108/01443331011085222

Noddings, N, 1984. *Caring: A Feminine Approach to Ethics and Moral Education* (University of California Press).

Owen, J R and Kemp, D, 2013. Social licence and mining: A critical perspective, *Resources Policy*, 38(1):29–35. https://doi.org/10.1016/j.resourpol.2012.06.016

Parliament of Australia, 2020. Inquiry into the destruction of 46,000-year-old caves at the Juukan Gorge in the Pilbara region of Western Australia, Joint Standing Committee on Northern Australia. Available from: <https://www.aph.gov.au/Parliamentary_Business/Committees/Joint/Northern_Australia/JuukanGorge>

Pereira, A, Buhmann, K and Fonseca, M, 2024. Ethics of meaningful stakeholder engagement in post-disaster mining remediation, in *The Cambridge Handbook of Business, Human Rights and the Environment* (eds: K Buhmann, C Sheppard and J Ford), pp 421–429 (Cambridge University Press).

Pereira, J J, 2021. Forgive us our sins: A critical perspective of Corporate Social Irresponsibility (CSiR), insights from the case of the Samarco mining dam collapse in Brazil, PhD thesis, Fundação Getulio Vargas.

Primo, P P B, Antunes, M N, Arias, A R L, Oliveira, A E and Siqueira, C E, 2021. Mining Dam Failures in Brazil: Comparing Legal Post-Disaster Decisions, *International Journal of Environmental Research and Public Health*, 18(21):11346. https://doi.org/10.3390/ijerph182111346

Prno, J and Slocombe, D S, 2012. Exploring the origins of 'social license to operate' in the mining sector: Perspectives from governance and sustainability theories, *Resources Policy*, 37(3):346–357. https://doi.org/10.1016/j.resourpol.2012.04.002

Roberts, J, 1996. From discipline to dialogue: Individualizing and socializing forms of accountability, *Accounting, Organizations and Society*, 21(1):27–46.

Sultana, F, 2015. Reflexivity, positionality and participatory ethics: Negotiating fieldwork dilemmas in International Research, *ACME: An International Journal for Critical Geographies*, 6(3):374–385. https://doi.org/10.14288/acme.v6i3.786

Thomson, I and Boutilier, R, 2011. The social license to operate, in *SME Mining Engineering Handbook* (ed: P Darling), 3rd edn, pp 1779–1796 (Society for Mining, Metallurgy and Exploration).

Tronto, J, 1993. *Moral Boundaries: A Political Argument for an Ethic of Care* (Routledge).

Tronto, J, 2013. *Caring Democracy: Markets, Equality and Justice* (NYU Press).

Temporal and spatial changes in riverine sediment geochemistry at the Ok Tedi Mine

G Brumm[1], B Gafie[2], E Kepe[3] and R Schumann[4]

1. Geologist, Environmental Geochemistry International, Castle Hill NSW 2154.
   Email: glen.brumm@geochemistry.com.au
2. ARD Manager, Ok Tedi Mining Limited, Tabubil, Western Province, PNG.
   Email: ben.gafie@oktedi.com
3. Environmental Manager, Ok Tedi Mining Limited, Tabubil, Western Province, PNG.
   Email: erizo.kepe@oktedi.com
4. Principal Environmental Geochemist, Environmental Geochemistry International, Castle Hill
   NSW 2154. Email: russell.schumann@geochemistry.com.au

INTRODUCTION

The Ok Tedi Au-Cu-Ag mine is located in the western province of Papua New Guinea (PNG) at the headwaters of the Ok Tedi. Steep terrain and high annual rainfall (10 000 mm) in the mining area preclude conventional waste rock storage dumps or a tailings storage facility (TSF). Attempts to construct a TSF at the commencement of mine operations resulted in collapse of the partially constructed facility following a seismic event. Subsequent to this, tailings from the flotation plant were discharged directly to the river system, a practice which continues today. Waste rock enters the river system via erodible waste rock dumps.. Riverine discharge of Ok Tedi mine waste rock and tailings are permitted under Section 13 of the Ok Tedi Environmental Management Act, which allows treated fine-grained waste and tailings to be placed into upper-Ok Tedi river valleys. In 2023, 16.5 Mt of treated tailings was discharged into the upper Ok Tedi and 65.5 Mt of waste rock was emplaced in dumps to the north and south of the mine which eroded into the river system.

Riverine sediment discharge is the most significant long-term environmental concern for the mine. To reduce environmental impact, Ok Tedi Mining Limited (OTML) have employed a variety of measures introduced over the past 20 years. These include river dredging at Bige, which commenced in 2001 to remove mine waste from the river system downstream of the mine, addition of limestone to failing waste rock dumps (2006) to mitigate acid rock drainage (ARD), tailings desulfurisation (2008), and addition of limestone to the processing plant mill (2010) to reduce the risk of ARD from tailings discharged to the river. Additionally, construction of an engineered waste rock dump (EWRD) commenced in early 2022, with the goal of reducing sediment loadings in the river system. In the future, construction of a second EWRD and a TSF downriver from the mine are being investigated.

OTML conduct an annual survey of sediment geochemistry, collecting samples from the Ok Tedi just below the mine to near the Fly River estuary over a distance of approximately 1000 km. These surveys provide a database containing information on both spatial and temporal changes to riverine sediment geochemistry. The data are used to determine the effectiveness of control measures in reducing the impact of mine waste on riverine environment and communities, and aid planning for future mitigation and closure strategies.

RIVERINE SEDIMENT SURVEY – RESULTS AND DISCUSSION

This paper summarises riverine sediment geochemical trends along the transport pathway of mine waste, from the Ok Tedi Mine site to the start of the Fly River Delta. Data covers a span of 23 years between 2001 to 2023. The data includes the ARD parameters total sulfur, Acid Neutralising Capacity (ANC) and Net Acid Production Potential (NAPP), and metals including copper, cadmium, lead and zinc in riverine and floodplain sediments.

Overall, data indicate that sediments in the Ok Tedi and Fly River systems continue to show the influence of sediments transported from the Ok Tedi mine. However, across the monitoring period 2001 to 2023, sulfur content, ANC and NAPP have improved substantially, with the vast majority of samples unlikely to produce ARD. Copper levels in the riverine sediments have likewise decreased significantly in all sections of the river system.

Figure 1 summarises data for the middle Fly River, showing relatively stable Total S, increasing ANC and decreasing NAPP, such that for the most recent period (2021–2023) almost all samples gave negative NAPP values indicating very low likelihood of riverine sediments in this reach of the river producing ARD. Sediment copper content has also decreased substantially between 2001 and 2023, albeit the most recent data indicate concentrations are still significantly above pre-mining concentrations. Temporal trends in the concentrations of other metals (Pb, Zn and Cd) are less pronounced, but, in the main, appear to remain above pre-mining concentrations. This pattern of decreasing ARD potential and decreasing metal concentrations in sediments sampled between 2001 and 2023 is repeated in samples collected from other sections of the river system downstream from the mine.

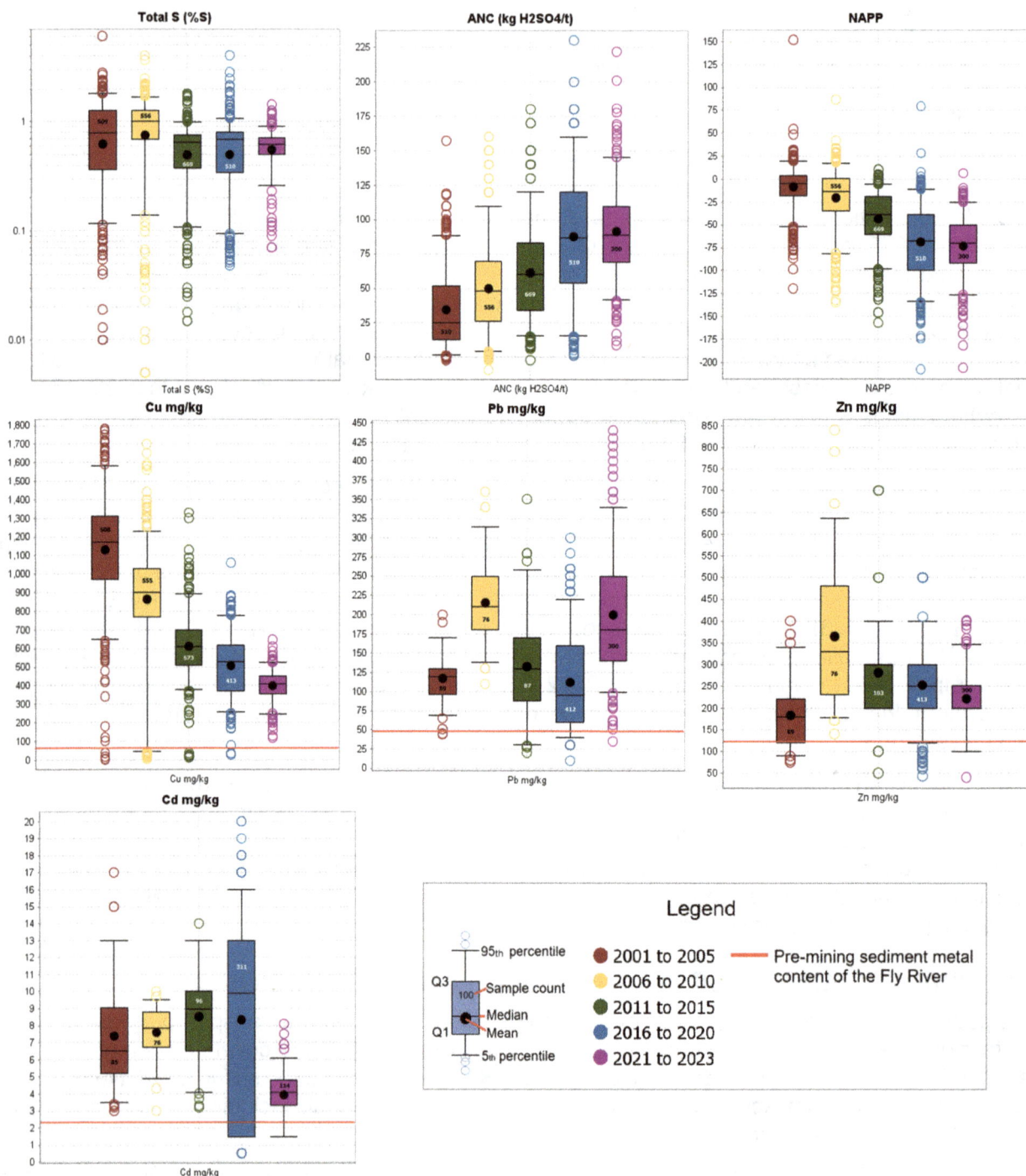

FIG 1 – Temporal geochemical trends for riverine sediments from the middle Fly River.

The substantial decrease in the ARD potential of riverine sediments can likely be attributed to the operational changes described in the introduction which OTML have introduced over the past 25 years. The addition of limestone to both failing waste rock dumps and ore during processing has led to a substantial increase in the ANC of mine wastes discharged to the river system, while desulfurisation of tailings has reduced their acid generating potential. Testing has demonstrated that sulfur in sediments in the Ok Tedi is predominantly associated with particle sizes >75 μm. Sediments in this size fraction are more efficiently removed by dredging operations conducted in the lower reaches of the Ok Tedi than smaller particles, resulting in further decreases in S content and ARD potential of sediments further downstream.

The decrease in Cu concentrations in sediments might also be attributed in part to desulfurisation of tailings, which removes not only pyrite (decreased Total S), but also residual chalcopyrite not recovered during bulk flotation (decreased Cu).

CONCLUSIONS

Sediment geochemical trends of the river system impacted by the Ok Tedi mine are influenced by multiple factors, including sediment deposition and dilution downriver by converging tributaries, and mine remediation and waste management procedures. The success of mine waste rock and tailings management practices are demonstrated by changes in riverine sediment geochemistry, including substantial improvements in copper concentrations and reduction in the ARD potential of river sediments between 2001 to present. While depositing mine waste rock and tailings into the Ok Tedi and Fly River system continues to be of major environmental concern, evaluation of riverine sediment data collected by OTML has shown that the measures they have undertaken have improved the environmental outlook for the river system as a whole. However, sediment survey results continue to show that riverine sediments as far down as just past the confluence of the Strickland and Fly rivers remain impacted by mine wastes (Figure 2). Future monitoring results will indicate whether the construction of EWRDs and installation of a TSF near Ningerum will also result in a significant decrease in adverse mine impact on sediment geochemistry in the downstream river system.

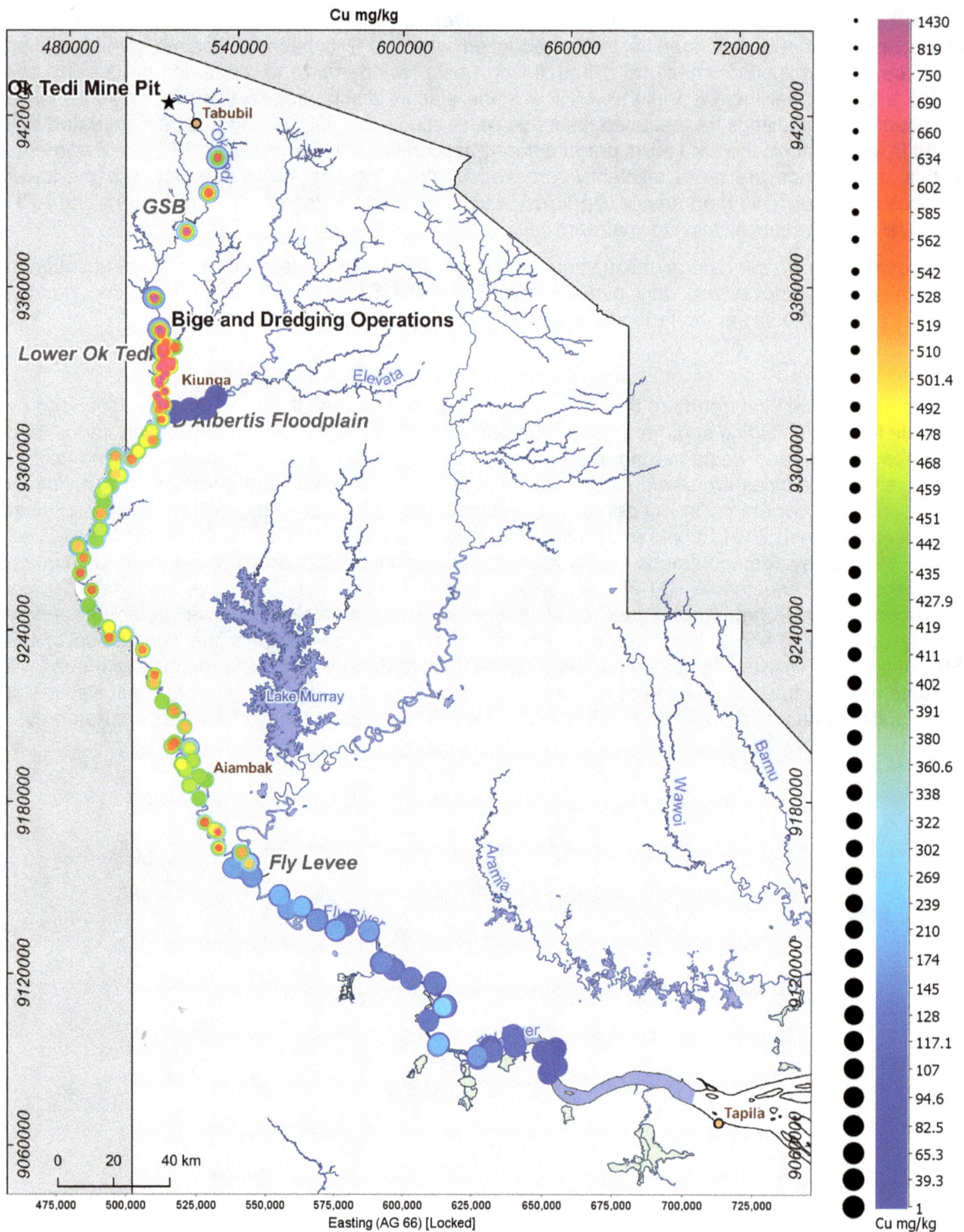

FIG 2 – Copper results for riverine sediment sampling along the Ok Tedi and Fly River systems. Scale to the right of the image shows data range. Data is compiled from 2021 through 2023.

Circularity in mining – innovation strategies for tailings as a resource

A Fernandez-Iglesias[1], F Vasconcelos[2] and H Oliveira[3]

1. Director, Sustainable Mining Portfolio, ArcelorMittal Global R&D, Spain.
   Email: ana.fernandez-iglesias@arcelormittal.com
2. Head of Mining R&D, ArcelorMittal Global R&D, France.
   Email: filipe.vasconcelos@arcelormittal.com
3. Research Engineer, ArcelorMittal Global R&D, Spain.
   Email: henrique-junio.oliveira@arcelormittal.com

ABSTRACT

The reuse of tailings presents significant opportunities and challenges for ensuring a circular economy within the mining industry. Iron ore is the upstream material for the steelmaking industry, and thus, enhancing circularity in mining operations can significantly contribute to achieving full circularity across the entire value chain. However, reusing mining waste, such as iron ore tailings, is fraught with complexities, technical challenges, and economic considerations. This paper examines the efforts and experiences of the research and development sector within the iron ore and steel industry, focusing on the establishment of circular solutions at mining sites across various continents.

The challenges of finding alternative uses for tailings include low availability of data, variability in tailings composition, difficulties in sampling, and challenges associated with remining tailings storage facilities or end of pipe streams. Additionally, identifying suitable local or regional applications and the lack of knowledge about potential applications of tailings further complicate the reuse process. Despite these challenges, the positive aspects of tailings circularity are significant. Reusing tailings can reduce liabilities and risks associated with tailings storage, as well as lower costs related to transport, pumping, dewatering, disposal, monitoring, and closure.

This paper highlights the benefits and obstacles encountered in the quest for sustainable tailings management. The findings underscore the importance of continued innovation, investment in research, and the development of robust frameworks to support the transition towards a more sustainable and circular mining industry. Forming partnerships with universities, research centres, other companies, and institutions is critical for overcoming the challenges associated with reusing iron ore tailings, as these collaborations facilitate the exchange of knowledge, innovative technologies, and resources, enabling the development of more effective and sustainable solutions. Addressing the technical and economic challenges associated with tailings reuse is crucial for paving the way for a more resilient and environmentally responsible industry, ultimately contributing to the global goals of sustainability and resource efficiency.

INTRODUCTION

Tailings, the residual materials produced during mineral processing, have long been considered an inevitable by-product of mining operations. However, the global push for sustainability and the advent of the circular economy are shifting perceptions of tailings from waste to resource. This transformation is critical for reducing the environmental footprint of mining and aligning the industry with global sustainability objectives.

The steel industry, as a major consumer of iron ore, has a vested interest in adopting circular approaches to tailings management. This paper explores research and development initiatives in this regard, highlighting how innovations in tailings reuse can drive circularity in mining and steel production.

BACKGROUND

The global volume of tailings is projected to increase substantially due to declining ore grades and the intensification of mining activities to meet demand for critical raw materials used in low-carbon and digitalisation technologies. As larger quantities of ore must be processed to extract the same amount of valuable metals, the generation of tailings is expected to continue growing, but due to the complexity of the issue and the diverse and extensive nature of the mining industry, rigorous scientific studies providing quantitative forecasts of future tailings generation growth are currently lacking. What is clear is that emerging processing technologies have enabled the exploitation of previously uneconomic deposits, further amplifying tailings output and stressing existing storage infrastructure.

To address this growing burden, the mining sector is advancing toward more sustainable and integrated tailings management frameworks. Primary metal entering the market should be produced sustainably, responsibly and with circularity principles built into operations and processes (ICMM, 2025a). Current research and industry practice are converging on circular economy principles, including the valorisation of tailings as secondary resources in construction materials, metal recovery, and land rehabilitation. These approaches aim not only to mitigate environmental liabilities but also to embed tailings management within a broader resource efficiency paradigm, aligning with regulatory expectations and long-term sustainability objectives.

MINING CIRCULARITY

Circularity in metals production – expanding the value chain

The concept of circularity in metals production has gained significant traction as industries seek to align with global sustainability goals and resource efficiency mandates. Metals, characterised by their high recyclability and enduring properties, are ideal candidates for circular economy frameworks. Integrating circular principles into the full life cycle of metals involves optimising the use of raw materials, minimising waste, and advancing the reuse of by-products like tailings, slag, and dust. These approaches not only reduce environmental impact but also foster economic benefits by creating secondary markets for repurposed materials. By extending circularity beyond manufacturing processes to encompass upstream mining operations, the metals industry can enhance its sustainability performance significantly.

Efforts to integrate circularity into the full value chain of metals production are being championed by influential organisations and alliances worldwide. The International Council on Mining and Metals (ICMM) advocates for sustainable practices that prioritise the reuse and valorisation of mining by-products, emphasizing the need for comprehensive circular frameworks in the sector. Similarly, the European Raw Materials Alliance (ERMA) supports initiatives aimed at enhancing resource efficiency and reducing dependence on virgin materials, particularly in the context of Europe's transition to renewable energy and green technologies. These organisations encourage collaboration between industry leaders, academia, and governments to address systemic challenges and scale innovative solutions. Their work highlights the importance of a unified approach to circularity, ensuring that mining activities are fully integrated into the sustainability strategies of metal production industries (ICMM, 2025b; ERMA, 2023).

The circular economy emphasises the reuse of waste as a resource, transforming traditional mining practices. Studies have highlighted the potential of tailings for value creation, such as extracting critical raw materials (Salminen et al, 2019). By implementing circular economy principles across the value chain, the metals industry has the opportunity to pave the way for resilient and future-proof operations that align with global sustainability priorities.

Visualising the material flows in metal production

This Sankey diagram showcases the flow of materials in a steel production process, highlighting how raw materials are converted into steel and the by-products streams generated at each stage. Data used for this approximation does not reflect a single mine, but an average of different iron ore mines from different countries and it has been developed for research and academic purposes.

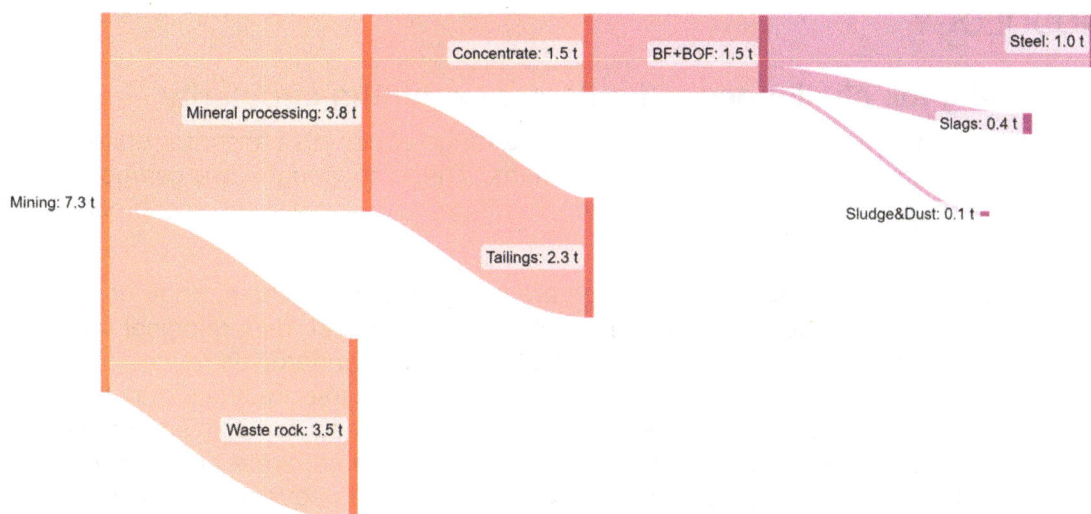

FIG 1 – Estimation of material flows for steel production including mining.

The right half of this diagram represents the steelmaking process, which generates approximately 0.5 tons of by-products, most of them slags from various stages such as blast furnaces, basic oxygen furnaces, and electric arc furnaces. For these materials, multiple market applications have been tested and are now well-established as examples of industrial symbiosis.

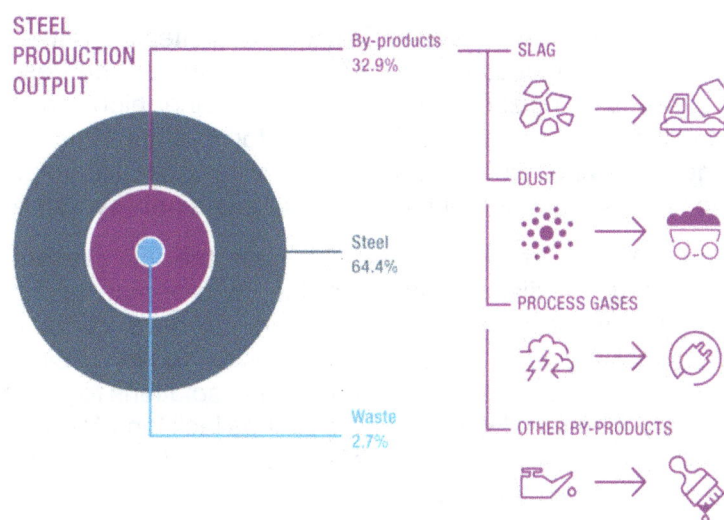

FIG 2 – Steel production output (World Steel Association, 2023).

These industrial circularity applications can even contribute to the decarbonisation of other sectors. ArcelorMittal highlights in its 2023 FactBook the reuse of 12.7 million tonnes (Mt) of blast furnace slag, with 9.2 Mt repurposed by the cement industry (avoiding 7.1 Mt of CO_2 emissions) demonstrating the transformative potential of leveraging industrial by-products.

This example underscores the vast opportunities for reusing or recycling mining waste streams in other industries, especially considering the substantial material flows generated in mining and processing operations. By capitalising on these streams, industries can significantly reduce environmental impacts, enhance circularity, and unlock economic benefits through sustainable practices. The synergy between industries, such as mining and construction, paves the way for innovative solutions to global sustainability challenges.

METHODOLOGY

Understanding the challenges to implement a mining circularity

Implementing mining circularity requires overcoming a range of complex and interrelated challenges that span technical, logistical, and regulatory domains. These challenges are compounded by the heterogeneous nature of tailings and the inherent difficulties in accessing and processing these materials for reuse.

- Variability in Tailings Composition: Tailings are composed of a diverse range of size fractions, from coarse particles to fine sediments, each with distinct mineralogical and chemical properties. This variability complicates the identification of standardised reuse applications and demands tailored processing approaches. For example, coarse fractions may be suitable for construction aggregates, while fine fractions could be used for soil amendments or the extraction of trace elements. The technical requirements to separate and characterise these fractions add complexity to circularity initiatives, necessitating advanced analytical techniques and specialised equipment to optimise their potential reuse.

- Low Availability of Data: A critical barrier to circular tailings management is the scarcity of reliable and comprehensive data sets on tailings composition and characteristics. Limited data availability hampers efforts to assess the feasibility of reuse applications and develop targeted solutions. Moreover, inconsistent methodologies for tailings sampling and characterisation across different mining operations exacerbate this challenge, creating gaps in understanding that hinder progress toward circular practices.

- Logistical Issues Due to Remote Locations: Many mining sites are situated in remote regions, far from potential markets or industries where tailings could be repurposed. This geographical isolation poses significant logistical challenges, including high transportation costs and difficulties in establishing efficient supply chains for tailings reuse. Furthermore, limited infrastructure in these areas often restricts the scalability of circular solutions and complicates efforts to integrate tailings management into broader sustainability frameworks.

- Using End-of-Pipe Tailings Versus Stored Tailings: tailings present distinct challenges depending on their stage of disposal and storage. End-of-pipe tailings, which are freshly generated during mineral processing, exhibit high moisture content that often requires immediate dewatering or drying to meet specifications for reuse. Their continuous production necessitates real-time collection, treatment, and storage solutions to prevent accumulation and operational disruptions. Additionally, end-of-pipe tailings lack long-term stabilisation, which can lead to segregation during handling or transport, complicating processing and repurposing workflows. The logistical complexity of aligning end-of-pipe tailings with reuse applications demands advanced management strategies, infrastructure development, and the integration of innovative drying and processing technologies.

FIG 3 – Different samples of iron ore tailings.

In contrast, stored tailings within tailings storage facilities (TSFs) pose their own unique challenges. These tailings, often dewatered using thickeners, paste thickeners, or filters before disposal, are stored in facilities with varying designs and degrees of complexity. Accessing stored tailings for reuse requires overcoming technical obstacles, such as the solidification and compaction that occur over time, which complicate excavation and processing efforts. Furthermore, the structural integrity and long-term stability of TSFs must be carefully managed during remining to avoid safety risks and potential environmental impacts, such as contaminant release or landscape alteration. The recovery of tailings from TSFs often demands specific moisture adjustments to match reuse applications, adding to the technical requirements. Addressing these challenges requires tailored excavation methods, risk assessments, and sustainable strategies to balance efficiency with environmental responsibility, ensuring that tailings from both sources are effectively integrated into circular mining practices.

- Remining Permits for Existing Tailings Dams: Regulatory hurdles are a significant impediment to mining circularity, particularly concerning remining permits for legacy tailings storage facilities. Obtaining these permits involves navigating complex and time-intensive approval processes, which often discourage investment and delay implementation. Furthermore, differing regulatory frameworks across jurisdictions create inconsistencies that complicate the standardisation of tailings reuse practices.

- Regulatory Barriers: As the mining industry moves toward adopting circularity principles for tailings management, a range of potential regulatory barriers can be anticipated. One significant challenge is the forecasted complexity of permitting processes for remining and reusing tailings. Current regulatory frameworks, tailored to conventional linear practices, are likely to lag behind the innovative approaches needed for circularity. Navigating intricate and protracted approval systems could discourage investment and delay progress. Furthermore, the absence of standardised international guidelines for tailings reuse may result in inconsistent requirements across jurisdictions. This inconsistency could complicate the global implementation of circular strategies, as mining companies struggle to align their operations with disparate regulatory expectations. Additionally, many regions are expected to continue classifying tailings as hazardous waste, which could impose strict limitations on handling, transportation, and application, thereby hindering their integration into sustainable reuse workflows.

Developing a systematic innovation approach

This strategy map illustrates a proposal of a systematic approach to develop circular mining solutions for tailings but could be applied to any other mining waste. The process begins with Information Gathering, which involves collecting critical data on tailings properties, such as composition, size fractions, production flows, etc. In parallel, this is complemented by a continuous Application Scouting, where feasible uses for the studied material are explored, targeting innovative applications that align with circular economy principles. These two foundational steps enable the detection of promising opportunities, depicted in the diagram as a green circle with a lightbulb icon. This milestone represents the identification of a project idea with potential environmental, economic, and industrial benefits.

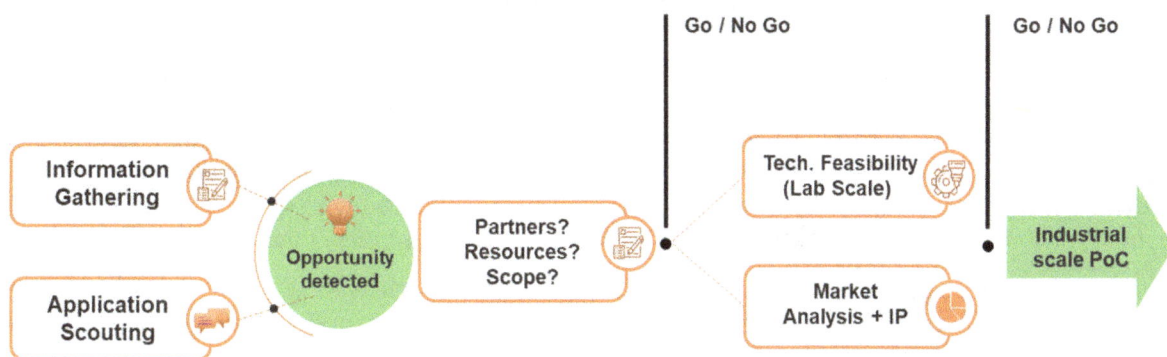

FIG 4 – Proposed approach for development of circular tailings applications.

Once an opportunity is identified, a comprehensive evaluation phase to assess the resources, partners, and scope necessary for the project's success is carried out. This phase involves identifying all the key elements to ensure that the technical, financial, and logistical requirements of the project are met. The evaluation phase culminates in a 'Go/No Go' decision point, where the project is either approved to advance or deemed impractical based on the findings.

Approved projects then proceed to Technical Feasibility Testing at Lab Scale, where the necessary laboratory test work is carried out with the target waste material in order to evaluate if it can be used as a substitute of the original raw material. In parallel, market analysis and intellectual property (IP) protection are addressed to ensure commercial viability and safeguard proprietary innovations. Another Go/No-Go checkpoint determines whether the project is ready for large-scale application.

If the project successfully passes all evaluation stages, it transitions to Industrial-Scale Proof of Concept (PoC). This final phase involves implementing and testing the circular solution on a larger scale, integrating it within existing mining operations. The progression from initial data collection to industrial application reflects a solid commitment to innovation and sustainability. By combining technical research, market readiness, and collaboration with strategic partners, R&D ensures that circular mining solutions not only address environmental challenges but also create lasting value across the mining and metals industry.

While the described strategy is effective in identifying potential opportunities for circularity in tailings management, it does not address all the challenges outlined. This methodology provides a structured approach for gathering data, scouting applications, and testing technical feasibility, enabling the recognition of promising projects. However, critical issues such as logistical barriers in remote locations, the variability in tailings composition, and the regulatory complexities surrounding remining permits remain only partially addressed. Additionally, challenges like the absence of standardised data sets and the financial and operational burdens associated with transitioning to circular practices require further solutions.

To overcome these unresolved challenges, a holistic approach is essential, one that combines joint innovation with active engagement of all key stakeholders. Collaborative efforts between mining companies, research institutions, policymakers, and local communities can drive the development of advanced technologies, standardised frameworks, and streamlined permitting processes. By working together, stakeholders can align objectives and resources to address these complexities comprehensively, accelerating the adoption of circularity in mining while fostering both environmental and economic sustainability.

RESULTS

The strategy successfully identified diverse applications for iron ore tailings by leveraging their high silica content for both advanced and foundational uses. Despite variations in size fractions and compositions caused by differences in ore mineralogy and beneficiation processes, the consistent presence of silica remains a key asset. Typically found as quartz or silicate minerals, this silica underpins numerous reuse possibilities, aligning with circular economy principles and transforming mining waste into valuable resources.

Among high-end applications, silica extracted from iron ore tailings can be purified for use in semiconductors, solar panels, and batteries. These industries require high-purity silica for its advanced optical, thermal, and electronic properties, enabling the development of cutting-edge technologies and delivering considerable economic value. On the other hand, construction-related uses, such as the production of cement, ceramic products, and geopolymers, demonstrate the potential to repurpose tailings in large-scale industrial applications. These uses demand less processing but still offer notable environmental benefits by reducing reliance on virgin materials. By employing innovative strategies, iron ore tailings can be efficiently repurposed across various industries, showcasing their versatility and alignment with sustainability initiatives.

The results emphasise that innovative strategies, combined with collaborative stakeholder engagement, play a critical role in unlocking the full potential of tailings reuse. By systematically identifying applications and addressing challenges such as composition variability and logistical constraints, the methodology laid the groundwork for effective repurposing of tailings. However, overcoming regulatory hurdles and refining technological processes remain necessary steps to realise the full scope of opportunities. Transforming tailings into both high-end and construction-related products demonstrates the capacity of mining waste to contribute to sustainable practices while advancing environmental and economic objectives.

REFERENCES

ArcelorMittal, 2023. FactBook 2023, Luxembourg: ArcelorMittal. Available from: <https://corporate.arcelormittal.com>

European Raw Materials Alliance (ERMA), 2023. Strategic Action Plan 2023–2030: Securing Critical Raw Materials for Europe, European Commission. Available from: <https://erma.eu>

International Council on Mining and Metals (ICMM), 2025a. Measuring Circularity Within Metals Value Chains, International Council on Mining and Metals. Available from: <https://www.icmm.com/en-gb/research/innovation/2025/measuring-circularity-within-metals-values-chains>

International Council on Mining and Metals (ICMM), 2025b. Tailings Management: Good Practice Guide [online], International Council on Mining and Metals. Available from: <https://www.icmm.com/tailings-management>

Salminen, J, Garbarino, E, Orveillon, G, Saveyn, H, Mateos Aquilino, V, Llorens González, T, García Polonio, F, Horckmans, L, D'Hugues, P, Balomenos, E, Dino, G, de la Feld, M, Mádai, F, Földessy, J, Mucsi, G, Gombkötő, I and Calleja, I, 2019. Recovery of critical and other raw materials from mining waste and landfills: State of play on existing practices (eds: G A Blengini, F Mathieux, L Mancini, M Nyberg, and H M Viegas), (JRC116131, EUR 29744 EN), Publications Office of the European Union, Luxembourg. https://doi.org/10.2760/600775

World Steel Association, 2023. Steel Facts 2023: Steel Recycling and Circularity, Brussels: World Steel Association. Available from: <https://worldsteel.org>

Applying GISTM to closed facilities – challenges and solutions

C Gimber[1], H McKay[2] and S Eagle[3]

1. Partner, ERM, Brisbane Qld 4000. Email: chris.gimber@erm.com
2. Consulting Director, ERM, Brisbane Qld 4000. Email: heather.mckay@erm.com
3. Consulting Director, ERM, Brisbane Qld 4000. Email: stefani.eagle@erm.com

INTRODUCTION

The Global Industry Standard on Tailings Management (GISTM) establishes a globally applicable framework for effective tailings management and governance across all stages of the facility life cycle (Global Tailings Review (GTR), 2020). Many organisations have chosen to implement GISTM across their portfolios, which often include a mix of planned, operational, and closed facilities. Applying GISTM to closed or legacy assets presents unique challenges, which are the focus of this paper.

SAFE CLOSURE

GISTM defines 'safe closure' of a tailings facility as not posing ongoing material risks to people or the environment. Following a safe closure designation, the facility is no longer deemed to be a tailings facility for the purposes of GISTM. As 'safe closure' relates primarily to the management or control of risks to people and the environment, a facility may be in a state of 'safe closure' without necessarily having achieved all the requirements for responsible closure (eg without having achieved post-closure land use objectives) (International Council on Mining and Metals (ICMM), 2025). Therefore meeting completion criteria and achievement of 'safe closure' do not necessarily happen together. It is possible to be in a state of 'safe closure' while not having met completion criteria. Conversely the reverse can also be true, especially in jurisdictions with poorly developed regulatory regimes or where there are narrow completion criteria.

A particularly complex issue arises when regulatory authorities are satisfied with the existing risk profile and are reluctant to authorise further work, yet 'safe closure' has not been fully achieved. Complying with the legislative regime is critical, and is compatible with GISTM, although in these situations may prolong or prevent the achievement of 'safe closure'. This situation has arisen on a TSF in France where the regulator has prevented further consultation with affected communities due to the low probability of failure.

KNOWLEDGE BASE

One of the key challenges that emerge for closed facilities is that there are often knowledge gaps in the design or construction practices. Commonly, the corporate knowledge that once existed has been lost through the passage of time. As a consequence, it is often necessary to conduct a historical review using information sources such as historical aerial photography, design drawings, operational records and conduct interviews with long-term or previous employees. It is usually necessary to undertake supplementary investigations, such as intrusive investigations (eg drilling), geophysics and/or remote sensing. It is important to prioritise information collection and data collection using a risk-based approach, so that the investigations are targeted and focused on the most material knowledge gaps. All of the TSFs reviewed by the authors, from a range of commodities and geographies, have had knowledge gaps that have needed to be addressed through supplementary investigations.

CLOSURE DESIGN

As commonly identified, towards the end of the facility life cycle, the number of options reduce and costs increase, at a time when the mining operation is not income generating. GISTM requires that closure design reduce the risk to As Low As Reasonably Practical (ALARP), and requires that all reasonable measures be taken with respect to 'tolerable' or acceptable risks to reduce them even further, until the cost and other impacts of additional risk reduction are grossly disproportionate to the benefit. There are a wide array of approaches used by practitioners to demonstrate ALARP has been achieved, ranging from qualitative to quantitative methodologies (International Council on Mining and Metals (ICMM), 2021). It is important to have a well-documented assessment, that has been approved by the Accountable Executive as required under GISTM.

MONITORING AND SURVEILLANCE

A tailings facility in the closure or post-closure phase may require even more monitoring and surveillance than it did in earlier phases of the life cycle. This may be a challenge especially when there are few or no personnel on-site, and often no supporting infrastructure such as power supply, communications or accommodation.

Many closed facilities lack functional monitoring equipment or up-to-date geotechnical, hydrological, and environmental data. In this case it is necessary to upgrade monitoring equipment and adapt to the changed site presence through the use of passive or autonomous systems, in combination with remote sensing techniques. Such equipment requires suitable design, ongoing maintenance and funding.

EVOLVING LANDSCAPE

The design or construction practices that were employed at old facilities often do not meet today's minimum standards, resulting in technical, environmental and social risks, despite complying with legal and regulatory obligations of the time. This may mean that further work is needed to adequately mitigate these risks and bring the facility into alignment with GISTM.

Closed facilities are not static and continue to change, as does the social, economic and environmental landscape within which they sit. Closure and post-closure periods can be very long (eg multiple decades) and recognising and responding to the changes over these extended periods is necessary, but challenging. The tailings facility itself can change too, for example by erosion or geotechnical instability caused by water movement, seismic activity or water infiltration. Tailings can oxidise, weather or release contaminants, posing geochemical risks that were not previously seen.

Over time, the social context around a tailings facility can shift dramatically, which may also impact on its consequence category. There may be: population growth; changing communities who are not aware of the risks from a nearby closed facility; communities may fall into a state of complacency; there may be changes in community expectations; and, there may be cultural and demographic shifts. Being connected to local communities can be increasingly challenging when sites have closed and there is limited on-site presence. Similarly, the sensitivities of engaging with communities on failure consequences and possible inundation for sites that have been inactive for many years needs consideration.

It is important that these changes are recognised and that the community stays engaged. This requires a robust plan, adequate resources and long-term view.

RESOURCING

ICMM member companies are required to implement GISTM, and many non-member companies have also publicly committed to implementation. GISTM implementation for closed sites requires ongoing investment for monitoring, auditing, corrective actions, reporting, community engagement and governance structures. Many closure provisions do not adequately cater for the ongoing trailing cost of GISTM implementation. It is important that such costs and resources are adequately planned and budgeted for.

CONCLUSIONS

There are many closed and legacy tailings facilities that will need to be brought into alignment with GISTM. The effort and resources to bring these facilities into alignment should not be underestimated – it will require funding, governance structures, community engagement, monitoring and accountability for many years, potentially many decades. While each site has its own unique circumstances and will apply GISTM in different ways, there are common themes that emerge to build and maintain the knowledge base, implement closure designs to achieve ALARP, monitor the physical and socio-economic context, and maintain connection with communities at risk. Sharing of case studies between industry peers will provide learnings and streamline the rollout of GISTM across closed and legacy assets.

REFERENCES

International Council on Mining and Metals (ICMM), 2021. *Conformance Protocols – Global Industry Standard on Tailings Management,* May 2021. Available from: <https://www.icmm.com/website/publications/pdfs/environmental-stewardship/2021/tailings_conformance-protocols.pdf?cb=21097>

International Council on Mining and Metals (ICMM), 2025. Tailings Management: Good Practice Guide [online], International Council on Mining and Metals. Available from: <https://www.icmm.com/tailings-management>

Global Tailings Review (GTR), 2020. Global Industry Standard on Tailings Management (GISTM) [online], Global Tailings Review. Available from: <https://globaltailingsreview.org/global-industry-standard/>

Sandy Ridge – a master-planned precinct with regional residue management enabling future material circularity

S Hosking[1], N Smith[2], N Blight[3] and M Gravett[4]

1. Chief Operating Officer, Tellus, Perth WA 6000. Email: steve.hosking@tellus.com.au
2. Managing Director and Chief Executive Officer, Tellus, Sydney NSW 2000.
   Email: nate@tellus.com.au
3. GM Studies, Tellus, Perth WA 6000. Email: nathan.blight@tellus.com.au
4. Technical Director, Resource Recovery and Waste Management, GHD, Perth WA 6000.
   Email: martin.gravett@ghd.com

INTRODUCTION

The Tellus Sandy Ridge facility is located 240 km north-west of Kalgoorlie in the Goldfields region of Western Australia. It is Australia's only licensed and continuously operating geological repository for permanent isolation of hazardous and radioactive waste.

In establishing the facility, Tellus negotiated a unique trust structure ensuring that when the facility ceases operation, the closure and 100-year monitoring period led by the WA Government will be fully funded. Tellus' assurance framework ensures that liability cannot revert to waste generators, or Australian taxpayers, while creating tangible value to generators by certifying balance sheet liability reduction. Tellus offers clients a permanent isolation certificate that enables them to remove hazardous waste disposal liability from their balance sheet under Australian Accounting Standards.

For materials with reuse potential, new developments at the Sandy Ridge facility will play a pivotal role in supporting circularity for minerals processing by-products and other residual materials through innovative management practices. These new developments are the basis of the proposed Sandy Ridge Sustainability Precinct.

MINSTORE

Central to the Sandy Ridge Sustainability Precinct model is 'MinStore', a regional residue storage solution that enables tailings management outsourcing across industries within a master-planned precinct. MinStore offers secure long-term storage, while also facilitating R&D into residue reuse and future backloading of fully characterised stored materials.

MinStore will be connected by rail, and comprise monocells (long-term storage facilities that receive only one type of solid material, or material from a single source) for minerals processing residues. Outsourcing of tailings/residue management enables distribution of establishment and operational overhead costs across multiple industry participants, reducing the cost burden for individual customers.

A key driver for the MinStore concept has been supporting the development of a viable Australian critical minerals industry and offering an outsourced materials management solution for minerals processing businesses. Establishment of residue storage monocells seeks to realise economy of scale and distribution of overheads to provide secure long-term storage of secondary materials that may become useful in the future. Developing MinStore progressively within the precinct and integrating the placement of monocells with waste cells will reduce the overall disturbance footprint.

The main development area set aside for MinStore is immediately east of the existing Sandy Ridge facility (shown in Figure 1).

FIG 1 – Tellus' Sandy Ridge facility.

Minerals processing facilities are often space-constrained and built in industrial estates or approved with minimal on-site storage of products and by-products. At times, legislation can also restrict disposal of minerals processing residues in disused mine voids that are not the source of the minerals being processed.

A key advantage of the MinStore concept over potentially cheaper, on-site alternatives at point of origin is the ability to store decades of by-product output from a minerals processing facility without impacting valuable space on constrained sites. In addition, potential impacts on sensitive receptors at point of origin will be negated. Finally, residue management is outsourced to a third party focused on identifying and developing reuse, blending and reprocessing opportunities to improve material use efficiency.

What makes MinStore different from a conventional tailings storage facility (TSF) or monocell design is the managed precinct model, which enables long-term efficient storage, monitoring and maintenance of by-product streams. Furthermore, development of individual single-material stacks within a master-planned precinct may offer scale-based operational cost efficiencies relative to a discrete, single customer TSF. Individual stacks can be capped with site-generated kaolin and protected against erosion by regular inspection and maintenance.

To recognise economies of scale for MinStore users, the site establishment, common infrastructure development, maintenance costs and operating overheads can be distributed across multiple customers, on a multi-year basis, reducing the unit cost per tonne for material managed or stored for individual customers.

An indicative layout in Figure 2 shows nine dry-stack tailings cells, each nominally sized to accommodate about 400 000 m^3 per annum for 25 years, and two smaller cells. Each material stack requires its own leachate pond to avoid cross-contamination of leachate from different residue materials. The annual material placement volume of 400 000 m^3 of residue for each large cell was derived from a scoping study commissioned by a potential MinStore customer, and the smaller cells were conceptually based on lower annual volumes of a different by-product, considered as part of a subsequent study commissioned by a second potential customer.

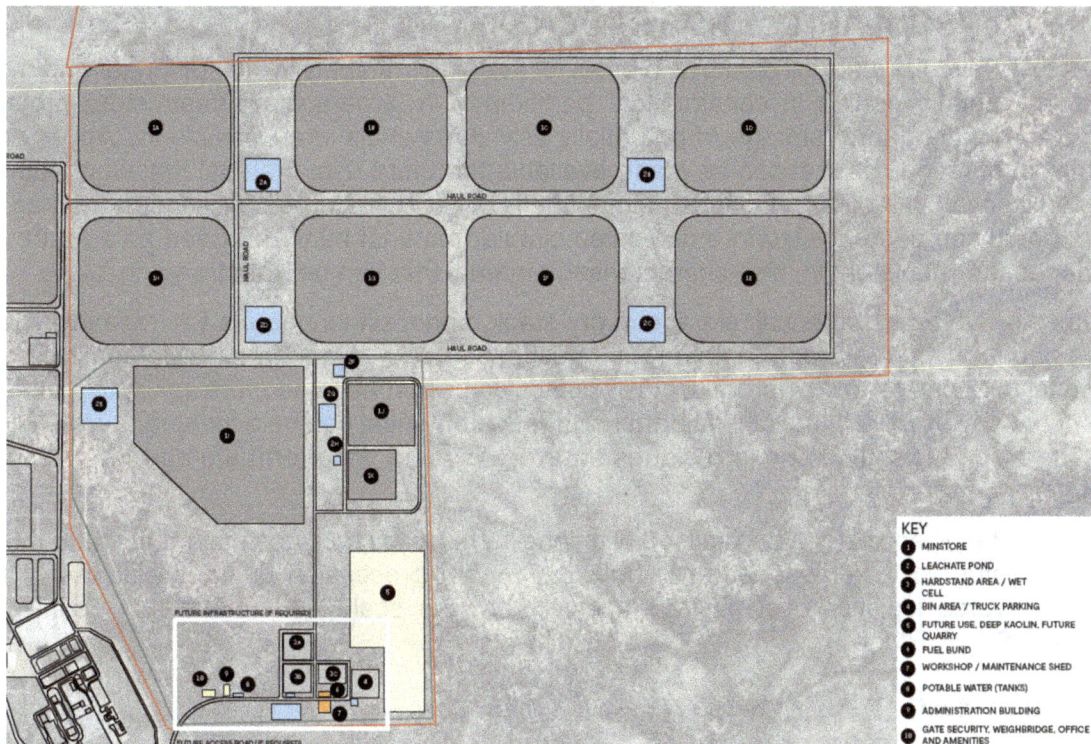

FIG 2 – Indicative layout of MinStore.

MinStore must be located in a suitable environment, so that if future beneficial uses of all stored materials are not found, MinStore can be safely closed and rehabilitated. The precinct lies within the Yilgarn craton that has been in place for about 250 million years, and the combination of a virtually flat plateau, cemented surface layers, dry geology and semi-arid conditions create a stable geomorphology.

The conceptual site model in Figure 3 highlights key characteristics of low rainfall, high evaporation, lack of a groundwater aquifer, lack of substantial trees and an extensive granite basement. Should materials need to remain at MinStore, all by-product storage installations would be capped at closure, and shaped to ensure stormwater run-off and minimise erosion.

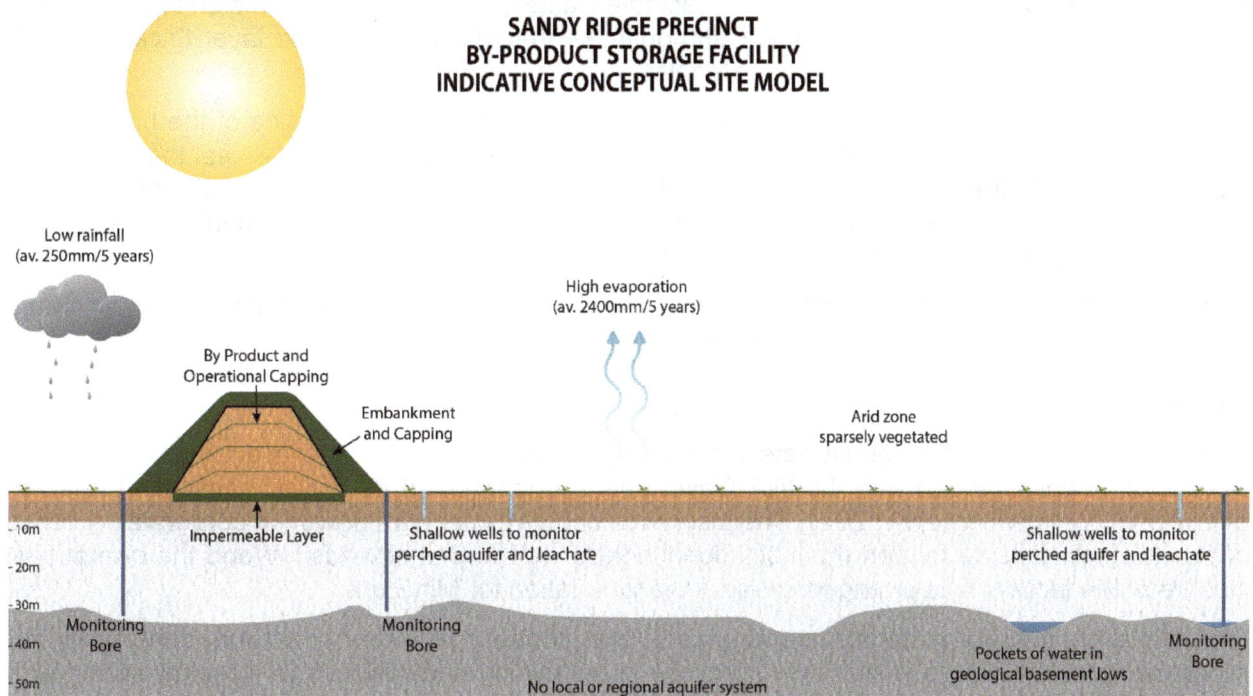

FIG 3 – Conceptual site model for MinStore.

MATERIAL CIRCULARITY AND TRACEABILITY

The concept of circular material banks as a basis for keeping materials in use for longer and reducing reliance on virgin materials in construction, particularly in large-scale infrastructure projects, is gaining traction. Long-term storage of potentially useful materials in accessible stockpiles enables the 'banking' of resources that can be made available for future use. Material that would otherwise be lost to disposal, particularly if disposed in a landfill or residue storage facility designed as a final disposal point, can be deposited in a dedicated circular material bank designed as a future quarry for homogenous, single origin, well-characterised material that can be 'withdrawn' in future.

For materials with reuse potential, storage in dry-stack monocells underpins full traceability, and an audit trail from origin to storage to final reuse destination can be generated with confidence. This may appeal to generators of by-product or residue streams that become more valuable in the future. That is not to say however, that only material with reuse potential can be accepted at MinStore. The facility can also efficiently manage residues that require safe, long-term storage in a remote site context.

Transport cost efficiencies can be realised with long-term rail capacity commitments, and if residue were to be returned to regions with rail hubs for further processing or beneficial use in future, significant reverse logistics cost savings accrue from the scale and regularity of consignment deliveries to Sandy Ridge.

RESEARCH AND DEVELOPMENT

Fundamental to MinStore's operating model is the incorporation of R&D trial facilities for material reuse investigation and blending of compatible materials to support material circularity.

Inclusion of a trial pad for evaluating the beneficial reuse potential of MinStore materials, either alone or in combination with other materials and products, will facilitate reuse feasibility evaluation. Ongoing R&D into material reuse will help unlock the latent value that would otherwise be lost in a contemporary mixed material tailings storage facility arrangement.

Single-source material monocells have the potential to become the resource banks of the future, with 'deposits' managed by Tellus under outsourced materials management contracts with the generators of the by-products under management. Future 'withdrawals' of traceable, geo-located, homogenous, quality-characterised recovered resources can be made available for large infrastructure projects on a reverse-logistics campaign basis.

Current R&D includes optimisation of the intractable waste cell capping profile design to improve the effectiveness of the capillary break layer and optimise material selection and layer thickness, with an emphasis on use of site-generated materials.

R&D is also evaluating use of a mineral processing residue to enhance sealing of the barrier layer beneath the capillary break layer for the intractable waste cells at closure. Given that the kaolin clay mined at Sandy Ridge can now be refined and sold into domestic and international markets, if the mineral processing residue can be utilised in the seal layer, this will reduce the quantity of virgin clay needed for capping at Sandy Ridge.

Further current R&D focus is blending Sandy Ridge kaolin with an industrial by-product to create bespoke products for use in the local mining industry.

CONCLUSION

The Sandy Ridge facility's robust safety case, and unique insurance and assurance framework, developed in collaboration with the WA Government, underpinned its social license and enabled operational commencement in 2021. The insurance and assurance framework incorporates a series of financial instruments to fund the institutional control period that extends beyond the operational closure of the facility. This arrangement sets the foundation for MinStore.

MinStore's centralised materials management hub concept provides a unique opportunity for progressing towards more sustainable practices in the mining and minerals processing sector, with material circularity built around leveraging the positive attributes of residual materials to create useful products with full traceability. Rail connectivity will enhance transport efficiency and enable backloading of recovered materials and products, unlocking the value of otherwise stranded waste materials.

The MinStore concept is not limited to WA, and other states with significant downstream processing could also benefit from MinStore.

Seeking highly reliable implementation of the Global Industry Standard for Tailings Management (GISTM)

L Howe[1], C Côte[2] and S Johnston[3]

1. PhD Candidate, Centre for Water in the Minerals Industry, Sustainable Minerals Institute, The University of Queensland, St Lucia Qld 4072. Email: layla.howe@uq.edu.au
2. Director for Leading for Higher Reliability Centre and Centre for Water in the Minerals Industry, Sustainable Minerals Institute, The University of Queensland, St Lucia Qld 4072. Email: c.cote@uq.edu.au
3. Sustainable Minerals Institute Industry Professor, Sustainable Minerals Institute, The University of Queensland, St Lucia Qld 4072. Email: susan.johnston@uq.edu.au

INTRODUCTION

The Global Industry Standard for Tailings Management (GISTM) is being progressively implemented by many mining companies, representing a potential turning point towards safer tailings storage. A few observers suggest the reliability of Tailings Storage Facility (TSF) performance primarily depends on improved technological solutions (Morgenstern, Vick and van Zyl, 2015). However GISTM requires companies to also place greater emphasis on organisational factors. These include consideration of organisational structures, performance incentives, training programs, greater collaboration between teams and knowledge sharing.

WHAT IS HRO THINKING?

High Reliability Organisations (HROs) routinely deliver their performance goals and remain free from catastrophic failures while operating in complex and hazardous environments (Cantu et al, 2021). Research seeking to understand how some entities sustained highly reliable performance led to the identification of common HRO principles (Weick and Roberts, 1993; Weick, Sutcliffe and Obstfeld, 1999). These include a collective mindset centred on a shared purpose and five hallmark characteristics: preoccupation with failure, reluctance to simplify, sensitivity to operations, commitment to resilience and deference to expertise (Weick and Sutcliffe, 2015). Over 30 years of research has shown many sectors have used HRO thinking to improve performance reliability across a wide variety of domains (Lekka and Sugden, 2011; Schaffer, Reynolds and Stringfield, 2012; Veazie, Peterson and Bourne, 2019; Oliver, Calvard and Potočnik, 2019). However, despite the high relevance of HRO thinking, these principles have yet to be applied in any substantive way to improve tailings management.

APPLYING HRO THINKING TO GISTM IMPLEMENTATION

GISTM contains 77 requirements that are grouped according to 14 principles of tailings management (International Council on Mining and Metals (ICMM), 2020; Hopkins and Kemp, 2021). The requirements integrate social, environmental, local economic and technical considerations to guide TSF management activities (ICMM, 2023). This paper explores how HRO thinking could be used to put GISTM requirements into practice with high reliability, using a theoretical framework (Figure 1).

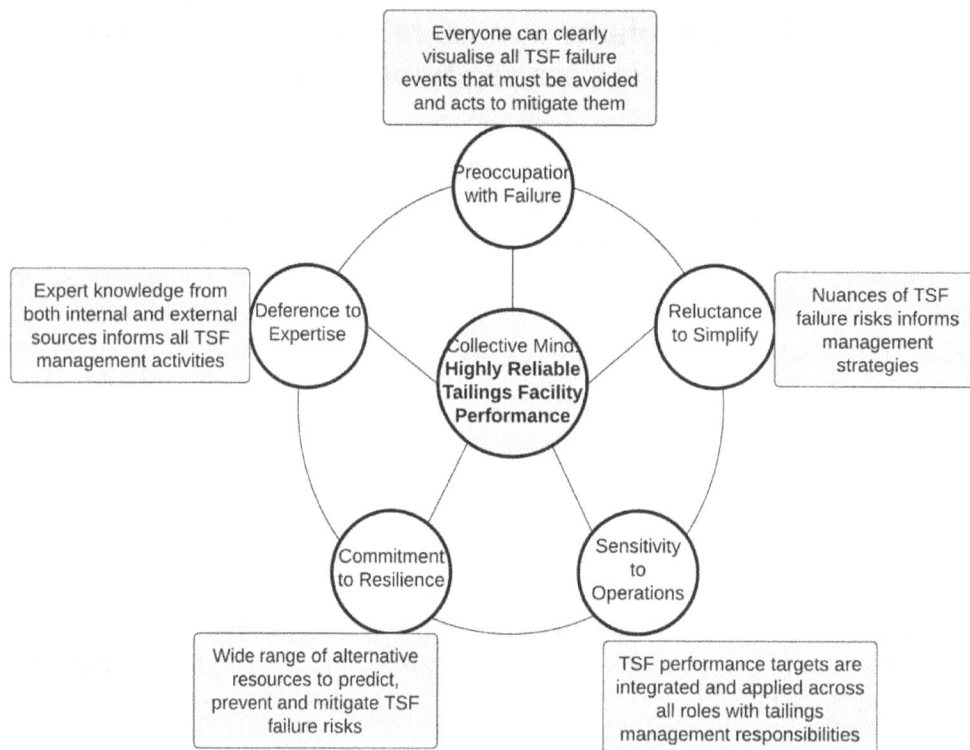

FIG 1 – HRO theory conceptualised in relation to tailings management implementation guided by the GISTM.

We analysed the content of GISTM by comparing the wording of the standard's goals and requirements with the six HRO characteristics. Outcomes of this comparison were reviewed in aggregate to identify major themes. We found that many GISTM requirements aligned with the types of organisational practices that HRO theory tells us support greater performance reliability. The GISTM requirements mostly aligned with preoccupation with failure, reluctance to simplify and deference to expertise. Key examples include:

- Proactively identifying tailings system errors through establishing appropriate monitoring methods and Quality Assurance/Quality Control processes (eg requirements 6.1, 6.2, 7.1, 7.2, 7.4).

- Sharing current information about progress towards tailings goals by regularly reviewing current data and confirming implementation aligns with design (eg requirements 7.5, 9.3, 11.3).

- Building a nuanced perspective of tailings system risks (eg requirements 1.1, 1.3, 2.1, 2.2, 5.1).

- Developing and integrating expertise in tailings governance processes (eg requirements 9.1, 9.2, 9.4, 11.1, 11.2).

These results are consistent with the established discourse on tailings management (Evans and Davies, 2020; McNab *et al*, 2023; Zare *et al*, 2024) highlighting that despite long-standing awareness of these practices, reliable implementation remains a challenge.

ORGANISATIONAL FACTORS NEEDING GREATER ATTENTION

Our review suggests there are three HRO principles that should receive more attention: collective mindset, sensitivity to operations, and commitment to resilience. Therefore, we urge that these organisational approaches are considered more deeply when putting GISTM requirements into action:

1. **Defining the shared purpose**: We found a limited focus on the key task of carefully stating a shared purpose for a TSF. Expressing a goal with ambiguous language can potentially cause confusion. It is important to consider the use of terms such as 'zero harm'. For instance, in

relation to environmental impacts, storing tailings in perpetuity inevitably results in permanent changes to the environmental values of a location. Implying zero environmental harm may introduce uncertainty about what is being aimed for. The way the shared goal is defined plays an important role in how it is understood by those working towards it. Each organisation needs to develop a clear, specific purpose for TSF management so it is tailored to individual business circumstances and can be understood by all those whose work impacts tailings.

2. **Embedding the shared purpose**: Although GISTM provides some approaches that may support putting a goal into action, it is also important to ensure the actions needed to operationalise the shared purpose are adapted to the organisation. The actions will be highly specific to each company and site. Leaders involved in tailings activities critically influence these actions. Anyone performing tailings activities will be affected by the commitment shown towards delivering on the organisational goals for a TSF. Important actions for sustaining commitment to the shared goals include actively seeking staff concerns related to TSF activities, and making sure reliability of TSF performance is considered a key priority in risk assessment and decision-making processes.

3. **Aiming for holistic TSF resilience**: Highly reliable TSF management requires resources to be made available for contingency strategies that are in addition to emergency preparedness activities, so TSF weaknesses can be contained before they escalate into larger or catastrophic failures. We found that while GISTM guides companies to prepare to respond to emergency scenarios, preparations are also needed to make sure there are adequate resources to fix unexpected issues that have not yet led to an emergency. For example, an erroneous TSF monitoring result warrants investigating to ensure it is resolved so a repeat of the same issue cannot happen again. It can be challenging to assess whether and to what extent additional resources need to be deployed. Reflecting on the shared purpose and associated organisational goals can provide direction. Other factors that could be considered include scoping training to optimise the collective capacity to respond to an unexpected problem, as well as sustaining processes that ensure corrective actions are verified.

Our results highlight that a greater awareness and commitment to these three organisational practices would assist with delivering all GISTM requirements with greater reliability.

ENHANCING THE RELIABILITY OF GISTM IMPLEMENTATION

The global mining sector is united by a common goal: to achieve safer, more sustainable operation of TSFs throughout the Life-of-mine. To achieve this goal, it is critical the full potential of GISTM is realised by reliably putting all the requirements into practice.

Our study demonstrates that applying HRO thinking can assist with achieving the full potential of GISTM. It provides insights about the additional approaches companies could use to enhance the reliability of their GISTM implementation, specifically by: (1) establishing clearer statements about their shared purpose for tailings management and their desired outcome(s) from implementing the GISTM; (2) taking specific actions to embed the shared purpose in their organisation; and (3) having resources available to manage unexpected problems before they escalate to enhance overall TSF resilience.

REFERENCES

Cantu, J, Gharehyakheh, A, Fritts, S and Tolk, J, 2021. Assessing the HRO: Tools and techniques to determine the high-reliability state of an organization, *Safety Sci,* 134:105082. https://doi.org/10.1016/j.ssci.2020.105082

Evans, R and Davies, M, 2020. Creating and retaining knowledge and expertise, in *Towards Zero Harm – A Compendium of Papers Prepared for the Global Tailings Review* (eds: B Oberle, D Brereton and A Mihaylova), Global Tailings Review, pp 150–161.

Hopkins, A and Kemp, D, 2021. Credibility Crisis Brumadinho and the Politics of Mining Industry Reform, CCH Australia Limited, Sydney, Australia.

International Council on Mining and Metals (ICMM), 2020. Global Industry Standard On Tailings Management. Available from: <https://globaltailingsreview.org/wp-content/uploads/2020/08/global-industry-standard-on-tailings-management.pdf>

International Council on Mining and Metals (ICMM), 2023. ICMM members to publish progress on implementing the Global Industry Standard on Tailings Management in August. Available from: <https://www.icmm.com/progress-implementing-gistm>

Lekka, C and Sugden, C, 2011. The successes and challenges of implementing high reliability principles: A case study of a UK oil refinery, *Process Saf Environ*, 89:443–451. https://doi.org/10.1016/j.psep.2011.07.003

McNab, L, Boshoff, J, Scampoli, L, Walsh, M and Cirillo, F, 2023. The Role of Corporate Governance: How to Develop a Global Tailings Division, in *Proceedings of the Tailings and Mine Waste 2023 Conference* (ed: J Goodwill), pp 1081–1090 (the University of British Columbia (UBC): Vancouver).

Morgenstern, N, Vick, S and van Zyl, D, 2015. Independent Expert Engineering Investigation and Review Panel: Report on Mount Polley Tailings Storage Facility Breach, Government of British Columbia, Canada.

Oliver, N, Calvard, T and Potočnik, K, 2019. Safe limits, mindful organizing and loss of control in commercial aviation, *Safety Sci*, 120:772–780. https://doi.org/10.1016/j.ssci.2019.08.018

Schaffer, E, Reynolds, D and Stringfield, S, 2012. Sustaining Turnaround at the School and District Levels: The High Reliability Schools Project at Sandfields Secondary School, *J Educ Stud Placed Risk (JESPAR)*, 17:108–127. https://doi.org/10.1080/10824669.2012.637188

Veazie, S, Peterson, K and Bourne, D, 2019. Evidence Brief: Implementation of High Reliability Organization Principles, Department of Veterans Affairs, US.

Weick, K E and Roberts, K H, 1993. Collective Mind in Organizations: Heedful Interrelating on Flight Decks, *Admin Sci Quart*, 38:357. https://doi.org/10.2307/2393372

Weick, K E and Sutcliffe, K M, 2015. *Managing the Unexpected: Sustained Performance in a Complex World* (John Wiley and Sons, Inc.: New Jersey).

Weick, K E, Sutcliffe, K M and Obstfeld, D, 1999. Organizing for High Reliability: Processes of Collective Mindfulness, *Res Organ Behav*, 1:81–123.

Zare, M, Nasategay, F, Gomez, J A, Far, A M and Sattarvand, J, 2024. A Review of Tailings Dam Safety Monitoring Guidelines and Systems, *Minerals*, 14:551. https://doi.org/10.3390/min14060551

Integrated waste landforms for tailings storage – a step towards the goal of zero harm

J Ranasooriya[1]

1. Geotechnical Consultant, Geotechnical Safety, Malaga WA 6944.
   Email: jay.ranasooriya@outlook.com

ABSTRACT

Tailings storage facilities (TSFs) are expected to perform without causing safety, health and environmental hazards during operation and after the end of their active life. The most common type of TSFs comprises embankment dams designed, constructed, and operated to meet the relevant guidelines, standards and regulatory requirements.

However, some of these structures have catastrophically failed at an alarming rate, causing not only severe safety, health and environmental disasters but also significant economic losses. In response to recent disasters caused by tailings dam failures, the mining industry, the regulators and other stakeholders have updated the existing guidelines and developed new guidelines and standards on tailings management. The main objective of these guidelines is to reach the ultimate goal of zero harm to people and the environment. Nonetheless, most of the guidelines and other relevant documents, for example, codes of practice and standards, focus on improving the design, construction, operation and closure of conventional tailings dams. They often overlook alternative approaches, such as Integrated Waste Landforms (IWLs), which integrate tailings dams with mine waste rock landforms.

Based on a review of the information available in the public domain, this paper briefly describes the causal factors of tailings dam failures and the advantages of IWLs that can be employed to reach the goal of zero harm by improving the longer-term stability of TSFs.

INTRODUCTION

The commodities produced by mining and ore processing are valuable and essential for all generations of human societies. However, for producing one tonne of most commodities, the industry, on average, produces about two tonnes of tailings, which are waste products usually discharged as slurry. For high-value commodities such as copper and gold, up to 98 per cent and over 99 per cent, respectively, of processed ore turns into tailings. In addition, several million tonnes of mine waste rock are removed at each mine to access orebodies and are stored separately in mine waste rock landforms. With the advances in extraction technology the low-grade orebodies, which are not currently mined, become economically viable. Hence, the rate of tailings discharge per one tonne of ore continually increases. Concurrently, the excavated volume of mine waste rock also increases.

Most tailings contain inherently toxic elements or added poisonous chemicals, or both. Some types of tailings may not contain toxic substances but the sheer volume could still pose floods and engulfment hazards. Hence, the tailings are usually deposited in tailings storage facilities (TSFs) which are expected to remain stable without causing safety, health and environmental hazards. The most common type of TSFs comprises embankment dams. These are engineered structures designed and constructed to meet the relevant guidelines, standards and legislative requirements. Nevertheless, during operation and after the closure, these structures have catastrophically collapsed at an alarming rate, causing deaths, adverse health impacts, property damage and environmental harm in different parts of the world (Azam and Li, 2010; Lyu et al, 2019; Dong, Deng and Wang, 2020; Lin et al, 2022; Piciullo et al, 2022; Stark et al, 2022). Due to tailings dam failures, the dam owners and operators also undergo severe financial losses in terms of loss of production, cost of site cleanup and remediation, compensation for affected parties, litigation expenses and loss of reputation.

In response to the tailings dam failure disasters, particularly the Brumadinho incident in January 2019, the mining industry, professional groups, volunteer organisations, and regulators in most jurisdictions have updated the existing guidelines and published new guidelines and standards on tailings management. The primary objective of these guidelines and standards is to reach the ultimate goal of zero harm to people and the environment, as emphasised in the Global Industry Standard on Tailings Management (GISTM) (Global Tailings Review (GTR), 2020).

However, most of the guidelines and standards focus mainly on improving the design, construction, operation and closure of conventional tailings dams without adequately considering other viable options, such as Integrated Waste Landforms (IWLs), which are structures formed by integration of dams with mine waste rock landforms. Based on a review of the information available in the public domain, this paper briefly describes the causal factors of tailings dam failures. It also outlines the advantages of IWLs that can be employed to reach the goal of zero harm while achieving a higher level of compliance with health and safety (HS) and environmental legislation applicable to tailings management.

LEGISLATIVE REQUIREMENTS

Health and safety (HS) legislation

In most jurisdictions of the world, the HS legislation requires the industry to minimise HS risks to a reasonably practicable level. In Australia, although the relevant legislation and the administrative framework may slightly differ in different states and territories, the law requires the elimination of HS risks so far as is reasonably practicable (SFAIRP), and if it is not reasonably practicable to eliminate HS risks, minimise those risks SFAIRP. This essentially means 'achieve zero harm to people'.

According to the Work (Occupational) HS Act of all jurisdictions in Australia, cost is a matter to be considered to ascertain what is 'reasonably practicable'. However, a low-cost option that provides less protection is unlikely to be considered a reasonably practicable means of eliminating or minimising HS risk (WorkSafe Victoria, 2007; Safe Work Australia, 2013; WorkSafe WA, 2021). Thus, if there is a better option for a comparable cost for eliminating or minimising HS risk, the owners and operators of TSFs must opt for the better option to fully comply with the HS law.

The legislative requirement in Australia conforms with the professional opinions of international tailings dam experts. For instance, a key recommendation of the Mount Polley expert panel (IEEIRP, 2015) is that '*safety attributes should be* evaluated *separately from economic considerations, and cost should not be the determining factor*'. Similarly, Roche, Thygesen and Baker (2017) recommended that '*the approach to tailings storage facilities must place safety first by making environmental and human safety a priority in management actions and on-the-ground operations*'.

Environmental legislation

The legislation applicable to environmental aspects of TSFs varies from one country to another, depending on the sensitivity of the local environment and the extent of the tailings production. Nonetheless, the primary objective of each jurisdiction is to eliminate or minimise the environmental harm from tailings. In Australia, environmental impacts from TSFs are regulated under State or Territory Government legislation. In all Australian jurisdictions more than one set of legislation could be applicable to tailings management. These include mining or mineral resources legislation and the environment protection legislation. Other applicable pieces of legislation cover water rights, heritage and land rights etc. The ultimate objective of all these legislations is to eliminate or minimise the damage to the environment.

TAILINGS STORAGE METHODS

Tailings dams

Depending on the topography, tailings dams can be constructed as cross-valley, side-hill and ring-dyke (paddock) dams. They are used mostly to store tailings slurry with a solid content of 25–40 per cent by mass. They can also be used for thickened or dewatered tailings storage.

Unlike water storage dams, which are usually constructed in a single phase, tailings dams are constructed in several stages with the first stage (starter dam) being sufficient only for a short period

(2 to 3 years) of tailings deposition. The starter dams are constructed using mine waste soil and rock from the mine excavation or using suitable soil borrowed from other sources in the area.

When the storage capacity of a starter dam is fully utilised, the dam is raised by either upstream, downstream, or centreline methods of raise construction as illustrated in Figure 1. The upstream raises are usually constructed using tailings borrowed from within the storage area. The downstream and centreline raises are constructed using mine waste or other suitable materials. Geoscience Australia (2023) shows that, in Australia, a vast majority of tailings dams are ring-dykes constructed on relatively flat ground.

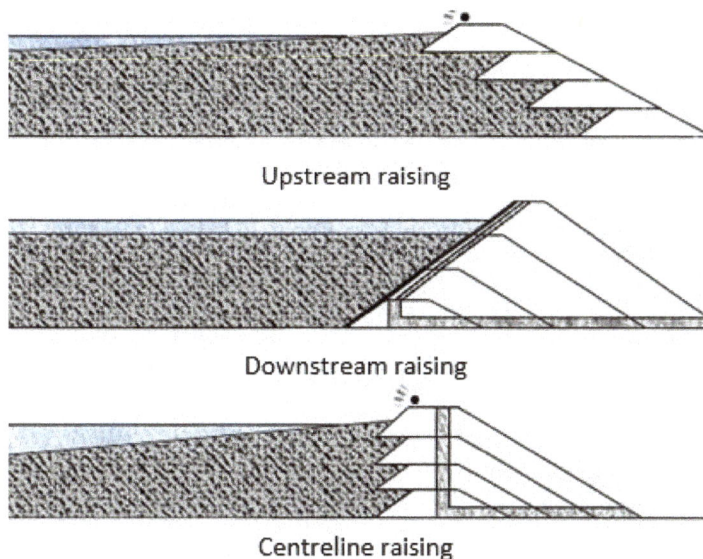

Upstream raising

Downstream raising

Centreline raising

FIG 1 – Tailings dam raising methods (after Vick, 1990).

Integrated waste landform (IWL)

IWL is usually a ring-dyke tailings dam fully enclosed and buttressed by a mine waste rock landform (Figure 2). The IWL concept was first introduced by a tailings dam consulting firm in the early 1990s and the first IWL was constructed in 1994 at the Bronzewing gold mine in Western Australia (WA) (Lane, 1998; Lacy, 2019). It is a TSF located inside a mine waste rock landform (Lane, 2018). It can be considered a two-cell storage system: a tailings cell and a mine waste rock cell surrounding the tailings cell.

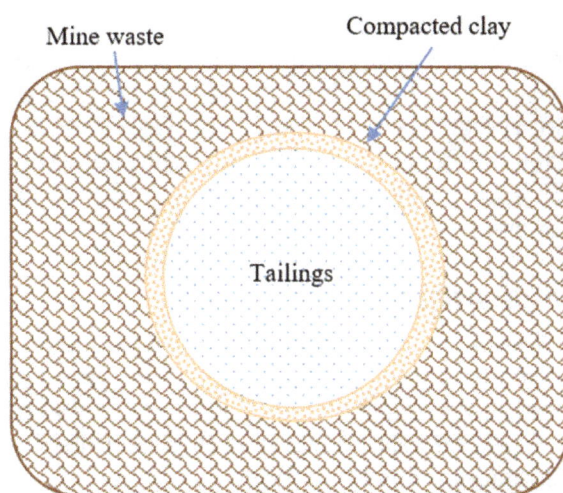

FIG 2 – Schematic plan view of IWL TSF.

In an IWL the tailings can be initially contained by a stand-alone ring-dyke starter dam, which is then buttressed and raised by coarse mine waste, similar to the downstream raising of ring-dykes (Figure 3). The inside face of the waste rock landform is lined with a zone of low permeability clay from a suitable source or tailings borrowed from within the storage. For the convenience of construction and operation, a circular shape is preferred for the tailings storage cell. Other shapes are also practicable.

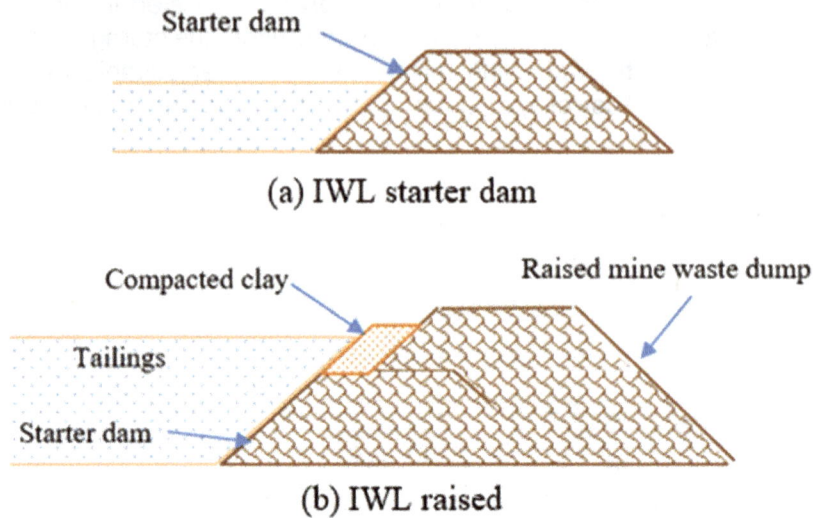

(a) IWL starter dam

(b) IWL raised

FIG 3 – IWL concept with starter dam: (a) starter dam; (b) raised by waste dump with a clay zone.

If mine waste rock is available well in advance of tailings production, an IWL can also be established by leaving a large enough circular void space within a waste rock landform and then constructing a low permeability material zone along the perimeter of the circular void space (Figure 4).

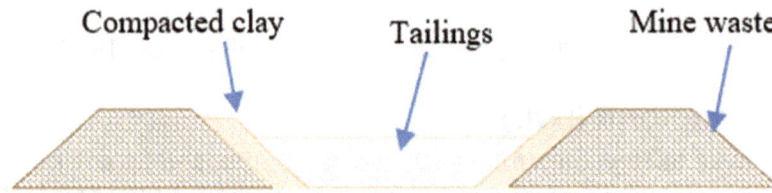

FIG 4 – IWL concept without a starter dam.

In both options, the buttressing waste rock landform eliminates the likelihood of tailings dam failure and catastrophic release of tailings. IWLs can be used for storage of slurry, thickened or paste tailings.

Since 1994, several other mines have adopted the IWL TSFs that can be operated efficiently with minimal risk of catastrophic failure and then rehabilitated to remain as stable landforms (Lane, 2004, 2013; Lacy and Barnes, 2006; Williams and Minard, 2010; Department of Foreign Affairs and Trade (DFAP), 2016; Lacy, 2019). Numerous other mines in Australia, especially the small to medium-scale mines, could have adopted the IWL concept to further improve the stability of ring-dyke TSFs and reduce the overall cost of site rehabilitation.

Other methods of tailings storage

Other methods of tailings storage include thickened tailings, paste tailings, dry stacking, and co-disposal with waste rock. These methods entail additional equipment for tailings thickening, paste making, dewatering to produce cake-like tailings, and mixing of coarse mine waste and tailings slurry for co-disposal. The expenses required for additional equipment discourage small to medium-scale mine operators with tight budgets from choosing these methods.

A low-cost option for tailings storage is in-pit disposal, for which a mined-out pit must be available before the commencement of ore processing. For new mining operations, this is an unlikely scenario. Hence, the small to medium-scale mine operators persist with conventional slurry tailings storage in TSFs comprising embankment dams.

TAILINGS DAM FAILURES

Causes of failures

Tailings dam could fail due to one or more causes depending on the conditions specific to the dam and the site. The International Commission on Large Dams (ICOLD, 2001) analysed 211 tailings dam incident data and identified eight causes of failure, ie overtopping (OT), slope instability under static loading (SI), earthquake-induced instability (EQ), weak foundation conditions (FN), seepage and internal erosion (SE), structural inadequacies (ST), external erosion (ER), and mine subsidence (MS) (Figure 5). ICOLD added that many failures were due to overtopping, slope instability, earthquakes, seepage, weak foundations, and external erosion. Due to the lack of sufficient data, the causes of the failure of some dams analysed by ICOLD remained unknown (UN).

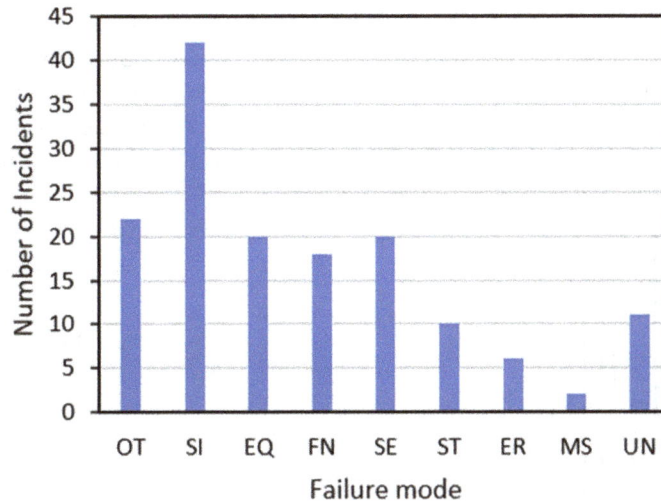

FIG 5 – Main causes of tailings dam failure (ICOLD, 2001) [(overtopping (OT), slope instability under static loading (SI), earthquake-induced instability (EQ), weak foundation conditions (FN), seepage and internal erosion (SE), structural inadequacies (ST), external erosion (ER), mine subsidence (MS), and unknown (UN)].

In ICOLD (2001) the other identified causes, that can be classified as human errors or omissions during the operation of dams, were: poor control of tailings beach development, deficiencies in dam maintenance, inadequate management of decant ponds, and rapid rate of dam raise construction.

Rico *et al* (2008) analysed 157 sets of tailings dam failure data and revealed that the most common cause of tailings dam failure is unusual rain followed by seismic liquefaction or earthquake-induced instability (EQ). Rico *et al* also observed other main causes of failure such as overtopping (OT), slope instability under static loading (SI), weak foundation conditions (FN), seepage and internal erosion (SE), structural inadequacies (ST), external erosion (ER), and mine subsidence (MS). They pointed out that unusual rain and snowmelt could lead to overtopping, excessive seepage, and foundation failure etc. Thus, the main causes of tailings dam failures identified by Rico *et al* are the same as those identified by ICOLD (2001).

Rico *et al* (2008) added that management and operation deficiencies such as poor control of tailings beach development, deficiencies in dam maintenance, presence of heavy equipment in unstable dams, inadequate management of decant pond and rapid rate of dam raise construction are also significant causes of tailings dam failures. These are literally the same as the other causes, which can be classified as operation stage human errors or omissions, identified by ICOLD.

A summary of the incident data reported by Rico *et al* (2008) is presented in Table 1 where UPS=upstream; DWN= downstream; CTL= centreline; MIX= mixed construction; ACT= dam active at the moment failure happened; INM= dam inactive at the moment failure occurred; ABN= abandoned; UNK= unknown).

TABLE 1

Number of incidents versus dam type and state (Rico *et al,* 2008).

Dam status	Method of dam raising				
	UPS	DWN	CTL	MIX	UNK
ACT	45	14	4	4	49
INM	2	1	0	0	2
ABN	9	2	2	0	6
UNK	13	0	0	0	4

From the reported cases with data on dam height, construction methods and dam status etc, Rico *et al* (2008) observed that the upstream method of dam raising accounted for 76 per cent of the failures, 83 per cent of failures occurred in active dams, 15 per cent in inactive dams, and 2 per cent in abandoned dams. Further, they found that approximately 77 per cent of the failures occurred in less than 30 m high tailings dams (Figure 6). This tallies with the ICOLD findings.

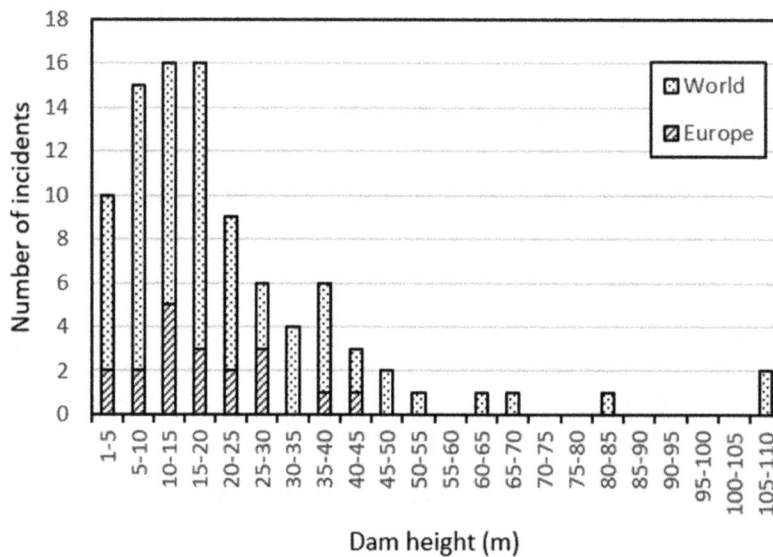

FIG 6 – Distribution of the number of incidents related to dam height (Rico *et al,* 2008).

A comprehensive review of 172 tailings dam failures from 1910–2018 by Lyu *et al* (2019) revealed a staggering 85 per cent of failures occurred in less than 45 m high tailings dams (Figure 7). The remaining 15 per cent of the dams that failed were 45 m to 60 m high. Figure 7 also shows that 73 per cent of failures occurred in less than 30 m high dams. This is consistent with the observation made by Rico *et al* (2008) that, 77 per cent of the failures occurred in less than 30 m high tailings dams. This does not imply that less than 30 m high dams are more likely to fail. It reflects the fact that most tailings dams are less than 30 m high and regardless of the low height they do fail.

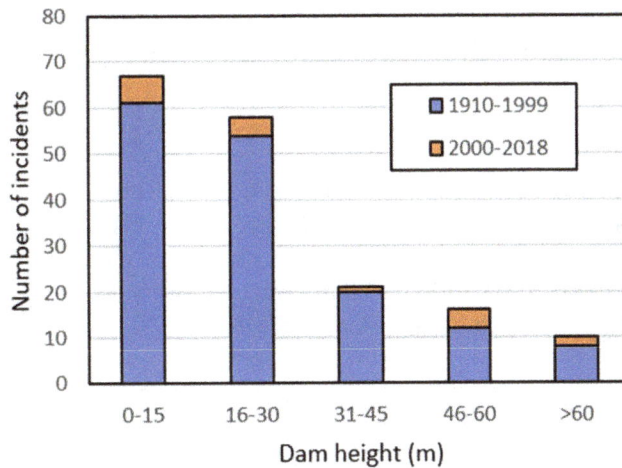

FIG 7 – Failure distribution by tailings dam height (Lyu *et al*, 2019).

Further, Lyu *et al* (2019) divulged that 59 per cent of failures occurred in upstream-raised dams (Figure 8). They also confirmed by providing examples the main causes of tailings dam failures are overtopping, slope instability, earthquake-induced instability, seepage and internal erosion, and weak foundations.

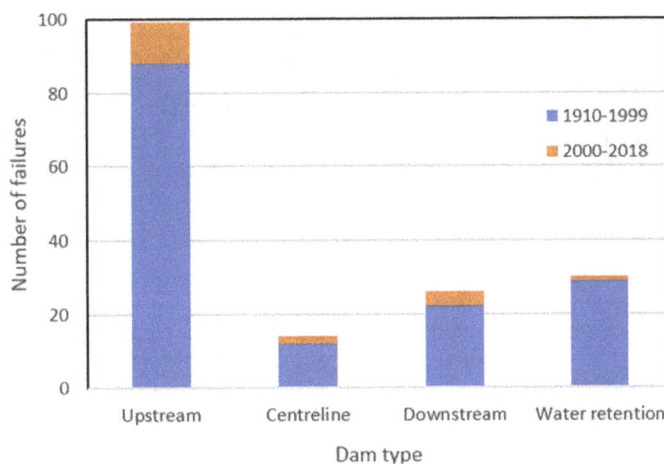

FIG 8 – Failure distribution by tailings dam type (Lyu *et al*, 2019).

Moreover, based on 342 tailings dam failure incidents, Lin *et al* (2022) confirmed that overtopping, slope instability, earthquake-induced failures, seepage and internal erosion, weak foundations, and structural inadequacies are the leading causes of tailings dam failures.

From a review of tailings dam incidents that occurred between 1915 to 2019, Piciullo *et al* (2022) observed that overtopping, static slope instability and dynamic liquefaction were the frequent causes of tailings dam failure as observed by several others including ICOLD (2001), Rico *et al* (2008), Lyu *et al* (2019), and Lin *et al* (2022). Piciullo *et al* also highlighted that the upstream-raised dams had the highest percentage of failure. Of the recorded failures, 82 per cent were in active tailings dams.

Stark *et al* (2022), from 325 tailings dam failure records confirmed that overtopping, earthquake-induced failures, and slope instability are the main causes of failure. Further, they observed that seepage, weak foundations, erosion and structural inadequacies are the other causes that lead to failures. Stark *et al* also showed that upstream raised tailings dams have a high percentage of failures per the number of tailings dams built in the world.

From the foregoing, it is clear that the most common causes of tailings dam failure are overtopping, slope instability under static loading, earthquake-induced instability, weak foundation conditions, seepage and internal erosion, and external erosion. Further, the upstream raised tailings dams accounted for the majority of the failures (Rico *et al*, 2008; Lyu *et al*, 2019; Piciullo *et al*, 2022). The available data also show that the tailings dam height is not necessarily a governing factor of failure. More than 70 per cent of failures occurred in less than 30 m high dams (Rico *et al*, 2008; Azam and Li, 2010; Lyu *et al*, 2019).

Consequences of tailings dam failures

Dong, Deng and Wang (2020) scrutinised 44 tailings dam failures and noted that 2138 lives were lost from 30 incidents that occurred between 1928 and 2019. Of this total, 1554 deaths occurred during the eight decades from 1928 to 2007, with an average rate of 20 deaths per annum. The remaining 554 deaths occurred from 2008 to 2019, with an average annual rate of 46 deaths, which is more than twice that of previous decades from 1928 to 2007 (Table 2). It is worth noting that two of the largest human tragedies resulting from tailings dam failures occurred in 2008 and 2019. Those two largest events at the Xinta tailings dam in China and the Vale tailings dam in Brazil led to at least 277 and 247 deaths, respectively.

TABLE 2

Annual average death rate (Dong, Deng and Wang, 2020).

Period	Total deaths	Annual death rate
1928–2019 (92 years)	2138	23
1928–2007 (80 years)	1554	20
2008–2019 (12 years)	554	46

Lin *et al* (2022) compiled 342 tailings dam failures globally from 1915 to 2021 and divided the period from 1947 to 2021 into three 25-year intervals (excluding 1915 to 1946 due to the lack of reliable data) to show the number of tailings dam failures in every 25 years (Table 3).

TABLE 3

Tailings dam failures 1947 to 2021 (Lin *et al*, 2022).

Period	No. of failures	Annual average
1947–1971	73	2.9
1972–1996	143	5.7
1997–2021	115	4.6

Using the same data set Lin *et al* (2022) show that from 1961 to 2020, five or more Very Serious TSF failures occurred in each of the six decades (Figure 9). They defined Very Serious failures as incidents causing more than 20 deaths and/or releasing 1 000 000 m³ of tailings, and/or release travel distance of 20 km or more. Whereas Serious failures are incidents resulting in the loss of life and/or release of 100 000 m³ of semi-solids. They emphasised the fact that '*Since the beginning of the 20ᵗʰ century, the frequency of tailings dam failures has been high worldwide*'.

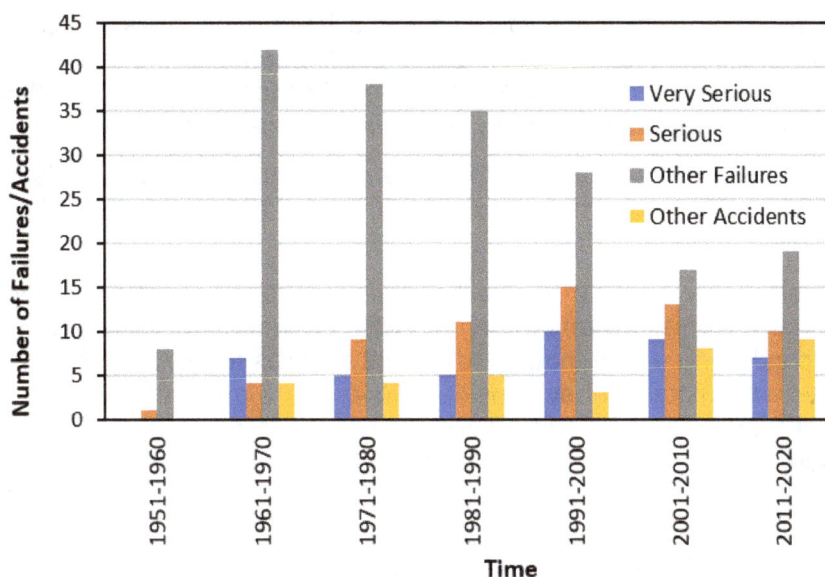

FIG 9 – Number of failures per decade (Lin *et al*, 2022).

Piciullo *et al* (2022) found that 257 tailings dams failed from 1915 to 2019 and released 250 million m³ of tailings into the environment. Out of the 257 failures, 17 per cent caused 2650 fatalities and 4 per cent caused more than 50 deaths.

TSF incidents since January 2019

The Brumadinho tailings dam incident in Brazil, which occurred in January 2019, highlighted the urgent need for improvements in tailings management within the mining industry. In response, the stakeholders launched a campaign that led to the issuance of GISTM in August 2020. Simultaneously, the existing guidance on tailings management were updated and, in some cases, new guidance documents were issued.

Since August 2020, most large mining companies have taken action to upgrade their existing TSFs. Those companies have formed Independent Tailings Review Bords (ITRB) and appointed experienced engineers to positions such as Engineer of Record (EOR) and Responsible Tailings Facility Engineer (RTFE) and empowered them in accordance with the GISTM principles. Some mining companies, especially small to medium-sized mine operators, are yet to upgrade their existing TSFs and appoint the EOR and RTFE due to budget restraints.

In the meantime, TSF failures continue to occur, leading to human casualties, property damage, and negative environmental impacts (WISE, 2025). Between February 2019 and August 2020, there were eight TSF failures, two of which resulted in a total of 183 fatalities. Moreover, from September 2020 to January 2025, there were 16 TSF failures, three of them causing a total of 90 deaths. A summary of these failures is provided in Table 4, which demonstrate that there was no noteworthy reduction in the annual death rate or the annual average failures since 2019 compared to those reported up to the Brumadinho tailings dam incident in January 2019.

TABLE 4

Tailings dam failures since 2019 (WISE, 2025).

Period	No. of failures	No. of deaths
Feb 2019 – Aug 2020	8	183
Sep 2020 – Jan 2025	16	90

Advantages of IWLs

Subject to detailed planning well in advance of the project commissioning, mining companies can attain several advantages from IWLs compared to conventional tailings dams, particularly upstream raised paddock dams. The planning should begin with a life-of-mine focus and cover: (a) scheduling and control of the mining sequence to meet material requirements for the tailings facility starter dam and subsequent lift construction; (b) the interaction between the mining fleet and the IWL construction team; and (c) final closure of the operation and site rehabilitation.

Some of the advantages of IWLs compared to conventional tailings dams are discussed in detail by Lacy and Lane (2007), Williams and Minard (2010), Lane (2017), Lacy (2019) and Mooder and McGreevy (2022).

The main advantage of IWLs is a very low likelihood of catastrophic failures causing deaths, injuries, property damage and adverse environmental impacts. Williams and Minard (2010) demonstrated this advantage by conducting a fault mode and effects analysis (FMEA) for the IWL at the Randalls Gold Project in Western Australia. They considered five failure mechanisms (ie, overtopping, slope instability under static and dynamic loading, internal erosion, progressive sloughing due to seepage, and external erosion) and determined that the likelihood of occurrence of the five failure mechanisms is negligible to low. The failure mechanisms considered by Williams and Minard are within the primary causes of tailings dam failures identified by ICOLD (2001), Rico *et al* (2008), Lyu *et al* (2019), Lin *et al* (2022), Piciullo *et al* (2022) and Stark *et al* (2022).

Further, Williams and Minard (2010) showed that the risk of a release of tailings from the Randalls IWL is low with an average risk rating of 3.7 in a scale of 1 (minimum) to 25 (maximum).

From a stability analysis of the Randalls IWL outer slopes under three loading conditions (ie, steady state static, operating base earthquake, and maximum design earthquake), Williams and Minard (2010) showed that the factor of safety of the outer slope is well above the minimum acceptable values for all three loading conditions. They added that '*major slope instability of the IWL embankment is unlikely to occur, even under the expected maximum earthquake loading conditions*'.

Taking the advantages mentioned above, several mines in WA (Mt McClure, Chalice, Mt Pleasant and Black Swan) and South Australia (Challenger, Prominent Hill and Kanmantoo) have successfully implemented the IWL method (Lane, 2017). In WA, the currently operating mines at Telfer, Dalgaranga, Gruyere, and Pilgangoora have successfully adopted the IWL method for tailings storage.

Other advantages of IWLs include the following:

- After the construction of the starter dam, the subsequent lifts require only the construction of the low permeability zone against the already placed waste rock landform. Thus, material requirements for lift construction will be much less than that required for conventional tailings dams. Materials for the low permeability zone may be borrowed from within the deposited tailings similar to the upstream raise construction in ring-dyke tailings dams.

- A reduction in the disturbed footprint, compared to that required to accommodate a separate TSF and a waste rock landform. This in turn reduces long-term environmental impacts and the cost of land clearance.

- Progressive rehabilitation as a single landform, instead of two separate landforms. The elimination of the need to rehabilitate a stand-alone TSF outer slope reduces the overall cost of rehabilitation. Further, a fully rehabilitated IWL will blend better into the natural landscape (Williams and Minard, 2010; Mooder and McGreevy, 2022).

- Elimination or significant reduction of the potential for rainfall run-off erosion and gullying in the tailings dam outer slopes. During heavy rainfall events, run-off erosion and gullying could occur on ring-dyke tailings dam outer slopes. The same rainfall could increase the pond size and elevate the phreatic surface within the dam. Deepening gullies and elevated phreatic surface could lead to slip failure on the outer slope of the dam. This is highly unlikely when the dam is fully buttressed by mine waste rock as in IWLs.

- Depending on the final design height of the surrounding mine waste rock landform the tailings cell can easily be raised to accommodate an increase in the tailings production.

- Tailings deposition can continue while the construction of the tailings cell perimeter embankment (low permeability zone) lift is in progress. This is not possible or difficult in most ring-dyke tailings dams.

- Since the tailings deposition surface area is continually increasing with each lift of the perimeter embankment (low permeability zone), a minor increase in tailings production can be easily accommodated without implementing significant modifications.

- The final cover design for the tailings surface can be water harvesting or shedding depending on the properties of the tailings. Water-shedding cover design can be limited to the tailings surface and the rest of IWL can be water harvesting if needed.

- IWLs can also be adapted for paste tailings, dry stacking, and co-disposal etc.

Disadvantages of IWLs

As with other tailings storage methods, IWLs also have several disadvantages. They are:

- From an environmental point of view, the main disadvantage is the perceived difficulties in the detection and management of seepage occurring under the mine waste rock surrounding the tailings cell. If seepage enters the area outside the perimeter of the tailings cell, the seepage may not be visually detectable until it emanates at the outer perimeter of the waste rock landform. This difficulty could be alleviated by incorporating appropriate seepage control measures into the WRL design and providing provisions for monitoring instrumentation and seepage recovery systems. This is a matter that the mining companies, TSF designers and regulators could jointly resolve.

- IWLs are viable only for projects where suitable mine waste materials are available in sufficient quantities within a reasonable distance to keep the haulage cost down.

- Even if sufficient quantities of suitable mine waste materials are to be excavated, the waste materials may not be available in time to construct the IWL for tailings storage. Detailed advanced planning is necessary to overcome this disadvantage.

- In hilly or steep terrains with frequent wet weather and weak foundation conditions, additional precautionary measures will be required to ensure the overall stability of the IWL. Otherwise, the entire IWL comprising the tailings and mine waste may become unstable.

- The IWL option may not be suitable for underground mining operations with a low volume of mine waste rock production (Lane, 2017).

DISCUSSION

The reported tailings dam incidents analysed by various researchers highlighted that the upstream raised dams are the most vulnerable to failure under several scenarios. This does not imply that all such dams would fail at some stage during their active life or after the closure. Nevertheless, the foremost concern is that several factors could initiate the collapse of such dams during operation or after the closure. Some of these factors are beyond human control. For instance, earthquakes and adverse weather. Other factors are human errors during planning, design, construction, operation, and closure stages, that are likely to recur even with the best efforts to prevent them.

IEEIRP (2015) underscored that the reliability of tailings dams is contingent on flawless execution in all phases, including planning, site investigation, data analysis, design, construction, operation, monitoring, regulatory actions, and risk management. They added that 'all of these activities are subject to human error'. As previously mentioned, Dong, Deng and Wang (2020) clearly recognised that there are still unresolved problems associated with tailings dams and that safety and environmental disasters from these dams are unavoidable. Roche, Thygesen and Baker (2017), in their Rapid Response Assessment report to UNEP, pertinently asked '… is there a way to reduce the risk of dam failure?'. An answer to this question is yes, consider IWLs where it is practicable.

The above answer is supported by Williams and Minard (2010) by showing that the likelihood of occurrence of the most common causes of tailings dam failures is negligible to low in IWLs. Further, the effect of operational stage human errors such as poor control of tailings beach development, deficiencies in dam maintenance, and inadequate management of decant ponds etc will be much less on IWLs than on conventional tailings dams. Hence, where possible, it is prudent to consider IWLs instead of standalone ring-dyke dams to comply with the HS Legislation.

It must be noted that some of the small to medium-scale mines operate within tight budgets and are sensitive to fluctuations in commodity prices. Downturns in commodity prices could compel low-budget mines to prematurely cease production for extended periods without attending to the required maintenance and closure works. In some cases, a sudden drop in commodity prices could lead the small mine operators to enter into voluntary administration or receivership, leaving their TSFs in a state of disrepair which is a burden to the general public and governing bodies. In an IWL, even if the closure works are delayed for an extended period without attending to maintenance work, the tailings containment perimeter dam is likely to remain in a better state than that of an upstream raised ring-dyke dam.

For these reasons, it is of paramount importance to avoid relying solely on dams, particularly when low-risk alternatives, such as IWLs, are available for comparable efforts and costs to achieve the ultimate goal of zero harm to people and the environment. If the planning and scheduling of construction are well coordinated, the construction and operation of IWLs can also be cost-effective. As emphasised by Mooder and McGreevy (2022), to achieve this objective, '*Mines should work toward creating a culture of collaboration, with ongoing, open communication among technical disciplines and the various operations, planning, reclamation, regulatory, and community relations teams*'.

It is not implied here that IWLs are the 'panacea' for tailings storage challenges at all mining operations. The point raised is that there are several mines where IWL concept could have been adapted to improve the stability of their tailings dams. Hence, it will be prudent for the health and safety regulators and the environmental regulators to encourage the industry to consider the IWL concept with a life-of-mine focus, before seeking regulatory approvals for tailings deposition.

CONCLUSIONS

It is well-known that tailings dams are vulnerable to failure under several scenarios, and the consequences of failure could be disastrous. The research on the behaviour of tailings as well as the new developments in design and construction quality have improved the performance of tailings dams. Despite these improvements, the literature confirms that the rate and the consequences of tailings dam failures remain largely unchanged during the last few decades. Catastrophic failure of tailings dams may occur due to natural causes or human errors. Hence, it is prudent to consider alternative methods of tailings storage that could eliminate or minimise the effect of natural causes and human errors that initiate catastrophic failure of tailings dams. This also supports the principle of precautionary approach for TSFs stability and it is a preferred option for social licence to operate.

There are several methods of tailings discharge and storage, with a low risk of catastrophic failure. Most of them are financially viable only for large, long-term mining operations. Small to medium-scale mining operations with tight budgets evade the new developments in tailings management and continue with slurry tailings storage in upstream raised dams. For such mining operations, an IWL will be a low-risk alternative to meet the goal of zero harm to people and the environment and to comply with the SFAIRP requirement of the Health and Safety legislation.

REFERENCES

Azam, S and Li, Q, 2010. Tailings dam failures: a review of the last one hundred years, *Geotechnical news*, 28(4):50–54.

Department of Foreign Affairs and Trade (DFAP), 2016. Tailings Management, Leading Practice Sustainable Development Program for the Mining Industry, Australian Government. Available from: <https://www.industry.gov.au/sites/default/files/2019-04/lpsdp-tailings-management-handbook-english.pdf>

Dong, L, Deng, S and Wang, F, 2020. Some developments and new insights for environmental sustainability and disaster control of tailings dam, *Journal of Cleaner Production*. https://doi.org/10.1016/j.jclepro.2020.122270

Geoscience Australia, 2023. Atlas of Australian Mine Waste, Geoscience Australia, Canberra, Australian Government Available from: <https://portal.ga.gov.au/persona/minewaste>

Global Tailings Review (GTR), 2020. Global Industry Standard on Tailings Management (GISTM) [online], Global Tailings Review. Available from: <https://globaltailingsreview.org/global-industry-standard/>

Independent Expert Engineering Investigation and Review Panel (IEEIRP), 2015. The Report on the Mount Polley Tailings Storage Facilities Breach, Vancouver, Province of British Columbia.

International Commission on Large Dams (ICOLD), 2001. Tailings dams–risk of dangerous occurrences, lessons learnt from practical experiences, *Bulletin 121*, International Commission on Large Dams, Paris, 155.

Lacy, H and Lane, J C, 2007. Integrated landforms and waste stream management, in Proceedings Planning for Mine Closure (Australian Centre for Geomechanics: Perth).

Lacy, H W B and Barnes, K L, 2006. Tailings storage Facilities – Decommissioning planning is vital for successful closure, in *Proceedings of the First International Seminar on Mine Closure*, pp 139–148 (Australian Centre for Geomechanics: Perth). https://doi.org/10.36487/ACG_repo/605_6

Lacy, H W B, 2019. Mine landforms in Western Australia from dump to landform design: review, reflect and a future direction, in *Proceedings Mine Closure 2019: the 13th International Conference on Mine Closure*, pp 371–384 (Australian Centre for Geomechanics: Perth). https://doi.org/10.36487/ACG_rep/1915_30_Lacy

Lane, J C, 1998. Innovative approaches to tailings disposal, paper presented to the Workshop on Future Directions in Tailings Management, Australian Centre for Mining Environmental Research, Kenmore, Australia, 1998.

Lane, J C, 2004. Integrated landforms–an alternative approach to tailings storage, paper presented to the Seminar on Mine Closure – Towards sustainable outcomes (Australian Centre for Geomechanics: Perth).

Lane, J C, 2013. In-pit tailings deposition (storage), paper presented to the Seminar on Advanced Tailings Management, 22 March 2013 (Australian Centre for Geomechanics: Perth).

Lane, J C, 2017. Integrated Waste Landforms Design, Construction, Operation and Management, paper presented to the Seminar on Management, Operation and Relinquishment of Tailings Storage Facilities, 27–28 April 2017 (Australian Centre for Geomechanics: Perth).

Lane, J C, 2018. Integrated Waste Landforms Design, Construction, Operation and Management, paper presented to the Goldfields Environmental Management Group Workshop, Kalgoorlie, Australia, 16–18 May 2018.

Lin, S Q, Wang, G J, Liu, W L, Zhao, B, Shen, Y M, Wang, M L and Li, X S, 2022. Regional Distribution and Causes of Global Mine Tailings Dam Failures, *Metals*, 12(6):905. https://doi.org/10.3390/met12060905

Lyu, Z, Chai, J, Xu, Z, Qin, Y and Cao, J, 2019. A Comprehensive Review on Reasons for Tailings Dam Failures Based on Case History, *Advances in Civil Engineering*, Volume 2019. Article ID 4159306. 18 pages, https://doi.org/10.1155/2019/4159306

Mooder, R B and McGreevy, J T G, 2022. Toward integrated mining landform design: recent tailings project examples from Canada, in Proceedings of Mine Closure 2022 (Australian Centre for Geomechanics: Perth). https://doi.org/10.36487/ACG_repo/2215_04

Piciullo, L, Storrøsten, E B, Liu, Z, Nadim F and Lacasse, S, 2022. A new look at the statistics of tailings dam failures, *Engineering Geology* 303:106657. https://www.sciencedirect.com/science/article/pii/S0013795222001429

Rico, M, Benito, G, Salgueiro, A R, Díez-Herrero, A and Pereira, H G, 2008. Reported tailings dam failures: A review of the European incidents in the worldwide context, *Journal of Hazardous Materials*, 152:846–852.

Roche, C, Thygesen, K and Baker, E (eds), 2017. Mine Tailings Storage: Safety Is No Accident, A UNEP Rapid Response Assessment, United Nations Environment Programme.

Safe Work Australia, 2013. How to determine what is reasonably practicable to meet a health and safety duty June 2013. Available from: <https://www.safeworkaustralia.gov.au/system/files/documents/2002/guide_reasonably_practicable.pdf>

Stark, T D, Moya, L and Lin, J, 2022. Rates and Causes of Tailings Dam Failures, *Advances in Civil Engineering 2022*, 21 p. https://doi.org/10.1155/2022/7895880

Vick, S G, 1990. *Planning, Design and Analysis of Tailings Dams*. http://dx.doi.org/10.14288/1.0394902

Williams, D A and Minard, T E, 2010. An environmentally and economically attractive integrated landform for the storage of tailings and waste at the Randalls Gold Project in Western Australia, paper presented to the Mine Waste 2010 Seminar (Australian Centre for Geomechanics: Perth). https://doi.org/10.36487/ACG_rep/1008_12_Williams

WorkSafe Victoria, 2007. How WorkSafe applies the law in relation to Reasonably Practicable, A guideline made under section 12 of the occupational health and safety act 2004. Available from: <https://content-v2.api.worksafe.vic.gov.au/sites/default/files/2018-06/ISBN-Reasonably-practicable-how-WorkSafe-applies-the-law-2007-11.pdf>

WorkSafe Western Australia (WA), 2021. How to determine what is reasonably practicable to meet a health and safety duty – Interpretive guideline: Department of Mines, Industry Regulation and Safety, Western Australia, 17 p. Available from: <https://www.wa.gov.au/system/files/2021-11/211101_GL_ReasonablyPracticable.pdf>

World Information Service on Energy (WISE), 2025. Chronology of major tailings dam failures. Available from: <https://www.wise-uranium.org/mdaf.html> [Accessed: 21 May 2025].

Cover performance and interactions between vegetation and metals at a tailings storage facility in New South Wales

T K Rohde[1], C Brownbil[2] and J Lang[3]

1. CEO, SGME Pty Limited, Windsor Qld 4030. Email: trohde@sgme.au
2. Environmental Scientist, SGME Pty Limited, Windsor Qld 4030. Email: cbrownbill@sgme.au
3. Ecologist, SGME Pty Limited, Windsor Qld 4030. Email: jlang@sgme.au

ABSTRACT

The Mine, based in central New South Wales, manages tailings within multiple tailings storage facilities (TSFs). Current geochemical assessments indicate that these tailings are chemically stable and likely capable of consuming acid, suggesting no risk of acid and metalliferous drainage (AMD). Nevertheless, X-ray fluorescence (XRF) analysis has determined that Aluminium, Magnesium, and Iron can become mobile under certain infiltration conditions, necessitating further examination of metal mobility to safeguard against environmental impacts. The rehabilitation process for the TSFs requires a growing medium that constrains or prevents metal movement while supporting the ultimate post-mining land use of grassland. This paper evaluates both the efficacy of placed mediums and direct tailings rehabilitation methods for TSF closure. Since 2015, initiatives have focused on establishing vegetation directly within the tailings, successfully demonstrating the direct seeding method. A sampling program conducted in 2024 analysed tailings up to a depth of 2 m and assessed vegetation, focusing on pH, electrical conductivity (EC), and total plant-available metal concentrations. Tailings displayed a consistently alkaline pH (median range 8.8–9.2) and generally low EC, which suggests that widespread salinity stress is unlikely. Nevertheless, there were deficiencies in total Nitrogen (below the detection limit of 150 mg/kg) and extractable Phosphorus (median <1 mg/kg below 0.25 m), which could hinder root growth and development. Total Copper levels were consistently elevated, with median values at or surpassing the threshold of 844 mg/kg, while maximum levels were above this threshold. Instances of total Arsenic and Mercury also surpassed established limits. The analysis of vegetation revealed a notable accumulation of Iron, Manganese, Boron, Molybdenum, and Arsenic in plant tissues, with multiple exceedances beyond the thresholds. Despite these challenging substrate conditions and elevated metal concentrations in the plants, field observations confirm that tolerant vegetation has been successfully established through direct seeding. Landscape Process Modelling at Multi-Dimensions and Scales (LAPSUS) projects low erosion rates over a century, provided that 50 per cent vegetation cover is preserved. This study concludes that direct seeding into unreactive, alkaline tailings with limited nutrients, particularly through the use of halophytes such as saltbush, presents a feasible and sustainable rehabilitation approach that does not require a placed growing medium. Continued monitoring of vegetation health and metal dynamics, along with an assessment of the environmental implications of metal accumulation, is essential for successful post-closure relinquishment.

INTRODUCTION

Location and operations

The Mine operates in central New South Wales (NSW), 27 km north-north-west of Parkes (Figure 1).

It has both underground and open cut mining, extracting from various orebodies utilising a combination of block and sub-level caving techniques. The Mine produces Copper (Cu) and Gold (Au), with waste rock stored in multiple waste rock dumps (WRDs) and tailings managed within five tailings storage facilities (TSFs).

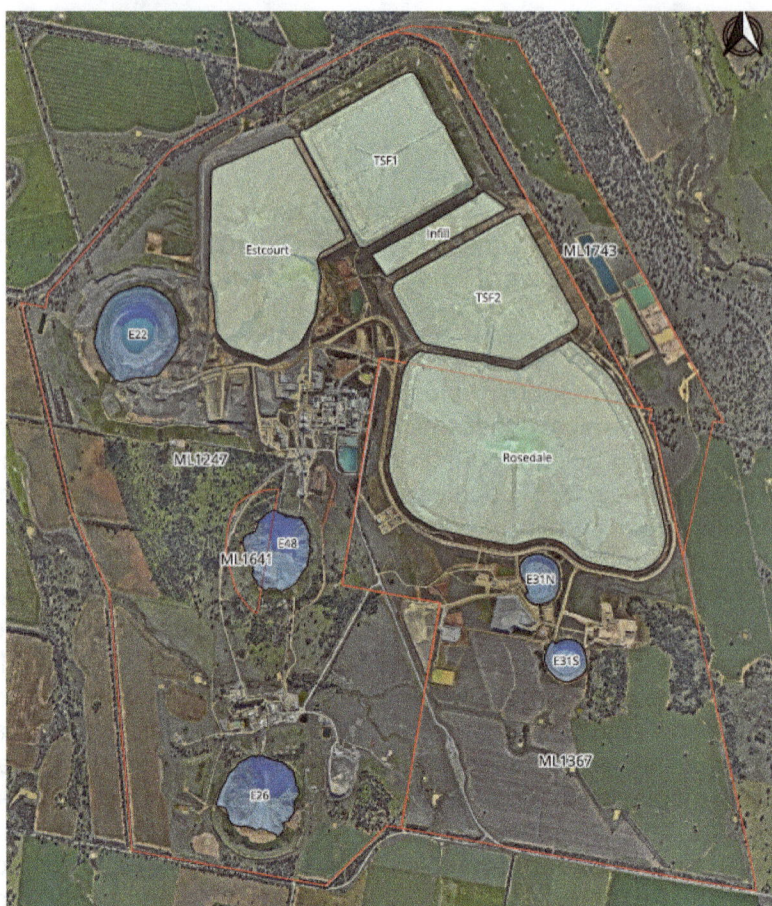

FIG 1 – Location of the Mine, central New South Wales, 27 km north-north-west of Parkes, with multiple tailings storage facilities (TSFs).

REVIEW

Climate

The climate of Central NSW is classified as temperate with no dry season, according to the Köppen-Geiger classification system (Bureau of Meteorology (BoM), 2024a, 2024b, 2024c). Historical and current rainfall and evaporation data (refer to Table 1) are collected from the nearest automatic weather station, which is located 30 km south-east of the Mine.

TABLE 1

Mean monthly and annual rainfall and evaporation.

Parameter	Jan	Feb	Mar	Apr	May	Jun	Jul
Mean rainfall (mm)	39.8	27.2	119.6	35.0	8.0	39.2	29.2
Mean evaporation (mm)	7.2	7.0	5.4	3.0	1.9	1.6	1.5
	Aug	**Sep**	**Oct**	**Nov**	**Dec**	**Annual**	
Mean rainfall (mm)	25.8	15.6	26.2	110.4	76.8	552.8	
Mean evaporation (mm)	2.2	4.3	5.8	6.1	7.6	1602.4	

The mean annual rainfall is 552.8 mm (see Table 1), with the wettest months typically occurring from November to March. Conversely, the driest months are from July to October.

Mean annual potential evaporation is 1602.4 mm (refer to Table 1), which is three times greater than the average yearly rainfall. Evaporation rates vary seasonally, with the highest rates occurring from December to February.

Geochemistry

Generally, post-mining land use risk is influenced by various factors, including geomorphology, the chemical reactivity of tailings, and the water quality that may result from infiltration, storage, capillary rise, or vegetation uptake.

The geochemical characterisation of tailings has been an ongoing effort, with the most recent assessment conducted by the University of Queensland (UQ) in 2023. This evaluation involved the analysis of 66 tailings samples collected to a maximum depth of 7 m. The samples were analysed using handheld X-ray fluorescence (XRF) to determine the total Sulfur (S) and Calcium (Ca) percentages. However, the analysis did not account for reactive sulfide (typically measured using Chromium (Cr) reducible Sulfur) or acid-neutralising capacity (ANC).

Acid-base accounting (ABA) assesses the balance between acid-generating and acid-neutralising processes. Values obtained from ABA include maximum potential acidity (MPA) and ANC. The difference between MPA and ANC is referred to as net acid-producing potential (NAPP). Classification criteria are as follows:

- If the ANC:MPA ratio is greater than or equal to 2, the material is at a very low risk of producing acid.

- If the ANC:MPA ratio is less than 2, the material is at a low risk of producing acid.

- If NAPP (Net Acid Producing Potential) is negative, the material may have sufficient ANC to prevent acid mine drainage (AMD).

- If NAPP is positive, the material may be capable of generating AMD.

The following equations have been used to calculate MPA, ANC, and NAPP:

$$MPA\ (kg\ H_2SO_4/t)) = \text{total S percentage} \times 30.6 \tag{1}$$

$$ANC\ (kg\ H_2SO_4/t) = ((\text{percentage Ca}/100) \times (1\,000\,000/40.08) \times 98.08))/1000 \tag{2}$$

$$NAPP\ (kg\ H_2SO_4/t) = MPA\ ((kg\ H_2SO_4/t) - ANC\ (kg\ H_2SO_4/t) \tag{3}$$

Where:

40.08   is the molar mass of calcium (g/mol)

98.08   is the molar mass of sulfuric acid (g/mol)

1 000 000   converts tonnes to grams

1000   converts grams to kilograms

Simplistically, the assumption that one mole of elemental Calcium neutralises one mole of Sulfuric acid leads to the ABA assessment depicted in Figure 2.

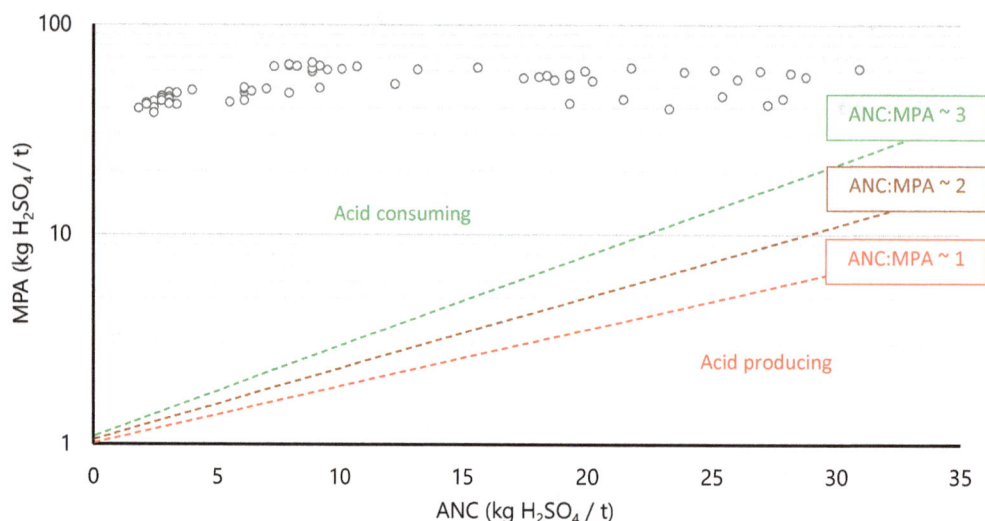

FIG 2 – Operations tailings reactivity.

Figure 2 shows that tailings are chemically unreactive and likely acid-consuming (AC) because their potential ANC:MPA ratio is greater than two. The AC potential of tailings has been reported in other geochemical investigations (CSIRO, 1997), which found AC potentials of 100–580 kg H_2SO_4/t. Guiton (2009) can supplement geochemical analysis and found the following metal water extract concentrations: Aluminium (4 mg/kg), Magnesium (830 mg/kg), Copper (1200 mg/kg), and Iron (25 000 mg/kg). It is noted that the water extract is not necessarily indicative of long-term capillary rise or potential for vegetation uptake.

Rehabilitation of the TSF

At rehabilitation, the five TSF's will form a single continuous structure and will have a final land use of grassland supported by a suitable growing medium. The growing medium will comprise one or more of the following: a combination of soil (topsoil, subsoil, or co-mingled; referred to as the soil herein), run-of-mine waste rock, or tailings with direct seeding. Research conducted on growing medium options is summarised in the following sections.

TSF growing medium trials

UQ TSF growing medium field trials, also known as the field trials, have been ongoing since 2008 (UQ, 2014) to determine a practical design that supports the final land use. Preliminary research suggests that the growing medium may need to be 1.8 m to 2.0 m deep to limit the potential for metal transport and accumulation by capillary rise, which could potentially impact vegetation. The field trials have investigated thinner growing medium thicknesses that could improve growing medium use efficiency. The field trials are summarised in Table 2.

TABLE 2

Growing medium options.

Design	A	B	C	D
Topsoil (m)	0.1	0.1	0.1	0.1
Waste rock (m)	-	0.4	0.4	0.9
Capillary break (m)	-	-	0.3	-
Total growing medium (m)	0.1	0.5	0.8	1.0

Percolation

A key outcome of the field trials has been the calculation of percolation using the volumetric water content (VWC) sensors buried in the growing medium and tailings. Stored infiltration was calculated daily by multiplying the change in VWC by the depth for the period from 2014 to 2017. The negative daily incremental change (-Δ) in stored water infiltration (-ΔSW) balances percolation daily, assuming that -ΔSW results from percolation (Figure 3). Figure 3 shows that cumulative percolation for the trial period corresponds to an annual rate of: trial A (267 mm), trial B (117 mm), trial C (100 mm), and trial D (20 mm). Groundwater recharge rates in Australia are variable and intrinsically linked to climatic and geological factors (Crosbie *et al*, 2010; Boas and Mallants, 2022; Lee *et al*, 2024). Determining groundwater recharge rates is essential for understanding how the growing mediums are performing. The average rate of groundwater recharge is generally calculated using the chloride mass balance method (CMB) for groundwater and rainfall. The CMB method has several generalised assumptions (Wood, 1999), including that chloride in groundwater is sourced from rainfall and is conservative within the system, that the flux exists as a steady state, and that there is no recycling. Lee *et al* (2024) used the CMB method to calculate a range of potential recharge rates for a temperate climate: low (2.6 mm/annum), medium (60 mm/annum), and high (522 mm/annum). A comparison of this study's results with the field trial percolation results indicates that percolation is medium to high, consistent with natural rates of groundwater recharge.

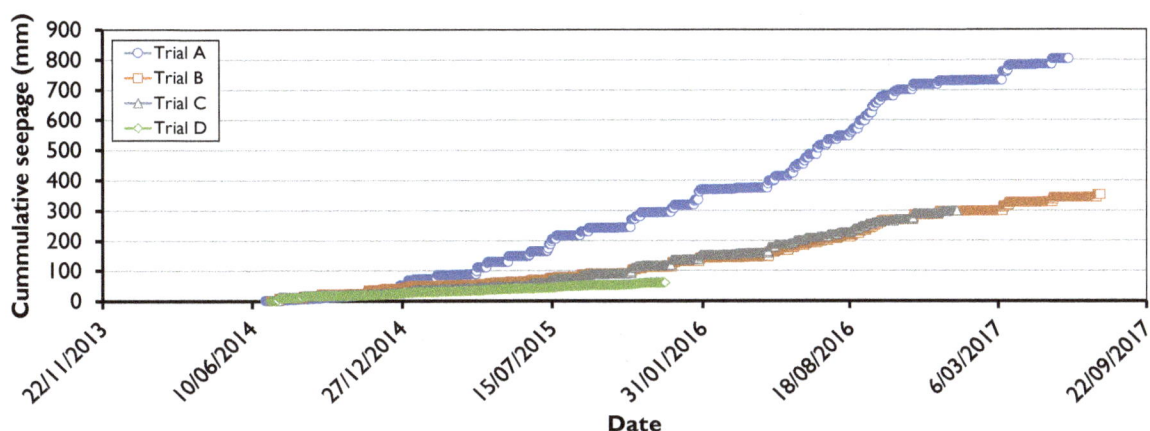

FIG 3 – Field trials – percolation.

Threshold criteria

In 2025, the field trials were sampled to develop threshold criteria (the threshold):

- Sub-samples were taken at depths based on the stratigraphy of the growing medium, and:
  - The field trial was analysed for pH and electrical conductivity (EC).
  - Total metal concentrations were analysed using handheld XRF for Cadmium (Cd), Lead (Pb), Arsenic (As), Chromium (Cr), Nickel (Ni), Mercury (Hg), Silver (Ag), and Copper (Cu).

- Vegetation sampling and laboratory analysis were conducted for:
  - Boron (B), Iron (Fe), Cu, Zinc (Zn), Molybdenum (Mo), Magnesium (Mg), Manganese (Mn), Cd, Pb, As, and Selenium (Se).

The resulting threshold (Table 3) was determined by adopting the highest value, or the 75th percentile from the field trials, in cases where an extreme outlier was present in the pooled trial results.

TABLE 3

Growing medium and vegetation thresholds.

Indicator	Units	Threshold
Growing medium		
pH	pH units	7.0–9.0
EC	dS/m (uS/cm)	Less than 2.0 (2000)
Total metals		
Cu	mg/kg	Less than 844
As	mg/kg	Less than 30
Cd	mg/kg	Less than 8
Hg	mg/kg	Less than 7
Vegetation		
Total metals		
Fe	mg/kg	Less than 1200
Cu	mg/kg	Less than 74
Mn	mg/kg	Less than 200
B	mg/kg	Less than 50
Mo	mg/kg	Less than 100
As	mg/kg	Less than 5

Direct seeding trials

Since 2015, vegetation has been established directly into tailings, demonstrating successful direct seeding and an effective dust control strategy. Over this period, a variety of salt-tolerant grasses, forbs, and chenopod species have colonised the TSF. The ongoing success of these species in colonising the tailings through continuous self-seeded succession has initiated a multi-year study into the potential for tailings to serve as a growing medium. Barley has been grown directly in the TSF to provide dust control; however, over the past five years, it has been deemed unnecessary for ongoing dust management due to the ground cover provided by established vegetation. In 2020, the TSF was sown with a seed mix of salt-tolerant species, resulting in increased ground cover density. Landloch (2021) conducted an assessment of the tailings in the south-east corner of the TSF. The focus was to assess differences between bare tailings (lacking vegetation), tailings with barley crops, and tailings that received nitro humus in 2016. The study found that establishing barley improves tailings as a growing medium by lowering salinity in the near-surface layer (the horizon) through increased leaching. The horizon then becomes suitable as a growing medium because the effective root zone is deeper. The Landloch study also found that the addition of litter or organic matter, such as nitro-humus, appears to enhance near-surface leaching of tailings, as well as nitrogen (N) and organic carbon. In 2021, a strip of chicken manure fertiliser (the organic strip) was placed on a section of the TSF at a location with poor saltbush establishment.

METHOD

Sampling locations

In 2024, a sampling program was completed in the TSF to collect samples of tailings (to 2 m below ground level, with samples taken at 0.25 m increments) and vegetation (salt bush – leaf (L), root (R), stem (S)) (Table 4 and Figure 4) to further assess the use of tailings as a growing medium.

TABLE 4

Sample locations.

Description	Location
Tailings samples	
Volunteer vegetation	16
Salt bush seeding	3, 4, 5
Organic strip	1, 2
Vegetation samples	
Volunteer vegetation	16 (L, R, S)
Salt bush seeding	11 (L, R, S)
Organic strip	10 (L, R, S)

FIG 4 – Sample locations.

Thresholds

Tailings fertility

An assessment of tailings fertility and post-mining land use potential was conducted in comparison to agricultural industry thresholds (Table 5). (Rayment and Bruce, 1984; Baker and Eldershaw, 1993; Department of Environment and Resource Management (DERM), 2011; Peverill, Sparrow and Reuter, 1999).

TABLE 5

Growing medium fertility thresholds.

Indicator	Units	Threshold
Total N	mg/kg	Greater than 1500
Extractable P (Colwell)	mg/kg	Greater than 10
Ammonia (NH$_3$)	mg/kg	-
Cation exchange capacity (CEC)	meq/100g	12.0–25.0
Total N	mg/kg	Greater than 1500
Vegetation available micronutrients		
Diethylenetriaminepentaacetic (DTPA) acid Cu	mg/kg	Greater than 0.5
DTPA Fe	mg/kg	Greater than 5
DTPA Mn	mg/kg	Greater than 2

Growing medium and vegetation thresholds

pH, EC, and total metal thresholds were derived from field trial sampling (Table 3).

LAPSUS

LAPSUS is a landform evolution model (LEM) that simulates the three-dimensional (3D) development of landscapes over time, used to estimate potential surface water run-off and predict resulting erosion and deposition processes on a landform scale. The primary inputs for LAPSUS include a digital elevation model (DEM), erosion characteristics, and climate data. Gradient and slope length vary across the TSF beaches and embankments, with the embankments also constructed from a range of materials.

Digital elevation model

A DEM grid size of 3 m was used for the TSF, as this resolution is suitable for the level of detail and size of the landform. The depth of the DEM was established to allow for erosion to occur to the depth of the landform, with a maximum depth of 28 m.

Erosion characteristics

A Water Erosion Prediction Project (WEPP) model was used to predict erosion rates, sediment yield, and run-off. An erosion range of 0.72–3.46 t/ha/a, representing the 75th quartile of the average erosion rates, was employed to semi-calibrate the LAPSUS model.

Other erosion input parameters used in LAPSUS are detailed in Table 6.

TABLE 6

LAPSUS input parameters.

Parameter	Input
m (exponent of overland flow)	2
n (exponent of slope)	1
p (multiple flow factor)	1
K (erodibility)	0.00081–0.00201
P (sedimentation)	0.0003

Each parameter in Table 6 contributes to the processes of water run-off, erosion, and deposition. The method employs formulas adapted to simulate spatial water erosion and sedimentation. (Schoorl, Sonneveld and Veldkamp, 2000). The adapted formulas are derived from the early works of Kirkby (1971) and Foster and Meyer (1972, 1975).

Climate

LAPSUS utilises annual rainfall, infiltration, and evaporation inputs to calculate the annual run-off volume. The model applied a constant value for each input: rainfall was set to 0.5528 m, equivalent to the average yearly rainfall of the region (Table 1), while infiltration and evaporation parameters were set to 0.2 m and 0.1 m, respectively.

RESULTS AND DISCUSSION

The following results and discussion present the findings for the TSF, including the organic strip, but exclude the UQ field trials, which were used as the threshold for comparison.

Tailings fertility

Results indicate that the fertility of tailings does not negatively impact the establishment of vegetation. Interestingly, incorporating fertiliser – especially a biologically active option – can enhance the effectiveness of using tailings as a growing medium. The distribution of total Nitrogen (Figure 5a) was consistently below the detection limit (150 mg/kg) across all depth intervals and is relatively low compared to the threshold of 1500 mg/kg. Although this deficiency could potentially restrict vegetation, it does not appear to affect established vegetation.

Extractable Phosphorus (Figure 5b) displays a stratified pattern with depth. The surface layer (0.00–0.25 m) exhibits the highest median value (4 mg/kg) and variability, with notably high outliers reaching 76 mg/kg, which surpasses the threshold of 10 mg/kg. Nonetheless, even at the surface, the median value remains below this threshold. Levels drop below 0.25 m (median less than 1 mg/kg) with minimal variability, indicating a deficiency throughout the profile that may limit root development and overall plant growth, particularly for species with deeper root systems. While specific surface patches may have adequate Phosphorus, a widespread limitation is evident; yet, it does not appear to hinder vegetation establishment and growth.

Ammonia (NH_3) (Figure 5c) levels are generally low, with most data points clustering at or below the detection limit (0.1–0.05 mg/kg). A few higher outlier values in the surface layer (up to 0.3 mg/kg) are exceptions. The consistently low values imply minimal readily available nitrogen in the form of Ammonia, reinforcing the notion that total Nitrogen could restrict vegetation.

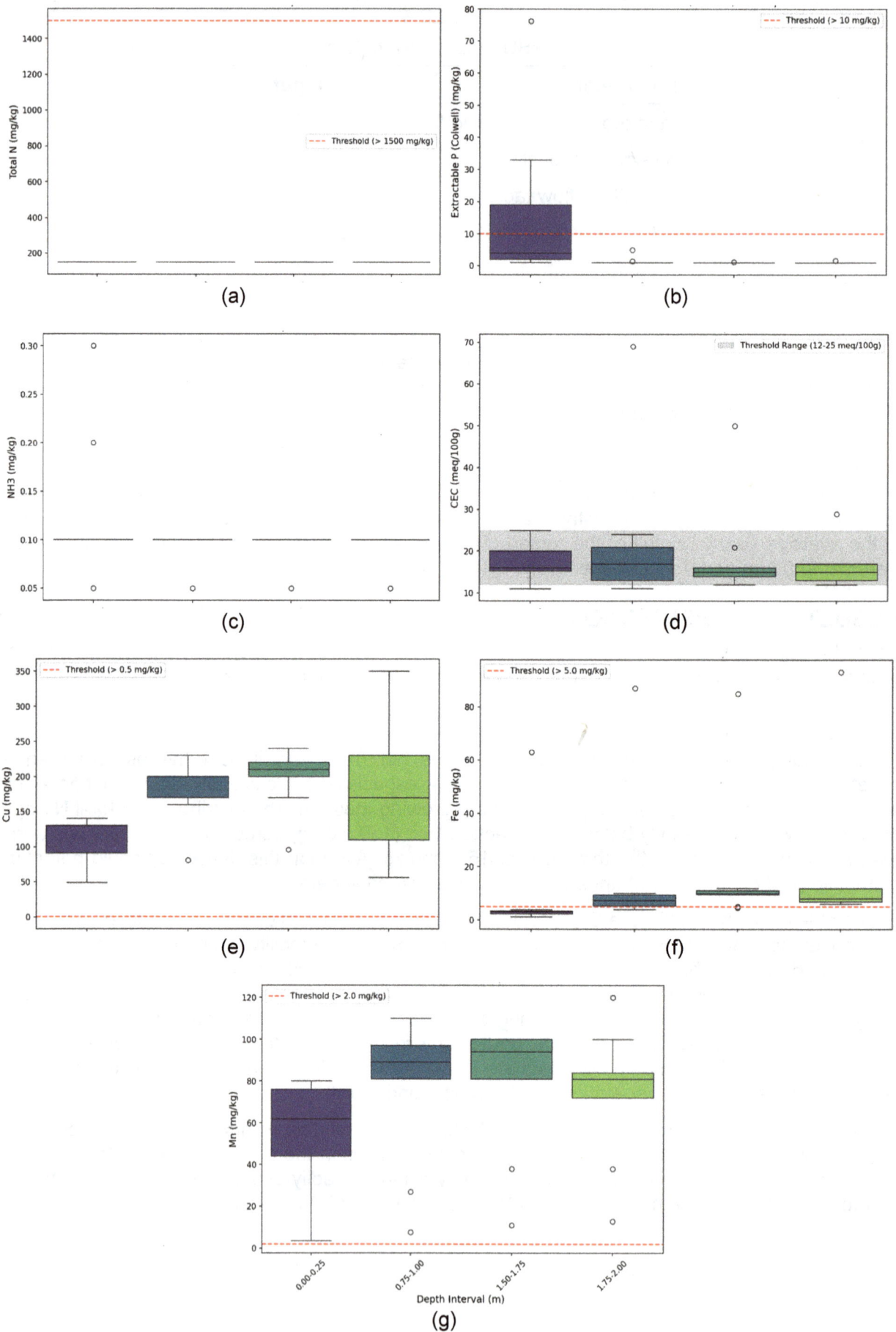

FIG 5 – Tailings fertility: (a) Total N; (b) Extractable P; (c) NH$_3$; (d) CEC; (e) Cu; (f) Fe; (g) Mn.

Cation exchange capacity (Figure 5d) generally falls within the acceptable range specified by the threshold (12–25 meq/100 g, grey band), particularly based on median values ranging from 15–16 meq/100 g. Variability peaks in the 0.75–1.00 m zone, where high outliers (50–70 meq/100 g) and a wider interquartile range indicate more dynamic conditions compared to deeper layers, characterised by tighter distributions. While the median values suggest a competent capacity for nutrient cation retention, the minimum values occasionally drop below the threshold (12 meq/100 g), especially at depths greater than 1.5 m. Overall, the CEC reflects a moderate potential for nutrient retention, which benefits vegetation; however, nutrient supply, including Nitrogen and Phosphorus, remains limited.

DTPA Copper (Figure 5e) concentrations are elevated across the profile compared to the threshold (greater than 0.5 mg/kg), with all observed values exceeding this, starting from a minimum of 49 mg/kg. A clear trend emerges, showing an increase in median concentration with depth, rising from 95 mg/kg to over 200 mg/kg at 1.50–1.75 m. Variability also rises with depth. The high DTPA Copper levels could lead to potential concerns for vegetation, inducing possible nutrient imbalances ('eg' with Iron and Manganese). DTPA Iron (Figure 5f) exhibits low median concentrations near the surface (3 mg/kg), which may dip below the threshold of 5.0 mg/kg. However, median levels increase to meet or exceed the threshold below 0.75 m (median 7–10 mg/kg). The data exhibit spatial heterogeneity, characterised by numerous high outliers at all depths (60–90 mg/kg). While median levels in the subsurface appear adequate, the exceptionally high DTPA Copper concentrations may lead to Iron deficiency, noting that Iron deficiency symptoms were not observed during the sampling program.

Lastly, similar to DTPA Copper, DTPA Manganese (Figure 5g) concentrations are notably high compared to the threshold (greater than 2.0 mg/kg), with the lowest observed value being 3.6 mg/kg. Median DTPA Manganese shows a considerable increase below the surface layer, with values rising from 45 mg/kg to a peak of 95 mg/kg at depths of approximately 1.50–1.75 m. Variability is moderate with depth, displaying several low outliers in the subsurface. The combination of high DTPA Copper and high DTPA Manganese suggests complex chemical conditions; however, these factors do not seem to negatively influence the establishment and growth of vegetation.

Growing medium and vegetation thresholds

Growing medium (the tailings)

The pH profile (Figure 6a) consistently exhibits alkaline characteristics, with median values around pH 8.9 at the surface (0.00–0.25 m) and a slight increase to stabilise between pH 9.1 and 9.2 below 0.75 m. Variability across all depths remains low. The median pH value exceeds the upper limit of the threshold (7.0–9.0) indicated by the grey band, suggesting that while the alkaline condition (pH greater than 9.0) can limit the availability of essential micronutrients like Iron, Manganese, Zinc, Copper, and Phosphorus, it does not completely hinder growth. Vegetation that can tolerate alkaline conditions, such as halophytes (saltbush), is establishing and growing under these conditions.

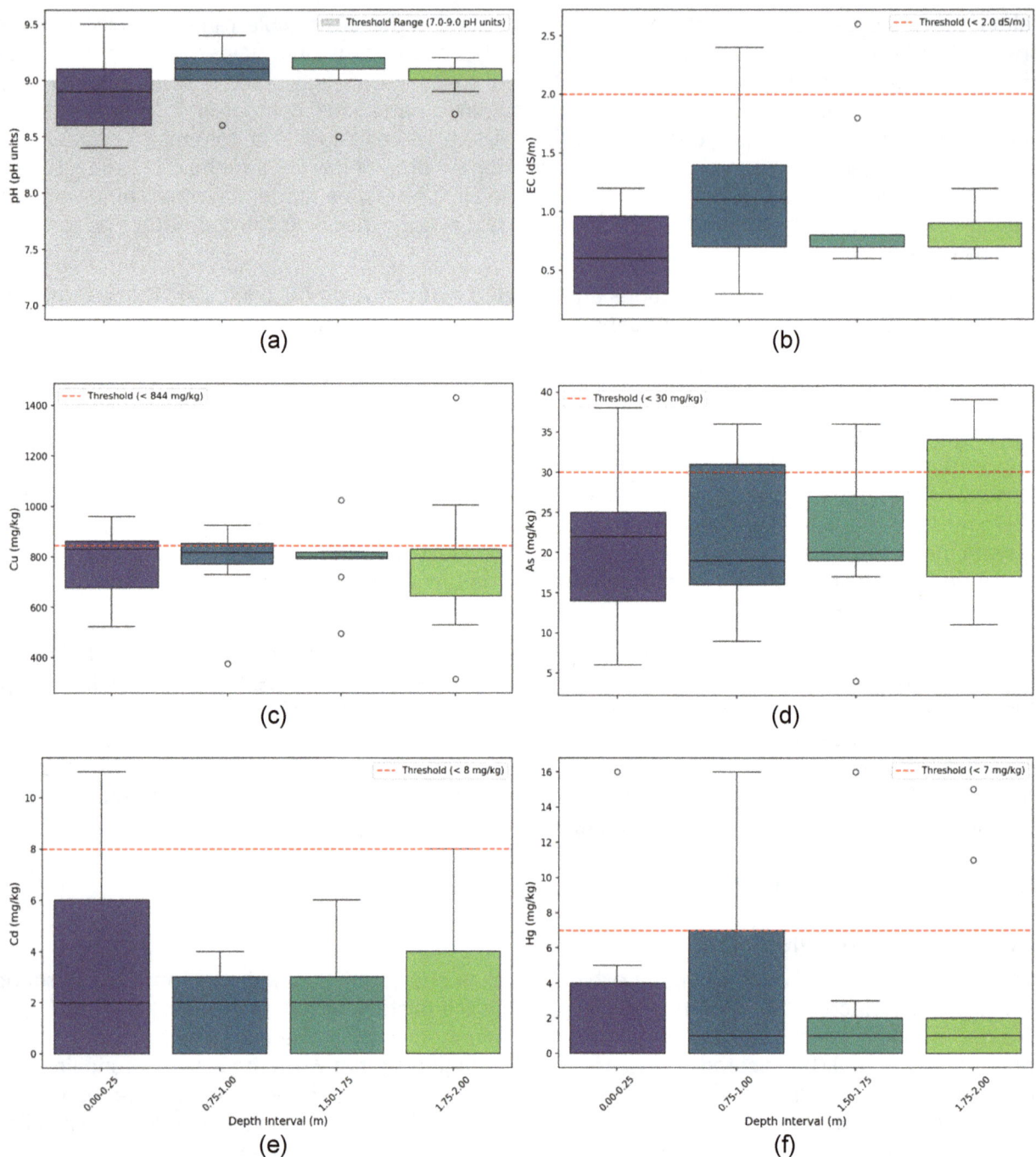

FIG 6 – Growing medium: (a) pH; (b) EC; (c) Cu; (d) As; (e) Cd; (f) Hg.

Electrical conductivity, an indicator of soluble salts, indicates generally low levels throughout the profile (Figure 6b). The median EC starts at its lowest at the surface (0.6 dS/m), peaks at 1.1 dS/m in the 0.75–1.00 m layer, and slightly declines to 0.7–0.8 dS/m in deeper layers. The most variability is seen in the 0.75–1.00 m zone, where the maximum measured value (2.4 dS/m) surpasses the threshold (less than 2.0 dS/m). Nonetheless, median values and the majority of data across all depths remain well below the salinity threshold, indicating that widespread salinity stress is not affecting vegetation.

Total Copper concentrations show a strong presence across all depths (Figure 6c). Median values vary from 800 mg/kg near the surface to above 815 mg/kg at depths of 0.75–1.75 m, before experiencing a slight decrease. Although these medians are just below the threshold (less than 844 mg/kg), the upper quartiles and maximum values consistently exceed this limit across all depth intervals, with the highest measurement (1430 mg/kg) found in the deepest layer. Despite these total concentrations and potential variations in vegetation availability due to high pH, this level of Copper reflects a high potential that could be toxic to some vegetation. Fortunately, established plants on the TSF appear to remain unaffected.

Total Arsenic concentrations display some variation with depth, with median values being highest near the surface (22 mg/kg) and in the deepest layer (27 mg/kg), while being slightly lower (19–20 mg/kg) in between (Figure 6d). Variability here is moderate. Although median concentrations remain below the threshold (less than 30 mg/kg) at all depths, the upper range frequently approaches or exceeds the threshold, particularly at the surface and below 1.5 m.

Total Cadmium concentrations are observed to be low throughout the profile (Figure 6e). Median values cluster between 0 and 2 mg/kg across all depths, below the 8 mg/kg threshold. Some variability does exist, with the maximum observed value (11 mg/kg) occurring at the surface (0.00–0.25 m) and exceeding this threshold. All other measurements beneath the surface remain within acceptable limits. Ultimately, based on these concentrations, Cadmium toxicity is not a constraint for vegetation.

Median total Mercury concentrations remain very low (0–1 mg/kg) at all depths, considerably below the 7 mg/kg threshold (Figure 6f).

Vegetation

Iron accumulation showed variation among the different parts of the vegetation across the four locations (Figure 7a). The highest median concentrations were found in leaves (4650 mg/kg), followed by stems (1340 mg/kg), and the lowest in roots (1190 mg/kg). When compared to the threshold of less than 1200 mg/kg, median concentrations in leaves and stems exceeded this limit, while the median concentration in roots was slightly below it. The maximum observed values were particularly high, notably in a stem sample (15 000 mg/kg at Location 16) and two leaf samples (8800 mg/kg at Location 11 and 7800 mg/kg at Location 10). These elevated Iron levels in vegetation tissue suggest a potential for Iron toxicity effects, noting that symptoms of toxicity were not observed during sampling.

In terms of Copper concentrations, all vegetation parts across the four locations remained below the threshold of less than 74 mg/kg (Figure 7b). The highest median concentration was found in leaves (28 mg/kg), while stems and roots recorded lower median levels (12–12.5 mg/kg). The maximum value recorded was 74 mg/kg in a stem sample. This trend suggests that, despite potentially high total Copper levels in the growing medium, the vegetation effectively limits Copper uptake, or its availability is low.

Manganese accumulation was present in all vegetation parts, often exceeding the threshold of less than 200 mg/kg (Figure 7c). While median concentrations were below or close to the threshold (highest in roots at 158 mg/kg), the maximum values consistently surpassed 200 mg/kg in all three vegetation parts: with a maximum of 310 mg/kg in leaves, 340 mg/kg in stems, and 390 mg/kg in roots. This finding suggests that many individual plants are actively accumulating Manganese, which may be beneficial if future analysis demonstrates that it is not recirculating back into the growing medium.

Boron uptake also frequently exceeded the threshold of less than 50 mg/kg, particularly in above-ground tissues (Figure 7d). Median leaf concentrations were near the threshold (47 mg/kg), with the maximum concentration (62 mg/kg) surpassing it. Although the median stem concentration was well below the threshold at 24 mg/kg, the maximum stem value reached 78 mg/kg, which is above the threshold. Median root concentrations remained well below the threshold, yet the maximum root value slightly exceeded it at 57 mg/kg.

Molybdenum uptake was highest in the leaves, with a median concentration of 71.5 mg/kg, which is still below the threshold of 100 mg/kg. However, maximum values ranged from 110 to 120 mg/kg, surpassing this threshold (Figure 7e). Stem and root tissues exhibited much lower concentrations, remaining comfortably below the threshold. While Molybdenum toxicity in vegetation is less common than that of other micronutrients, accumulation levels above the threshold, particularly in leaves, could potentially disrupt other nutrient metabolic processes.

Arsenic accumulation was also observed, with median concentrations recorded at 39 mg/kg in leaves, 7 mg/kg in stems, and 5.5 mg/kg in roots, all meeting or exceeding the threshold of less than 5 mg/kg (Figure 7f). Maximum values were even higher, especially in leaves (90 mg/kg) and stems (60 mg/kg).

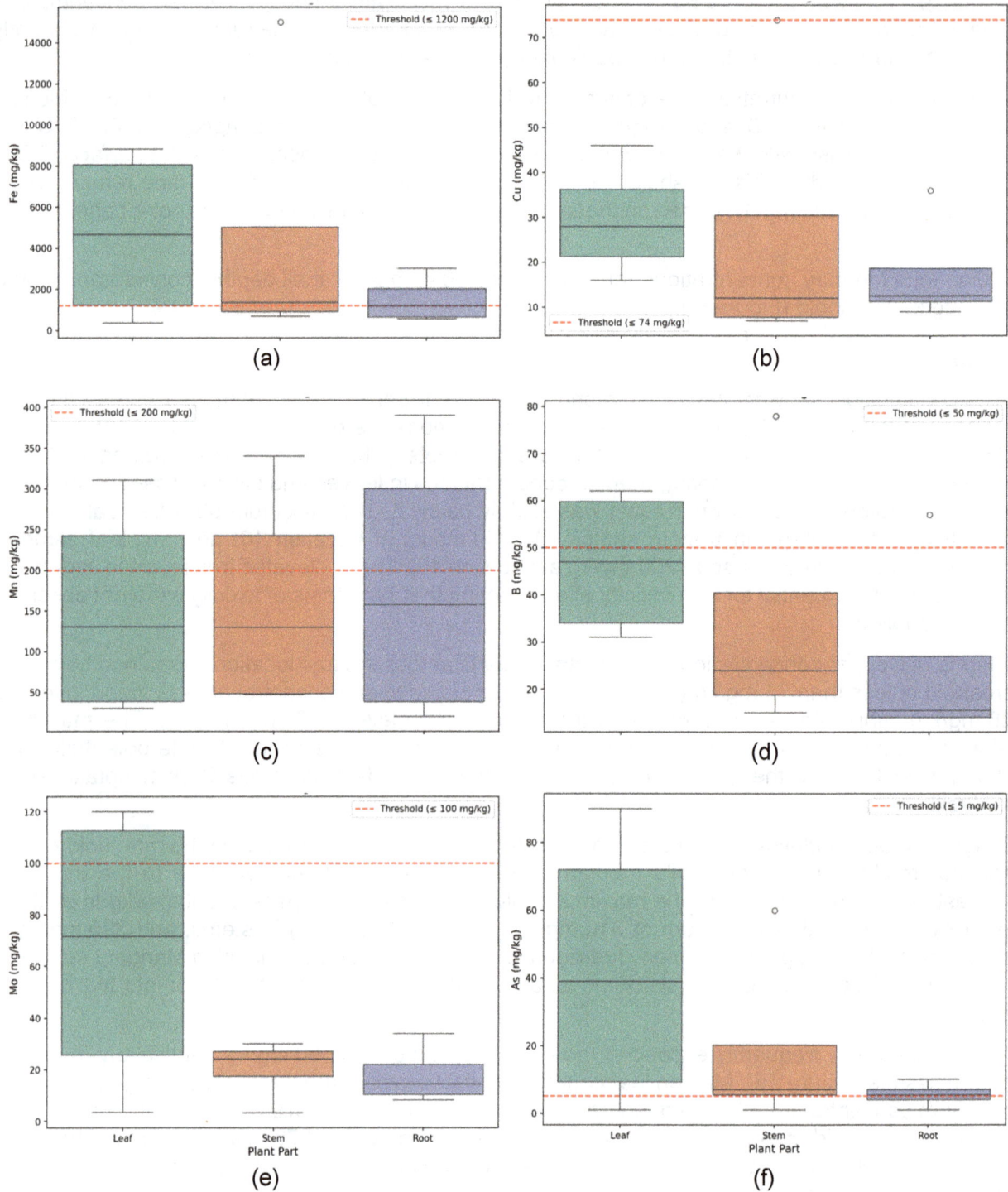

FIG 7 – Vegetation: (a) Fe; (b) Cu; (c) Mn; (d) B; (e) Mo; (f) As.

LAPSUS

The LAPSUS 3D landform render after 100 years is shown in Figure 8. It indicates that the upper surface of the TSF will experience some erosion when tailings are used as a growing medium; however, the erosion will create gullies less than 0.5 m deep, assuming the vegetation cover can be maintained at 50 per cent. Noting that vegetation covers less than 50 per cent will increase the likelihood and severity of erosion.

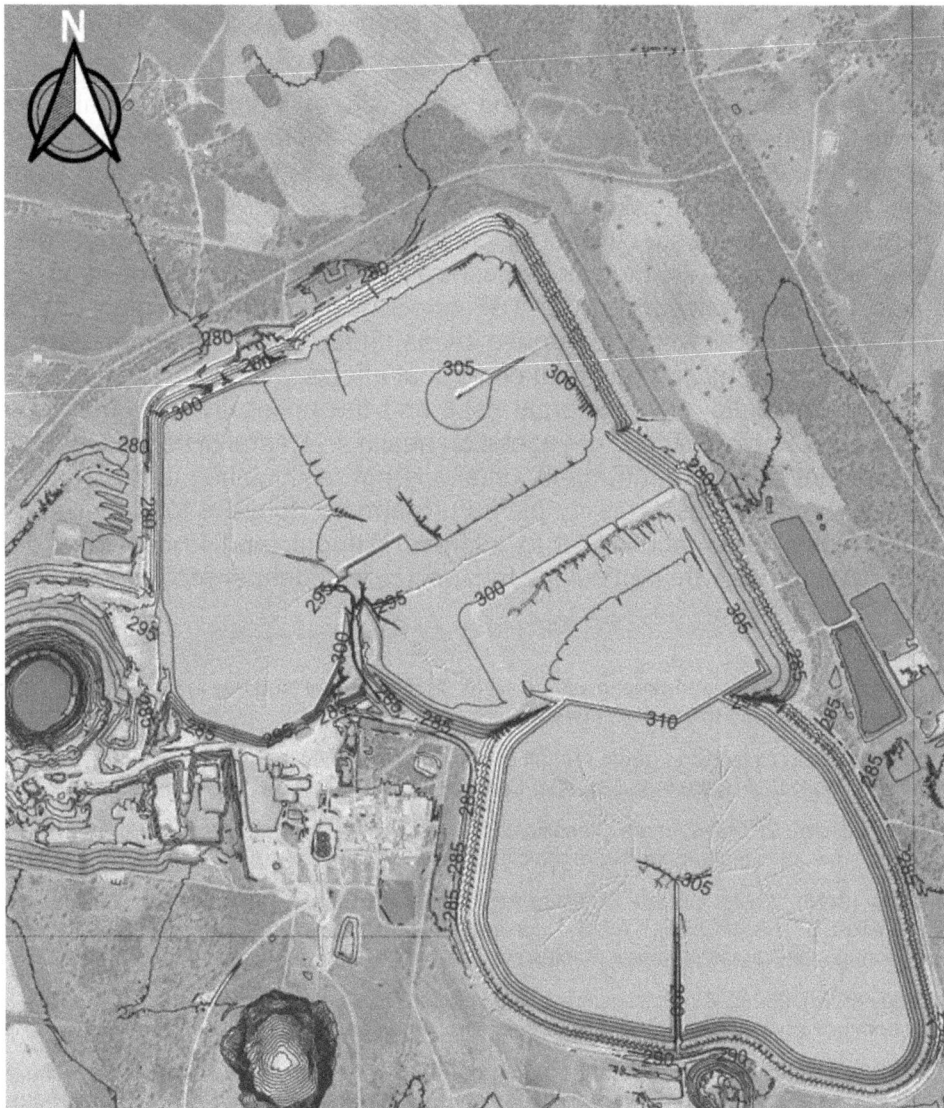

FIG 8 – Rehabilitated landform after 100 years.

CONCLUSIONS

This study identified physicochemical characteristics of the tailings, revealing a challenging substrate for plant growth primarily characterised by alkaline pH, severe deficiencies in nitrogen and phosphorus, and elevated levels of metals such as Copper, Manganese, and Arsenic. Despite these challenges, the rehabilitation strategy employed has effectively advanced the TSF toward its post-mining land use goals.

Direct seeding into the tailings is the proposed method for the TSF closure and rehabilitation. This strategy has proven successful in establishing resilient vegetation, particularly species such as saltbush, which thrive under the site conditions. The growth of this vegetation is vital for stabilising the landform surface. Additionally, incorporating organic fertilisers improves the effectiveness of the tailings as a growing medium and enhances ground cover, which is crucial for controlling erosion risk.

The findings of this research directly influence future TSF closure plans by illustrating that installing a growing medium over the tailings is unnecessary, given the effectiveness of direct seeding and the resulting geomorphic stability. The utilisation of the LAPSUS model forecasts low erosion rates across the constructed landform under present conditions. Importantly, modelling suggests that if the current state of the landform, including the established vegetation cover, is preserved, the TSF is expected to remain stable with low erosion rates for at least 100 years. This anticipated long-term geomorphic stability, paired with the successful establishment of target vegetation species directly into the tailings surface, indicates that the landform is effectively progressing towards a sustainable post-mining land use.

While the rehabilitation strategy is demonstrating effectiveness in creating a stable and functional landscape, despite the inherent difficulties of the tailings, several metal threshold exceedances were noted in the vegetation. Specifically, median Iron concentrations in leaves and stems exceeded acceptable levels, and maximum measurements for Manganese, Boron, and Molybdenum in various plant parts also exceeded their respective thresholds. Additionally, Arsenic accumulation was noted in leaves, stems, and roots, with median concentrations meeting or surpassing the threshold. Although the established vegetation on the TSF appears unaffected by high total Copper, and despite elevated DTPA Copper and Manganese levels, these heightened metal concentrations in vegetation raise concerns regarding potential ecological effects. For post-closure relinquishment, a formal assessment of potential environmental risks and the necessity for adaptive management strategies, including ongoing monitoring of vegetation health and metal dynamics, as well as targeted vegetation management to ensure that metal uptake is not recirculating into the growing medium, would enhance the long-term sustainability plan. Furthermore, future efforts could aim to develop more efficient and sustainable techniques to address Nitrogen and Phosphorus deficiencies to further promote vegetation growth, particularly for species with deeper roots systems.

REFERENCES

Baker, D E and Eldershaw, V J, 1993. Interpreting soil analyses for agricultural land use in Queensland, Information Series QO93014, Queensland Department of Primary Industries, Brisbane.

Boas, T and Mallants, D, 2022. Episodic extreme rainfall events drive groundwater recharge in arid zone environments of central Australia, *Journal of Hydrology: Regional Studies*, 40. https://doi.org/10.1016/j.ejrh.2022.101005

Bureau of Meteorology (BoM), 2024a. Climate Classification Maps [online], Commonwealth of Australia. Available from <http://www.bom.gov.au/climate/maps/averages/climate-classification/?maptype=kpn> [Accessed: 13 April 2025].

Bureau of Meteorology (BoM), 2024b. Climate Statistics for Australian Locations: Monthly Climate Statistics — Summary Statistics PARKES AIRPORT AWS [online], Commonwealth of Australia. Available from <http://www.bom.gov.au/climate/averages/tables/cw_065068_All.shtml> [Accessed: 13 April 2025].

Bureau of Meteorology (BoM), 2024c. Evapotranspiration calculations [online], Commonwealth of Australia. Available from: <http://www.bom.gov.au/watl/eto/tables/nsw/parkes_airport/parkes_airport.shtml> [Accessed: 13 April 2025].

Crosbie, R S, Jolly, I D, Leaney, F W and Petheram, C, 2010. Can the data set of field-based recharge estimates in Australia be used to predict recharge in data-poor areas?, *Hydrology and Earth System Sciences*, 14(10):2023–2038.

CSIRO, 1997. Characterisation of TSF1, Final Report of Stage 2, Mine site Rehabilitation Research Program.

Department of Environment and Resource Management (DERM), 2011. Protecting Queensland strategic cropping land Guidelines for applying the proposed strategic cropping land criteria [online]. Available from <https://www.parliament.qld.gov.au/Work-of-the-Assembly/Tabled-Papers/docs/5311t5265/5311t5265.pdf> [Accessed: 13 April 2025].

Foster, G R and Meyer, L D, 1972. A closed-form soil erosion equation for upland areas, in *Sedimentation: Symposium to Honor Professor H A Einstein* (ed: H W Shen), pp 12-1–12-19 (Colorado State University).

Foster, G R and Meyer, L D, 1975. Mathematical simulation of upland erosion by fundamental erosion mechanics, in *Present and Prospective Technology for Predicting Sediment Yields and Sources: Proceedings of the Sediment-Yield Workshop*, pp 190–207 (US Department of Agriculture, Agricultural Research Service).

Guiton, S P, 2009. The Geochemistry of Advanced Weathering in Mine Tailings, New South Wales, School of Earth Sciences.

Kirkby, M J, 1971. Hillslope process-response models based on the continuity equation, in *Slopes, forms and processes, Transactions of the IBG* (ed: D Brunsden), Special publication, pp 15–30 (Royal Geographical Society: London).

Landloch, 2021. Material Characterisation Study: Material Characterisation Program for Rehabilitation.

Lee, S, Irvine, D J, Duvert, C, Rau, G C and Cartwright, I, 2024. A high-resolution map of diffuse groundwater recharge rates for Australia [online], *Hydrology and Earth System Sciences*, 40(7):1771–1790. https://doi.org/10.5194/hess-28-1771-2024

Peverill, K I, Sparrow, L A and Reuter, D J, 1999. *Soil Analysis: An Interpretation Manual* (CSIRO Publishing: Victoria).

Rayment, G E and Bruce, R C, 1984. Soil testing and some soil test interpretations used by the Queensland Department of Primary Industries, Queensland Department of Primary Industries Information Series QI84029, Queensland Department of Primary Industries, Brisbane.

Schoorl, J M, Sonneveld, M P W and Veldkamp, A, 2000. Three-dimensional landscape process modelling: The effect of DEM resolution, *Earth Surface Processes and Landforms*, 25:1025–1034.

University of Queensland (UQ), 2014. Rehabilitation Strategies for Tailings Storage Facilities — Planning for Closure – Stage 4: Field trial for testing of cover systems construction report, Centre for Mined Land Rehabilitation, the University of Queensland.

University of Queensland (UQ), 2023. New South Wales Mine Reuse Project: Cu-Au Porphyry Mine Final Report, Department of Regional NSW.

Wood, W W, 1999. Use and misuse of the chloride-mass balance method in estimating ground water recharge, *Ground Water*, 37(1):2–3.

Ok Tedi Mining Limited life-of-mine studies

N Subul[1], J Yalomba[2] and J Isarua[3]

1. Snr Mining Engineer – LTP, Ok Tedi Mining Limited, Tabubil Western Province 332, Papua New Guinea. Email: nongkas.subul@oktedi.com
2. Snr Mining Engineer – STP, Ok Tedi Mining Limited, Tabubil Western Province 332, Papua New Guinea. Email: john.yalomba@oktedi.com
3. Snr Mining Engineer – STP, Ok Tedi Mining Limited, Tabubil Western Province 332, Papua New Guinea. Email: joe.isarua@oktedi.com

ABSTRACT

Ok Tedi Mining Limited (OTML) operates one of the largest open cut copper and gold mines in the Southern Hemisphere. Located at the remotest part of PNG's Western Province, mine supplies are sourced by sea and river to Kiunga Port, then the road 140 km up to the mine. The business has future plans to improve its current supply chain by providing route options to Indonesia by road network, as well as other ports via road network in country. OTML has embarked on mine life extension from 2033 to 2050 adding operating years to the mine.

Waste and Tailings Management form a large part of the mine life extension, and the company has already progressed studies for two Engineering Waste Rock Dumps (EWRD) and a Tailings Storage Facility (TSF). The two EWRD set-ups are both at the northern and southern end of the mine respectively whilst TSF is being looked at potentially along the Kiunga Tabubil Highway. Currently the mine operates an erodible waste dump approach where waste rocks are dumped into the river system and reclaimed at Bige complex. With seven pushbacks, scheduling challenges are common especially bridging gaps in ore and metal profile. The life-of-mine (LOM) designs are done on-site, whilst pit optimisation is sourced out to consultants to deliver results. Blending is key in results delivery of a high tonnage low head grade operation.

Scope for this abstract is on OTML LOM, EWRD and TSF for a sustainable and reputable operation. The optimisation results are leaning towards a potential relocation of the current office and maintenance complex. With strong performances recently, the business aims to acquire other mines to bring value up to USD 2 billion from USD 1.3b billion at present. First five years are formative years for the business that will determine the mine life 2050 and beyond.

BACKGROUND

Since 1984 Ok Tedi open cut mine has been producing both copper and gold with annual mine targets reaching 110 Mt total material movement, 95 kt copper production and 8 kt gold generating billions of dollars in revenues. The operation is purely a pushback mine that has several pushbacks and controlled through mine planning process by quarterly schedules, rolled out in the monthly planning space. This paper will discuss more about Ok Tedi Mining Limited (OTML) operation, erodible dumping and Bige dredging set-up, business strategy into 2050, EWRD 1 and EWRD 2 and TSF.

The current mining practice (Isarua and Elit, 2023) uses an erodible dump mechanism whereby dumping of mine waste is through pit edge and transferred down through Fly River system. Initial TSF programme in 1984 was put to a stop when the dam failed in 1987 forcing the government then to permit dumping into the river system. This was the practice since then. The government upon set-up of Bige allowed for the dumping to occur where waste material reclamation was happening in Bige.

With mine life extension to 2032, the key enabler was the EWRD 1 which was to store waste on the land, north of the current pit. In 2020 the concept of EWRD 1 was successfully introduced and managed internally through a dedicated project team. After change of management in 2023, OTML embarked on a mine life extension project to 2050 and key part to this are the strategic projects in EWRDs and TSF. Figure 1 shows mine pit areas and the locations of EWRD1, EWRD2 and TSF. Also note that both EWRD 2 and TSF are at the feasibility study state, hence no visible associated infrastructure shown.

FIG 1 – Showing mine pit areas with reference to EWRD1, EWRD2 and TSF locations.

The Bige dredging set-up is a waste management system where the mine waste deposited into the river system and transferred down is dredged out to the stockpiles. The Picon (pyrite) stockpiles are also set and buries the tails from the mine. The waste materials get stockpiled onto the land as an overturned Fubilan mine pit whilst the tails are covered. This operation is a 24/7 operation with its own camping facilities and management structure under the Environment Department. The site is approximately 88 km south from the mine and has both west and east banks that cater for the waste storage. The pictures in Figure 2 show both the west and east bank stockpiles with dredging barge as shown (left); and (right) shows the six lifts of the eastern bank stockpile with rehabilitation methods applied to restore the environment features.

FIG 2 – Bige dredging set-up showing both west and east banks of stockpile with dredging barge (left); and close up view on East bank Stockpile (right).

BUSINESS STRATEGY

The business strategy (Bong, 2024) anticipates the first five years of full production and high value operation from 2025 to 2030. This is labelled as productive and formative years where there is sufficient cash to actually complete the strategic projects being put forward to realise the Ok Tedi 2050. From 2031 onwards to 2050, all strategic projects would have already been completed and the need for more people will no longer be required, although revenue is expected to be higher. As shown by the graph in Figure 3, in the first five years the production margin is USD +300 M squared off by the USD -300 M from capital projects. Sustainable loans are also being considered to cushion the effects of low cash flow with discussions with Japan International Cooperation Agency (JICA) and other development partners underway.

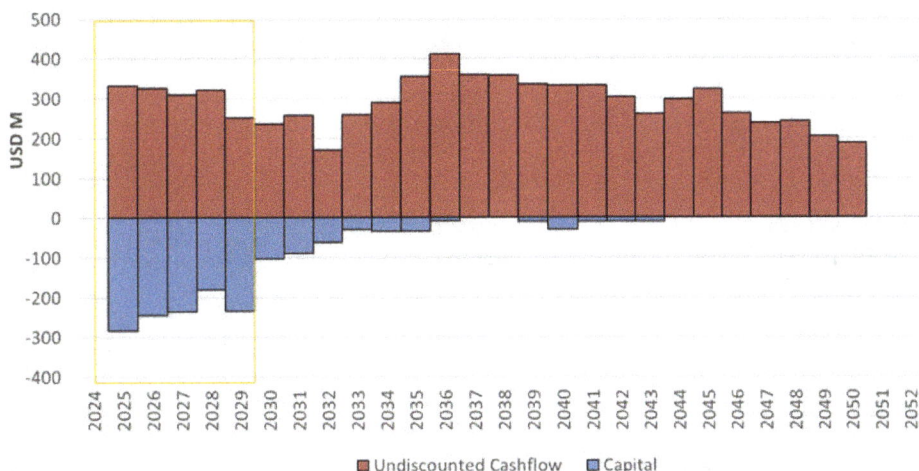

FIG 3 – Cash flow assumptions into 2050 for both Cu and Au.

With the need for people subsiding post-2030 and the corporate goal to reach a USD 2 billion company status, OTML management now plans to go into acquisition of other projects that will elevate the value to expected target. The three prospects into consideration are the Woodlark and Misima projects, Mt Kare project and one of the mines in Queensland. This plan will see OTML distributing its workforce to these locations and improve its value to expectation. The potential of underground (UG) mine is also being studied with additional revenue of USD 3 billion expected from prospects identified.

OTML has extended its mine life to 2050. The strategy deployed is the corporate social responsibility over profitability. In corporate social responsibility, it's about the people where some profit is made and keeps the business afloat at the same time job opportunities are created and employment sustained during the life-of-mine. Profitability is when the business strategy focuses on Net Present Value (NPV) which means to get more value out while the price is high and once reserves are exhausted, the business closes the mine. This was the strategy deployed previously whereby the mine life was expected to end by 2032. With Mr Ilimbit' s appointment as Managing Director (MD) in 2023, this strategy has changed since and the renewed focus is on corporate social responsibility with business opportunities to the locals. Local business partners are now engaged in spin off benefits better than before with NPV arrangements.

Key enablers to the mine life extension to 2050 are the EWRD1, EWRD2, TSF and Underground (UG) mining, focusing on growth, energy and waste management, environment and social responsibility, governance and risk. Figure 4 shows the 2050 growth map for the business. UG is being mapped to progress UG opportunities in reserves and eventual mining. Close consultation with consultants in all areas of strategic projects are evident during the initial stages of the project. The objective is to eventually transfer those knowledge base to the employees as part of the business strategy on its workforce.

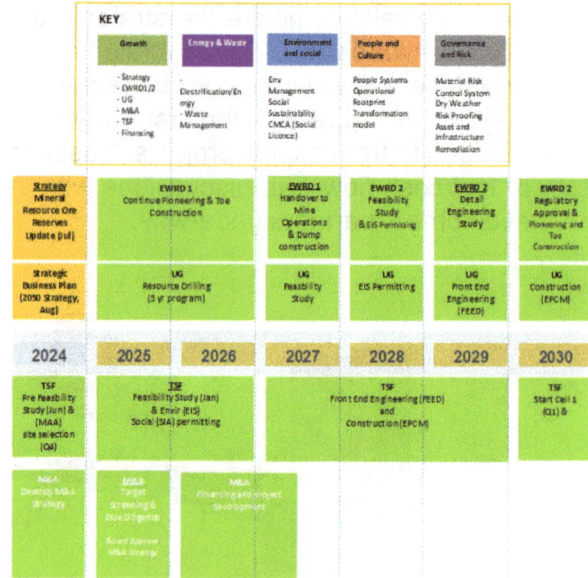

FIG 4 – OTML 2050 strategy encompassing key enablers to achieving 2050 mine life.

LIFE-OF-MINE (LOM)

From the business strategy and studies, practical pit designs are generated to determine the practical aspect of the Life-of-mine (LOM). The depths of the pit and pushbacks quantify that there are sufficient reserves to continue mine life to beyond 2050. For now, the approved designs target 2050 as the end of mine life approved by the Executive Leadership Team (ELT). The site team produces the designs whilst the business strategy work is contracted to consultants.

The next office pushback and UG reserves will see a mine life that is beyond 2050. Current pit bottom sits around 1303RL with ultimate pit bottom designed at 1093RL. These practical designs are then put through LOM schedules using Whittle from which a high-level output is produced and communicated to all stakeholders. As it stands, the Whittle schedules are telling that mine life will end in 2050. These schedules are annually adjusted as new findings come in. In the event where there is absence in geotechnical data, more drilling is slotted in the day-to-day planning to enable data update.

At the studies space, UG options are also being looked at post the pit limit to see if there are opportunities left by the pit limit. Yes, with GC model graded at 0.30 per cent Cu, there are several ore deposits that can be mined using the underground option. These are being put together for consideration. Figure 5 is a plan view of the LOM pit limit with pink colour showing 0.30 per cent Cu shell left behind by the pit limit. Figure 6 shows the cross-section view with orebody which can be mined using a UG method.

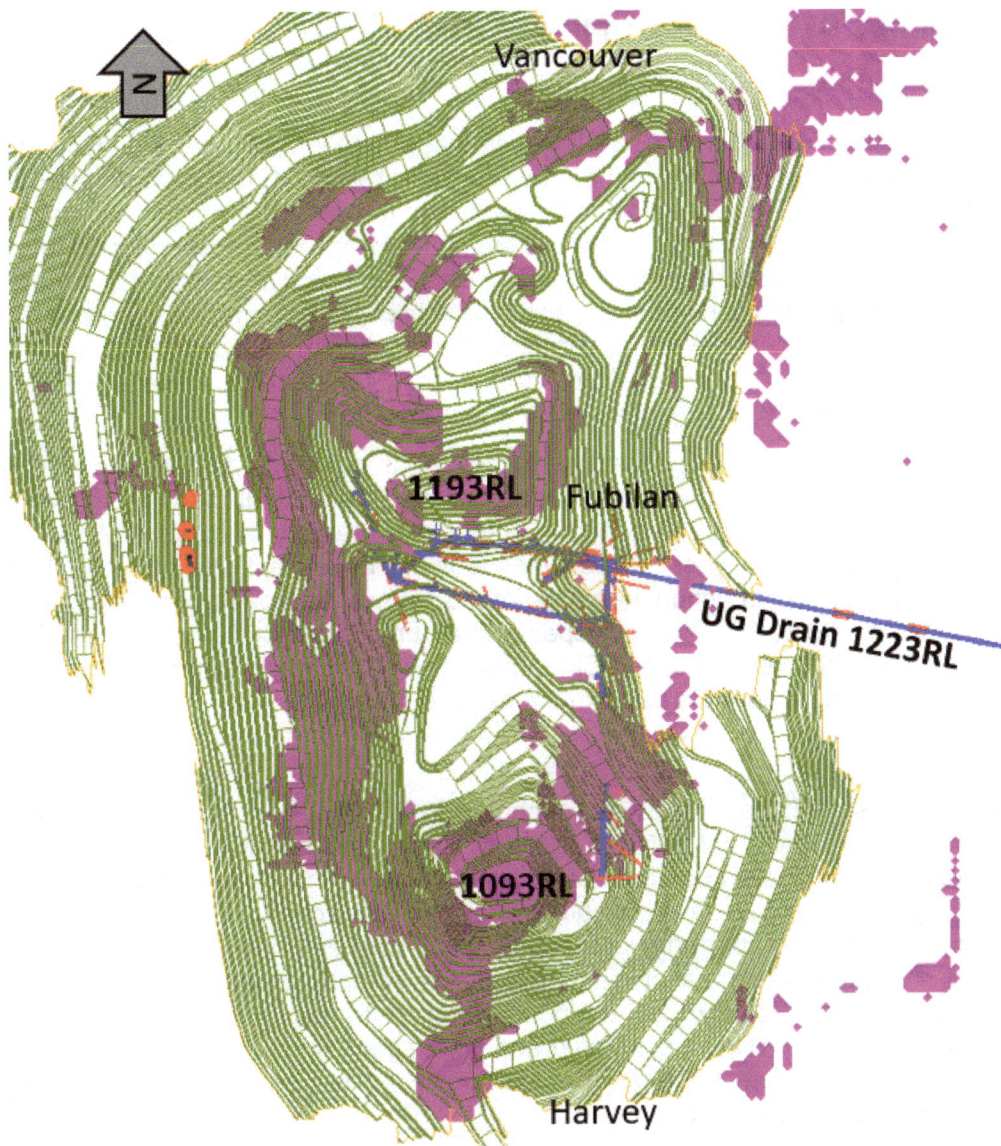

FIG 5 – Showing plan view of the LOM pit with remaining ore reserves in pink.

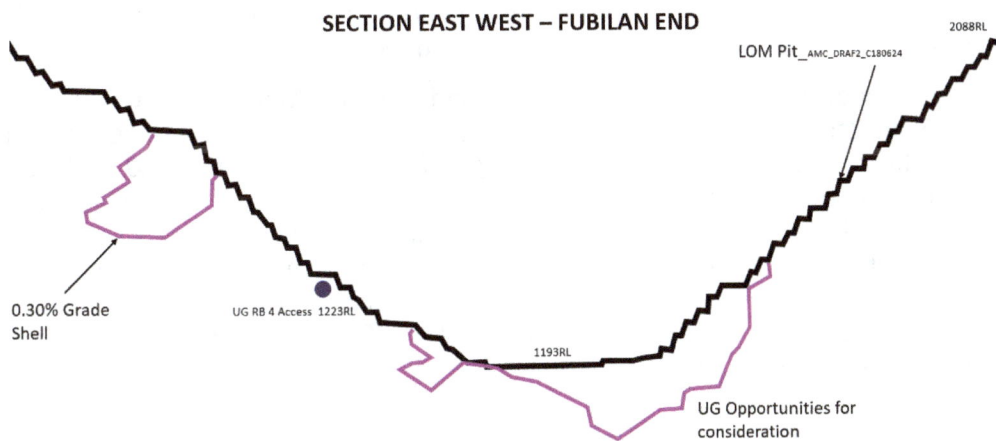

FIG 6 – Cross-section showing UG opportunity post 1193RL from the cross-section provided. Ultimate pit limit at 1093RL.

ENGINEERED WASTE ROCK DUMP (EWRD)

EWRD 1

The Engineering Waste Rock Dump (EWRD1) was introduced in 2021 to store waste on the land. This was to minimise waste dumping into the river system by way of dumping on the northern end of the pit gully at Sulphide Creek with toe of the design at the crest of Wellington orebody. This will enable the dumping of waste rock at the same time allowing access construction to the Wellington orebody drilling site. Due to scarcity in the limestone for construction phase, the project has been slow. Blocky siltstone is being considered based on the recommendation categories to supplement construction phase. After the toe drain construction is complete, the dump will be built in 50 m lifts with systemic drainage set-up and Non-Acid Forming (NAF) material as cover whilst the Potential Acid Forming (PAF) material will be placed within the bench.

Challenges of varying truck cycles, material scarcity, delayed critical supplies to site, unplanned equipment downtime are reversed with improved equipment rates, high bench turn overs, versatile design amendment and implementation and segregated placement of rocks whilst waiting for the exposure of categorised rock types. These mechanisms have kept the project afloat since its inception in 2021. With construction heading into unknown territories, ongoing Geotech radar monitoring are carried out to monitor the slopes of the dump being built. Recent introduction of six Xuzhou Construction Machinery Group (XCMG) truck fleet is also improving waste rock movement to the project site with more fleet expected in the next 18 months.

EWRD 2

From the high-level schedules, EWRD2 is expected in 2028 with limestone inventory checks in progress. Limestone is usually found both in the southern and northern part of the pit, the main challenge has been the timing to access this quality rock type, which was affecting EWRD1 progress. The current thought process is to have the southern pit end limestone for EWRD2. Also, in the picture to be considered is the UG opportunity presenting at the Harvey Creek, below where EWRD2 will be constructed. The site challenge at present is limestone scarcity where model predicted limestone is not presenting and that is causing an engineering misery to continuity of current EWRD1. Since 2021, backup dumping option was a solution giving time, space and funding flexibilities for EWRD1. It's a business strategy to store waste on the land hence will still be developed including reducing limestone dumping into the river system.

With EWRD2 project, no landowner relocation is expected with this construction since it's within a mine site although it may affect the orebodies at depth when we have to mine. This will mean doing offsets and then providing designs that will allow mining both open pit and underground. The location of the EWRD2 encompasses five erodible dump edges used at present.

In PNG, this is the third EWRD in the country after Hidden Valley Mine and EWRD1 set-up and will mimic the standards and design from EWRD1 where the designs are 50 m Lifts, adequate berms and toe drain systems that allow for adequate storage of waste rock in NAF and PAF configuration. The construction methodology is similar where 10 m lifts are completed, drainage system set before the next lift happens. This is crucial for safe lifts. The learnings from EWRD1 will help improve the second project and that will be a huge success for the business in terms of total waste management. The capacity expected is at least ten years (88 Mt waste storage) minimum that will support the active part of the waste mining to uncover ore from 2030 to 2040 period. Figure 7 shows a simplified EWRD concept in design and construction by lift and toe drain. Surface drainage is managed on bench.

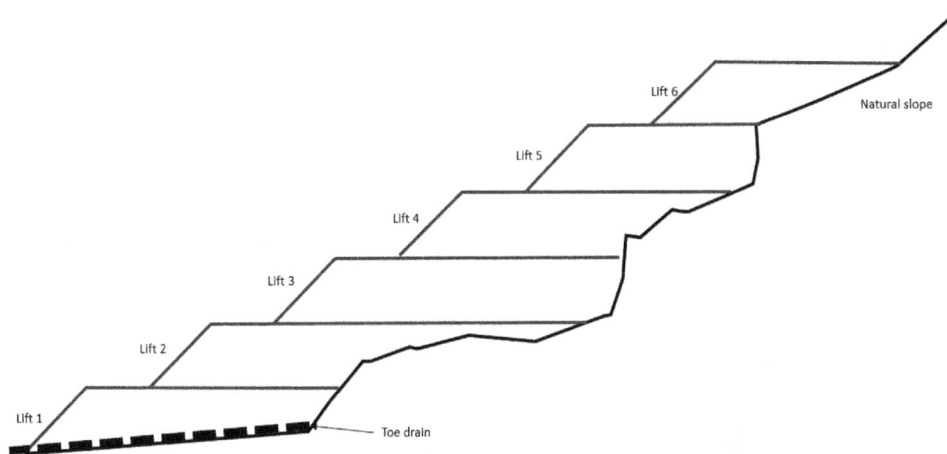

FIG 7 – Simplified EWRD design and construction.

TAILINGS STORAGE FACILITY

TSFs are a storage facility that stores tailings waste on the land, it is a designed, engineered and constructed structure to permanently store tailings (BHP, 2024). The primary purpose of a TSF is to safely contain tailings by allowing the sands to settle consolidate and dewater. Same document indicates that the three different types of TSFs are surface facilities, in-pit facilities and dry stack facilities. OTML has made a huge strategic business call to store tailings waste on the land instead of the current piping of tails into the river system, to minimise tails into the river system with legacy issues of dieback and other environmental damages the mine has incurred over many years.

The current practice is such that the mill processing plant has a thickener tank that performs the function of TSF before the tails discharge to the environment. Just like the flotation circuits where gangues are separated from the ore, the tailings storage plant at the mill performs similar function by reducing the pyrite content in the tails before sending the tails to the environment through pipes at Gandagai. The catch is bulk of these tails after processed are sent down the river system and then get pumped back onto the land at the Bige complex. The pH level is usually set to 7 before the discharge where limestone is used as a neutralising agent. The plan is to expect TSF established within the next five years with the due date for completion of the feasibility study by end of quarter two next year. Currently the TSF sessions are managed by the consultants and OTML provides the technical support required to complete the feasibility studies. Objective of TSF is to ensure all tails are piped down to Ok Birim TSF site and store all tails on the land without any spillage and piping into the river system. This will eliminate current practice of mill tails into the river system.

OTML engaged KCB in 2013 to complete a Pre-Feasibility Study (PFS) on the TSF options. Ok Birim site was selected after thorough assessment and selection criteria. The site is less than 100 m west of Ok Birim River, covering some tributaries. The thought process initially was to test the viability of the TSF on-site which the study was completed, and idea shelved. New OTML management, since taking over in 2023, has embarked on the study this time shifting the site south-west towards higher land, away from Ok Birim River. Initial PFS (Williams, 2024) had five stages of storing the tails. Stage 1 had picon impoundment extents with cyclone overflow extent shown. This really is the initial foundation of tailings placement with conceptual flows into the environment. Stage 2 had similar set-up of stage 1 with addition of operating pond that has discharge into the river system. Stage 3 had extended version of stage 2 with addition of reclaimed tailings surface. Stage 4 had similar set-up of stage 3 will addition of increased set-up of reclaimed tailings surface. Stage 5 had the ultimate outlook of the TSF. As mentioned earlier – the initial study aimed at nullifying installation of the TSF with placement close to the Ok Birim River. Figure 8 shows initial PFS layout of Ok Birim TSF.

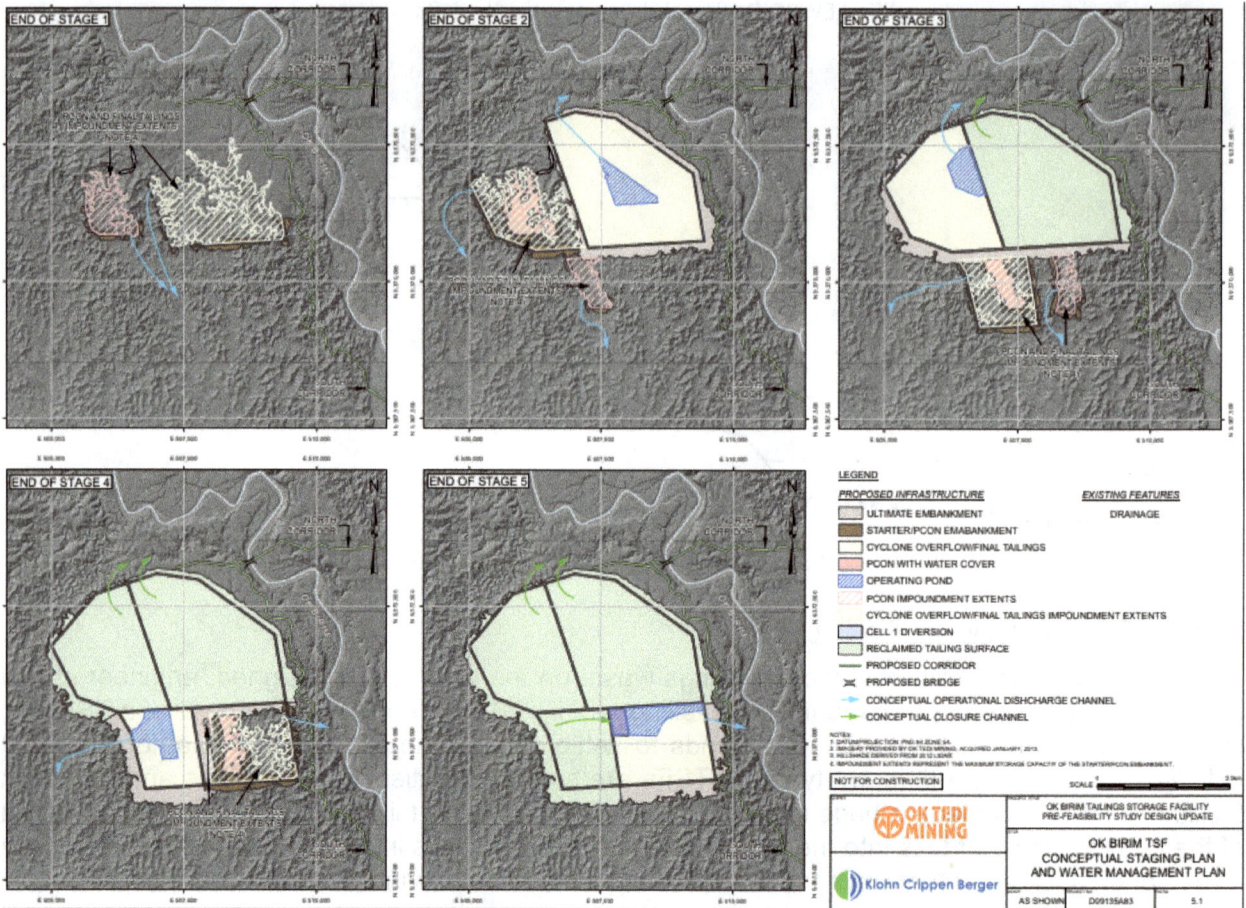

FIG 8 – Different stages of TSF stages at Ok Birim Area completed in 2014.

The proposed TSF site south-west of Ok Birim will encompass the two villages of Putmabin and Kungembit both within the boundary of Tailings Storage Facility. Relocation exercise is being discussed at high level and Community Relations (CR) will be engaged to actually carry out formal awareness once feasibility study details are completed, presented and approved by ELT. The CR team is doing initial awareness at present to ensure the relocation happens smoothly. The plan is to relocate the two villages of Putmabin and Kungembit to a nearby station in Ningerum where the company provides all facilities for the relocation including house, compensation and increased utility services capacity.

Generally, the broad types of TSF construction methodology are the upstream construction, the downstream construction and centreline construction. This choice is usually depended on the location, geology, seismicity, climatic condition, construction material and the nature of the tails. Given Ok Tedi mine is in the tropical region, TSF selected will include all conditions of the tropics. From the life cycle of tailings storage facilities which cover site selection and design, construction, operations, transition, closure and landform, OTML is in the site selection and design stage with lidar data being crucial. Lidar data has been recently completed, with updated TSF design to be completed together as well as the pipeline routes with June 2026 as the deadline for Feasibility Study and construction commencement in 2027.

CONCLUSION

After 40 years in operation, Ok Tedi Mining Limited has come a long way in managing its own challenges in the midst of decorated successes and have recently embarked on sustainable mining practice that is environmentally friendly and promote employment opportunities for the Papua New Guineans and spin off benefit for the local communities. OTML has made a serious investment in Tailings Storage Facility and Engineered Waste Rock Dumps to ensure the Life-of-mine realises the year 2050 target as embedded in the recent Strategic Business Plan.

REFERENCES

Bong, A, 2024. 5 Year Strategy Session, Strategic Business Development and Optimisation (SBD&O) BU, Business strategy, Ok Tedi Mining Limited document.

BHP, 2024. What are tailings and TSF? [online]. Available from: <https://www.bhp.com/sustainability/tailings-storage-facilities/what-are-tailings-storage-facilities>

Isarua, J and Elit, S, 2023. In-pit waste rock dumping as a mitigating strategy for waste rock management, in *Proceedings of Mine Waste and Tailings Conference 2023*, pp 198–205 (The Australasian Institute of Mining and Metallurgy: Melbourne).

Williams, T, 2024. Ok Birim Tails Storage Facility Capital and Sustaining Cost Estimate Prefeasibility Study, 19 p.

ESG aspects of tailings management in Australia – aligning two global standards

K Vershinina[1], L Rollin[2] and H Thomson[3]

1. Principal Consultant (ESG), SRK Consulting (Australasia) Pty Ltd, Brisbane Qld 4000. Email: kvershinina@srk.com.au
2. Senior Consultant (ESG), SRK Consulting (Australasia) Pty Ltd, West Perth WA 6005. Email: lrollin@srk.com.au
3. Principal Consultant (Tailings Engineering), SRK Consulting (Australasia) Pty Ltd, West Perth WA, 6005. Email: hthomson@srk.com.au

ABSTRACT

The mining industry, both in Australia and globally, strives to minimise risks associated with mine project development, with tailings management being a key area of public and regulatory scrutiny. While compliance with national and state legislation governing the design, operation, and closure of tailings storage facilities is mandatory, leading international best practices extend beyond regulatory requirements to ensure the highest standards of safety, environmental stewardship, and social responsibility. Numerous global and Australian mining companies elect to follow international practices to enhance and strengthen their management approaches, reduce reputational risks, and improve relationships with stakeholders.

The Global Industry Standard on Tailings Management (GISTM or the Standard; Global Tailings Review (GTR), 2020) and the Towards Sustainable Mining (TSM) Tailings Management Protocol (TSM TMP or the Protocol; Mining Association of Canada, 2023) are internationally recognised frameworks that guide best practices in Australia. GISTM was developed as a global benchmark for tailings governance, while the Minerals Council of Australia (MCA), in collaboration with the Mining Association of Canada (MAC), has adapted and adopted MAC's TSM TMP to align with the Australian mining context.

It is not unusual for medium to major mining companies and investor groups (including those operating in Australia) to commit to implementing both the Standard and the Protocol. Junior companies, as members of the MCA, may also align their tailings storage facility (TSF) governance with international good practice to unlock funding or meet the expectations of downstream markets, but best practice is now widely considered to be GISTM. Although both the International Council of Mining and Metals (ICMM) and Canadian TSM are currently collaborating on the Consolidated Mining Standard Initiative (2025)—which aims to merge various voluntary responsible mining standards into one global standard—the new consolidated standard continues to refer to both tailings management frameworks as the preferable management tools.

Both frameworks share a common objective – to minimise risk and enhance accountability while addressing the key environmental, social, governance (ESG), and technical aspects of tailings management. However, conforming with both standards concurrently can be challenging for mining companies as differences in structure, implementation, and reporting requirements between the standards adds complexity. Given that regulatory compliance remains obligatory, understanding the synergies and distinctions between these frameworks is essential to streamline conformance efforts, reduce duplication, and enhance operational efficiency.

To understand the application of the Standard and the Protocol it is important to recognise the background and the drivers behind their development.

The GISTM was developed collaboratively with Principles for Responsible Investors (PRI, investors), United Nations Environment Programme (UNEP) and ICMM (industry body, mining companies and associations) to strengthen existing practices through the integration of social, environmental and technical considerations throughout the full life-cycle of a TSF. It is a governance standard that supplements, rather than replaces, existing technical TSF management standards.

The Standard was published in August 2020 as a response to the unprecedented loss of life following the catastrophic Brumadinho TSF failure in 2019, when 259 people were confirmed dead and 11 people remain missing. The erosion of trust in mining companies (including major companies such as Vale and other ICMM members) as a result of the incident required immediate action from the industry but with oversight from trusted bodies. As of January 2024, over half of the mining sector (by market capitalisation) was committed to implement the Standard (The Church of England, 2024).

The Protocol is part of the Towards Sustainable Mining initiative that began as a project of the MAC in 2004 and was the first initiative globally to require site-specific reporting with external verification. As of April 2025 there are eight other national mining associations outside of Canada that have adopted the TSM initiative (MAC, 2025) including MCA. The TMP, among other TSM protocols and guides, frames the TSM commitments to responsible mining. The Protocol is applied to TSFs in commercial production and to inactive TSFs in the closure and post-closure phases of the life cycle as well as to facilities on long-term care and maintenance.

For both the Standard and Protocol, the technical expertise, engineering excellence and accountability for TSF management are fundamental requirements and cannot be compromised. At the same time the objectives of the documents of achieving 'zero harm to people and the environment' (for GISTM) and 'zero catastrophic failures... and no significant adverse effects on the environment or human health' (for the Protocol) are multidisciplinary in nature and require much broader consideration of ESG factors.

However, while the Protocol refers to other TSM protocols for some components (with four topics covered – Indigenous and Community Relationships, Climate Change, Water Stewardship, and Crisis Management and Communication Planning), all conformance indicators for the Protocol itself are focused on technical and governance aspects to ensure effective and safe function of the TSF during the commercial production stage. In terms of ESG factors, the GISTM goes beyond the engineering and technical excellence and requires consideration of topics that are outside the purely technical parameters, such as human rights, an integrated knowledge base that comprises surrounding environmental and social characteristics, and disclosure of the information to public.

Conformance to both the Protocol and GISTM can be demonstrated by self-assessment and through third-party validation. However, self-assessment for GISTM must address all requirements of the Standard, whereas the TSM TMP requires conformance with five technical governance indicators. With the similar requirements to the engineering and operational procedures control in the Protocol and Standard, the conformance to both frameworks require an additional effort to meet the GISTM broader ESG expectations. Some areas of the differences are summarised in Table 1.

TABLE 1

Comparison of some ESG requirements between the GISTM and TSM.

Item	GISTM	TSM
Life cycle	Applied directly to all existing and to be built facilities, from design to post-closure. Alternative analysis for new facilities.	Applied to commercial production, closure and post-closure, care and maintenance. The supporting tailings guide can be used for new facilities.
Application	To an individual TSF	To a mine site as a whole (collectively to several TSFs)
Spatial application	Considers areas beyond the TSF footprint and immediate surrounding the TSF including community, biodiversity, and water impacts.	The focus is on the facility and referring to the local community and authority as stakeholders (criteria provided in another protocol).
Expertise areas for conformance	Requires a holistic approach using the ESG-based interdisciplinary integrated knowledge base. The consequence classification matrix considers environmental, health, social and cultural, and infrastructure and economic incremental losses for the dam failure consequence classification.	The Protocol itself is based on the technical and governance criteria with reference to other protocols for specific areas.
Affected communities	Four requirements under Principle 1 including human rights due diligence, general requirements for engagement, Free, Prior and Informed Consent and Indigenous communities, grievance mechanism	Reference to another TSM protocol (with separate conformance assessment).
Public disclosure	Publish and regularly update information related to the TSF.	No requirements

REFERENCES

Church of England, The, 2024. The Investor Mining and Tailings Safety Initiative [online], The Church of England. Available from: <https://www.churchofengland.org/about/leadership-and-governance/national-church-institutions/church-england-pensions-board/pensions> [Accessed: 16 April 2025].

Consolidated Mining Standard Initiative, 2025. Home page, Consolidated Mining Standard Initiative. Available from: <https://miningstandardinitiative.org> [Accessed: 30 May 2025].

Global Tailings Review (GTR), 2020. Global Industry Standard on Tailings Management (GISTM) [online], Global Tailings Review. Available from: <https://globaltailingsreview.org/global-industry-standard/>

Mining Association of Canada (MAC), 2023. Toward Sustainable Mining – Canada, Tailings Management Protocol, ver March 2023 [online], The Mining Association of Canada. Available from: <https://mining.ca/wp-content/uploads/2024/03/Tailings-Management-Protocol-2023-03-09-ENG.pdf> [Accessed: 30 June 2025].

Mining Association of Canada (MAC), 2025. International Uptake [online], The Mining Association of Canada. Available from: <https://mining.ca/towards-sustainable-mining/international/> [Accessed: 16 April 2025].

Post-mining land use

Agro-climatic suitability of central Queensland mining region for Pongamia Plantation

M Anzooman[1], M Shaygan[2], N McIntyre[3], P deVoil[4], J Lane[5] and D Rodriguez[6]

1. Post Doctoral Research Fellow, Centre for Water in the Minerals Industry, The University of Queensland, St Lucia Qld 4072. Email: m.anzooman@uq.edu.au
2. Research Fellow, Centre for Water in the Minerals Industry, The University of Queensland, St Lucia Qld 4072. Email: m.shaygan@uq.edu.au
3. Professorial Research Fellow, Centre for Water in the Minerals Industry, The University of Queensland, St Lucia Qld 4072. Email: n.mcintyre@uq.edu.au
4. Principal Farming Systems Modeller, Queensland Alliance for Agriculture and Food Innovation, The University of Queensland, St Lucia Qld 4072. Email: p.devoil@uq.edu.au
5. Senior Research Fellow, UQ Gas and Energy Transition Research Centre, The University of Queensland, St Lucia Qld 4072. Email: j.lane@uq.edu.au
6. Professorial Research Fellow, Queensland Alliance for Agriculture and Food Innovation, The University of Queensland, St Lucia Qld 4072. Email: d.rodriguez@uq.edu.au

ABSTRACT

Pongamia trees, known for their resilience to diverse soil types and climatic conditions, offer potential for bioenergy, land rehabilitation and carbon sequestration. This study examines the agro-climatic suitability of the Central Queensland mining region for Pongamia plantations by analysing climate, soil properties, and topography. Climate analysis, based on 50 years of rainfall and temperature data (1960–2020), revealed annual rainfall ranging from 250 to 1400 mm, with a 10-year average of 500–800 mm. Rainfall was concentrated in 4–13 days/month, while temperatures fluctuated between -1.1°C and 44°C. Frost days were minimal, averaging three annually. These conditions are favourable for Pongamia growth, provided supplemental irrigation is applied during the establishment phase and extended drought condition. Soil assessments, utilising spatial data from the CSIRO and Queensland Spatial Catalogue and 50 field survey points, identified Vertosols and Sodosols as the dominant soil types. Soil physical properties showed plant available water ranged from 5–30 mm in the topsoil (0–30 cm) to 20–55 mm in the subsoil. Bulk density values increased from 1.28 g/cm^3 at the surface to over 1.66 g/cm^3 at depth, necessitating land preparation to reduce compaction. Soil texture varied, with sandy topsoil (29–83 per cent sand, 8.9–16.3 per cent silt, 3–42 per cent clay) transitioning to more clay-rich subsoil, which improves water retention. These physical properties align with Pongamia's adaptability but highlight the necessity for tailored land management practices. Chemical properties, including pH (4.6–7.5 at surface), organic carbon (1.3–2.1 per cent in topsoil), and cation exchange capacity (11–90 meq/100 g), were within acceptable ranges for Pongamia establishment. However, nutrient deficiencies in deeper layers necessitate fertilisation and soil amelioration to optimise growth. Topographic analysis revealed predominantly flat terrain (<2° slope), conducive to plantation establishment with minimal leveling required. This study demonstrates the potential of Central Queensland mining region for Pongamia plantation, with strategic management interventions to address specific soil and climate challenges.

INTRODUCTION

Pongamia pinnata (Pongamia) is a leguminous tree native to India and South-east Asia, with a growing presence in Australia (Degani *et al*, 2022). It is a highly adaptable species, tolerating a broad range of climatic conditions, including temperatures from light frost up to 50°C (Islam *et al*, 2021). Pongamia thrives in regions receiving annual rainfall between 500 and 2500 mm, provided that soil moisture is adequate during the establishment phase (Islam *et al*, 2021). The tree is known for its ability to grow in diverse soil types, including sandy, rocky, heavy clay, alkaline, and saline soils (Usharani, Naik and Manjunatha, 2019). However, optimal growth is observed in well-drained sandy-loam soils with moderate moisture content (Degani *et al*, 2022).

Due to its environmental resilience and ability to grow in marginal soils, Pongamia is increasingly recognised for its potential in agroforestry, biofuel production, land rehabilitation, and carbon sequestration (Odeh, Tan and Ancey, 2011). Mature Pongamia trees (approximately 15 years old) can sequester 45–50 kg of carbon per tree per annum (Prasad, 2014). Additionally, the species exhibits strong land amelioration potential, making it suitable for rehabilitating degraded or mined lands. Notably, Pongamia has demonstrated the capacity to establish on mine waste materials (tailings) with high copper concentrations—up to 6000 mg kg^{-1} (Kumar, Mehta and Hazra, 2009).

Pongamia is also considered a promising biodiesel feedstock due to its high seed oil content, which is approximately 40 per cent. Reported oil yields range from 1500 to 2000 kg/ha under optimal growing conditions (Fu *et al*, 2011; Scott *et al*, 2008) further reinforcing its viability as a sustainable biofuel source.

Understanding the environmental factors influencing Pongamia's growth is crucial for expanding its cultivation. Tree growth is primarily influenced by climatic conditions and soil properties (Vlam *et al*, 2014). Temperature and rainfall are critical determinants of tree growth, with higher temperatures often negatively affecting biomass accumulation. The specific soil characteristics also play a key role in plant establishment, as different soil types vary in their water-holding capacity, nutrient availability, and aeration properties (Pratiwi, 2000). The ability of plants to adapt to new environments is strongly influenced by genetic factors and local environmental conditions.

Several studies have explored Pongamia's adaptability to diverse environmental conditions, highlighting its resilience in saline, drought-prone, and nutrient-deficient soils (Scott *et al*, 2008; Saxena, 2002). Research has demonstrated that Pongamia can tolerate saline conditions and thrive in dry, marginal lands with poor soil quality, including sandy and rocky substrates. For instance, a study by Terviva (2020) found that Pongamia trees grown in arid and semi-arid regions with highly saline soils (electrical conductivity exceeding 8 dS/m) exhibited sustained growth and biomass production, making it a promising species for afforestation in degraded landscapes. Similarly, in drought-prone regions, Pongamia has been shown to withstand prolonged dry spells and extreme heat, tolerating temperatures up to 50°C due to its deep taproot system that enables access to deeper soil moisture reserves (Leksono *et al*, 2021). A study conducted in India found that Pongamia seedlings had a survival rate of over 80 per cent in regions receiving less than 500 mm of annual rainfall, indicating its suitability for low-rainfall environments (Arpiwi, Muksin and Song, 2023). Moreover, research on soil fertility has revealed that Pongamia can establish and grow in soils with low nutrient availability, particularly those deficient in nitrogen and phosphorus, two essential macronutrients for plant growth. A study assessing naturally growing Pongamia trees in Indonesia found that despite growing in soils with low organic matter content (<1.5 per cent) and phosphorus levels below 10 mg/kg, the trees exhibited stable growth rates, suggesting an efficient nutrient-use strategy and possible associations with nitrogen-fixing bacteria (Leksono *et al*, 2021). These findings collectively highlight Pongamia's potential for cultivation in challenging environments, making it a valuable species for land rehabilitation, agroforestry, and bioenergy production in regions with marginal soils.

Despite the promising attributes of Pongamia, limited research has been conducted on its establishment and growth potential in Queensland, Australia. There is a need for a comprehensive assessment of the region's agro-climatic conditions to determine the feasibility of large-scale Pongamia plantations. This study aims to evaluate the suitability of Central Queensland's mining region for Pongamia cultivation and seed yield by analysing climate parameters, soil properties, and topography. By comparing local environmental conditions with known requirements for optimal Pongamia growth, this research will provide critical insights into potential constraints and necessary land management practices. The findings will contribute to the broader understanding of Pongamia's adaptability and its potential role in sustainable land management and biofuel production in Australia.

METHODS

Study region and climate analysis

The study area is situated in the semi-arid inland regions of Central Queensland, Australia (Figure 1). Fifty years of rainfall data (1960–2020) was collected from the SILO website (SILO grid point Latitude: -22.00, Longitude: 148.05) (Queensland Department of Environment and Science, 2023). The maximum and minimum daily temperature data (1930–2020) were also collected from the SILO website. Air temperature was used as an indicator of frost days, with days having temperatures below 0°C considered as frost days (Bureau of Meteorology, 2023). The rainfall, temperature and frost time series were plotted. The frequency of occurrences for precipitation and temperature data were also analysed. The analysed climatic conditions were compared with the optimal climatic conditions for the growth of Pongamia trees.

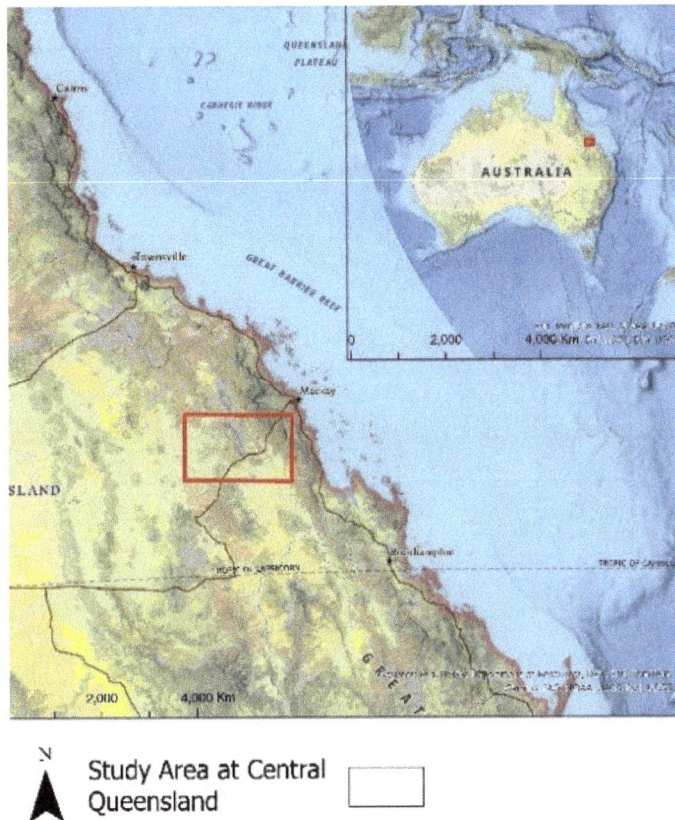

Study Area at Central Queensland

FIG 1 – Location of the study area in Central Queensland.

Land assessment

A literature review was conducted to understand the optimal conditions for growth of Pongamia trees in terms of soil conditions. Soil spatial data (soil properties and soil types) of the study area were collected from public domain sources including CSIRO Access Portal (CSIRO, 2022) and Queensland Spatial Catalogue (QSpatial, 2019). The topography was also obtained from Queensland Spatial Catalogue (QSpatial, 2019) (The spatial data were analysed using ArcGIS Pro (version 3.3). To verify the analysed spatial data, they were compared with the land survey data (50 sampling points) conducted in the study area. The selected mine sites are representative of the broader Central Queensland mining region in terms of soil types, climatic conditions, and land use history. The analysed soil spatial data were compared with the optimal soil conditions for growing Pongamia trees. The studied soil physical properties included soil bulk density (BD), soil texture (sand, silt, and clay content) and plant available water. The studied soil chemical properties included soil pH, Electrical Conductivity (EC)/salinity, organic carbon content, total nitrogen, total phosphorus and cation exchange capacity (CEC). Soil types of the study area were analysed using the Australian Soil Classification (Isbell, 2016).

RESULTS

Evaluation of climate

The analysis of rainfall data indicated that, over 1960–2020, the annual rainfall at the study area varied between 250 and 1400 mm, and the running ten year average ranged between 500 to 800 mm (Figure 2a). The annual average number of rainy days in the study area fluctuated between four and 13 days per month, with a running ten year average between six to eight days per month. Annual rainfall is in the range 600 to 800 mm in 32 per cent of years, and in the range 400 to 600 mm in

29 per cent of years (Figure 2a). The monthly total rainfall followed a strong seasonal pattern, peaking from January to March (median ~90–110 mm) with high variability (50–150 mm, Figure 2b). The driest period occurred from June to September (median ~10 mm, range 5–30 mm), with occasional extreme rainfall events. Rainfall began increasing in October (~40 mm), reaching ~70 mm in December.

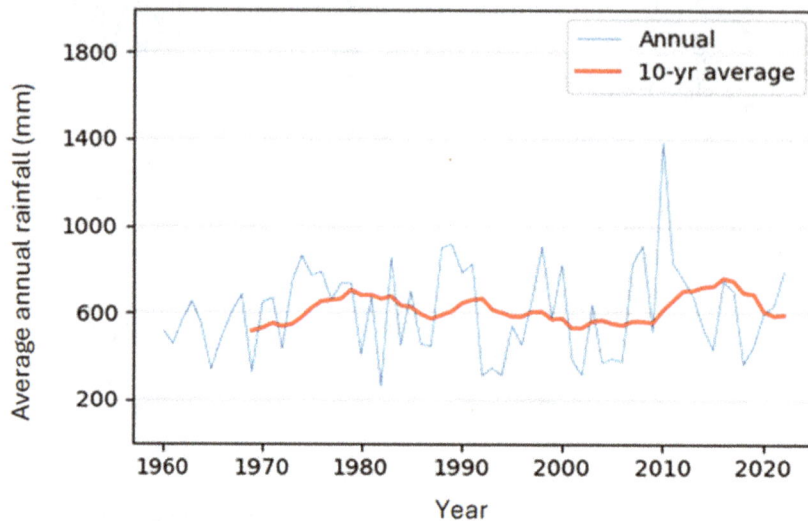

(a)

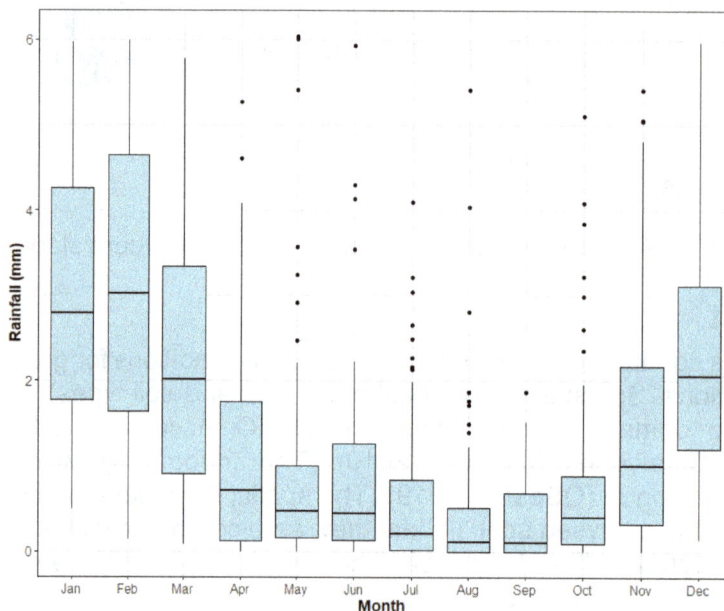

(b)

FIG 2 – Rainfall at the study area from 1960 to 2020: (a) rainfall amount; and (b) monthly rainfall distribution represented as boxplots. The boxes indicate the interquartile range (IQR), with the horizontal line inside each box representing the median monthly rainfall. The whiskers extend to 1.5 times the IQR, while the black dots represent outliers—rainfall values that exceed the whisker range.

The maximum daily temperature of the study area ranged from 17°C to 44°C, while the minimum daily temperature varied between -1.1°C and 28°C (Figure 3a). Monthly temperature showed that maximum temperatures peaked from November to February (~35°C, range 30–42°C) and were lowest in June and July (~25°C, range 20–30°C). Minimum temperatures followed a similar trend, with the coldest period in June and July (~10°C, range 0–15°C) and the warmest in January and December (~20°C, range 15–25°C, Figure 3b). The annual number of frost days (between June and August: wintertime) were between one and nine days, with the 10-year average of three frost days.

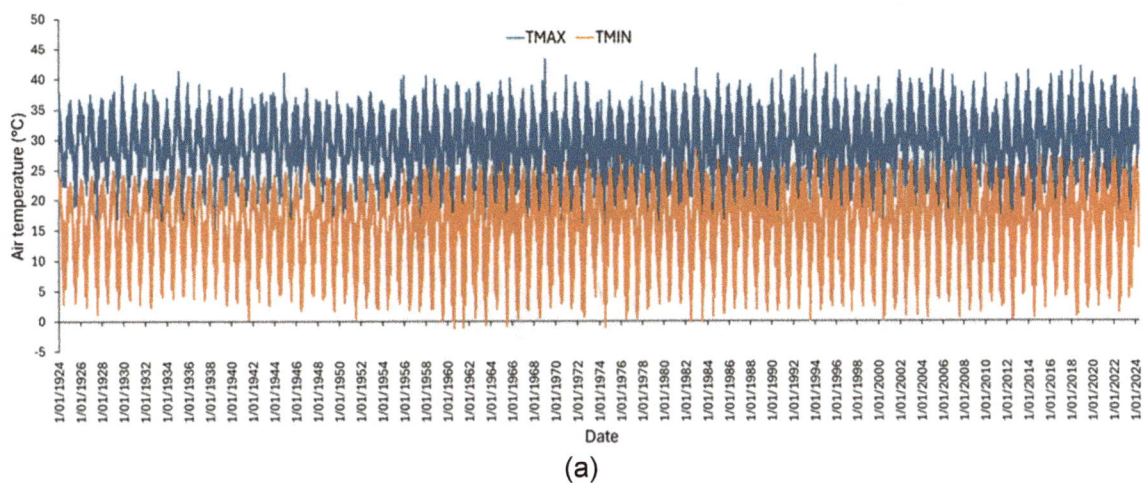

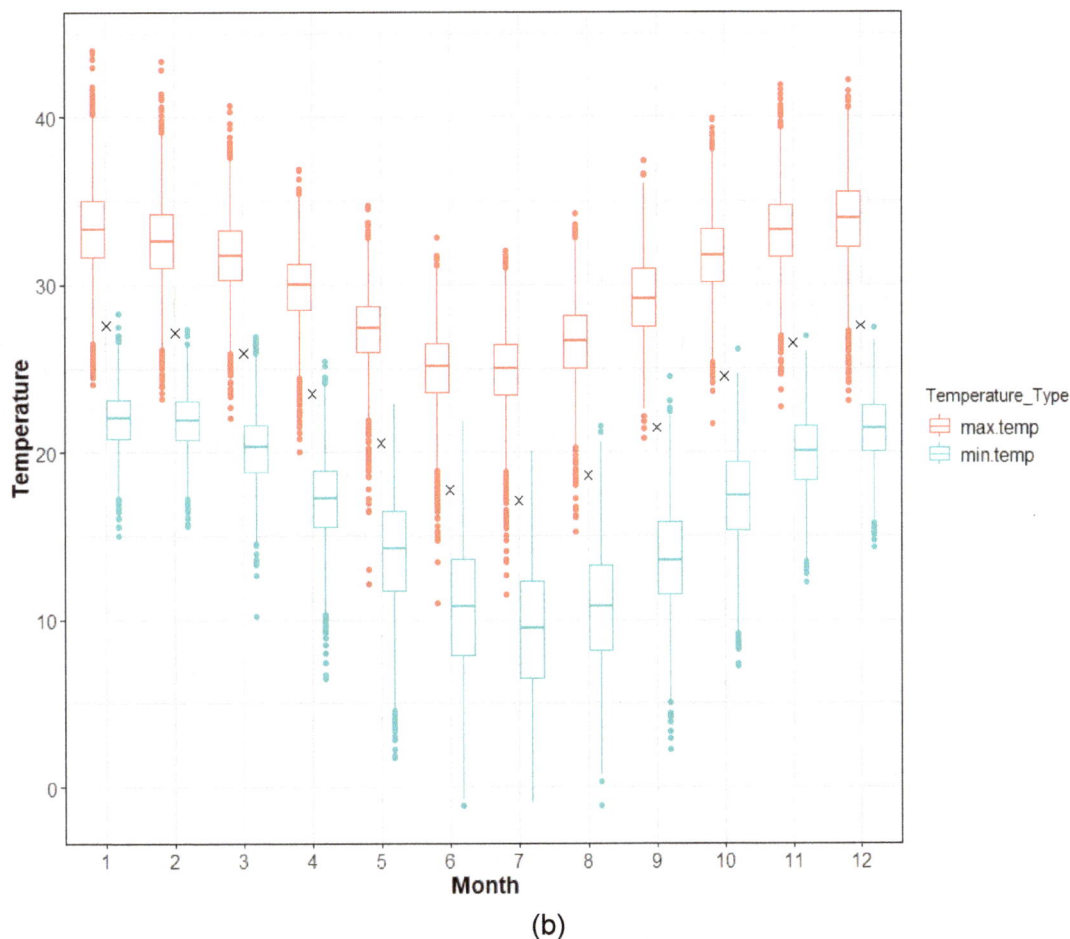

FIG 3 – Temperature at the study area from 1960 to 2020: (a) Daily air temperature at the study area; and (b) Monthly variation of maximum and minimum temperatures represented as boxplots. The boxes indicate the interquartile range (IQR), with the horizontal line inside each box showing the median. The whiskers extend to 1.5 times the IQR, and points beyond the whiskers represent outliers. The black crosses (×) represent the mean temperature for each month.

Evaluation of soil properties

Based on the Australian Soil Classification, the study area predominantly had two soil types: Vertosols and Sodosols. The land surface of the study area had mostly less than 2° (3.49 per cent) slope (Figure 4). Some small areas of the area the area had higher slopes, mainly between 2° (3.49 per cent) and 7° (12.28 per cent) (Figure 4).

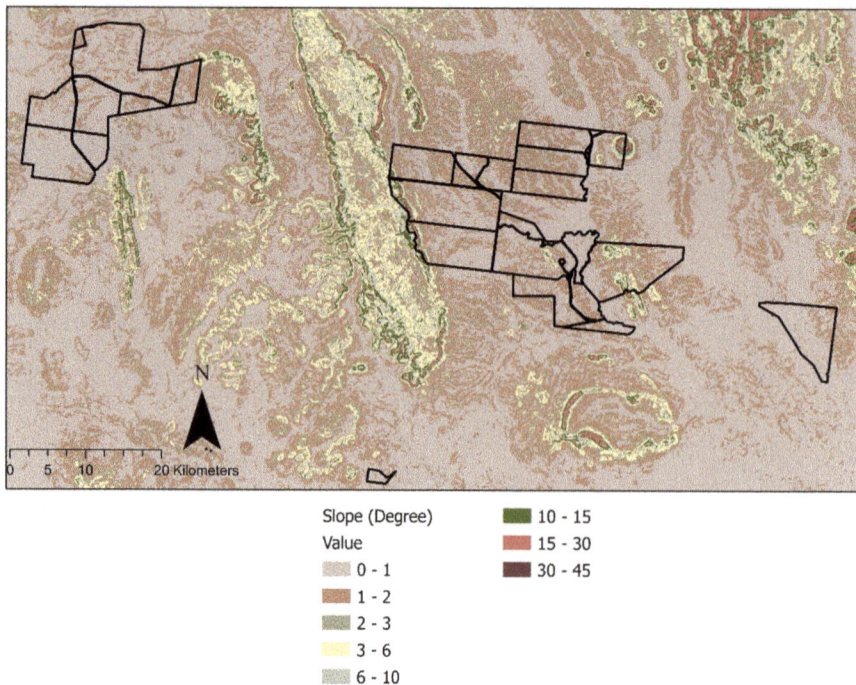

Slope (Degree)
Value
- 0 - 1
- 1 - 2
- 2 - 3
- 3 - 6
- 6 - 10
- 10 - 15
- 15 - 30
- 30 - 45

FIG 4 – Slopes at the study area; the black lines indicate the boundary of the study area.

The soil salinity spatial data were only available for the north-west parts of the area. Therefore, the soil salinity for the whole study area could not be analysed. The available data indicated low electrical conductivity (less than 0.37 dS/m) for the soil profiles in the area. The pH of surface soil (0–5 cm) mainly ranged between 6.1 and 7.5 (Figure 5). Below the surface, the pH values remained relatively constant, between 4.6 and 7.5, up to a depth of 60 cm (Figure 5). The pH values increased to up to 10.5 at the deeper depths before returning to <7.5 pH at 100 to 200 cm depths (Figure 5).

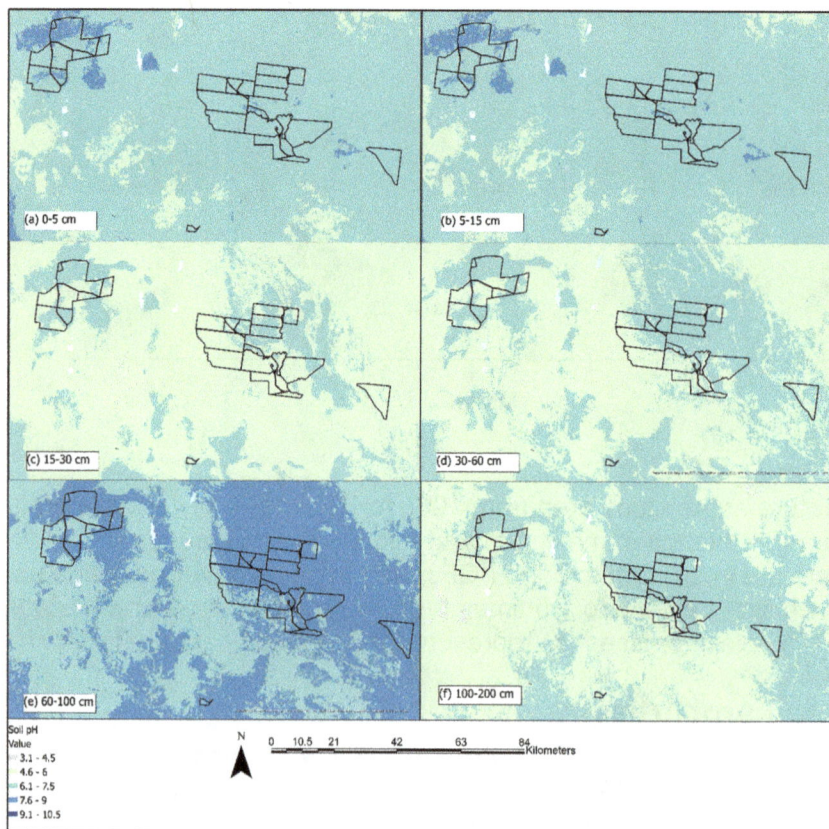

Soil pH
Value
- 3.1 - 4.5
- 4.6 - 6
- 6.1 - 7.5
- 7.6 - 9
- 9.1 - 10.5

FIG 5 – Soil pH at depths: (a) 0–5 cm; (b) 5–15 cm; (c) 15–30 cm; (d) 30–60 cm; (e) 60–100 cm and (f) 100–200 cm; the black lines indicate the boundary of the study area.

The Life of Mine | Mine Waste and Tailings Conference 2025 | Brisbane, Australia | 29–30 July 2025

The soil organic carbon, which is correlated with soil organic matter, decreased with increasing soil depth. The soil organic carbon content varied between 1.3 per cent and 2.1 per cent up to a depth of 15 cm. It then decreased to between 0.1 per cent and 0.7 per cent at deeper depths. Soil cation exchange capacity (CEC) did not change with depth. Widespread CEC values, mainly between 11 and 90 meq/100 g, were observed across the study area. Total N content decreased with increasing the depth of soil profiles. The total N in topsoil (0–60 cm) mainly varied between 0.031 per cent and 0.15 per cent. The total N dropped to 0.017 per cent in subsoil. Like total N, total P content decreased with increasing depth. In the topsoil (0–30 cm), total P content ranged from 0.026 per cent to 0.073 per cent. However, the total P content reduced to 0.007 per cent in subsoil.

The soil bulk density increased with increasing depth (Figure 6). The surface soil (0–5 cm) had bulk density values between 1.28 g/cm³ and 1.45 g/cm³ (Figure 6). The soil bulk density values were predominantly between 1.39 g/cm³ and 1.66 g/cm³ up to a depth of 100 cm (Figure 6). The values then increased to less than 1.66 g/cm³ at the depth of 200 cm (Figure 6). The plant available water increased with increasing depth. The topsoil was possibly drier due to evaporation and/or evapo-transpiration compared to subsoil. In the study area, the plant available water of topsoil (0–30 cm) varied from 5 mm to 30 mm while the subsoil plant available water ranged between 20 mm and 55 mm. The soil had high sand content, particularly for topsoil (0–30 cm). The topsoil sand content ranged between 29 per cent and 83 per cent. However, the subsoil sand content varied from 29 per cent to 65 per cent. There were no notable changes among silt contents at different depths of the soil profile, and the silt content was predominantly between 8.9 per cent and 16.3 per cent. The clay content of the topsoil (0–30 cm) ranged from 3 per cent and 42 per cent, and the clay content increased with increasing depth.

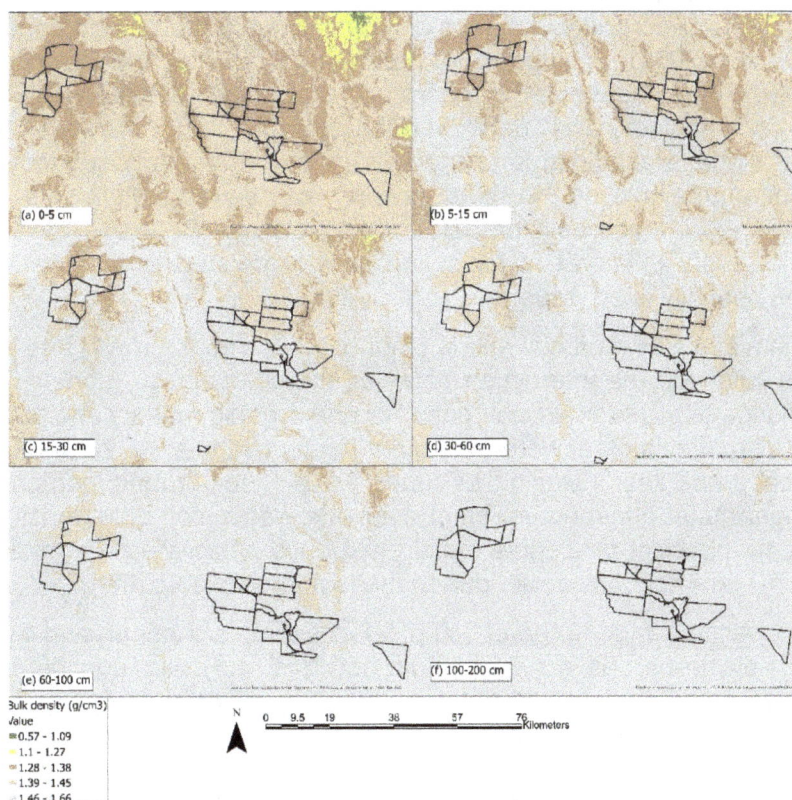

FIG 6 – Soil bulk density at depths: (a) 0–5 cm; (b) 5–15 cm; (c) 15–30 cm; (d) 30–60 cm; (e) 60–100 cm and (f) 100–200 cm; the black lines indicate the boundary of the study area.

DISCUSSION

Climate suitability

Pongamia trees typically grow in humid tropical and subtropical climates but demonstrate adaptability to various environmental conditions (Degani et al, 2022). Pongamia trees can grow in regions with annual rainfall ranging from 500 to 2500 mm (Islam et al, 2021). Optimal growth rates are generally observed in regions receiving a minimum annual rainfall of 500 to 800 mm (Murphy *et al,* 2012). This suggests that our study area, which has 500 to 800 mm rainfall per annum, can support growth of Pongamia trees. However, irrigation at the early stage of growth (establishment) is recommended to ensure optimal development (Murphy *et al,* 2012).

Islam *et al* (2021) noted Pongamia trees can tolerate minimum temperatures from 0°C to 16°C and maximum temperatures from 27°C to 50°C. Aminah and Syamsuwida (2019) reported that Pongamia trees grow in climates characterised by warm, extended summers with day-time temperatures ranging from 30 to 40°C, and mild night-time temperatures not dropping below 17°C. Degani *et al* (2022) also mentioned Pongamia can tolerate a range of climatic conditions, from light frost to 50°C. Our study indicated a very low occurrence of air temperatures below 0°C and above 42°C for the study area (Figure 6). This suggests temperature suitability of the study area for planting Pongamia trees.

Several studies (Usharani, Naik and Manjunatha, 2019; Murphy *et al,* 2012; Degani *et al,* 2022) reported that mature Pongamia trees can endure light frosts, but they require a period of no frost lasting between two to six months to maintain optimal growth. In this study, the number of annual frost days (minimum one day and maximum nine days) also recommends that the climatic conditions of the study area are suitable for growing Pongamia trees. However, as Duke (1983) mentioned a period (two to six months) without frost is required for achieving an optimal growth.

Suitability of topography and soils

Soil topography significantly affects land preparation for Pongamia tree plantations as the slope and direction of the slope impact the accessibility for heavy machinery. Topography also influences how the land will be segmented into paddocks, thus dictating tillage patterns and irrigation and drainage requirements (Anderson, 2009). A slope of <3 per cent (<1.72°) is preferred for agricultural land (Ghazanshahi, 2008). Slopes between 3 per cent and 6 per cent (between 1.72° and 3.43°) allow easy access for heavy machinery and operation (Ghazanshahi, 2008). However, levelling would be required to create a suitable condition for surface irrigation. This study suggests that most of the study area would not require leveling. However, leveling may be required in some areas due to land heterogeneity or selected irrigation system. Both Vertosols and Sodosols have potential to grow Pongamia trees. However, amelioration for saline soils is needed to achieve optimal growth on Sodosols or wherever salinity is an issue.

Soil physics and chemistry play a crucial role in plant establishment, growth and yield. Soil physico-chemical properties influence the interaction between the soil solution (comprising soil water and nutrients), soil exchange sites (such as clay particles and organic matter) and plant root (Singh and Sainju, 1998). This also affects plant water and plant health (Sahu *et al,* 2019). The most important soil physico-chemical properties, including EC (salinity), pH, soil organic carbon, cation exchange capacity, total nitrogen, total phosphorus, plant available water, soil bulk density and soil texture, were assessed. Some physical properties mainly hydraulic parameters (eg hydraulic conductivity and pore distribution) were not considered due to the lack of available data.

Pongamia can grow well in a wide range of soil pH. For example, Aminah and Syamsuwida (2019) reported that Pongamia trees can grow in acidic soil (pH: 4.3) although neutral to alkaline pH conditions prevail in Pongamia's natural habitats. Masilamani *et al* (2002) mentioned that under strongly alkaline conditions (>10.5), Pongamia germination significantly reduces because the alkalinity limits nutrient availability. While the surface soil (0–5 cm) of the study area has a suitable pH, amelioration may be required at deeper depths of the soil profile to increase growth and yield. Masilamani *et al* (2002) noted that germination and shoot length of Pongamia are affected in soils with CEC <15 meq/100 g. This suggests the soil CEC values in the study area are suitable. Soil organic carbon is related to soil organic matter (soil organic matter= soil organic carbon × 1.72) (Pribyl, 2010). The soil organic matter content of agricultural land is typically between 1 per cent and 6 per cent, depending on the type of agricultural land (Ghazanshahi, 2008). While the organic matter content of the study area, particularly topsoil, was within the suitable range for agriculture, and while Pongamia trees are nitrogen fixing plants, application of chemical fertilisers to promote establishment would be required. To determine the amount and type of chemical fertilisers, localised soil sampling and then soil characterisation are required.

The study area has high soil bulk density (>1.28 g/cm^3). This high bulk density can inhibit root growth (Marshall, Holmes and Rose, 1996). Thus, an appropriate land preparation strategy and addition of amendment to reduce soil bulk density to at least a depth of 100 cm is recommended. This would allow growing roots to easily access the higher plant available water contents observed in the deeper soil profile. The study area has a range of soil texture, including loamy sand, sandy-loamy and sandy clay loam. Usharani, Naik and Manjunatha (2019) reported that Pongamia trees can grow in a wide range of soil textures, including sandy, heavy clay, and rocky soils. However, the only limitation is germination as the germination is low on extremely sandy soil with low water-holding capacity (Degani et al, 2022). This is unlikely to create an issue for planting Pongamia trees in the study area as no extremely sandy soil was mapped.

CONCLUSION

This study suggested that the climatic conditions of the study area are suitable for a Pongamia plantation. However, variation in precipitation suggests that applying irrigation during the early stages of Pongamia growth (plant establishment) would be required. The irrigation would be also required in extreme dry years even following establishment. A period (two to six months) without frost (equivalent to minimum overnight temperatures not going below 0°C) is also required for achieving an optimal yield. The soils of the study area are generally suitable in terms of pH. However, the plantation area would require fertilisers (organic and chemical fertilisers) to improve soil organic matter content and nutrients. The plantation area would also benefit from land amelioration if salinity and sodicity issues emerge. An appropriate land preparation strategy (such as soil ripping and addition of amendments) needs to be conducted for reducing soil bulk density as well as levelling to reduce the land slope where needed. These conclusions are based on spatial data that only provide an overview of soil/land conditions. Thus, localised soil sampling and soil characterisation of the area of interest are required to provide better knowledge of required management practices before planning and design of a plantation.

In the next phase of this study, the Agricultural Production Systems Simulator (APSIM) model was used to simulate Pongamia growth and seed yield under the analysed climate and soil conditions. The model results suggest that the region is potentially suitable for Pongamia cultivation, although yield variability is expected due to environmental factors. Notably, Vertosols were predicted to support better growth than Sodosols, likely due to their higher cation exchange capacity, which enhances nutrient availability.

REFERENCES

Aminah, A and Syamsuwida, D, 2019. Natural growing site and cultivation of Pongamia (*Pongamia Pinnata (L.) Pierre*) as a source of biodiesel raw materials, *IOP Conference Series: Earth and Environmental Science*, 2019:012050 (IOP Publishing).

Anderson, G, 2009. The impact of tillage practices and crop residue (stubble) retention in the cropping system of Western Australia, Citeseer.

Arpiwi, N L, Muksin, K and Song, A N, 2023. Drought Stress Decreases Morphophysiological Characteristics of *Pongamia pinnata (L.) Pierre* a Biodiesel Tree, *Pakistan Journal of Biological Sciences*, 26.

Bureau Of Meteorology, 2023. Annual and monthly potential frost days [online]. Available from: <http://www.bom.gov.au/climate/map/frost/what-is-frost.shtml> [Accessed: 15 September 2023].

CSIRO, 2022. Soil and Landscape Grid National Soil Attribute Maps – Australian Soil Classification Map (3" resolution), Release 1, 28/10/2022 ed.

Degani, E, Prasad, M, Paradkar, A, Pena, R, Soltangheisi, A, Ullah, I, Warr, B and Tibbett, M, 2022. A critical review of Pongamia pinnata multiple applications: from land remediation and carbon sequestration to socioeconomic benefits, *Journal of Environmental Management*, 324:116297.

Duke, J A, 1983. Handbook of Energy Crops, Purdue University, Center for New Crops and Plants Products.

Fu, S J, Summers, T J, Morgan, S Q and Turn, W K, 2021. Fuel properties of pongamia (Milletia pinnata) seeds and pods grown in Hawaii, *ACS Omega*, 6(13):9222–9233.

Ghazanshahi, J, 2008. Soil and interaction with agriculture, Aeeizeh, Tehran, Iran.

Isbell, R, 2016. *The Australian soil classification* (CSIRO Publishing).

Islam, A K M A, Chakrabarty, S, Yaakob, Z, Ahiduzzaman, M and Islam, A K M M, 2021. Koroch (Pongamia pinnata): A promising unexploited resources for the tropics and subtropics, *Forest Biomass-From Trees to Energy*, IntechOpen.

Kumar, S, Mehta, J and Hazra, S, 2009. In vitro studies on chromium and copper accumulation potential of Pongamia pinnata (L.) Pierre seedlings, *Bioremediation, Biodiversity and Bioavailability*, 3:43–48.

Leksono, B, Rahman, S A, Larjavaara, M, Purbaya, D A, Arpiwi, N L, Samsudin, Y B, Artati, Y, Windyarini, E, Sudrajat, D J and Aminah, A, 2021. Pongamia: A possible option for degraded land restoration and bioenergy production in Indonesia, *Forests*, 12:1468.

Marshall, T J, Holmes, J W and Rose, C W, 1996. *Soil physics* (New York: Cambridge University Press).

Masilamani, P, Annadurai, K, Saravanapandian, P and Nadu, T, 2002. Effect of soil pH and organic matter on germination and growth attributes of Pungam (Pongamia Pinnata Roxb), *Indian Journal of Forestry*, 25:415–419.

Murphy, H T, O'Connell, D A, Seaton, G, Raison, R J, Rodriguez, L C, Braid, A L, Kriticos, D J, Jovanovic, T, Abadi, A, Betar, M, Brodie, H, Lamont, M, Mckay, M, Muirhead, G, Plummer, J, Arpiwi, N L, Ruddle, B, Saxena, S, Scott, P T, Stucley, C, Thistlethwaite, B, Wheaton, B, Wylie, P and Gresshoff, P M, 2012. A Common View of the Opportunities, Challenges and Research Actions for Pongamia in Australia, *BioEnergy Research*, 5:778–800.

Odeh, I O, Tan, D K and Ancev, T, 2011. Potential suitability and viability of selected biodiesel crops in Australian marginal agricultural lands under current and future climates, *Bioenergy Research*, 4:165–179.

Prasad, M V R, 2014. Genetic Enhancement of Pongamia Pinnata for Bio-Energy, in *Recent Advances in Bio-Energy Research* (eds: S Kumar, A K Sarma, S K Tyagi and Y K Yadav), pp 380–388 (Sardar Swaran Singh National Institute of Renewable Energy, Kapurthala, India).

Pratiwi, B M, 2000. The relationship between soil characteristics and species diversity in Tanjung Redep, East Kalimantan, *Journal of Forestry and Estate Research*, 1:27–23.

Pribyl, D W, 2010. A critical review of the conventional SOC to SOM conversion factor, *Geoderma*, 156:75–83.

Qspatial, 2019. Queensland Ortho Dem Data, 26/03/2019 ed.

Queensland Department of Environment and Science, 2023. SILO-Australian Climate Data [online], Queensland Government. Available from: <https://www.longpaddock.qld.gov.au/silo/> [Accessed: 15 May 2020].

Sahu, P K, Singh, D P, Prabha, R, Meena, K K and Abhilash, P, 2019. Connecting microbial capabilities with the soil and plant health: Options for agricultural sustainability, *Ecological Indicators*, 105:601–612.

Saxena, N C, 2002. Environmental benefits during the growth phase of Jatropha and Pongamia, International Energy Initiative.

Scott, P T, Pregelj, L, Chen, N, Hadler, J S, Djordjevic, M A and Gresshoff, P M, 2008. Pongamia pinnata: an untapped resource for the biofuels industry of the future, *Bioenergy Research*, 1:2–11.

Singh, B and Sainju, U, 1998. Soil physical and morphological properties and root growth, *HortScience*, 33:966–971.

Terviva, 2020. Sustainability Report, Terviva, United States of America.

Usharani, K, Naik, D and Manjunatha, R, 2019. Pongamia pinnata (L.): Composition and advantages in agriculture: A review, *J Pharmacogn Phytochem*, 8:2181–2187.

Vlam, M, Baker, P J, Bunyavejchewin, S and Zuidema, P A, 2014. Temperature and rainfall strongly drive temporal growth variation in Asian tropical forest trees, *Oecologia*, 174:1449–1461.

Renewable energy as post-mining land use in Queensland

C Côte[1], P Bolz[2] and J Robertson[3]

1. Director, Centre for Water in the Minerals Industry, Sustainable Minerals Institute, The University of Queensland, St Lucia Qld 4072. Email: c.cote@uq.edu.au
2. Research Fellow, Centre for Water in the Minerals Industry, Sustainable Minerals Institute, The University of Queensland, St Lucia Qld 4072. Email: p.asmussen@uq.edu.au
3. Lecturer, Griffith Law School, Griffith University, Nathan Qld 4111. Email: jacqui.robertson@griffith.edu.au

INTRODUCTION

A baseline assessment of post-mining land use (PMLU) in Queensland concluded that in the Bowen Basin, renewable energy produced by solar and wind had vast potential, but the pathway to implementation was not clear. This project aimed at examining how renewable energy projects could be set-up on a mining lease and be accepted as a productive post-mining land use by regulators and stakeholders. It comprised a review of legal instruments and planning requirements, an assessment of the potential for renewable energy production at three case study sites, in terms of available land and associated generation capacity, and cost-benefit analyses.

REGULATORY PATHWAYS

The project analysed the legislative instruments that will be triggered when planning to install a solar or wind farm facility on mining tenure in the Isaac Regional Council and Central Highlands Regional Council local government areas. The process for obtaining the required approvals can take four different paths depending on the miner's interests in the land upon which the facility is to be operated, and the tenure of that land. For all pathways, numerous approvals are required and there is no entity coordinating them, at either Federal or State level.

Even though it is potentially possible under the current regulatory framework to establish a renewable energy facility as a PMLU, it will require extensive time, effort and expense, mostly because the Progressive Rehabilitation and Closure Plan (PRCP) and potentially the Environmental Authority will need to be amended. A mine could simply rehabilitate the land to, for example, a grazing or native ecosystem land use, and then relinquish the land for it to be developed as a renewable energy facility at that later time. By simply waiting until relinquishment, no changes to the PRCP and EA are necessary. There would need to be significant benefit, such as a high economic return, to overcome the complexity of establishing the renewable energy facility earlier as a PMLU, rather than at a later date, after relinquishment.

CASE STUDIES – POTENTIAL GENERATION CAPACITY

A multi-criteria analysis was conducted to determine which mines would be most representative of the Bowen Basin mining sector and could be used as case studies. Criteria included location and local government area, extent of disturbed land, exposure to solar radiation and wind, and distance to energy transmission networks. Three coalmines were selected, with both underground and open cut workings. Land use information was compiled from the mines' PRCP. Each parcel of disturbed land was rated for its suitability for installation of renewable energy equipment. For instance, the types of land use with greatest suitability for installation of solar panels or wind turbines are cleared area, laydown yard, building, waste rock dump, co-disposal area and tailings storage facility. Wind turbines can also be installed along haul roads or conveyor belts.

In Queensland, four hectares per MW are required for installation of large-scale solar farms. Wind turbines have a generation capacity of 4.5 MW and require 200 hectares of land, to allow sufficient separation between the blades (Department of Energy and Climate, 2022). It was assumed that solar panels would operate at full capacity during 20 per cent of the time, and wind turbines 45 per cent of the time (this is called the capacity factor; these adopted values are conservative estimates for Queensland). Using these assumptions, the extent of disturbed land suitable for installation of renewable energy facilities was converted into potential generation capacity (MW) and electricity produced per annum (MWh).

Results are summarised below and compared with the Queensland renewable energy targets that were in place at the time of the project (10 000 MW of new large-scale solar capacity, 12 200 MW of new wind generation capacity).

TABLE 1

Potential energy generation from disturbed land on a mine site, based on the results from the case studies.

Potential generation of solar energy (MW)	Contribution to target	Potential generation of wind energy (MW)	Contribution to target
80–700	0.8–7%	4–60	0.03–0.5%

There is great potential for individual mines to contribute to the target for solar energy. Most utility-scale wind farms have very large generation capacity, greater than 500 MW. It is unlikely that there will be sufficient disturbed land on a mine site to support installation of a wind farm, but several mines could collaborate to combine their available land.

FINANCIAL BENEFITS

The aspects that govern the financial benefits of renewable energy production are capital costs, access to energy markets, and the ability to monetise the energy production through access to the Australian electricity market. Information was compiled from several sources, including: The Australian Renewable Energy Agency (ARENA), the Queensland Government 'Electricity generation map' (<https://electricity-generation-map.epw.qld.gov.au>), the Australian Energy Market Operator (AEMO) and the National Renewable Energy Laboratory of the U.S. Department of Energy, Office of Energy Efficiency and Renewable Energy. This information was used to derive several financial indicators of solar and wind plants: Net Present Value (NPV), Return on Investment (RoI), and payback period.

Based on the calculated value of these financial indicators, utility-scale wind energy production has a much greater potential to return financial benefits (positive NPV, with payback period around five years) for a generation capacity greater than 200 MW. However, results from the case studies showed that there is not enough disturbed land on a single mine to establish more than 200 MW of generation capacity. Solar plants offer fewer financial benefits: the NPV is only positive over a 20-year period if electricity can always be sold at the highest spot price ($200/MWh), which is highly unlikely. Currently, there is no financial incentive to overcome the complexity of obtaining approval for establishing a solar energy facility as a PMLU.

CONCLUSIONS

If renewable energy generation is indeed a PMLU that stakeholders support, reducing the requirements associated with amending the PRCP and EA for such a novel PMLU may assist, as it would overcome the current lack of incentives, at least for solar energy. There is a requirement to engage with regulators to discuss how regulatory frameworks could be updated to encourage adoption of renewable energy as post-mining land use, and to consider closure completion criteria adapted to the installation of renewable energy facilities, particularly in relation to landform design intent. In addition, the availability of financial schemes to support installation of solar plants should be investigated.

In response to this project, the Queensland Resources Council obtained funding from the Queensland government Regional Economics Futures Fund, which will seek to implement the above recommendations.

ACKNOWLEDGEMENTS

This project was funded by ACARP.

The authors wish to thank the representatives from industry, and local and state government who gave their time to discuss the issues outlined in this paper. In particular, representatives from Central Highlands Regional Council, Isaac Regional Council, Department of Environment, Tourism, Science and Innovation, Department of Natural Resources and Mines, Manufacturing and Regional and Rural Development, Department of Energy and Public Works, Department of State Development, Infrastructure, Local Government and Planning, and Office of the Queensland Mine Rehabilitation Commissioner. Environmental teams from the three case study sites also provided extensive support by compiling and submitting their PRCP data.

REFERENCE

Department of Energy and Climate, 2022. Queensland SuperGrid Infrastructure Blueprint, Energy and Jobs Plan – Power for generations, September 2022, Department of Energy and Climate, Queensland Government. Available from: <https://www.epw.qld.gov.au/__data/assets/pdf_file/0030/32988/queensland-supergrid-infrastructure-blueprint.pdf>

Standards and reporting

Acceptable erosion limits for constructed mine landforms – a literature review

G Maddocks[1] and A M Khalifa[2]

1. Technical Director – SLR Consulting Australia Pty Ltd, Brisbane Qld 4000.
   Email: gmaddocks@slrconsulting.com
2. Principal Consultant (CPSS), SLR Consulting Australia Pty Ltd, Brisbane Qld 4000.
   Email: akhalifa@slrconsulting.com

ABSTRACT

Constructed mine landforms are rehabilitated with the intent to ideally provide a beneficial use that may include returning the land to pre-mine vegetation, using the rehabilitated land for forestry, grazing or cropping, or in some cases, using the land for alternate uses such as solar farms or hydro-pump back opportunities. In all cases and particularly in cases where the constructed mine landforms contain geochemically reactive material the surface material on the landform must remain geotechnically and erosionally stable.

The establishment of performance targets for rehabilitated landforms provides a measurable way of determining if the design objectives and design criteria have been achieved. In Queensland, the performance targets are also framed as acceptable erosion rates.

Defining a one-size fits-all performance target for erosion rates or sediment yields is problematic due to the variability in landform design, material properties, and climatic conditions. This review evaluates published erosion benchmarks and the applicability of modelling tools—such as RUSLE, WEPP, MINErosion, GEOWEPP, SIBERIA and CAESAR$_{LISFLOOD}$ in setting site-specific erosion performance targets. Empirical estimates, including those from Lu *et al* (2003), suggest average erosion rates of 4–5 t/ha/annum across continental Australia. These values are often adopted as fixed targets, yet the original research cautions against their use at site scale due to landscape variability and model limitations. Similarly, the notion of aligning erosion rates with natural soil formation (<4 t/ha/annum) oversimplifies the dynamic nature of rehabilitated landforms and post mine land use.

Process-based models, particularly when calibrated using site-specific data, offer a more robust approach. WEPP and MINErosion allow for annual and storm-scale erosion assessment, while GEOWEPP and SIBERIA provide long-term landform evolution evaluation. Together, these tools support the development of performance-based erosion thresholds that evolve with rehabilitation progress. The review recommends defining acceptable erosion rates as site-specific ranges, not fixed values and ensuring outcomes are scientifically defensible and operationally realistic within post-mining landscapes.

INTRODUCTION

Establishing 'acceptable' performance targets for constructed mine landforms is an important component of mine closure planning and long-term rehabilitation success. These targets serve as benchmarks for assessing landform stability, long-term sediment yield, and environmental hazards.

However, defining what constitutes an 'acceptable' rate of erosion remains contentious, with significant variability in approaches adopted across jurisdictions, model interpretations, and environmental contexts. While some regulators reference broad-scale empirical averages such as those published by Lu *et al* (2003), others advocate for site-specific, process-informed thresholds grounded in model outputs and reference conditions that are linked to the life-of-mine plan and progress through engineering stages of development ie advancing from concept, to pre-feasibility and feasibility studies and then into detailed design and implementation during operations.

This review synthesises the current scientific understanding and regulatory practice concerning erosion rates applicable to constructed mine landforms, with a particular focus on Queensland and New South Wales (NSW), Australia. The review also draws comparisons with regulatory frameworks and scientific thresholds employed internationally, including in the United States, Europe, and

Canada. It evaluates how physically based erosion models, such as WEPP, MINErosion, GEOWEPP and SIBERIA, inform erosion assessment and how erosion rates can be defined within a scientifically and practically defensible framework.

ACCEPTABLE EROSION LIMITS IN AUSTRALIAN CONTEXTS

In situ geological units that underlie natural landforms have been eroding for millennia and are in a steady state of equilibrium. Constructed mine landforms can be constructed from several major stratigraphic and or lithological units that include fresh rock and partially weathered to extremely weathered units in the saprolith that transition to subsoil and topsoil in the pedolith.

Constructed mine landforms are physically and geochemically highly heterogenous and it may be unreasonable to assume these landforms that are constructed over 10 to 50 years and rehabilitated over 5 to 10 years can reflect natural rates of erosion that take millennia to reach equilibrium.

Queensland and New South Wales

In Queensland, historical reference values for acceptable erosion ranged from 12 to 40 t/ha/annum, particularly in rehabilitation targeting low intensity grazing or native vegetation (Williams, 2001). More recently, state regulators have adopted average annual soil loss limits of 5 t/ha/annum and maximum thresholds of 10 t/ha/annum, derived from national-scale modelling by Lu *et al* (2003) using RUSLE. These values are increasingly applied across rehabilitation and mine closure assessments, often without adequate consideration of site-specific variability or the limitations inherent in empirical models.

In NSW, similar figures are frequently cited, with reference to tolerable soil loss rates from the agricultural sector. Although NSW has not codified specific numeric thresholds for mine rehabilitation, erosion targets often default to <10 t/ha/annum in regulatory correspondence, especially for post-mining landforms supporting grazing or native bushland.

Critique of uniform thresholds

Lu *et al* (2003) explicitly cautioned against applying national erosion maps and values (eg 5–6 t/ha/annum average) to site-specific assessments. Their modelling demonstrated high spatial variability, with more than 28 per cent of the continent predicted to erode at <0.5 t/ha/annum and only 8 per cent at > 10 t/ha/annum. Moreover, they noted that the maps provide erosion potential (not sediment yield) and must be adjusted for sediment delivery ratios (SDRs). Importantly, their modelling was based on RUSLE, a model that is frequently criticised by some regulators for oversimplifying complex erosion dynamics and excluding gully formation or depositional processes.

Broader land use comparisons further highlight the limitations of applying a uniform erosion threshold to rehabilitated mine landforms. Lu *et al* (2003), using a national-scale application of RUSLE, reported that erosion rates vary substantially across Australian land uses. Sugarcane cultivation, for instance, had the highest predicted average erosion rate at 16.1 t/ha/annum, with observed values reaching up to 227 t/ha/annum under conventional tillage systems. Grazing on residual/native pastures was associated with mean erosion rates of 5.4 t/ha/annum, while improved pastures showed lower averages around 1.1 t/ha/annum. Cropping systems (eg cereals and oilseeds) averaged between 2.7 and 3.2 t/ha/annum but were estimated to accelerate erosion 18–30 times above background levels.

These figures demonstrate that erosion rates in some managed agricultural landscapes—such as sugarcane cultivation (with mean rates of 16.1 t/ha/annum and observed rates exceeding 200 t/ha/annum)—can far exceed the conservative thresholds now being proposed for mine rehabilitation (typically 5 to 10 t/ha/annum). However, other land uses such as improved pastures and cereal cropping often show lower erosion rates, typically ranging from 1.1 to 3.2 t/ha/annum. While these values are numerically below the 5–10 t/ha/annum mine rehabilitation threshold, they still represent accelerated erosion compared to natural background levels, often estimated at <1 t/ha/annum. The key point is not that all agricultural erosion exceeds mine rehabilitation thresholds, but that acceptable erosion rates vary widely across land uses and contexts. This reinforces the need for mine site erosion limits to be grounded in landform-specific characteristics and local analogues, rather than applying a one-size-fits-all threshold that may be more conservative than necessary or less protective than intended.

This contrast reinforces the need for erosion limits to be contextualised, rather than standardised across all post-mining sites. Applying a fixed threshold of 5–10 t/ha/annum, as currently advocated by some regulators, would impose stricter performance criteria on constructed landforms than those applied to long-established agricultural systems. A more pragmatic and defensible approach involves benchmarking acceptable erosion rates against analogous land uses and incorporating modelled site-specific data, rather than relying solely on national averages or generalised background rates.

Grigg, Mulligan and Worrall (2001) proposed a criterion for post-mining landforms whereby erosion should not exceed the natural soil formation rate, estimated as <4 t/ha/annum. However, such rates are often orders of magnitude below the erosion rates occurring on newly constructed landforms composed of unconsolidated spoil.

TABLE 1

Estimated soil loss rates and rates of erosion acceleration since European settlement from major land used categories (after Lu *et al*, 2003).

Landuse group	Approx. total area $(km^2 \times 10^3)$	Total erosion (Mt/year)	Av. erosion rate (t/ha.year)	Rate of acceleration (ratio of current to natural rates)
Forest	277	26	1.0	1
Woodland	2180	726	2.8	1
National park	184	129	7.4	1
Residual/native pastures	4232	2304	5.4	3
Plantation	146	17	2.1	4
Improved pastures/legumes	200	19	1.1	5
Cereals excluding rice	193	63	2.7	18
Oilseeds	4	2	3.2	30
Other agricultural lands	16	26	11.1	33
Sugarcane	5	10	16.1	33

INTERNATIONAL PERSPECTIVES ON ACCEPTABLE EROSION RATES

United States

In the US, tolerable soil loss (T-value) is commonly defined in agricultural systems as 4.5 to 11.2 t/ha/annum, depending on soil depth and use. The USDA-NRCS defines T-values based on maintaining long-term soil productivity. However, these figures are not directly transferable to post-mining contexts, where landscape function and stability are more critical than agricultural yield. In surface mining reclamation, the Office of Surface Mining Reclamation and Enforcement (OSMRE) mandates that landforms be stable, support designated land use, and not contribute to off-site sedimentation. However, numerical erosion rate thresholds are not universally prescribed and are instead evaluated through performance-based standards and post-reclamation monitoring.

Europe

European Union directives, such as the Soil Thematic Strategy and the Common Agricultural Policy (CAP), suggest tolerable erosion rates of 1 to 6 t/ha/annum. However, like the US, these apply primarily to agricultural soils and are not extended to post-mining landscapes. In the UK, soil erosion thresholds for reclaimed land are determined case-by-case, with regulators favouring evidence-based modelling over static benchmarks.

Canada

Canadian frameworks vary by province. In Alberta and British Columbia, erosion control is governed by performance criteria rather than fixed limits. Thresholds are often guided by land capability classifications and empirical erosion data, with typical tolerable ranges from 2 to 6 t/ha/annum for disturbed lands. Erosion modelling using tools like RUSLE, SIBERIA, and site-specific hydrological modelling is common practice.

MODELLING-BASED APPROACHES TO EROSION THRESHOLD DEFINITION

Role of physically based models

Models like WEPP and MINErosion enable the simulation of sediment detachment, transport, and deposition over short time scales (storm to annual), accounting for slope length, material erodibility, and vegetation cover. In contrast, SIBERIA provides long-term landform evolution projections, simulating gully incision, slope retreat, and sediment redistribution over decades to centuries.

These models offer a pathway to replace generic numerical targets with probabilistic or scenario-based outcomes. For example, erosion rates can be expressed as a range (eg 15–50 t/ha/annum during early establishment, declining to <10 t/ha/annum after stabilisation), linked to vegetation cover, slope geometry, and rainfall erosivity. This approach enables defensible, site-specific criteria aligned with landform design and closure objectives.

Benchmarking against reference sites

An increasingly accepted method is to benchmark acceptable erosion rates against the performance of analogous, stable landforms in the same region—for instance, undisturbed hillslopes or successful older rehabilitation. This empirical calibration allows for a context-sensitive threshold grounded in measurable outcomes rather than fixed assumptions.

DISCUSSION AND CONCLUSION

This review highlights the limitations of applying fixed, universal erosion thresholds to constructed mine landforms. Although numeric erosion targets such as 5–10 t/ha/annum are frequently cited in regulatory settings, particularly in Queensland and New South Wales, these figures are derived from broad-scale empirical modelling and are not intended to be used as compliance benchmarks for highly variable, site-specific conditions. The application of such thresholds, without accounting for material heterogeneity, climate variability, slope geometry, vegetation cover, and evolving land use, risks misrepresenting actual erosion potential and may impose unrealistic or unnecessarily conservative performance expectations.

Constructed mine landforms, particularly in their early stages of establishment, experience erosion dynamics that differ markedly from natural landscapes or long-managed agricultural systems. These landforms are often comprised of unconsolidated materials with limited surface armouring and require time to develop physical and biological stability. Therefore, performance targets must reflect the temporal evolution of landform stability and allow for higher erosion rates during the initial post-construction phase, progressively declining as vegetation becomes established and hydrological function improves.

A key outcome from this review is the recommendation to define acceptable erosion limits as performance-based ranges that are derived from site-specific modelling, supported by empirical data, and benchmarked against regional and local reference conditions. Process-based erosion models, such as WEPP, MINErosion, and SIBERIA, offer a robust framework to simulate sediment detachment, transport, and deposition over various temporal and spatial scales. These models allow for the prediction of erosion under different landform configurations, vegetation scenarios, and climatic inputs, making them valuable for evaluating landform design options and quantifying potential short-term stability and long-term landform evolution.

SLR also advocate for the incorporation of direct erosion measurements using rainfall simulators, to quantify analogue erosion rates on nearby stable or rehabilitated landforms with similar material and slope characteristics. These analogue benchmarks provide critical calibration data for models and offer a pragmatic, evidence-based method for defining acceptable erosion targets. When used alongside modelling, rainfall simulator trials can significantly improve confidence in performance assessments by generating high-resolution data under controlled and repeatable conditions.

Additionally, benchmarking acceptable erosion rates against stable natural analogues or mature rehabilitation in the same region allows for a more defensible interpretation of landform function. This approach recognises that erosion is a natural process and that some degree of soil loss is inevitable and acceptable, provided it does not compromise landform stability, long-term sustainability, or off-site environmental values.

Ultimately, acceptable erosion limits should not be defined by static one size fits all numerical values adopted from agricultural policy or national erosion averages. Instead, they should represent dynamic, scientifically justified performance targets that evolve from concept level studies through to detailed design phases over the course of the mine life.

Integrating measured data from the range of cover materials that will be used on a constructed mine landform with field measurements (including rainfall simulation), and post-rehabilitation monitoring enables a more adaptive and transparent approach to evaluating landform stability. This will ensure that post-mining landscapes are designed and managed to be resilient, self-sustaining, and aligned with agreed land use outcomes.

REFERENCES

Grigg, A H, Mulligan, D R and Worrall, R, 2001. Developing a framework for evaluating rehabilitation success on mine sites in Australia, Proceedings of the 26th Annual Minerals Council of Australia Environmental Workshop, October 2001, Newcastle, NSW, Minerals Council of Australia, Canberra.

Lu, H, Prosser, I P, Moran, C J, Gallant, J C, Priestley, G and Stevenson, J G, 2003. Predicting sheetwash and rill erosion over the Australian continent, Australian Journal of Soil Research, 41(6):1037–1062. https://doi.org/10.1071/SR02157

Williams, D J, 2001. Prediction of erosion from steep mine waste slopes, Environmental Management and Health, 12(1):35–50 (MCB University Press).

Author index

www.ingramcontent.com/pod-product-compliance
Lightning Source LLC
Chambersburg PA
CBHW061103210326

41597CB00021B/3967